Nonlinear Numerical Methods and Rational Approximation II

Mathematics and Its Applications

Volume 296

Nonlinear Numerical Methods and Rational Approximation II

edited by

Annie Cuyt
Department of Mathematics and Computer Science,
University of Antwerp,
Antwerp, Belgium

KLUWER ACADEMIC PUBLISHERS
DORDRECHT / BOSTON / LONDON

A C.I.P. Catalogue record for this book is available from the Library of Congress.

ISBN 0-7923-2967-8

Published by Kluwer Academic Publishers,
P.O. Box 17, 3300 AA Dordrecht, The Netherlands.

Kluwer Academic Publishers incorporates
the publishing programmes of
D. Reidel, Martinus Nijhoff, Dr W. Junk and MTP Press.

Sold and distributed in the U.S.A. and Canada
by Kluwer Academic Publishers,
101 Philip Drive, Norwell, MA 02061, U.S.A.

In all other countries, sold and distributed
by Kluwer Academic Publishers Group,
P.O. Box 322, 3300 AH Dordrecht, The Netherlands.

Printed on acid-free paper

Printed in the Netherlands

TABLE OF CONTENTS

EDITOR'S PREFACE

These are the proceedings of the international conference on "Nonlinear numerical methods and Rational approximation II" organised by Annie Cuyt at the University of Antwerp (Belgium), 05–11 September 1993. It was held for the third time in Antwerp at the conference center of UIA, after successful meetings in 1979 and 1987 and an almost yearly tradition since the early 70's. The following figures illustrate the growing number of participants and their geographical dissemination. In 1993 the Belgian scientific committee consisted of A. Bultheel (Leuven), A. Cuyt (Antwerp), J. Meinguet (Louvain-la-Neuve) and J.-P. Thiran (Namur).

The conference focused on the use of rational functions in different fields of Numerical Analysis. The invited speakers discussed "Orthogonal polynomials" (D. S. Lubinsky), "Rational interpolation" (M. Gutknecht), "Rational approximation" (E. B. Saff), "Padé approximation" (A. Gonchar) and "Continued fractions" (W. B. Jones). In contributed talks multivariate and multidimensional problems, applications and implementations of each main topic were considered. To each of the five main topics a separate conference day was devoted and a separate proceedings chapter compiled accordingly. In this way the proceedings reflect the organisation of the talks at the conference. Nonlinear numerical methods and rational approximation may be a narrow field for the outside world, but it provides a vast playground for the chosen ones. It can fascinate specialists from Moscow to South-Africa, from Boulder in Colorado and from sunny Florida to Zürich in Switzerland.

The organizer and the scientific committee would like to thank the "Nationaal Fonds voor Wetenschappelijk Onderzoek (NFWO)", the "Fonds National de la Recherche Scientifique (FNRS)" and the "University of Antwerp (UIA)" for the financial support as well as IBM, ASLK, and MIVA for logistic support. Thanks are also due to the "Congrescentrum UIA" for the hospitality and technical know-how, to P. Janssens and B. Verdonk for 24-hour standby, to my daughter A. Van Soom and my husband U. Van Soom, to "Kluwer Academic Publishers" for their involvement, before, during and after the conference, and to the referees for their cooperation in producing journal quality proceedings. Without all these people the conference wouldn't have been possible. Last but not least I want to thank the participants: we very much enjoyed meeting all of you !

Annie Cuyt*

* Research Director NFWO, Universiteit Antwerpen (UIA), Dept WIS–INF, Universiteitsplein 1, B-2610 Wilrijk-Antwerp, Belgium, Annie.Cuyt@wins.uia.ac.be

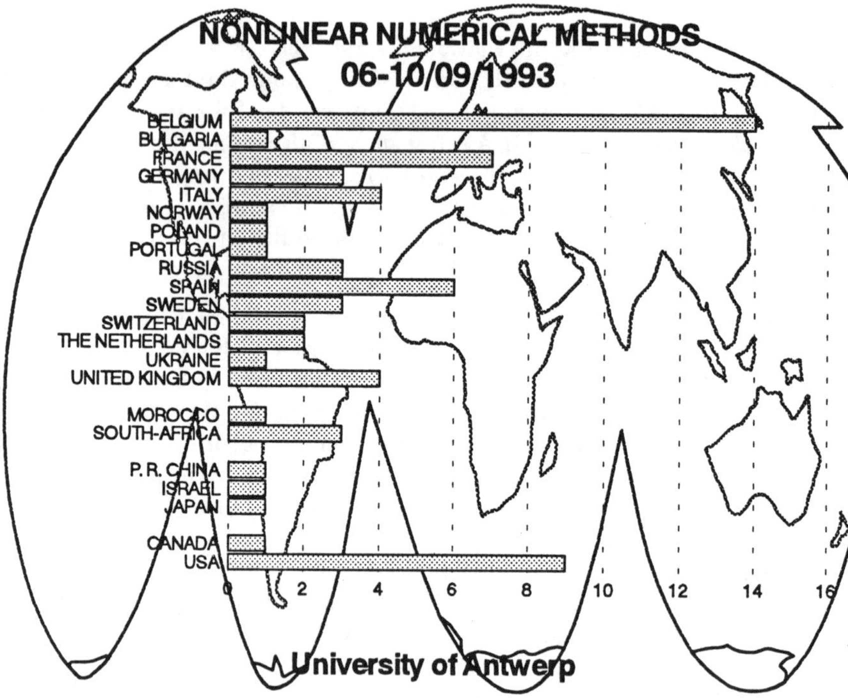

NONLINEAR NUMERICAL METHODS AND RATIONAL APPROXIMATION

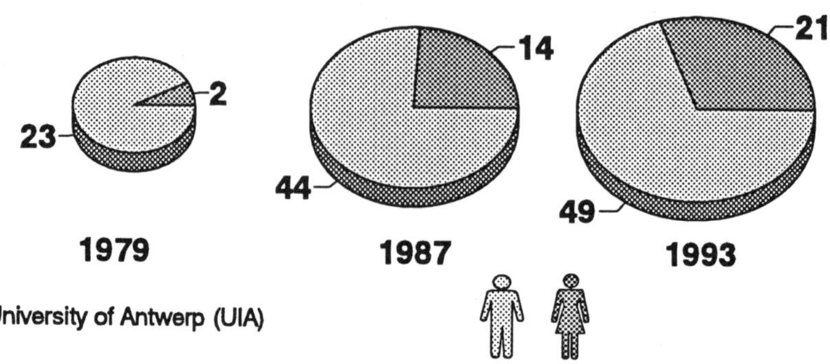

University of Antwerp (UIA)

LIST OF PARTICIPANTS

Jilali Abouir
1/17 Residence Comtesse de Segur
F-59790 Ronchin, France

Carlo Alabiso
Dipto di Fisica, Universita di Parma
V.le delle Scienze, 43100 Parma, Italy
ALABISO@VSPR2.CINECA.IT

Alexander Aptekarev
Keldysh Inst. of Appl. Mathematics
Miusskaya Square 4, 125047 Moscow A-47, Russia
aptekaa@refal.msk.su

George A. Baker Jr.
Group T11, MS B262, Los Alamos Nat. Lab.
87545 Los Alamos, New Mexico, USA
gbj@griffin.Lanl.GOV

Laurent Baratchart
INRIA
2004, Route des Lucioles B.P. 93, 06902 Sophia-Antipolis Cdx, France
baratcha@sophia.inria.fr

Bernhard Beckermann
Laboratoire ANO, U.F.R.-IEEA-Informatique, Universite de Lille 1
F-59655 Villeneuve D'Ascq, France
bbecker@omega.univ-lille1.fr

Jean-Paul Berrut
Departement de Mathématique, Universite de Fribourg
Perolles, CH-1700 Fribourg, Suisse
BERRUT%CFRUNI51.bitnet@cc1.kuleuven.ac.be

Luca Billi
SISSA
via Beirut 4, I-34013 Miramare, Italy
billi@vs4te6.bo.infn.it

Cathy Bonan-Hamada
Dept of Mathematics, University of Colorado, Campus Box 395
CO 80304 Boulder, Colorado, U.S.A.
bonan@euclid.Colorado.EDU

Claude Brezinski
Laboratoire ANO, U.F.R.-IEEA-Informatique, Universite de Lille 1
F-59655 Villeneuve D'Ascq, France
brezinsk@omega.univ-lille1.fr

Adhemar Bultheel
Dept. Computerwetenschappen, Katholieke Universiteit Leuven

Celestijnenlaan 200A, B-3001 Heverlee, Belgium
Adhemar.Bultheel@cs.kuleuven.ac.be

Alicia Cachafeiro
Dept. Mathematica Aplicada, Universidad de Vigo
S-36280 Vigo, Spain
acachafe@dtc.uvigo.es

Claudine Chaffy-Camus
LMC INPG
46, av. Felix Viallet, 38031 Grenoble Cedex, France
chaffy@imag.fr

Alan K. Common
Mathematical Institute, University of Kent
CT2 7NF Canterbury, U.K.
akc@ukc.ac.uk

Cathleen Craviotto
Dept of Mathematics, University of Colorado, Campus Box 395
CO 80309 Boulder, Colorado, USA
craviott@euclid.Colorado.EDU

Annie Cuyt
Dept Mathematics & Computer Science, University of Antwerp (UIA)
Universiteitsplein 1, B-2610 Antwerp-Wilrijk, Belgium
cuyt@wins.uia.ac.be

Rudnei D. da Cunha
Computing Laboratory, University of Kent
CT2 7NF Canterbury, Kent, U.K.
trh@ukc.ac.uk

Kathy Driver
Department of Mathematics, University of Witwatersrand, Private Bag 3
2050 Wits, Johannesburg, South Africa
036KAD@witsvma.wits.ac.za

Jan Gelfgren
Department of Mathematics, University of Umeå
S-90187 Umeå, Sweden

Andrey A. Gonchar
Steklov Mathematical Institute, Academy of Science of USSR
GSP1 Ul. Valvilova 42, 117966 Moscow, Russia

Pablo Gonzalez-Vera
Dpto Analisis Matematico, Facultad de Matematicas, Universidad de La Laguna
38271 La Laguna, Canary Islands, Spain

Peter R. Graves-Morris
Department of Mathematics, University of Bradford
BD7 1DP Bradford, West Yorkshire, U.K.
P.R.Graves-Morris@bradford.ac.uk

Philip Gustafson
Dept of Mathematics, Washington State University, Campus Box 3113
WA 99164 Washington, Washington, USA

Martin Gutknecht
IPS Supercomputing, ETH-Zurich, ETH-Zentrum
CH-8092 Zurich, Switzerland
mhg@ips.id.ethz.ch

Erik Hendriksen
Fac. Wiskunde en Informatica, Universiteit Amsterdam
Plantage Muidergracht 24, 1018 TV Amsterdam, Nederland
erik@fwi.uva.nl

Marie-Paule Istace
Dept. de Mathématique, Fac. Univ. Notre-Dame de la Paix
Rempart de la Vierge 8, B-5000 Namur, Belgium
mis@math.fundp.ac.be

William B. Jones
Dept of Mathematics, University of Colorado, Campus Box 395
CO 80309 Boulder, Colorado, U.S.A.
wjones@euclid.Colorado.EDU

Valeri Kaliaguine
Kstovo, Nizhniy Novgorod
606200 Talalushkina 12-12, Russia

Ralitza Kovacheva
Institute of Mathematics, Bulgarian Acad. Science
PO Box 373, 1090 Sofia, Bulgaria
rkk@bgearn.bitnet

Khrystyna Kuchmins'ka
Inst. Appl. Problems in Mech. and Math., Ukrainian Academy of Science
3-6 Naukova Street, 290601 L'viv, Ukraina
kalyniak%ippmm.lviv.ua@LITech.Lviv.UA

Arno Kuijlaars
Universiteit van Amsterdam
Plantage Muidergracht 24, 1018 TV Amsterdam, Nederland

Eli Levin
Max Rowe Educational Center, The Open University of Israel
P.O.Box 39328 Ramat Aviv, 61392 Tel Aviv, Israel
levin@math.tau.ac.il

Paul Levrie
KIHA
Salesianenlaan 30, 2660 Hoboken, Belgium
Paul.Levrie@cs.kuleuven.ac.be

Jia-Liang Li
Dept. Mathematics, Central China Normal University
430070 Wuhan Hubei, PR China

Doron S. Lubinsky
Departement of Mathematics, University of Witwatersrand
PO Wits 2050, Johannesburg, South-Africa
036DOR@witsvma.wits.ac.za

Alphonse Magnus
Institut de Mathématique, Universite Catholique de Louvain
Chemin du cyclotron 2, B-1348 Louvain-La-Neuve, Belgium
magnus@anma.ucl.ac.be

Francisco Marcellan
Departemento Ingenieria, Universidad Carlos III de Madrid
Avda Mediterraneo 20, 28913 Leganes-Madrid, Spain
pacomarc@ing.uc3m.es

David Masson
Department of mathematics, University of Toronto
M5S 1A1 Toronto, Ontario, Canada
masson@math.toronto.edu

Daniel M. Matjila
Fac. Maths & Nat. Sciences, University of the North
Private Bag X1106, 0727 Sovenga, South Africa

Ana C. Matos
Grupo de Matematica Aplicada, Universidade do Porto
R. Das Taipas 135, 4000 Porto, Portugal
ammatos@nccup.ctt.pt

Jean Meinguet
Institut de Mathématique, Universite Catholique de Louvain
Chemin du cyclotron 2, B-1348 Louvain-La-Neuve, Belgium
Meinguet@anma.ucl.ac.be

Olav Njåstad
Institute of Mathematics, University of Trondheim - NTH
N-7034 Trondheim, Norway
njastad@imf.unit.no

Teresa E. Perez-Fernandez
Dpto Matematica Aplicada, Universidad de Granada
18071 Granada, Spain

Miguel Angel Piñar-Gonzalez
Dpto Matematica Aplicada, Universidad de Granada
18071 Granada, Spain
mpgonzalez@ugr.es

Michela Redivo-Zaglia
Dpto di Elettronica e Informatica, Universita di Padova
Via Gradenigo 6A, 35131 Padova, Italy
ELEN@ELETT1.DEI.UNIPD.IT

Maria Luisa Rezola
Dpto Matematicas, Universidad de Zaragoza

50009 Zaragoza, Spain

David E. Roberts
Mathematics Department, Napier College
Colinton Road 219, EH14 1DJ Edinburgh, Scotland, U.K.
SDZ511@PRIME.NAPIER.ac.uk

André Ronveaux
Facultés des Sciences, Département de Physique, FUNDP
Rue de Bruxelles 61, B-5000 Namur, Belgium

Milvia Rossini
Dpto di Matematica, Universita di Milano
Via C. Saldini 50, 20133 Milano, Italy
bozzini@vmimat.mat.unimi.it

Catarina Rudälv
Department of Mathematics, University of Umeå
S-90187 Umeå, Sweden

Paul Sablonnière
Laboratoire LANS, INSA Rennes
Avenue des Buttes de Coesmes, F 35043 Rennes Cedex, France
sablonni@perceval.univ-rennes1.fr

Ed B. Saff
Department of Mathematics, University of South Florida
FL 33620 Tampa, Florida, USA
esaff@szego.math.usf.edu

Tetsuya Sakurai
Inst. Information Sc. and Electronics, Tsukuba University
305 Tsukuba, Japan
sakurai@is.tsukuba.ac.jp

Ahmed Salam
Laboratoire ANO, UFR IEEA - M3, Université de Lille I
F-59655 Villeneuve d'Ascq, France

Geert Schelfhout
ESAT Laboratory, Dept Electrical Engineering KUL
K. Mercierlaan 94, B-3001 Heverlee, Belgium
Geert.Schelfhout@esat.kuleuven.ac.be

Ann Sinap
Departement Wiskunde, Katholieke Universiteit Leuven
Celestijnenlaan 200B, B-3001 Heverlee, Belgium

Herbert Stahl
TFH-Berlin/FB2
Luxemburger Str. 10, D-1000 Berlin 65, Germany

Franek Szafraniec
Instytut Matematyki, Uniwersytet Jagiellonski
Reymonta 4, PL 30059 Krakow, Poland
fhszafra@IM.UJ.EDU.PL

Jean-Pierre Thiran
Dept. de Mathématique, Fac. Univ. Notre-Dame de la Paix
Rempart de la Vierge 8, B-5000 Namur, Belgium

Suzanne Thiry
Dept. de Mathématique, Fac. Univ. Notre-Dame de la Paix
Rempart de la Vierge 8, B-5000 Namur, Belgium

W.J. Thron
Dept of Mathematics, University of Colorado, Campus Box 395
80309 Boulder, Colorado, U.S.A.

Walter Van Assche
Departement Wiskunde, Katholieke Universiteit Leuven
Celestijnenlaan 200B, B-3001 Heverlee, Belgium
walter%twi%wis@cc3.KULeuven.ac.be

Marc Van Barel
Dept. Computerwetenschappen, Katholieke Universiteit Leuven
Celestijnenlaan 200A, B-3001 Heverlee, Belgium
Marc.VanBarel@cs.kuleuven.ac.be

Jeannette Van Iseghem
U.F.R. IEEA-Informatique, Universite de Lille 1
F-59655 Villeneuve d'Ascq, France
jvaniseg@ano.univ-lille1.fr

Brigitte Verdonk
Dept Mathematics & Computer Science, University of Antwerp (UIA)
Universiteitsplein 1, B-2610 Antwerp-Wilrijk, Belgium
verdonk@wins.uia.ac.be

Hans Wallin
Department of Mathematics, University of Umeå
S-90187 Umeå, Sweden

Ernst J. Weniger
Inst. Phys. & Theor. Chemie, Universitaet Regensburg
Universitaetsstrasse 31, D-93040 Regensburg, Germany
weniger@vax1.rz.uni-regensburg.dbp.de

Franck Wielonsky
INRIA
2004, Route des Lucioles B.P. 93, 06902 Sophia-Antipolis Cdx, France

Jet Wimp
Dept. of Mathematical Sciences, Drexel University
19104 Philadelphia, Pennsylvania, U.S.A.
jwimp@mcs.drexel.edu

Nancy Wyshinski
Dept of Mathematics, Trinity College
CT 06106 Hartford, U.S.A.
nwyshin@starbase.trincoll.edu

WELCOME

It is my pleasure to welcome the participants of the international conference on "Nonlinear Numerical Methods and Rational Approximation" which is organised for the third time at our University. I am glad that the scientific committee chose our University to host the conference. This meeting gives me the opportunity to say a few words about the University of Antwerp and to express some outsider's views on the topic of the congress.

The University of Antwerp consists of three cooperating centres: RUCA, UFSIA and UIA. They are governed by a common board and were respectively founded in 1852 and 1971. The University of Antwerp has approximately 8500 students which makes it the third largest in Flanders. Our library contains more than 1.3 million volumes and about 7000 journals. The operating language is Dutch, but some post-graduate programmes are taught in English, while the internationalisation of scientific research has made a knowledge of Dutch unnecessary for scientific work.

The general topics of interest of this audience are numerical analysis, approximation theory and applied mathematics. Nonlinear numerical methods and rational approximation may be a narrow field for the outside world, but it provides a vast playground for the chosen ones. It can fascinate specialists from Moscow to South Africa, from Boulder in Colorado and from sunny Florida to Zürich in Switzerland. People from all over the world meet in Antwerp, with a common interest that stretches across all the boundaries of countries and across all the boundaries of mathematics.

Mathematicians have always had the connotation of being eccentric. It is quite common that a painful silence occurs during a conversation when one of the participants has confessed to be a mathematician, as if he speaks another language... And they do: mathematicians do speak their own language. Henri Poincaré once said "Mathematics is the art of giving the same name to different things". But mathematicians are creative enough to invent their proper words too. Who would ever have been so clever to translate only half of a German word into English? Who else, but mathematicians would ever see the logic of the existence of a word like "eigenvalue"? Mathematicians also invent words or give them a special meaning. Words like fractal, wavelet or multigrid for instance belong to the vocabulary of every mathematician nowadays, but if you asked the man in the street, the diversity of possible definitions you would get for such terms would certainly diverge towards infinity. So far for the linguistic note.

From "A History of Scientific Computing" edited by Nash, I quote a precise description of what numerical analysis is all about:

"The numerical analyst of today often finds himself in a position in which he does not know what he is doing, and worries whether what he says is true. This undesirable state of affairs is caused, to a large extent, by present day methods of computation on high-speed electronic computers. For, most frequently, only the final result from a computation is read out from the computer's memory and most if not all the intermediate computations are performed more or less blindly. Thus, the numerical analyst may only become aware of a pathological situation that has developed if it renders the final result obviously incorrect."

I am convinced that many types of mathematicians are represented here at the conference. This is certainly a good basis for fruitful collaboration. The major topics of the conference, whether you call it numerical analysis, computer science, applied mathematics, scientific computing, or whatever you think appropriate, are reaching maturity. This, one can tell by the fact that several of the founders of the discipline are no longer with us. I can cite names like Wilkinson, Werner, Collatz, Henrici and recently Householder with whom some of you may have collaborated or whom you may have met at conferences like these. But mathematics is as old as humanity and it will probably exist as long as humanity exists. It is like a chameleon always changing colours and always posing new challenges to new generations. May this conference be a major step forward in its development. I wish all of you a successful meeting!

F. Adams*

* Rector, Universiteit Antwerpen (UIA), Universiteitsplein 1, B-2610 Wilrijk-Antwerp, Belgium

ORTHOGONAL

POLYNOMIALS

ZEROS OF ORTHOGONAL AND BIORTHOGONAL POLYNOMIALS: SOME OLD, SOME NEW

D.S. LUBINSKY
Department of Mathematics, University of the Witwatersrand, P.O. Wits 2050, Rep. of South Africa.

Abstract. We briefly review some old and new results on zeros of orthogonal polynomials for weights on $(-\infty, \infty)$. In addition, we discuss the zeros of some special biorthogonal polynomials.

AMS(MOS) Classification: Primary 42C05, Secondary 33E99.

Key words: : Zeros of orthogonal polynomials, Spacing of Zeros, Freud Weights, Biorthogonal polynomials.

1. INTRODUCTION

Let $W := e^{-Q}$, where $Q : I\!R \to I\!R$ is measurable, everywhere finite, and satisfies

$$\lim_{|x|\to\infty} Q(x)/\log|x| = \infty.$$

The latter condition ensures that W decreases so rapidly at $\pm\infty$ that all the moments

$$(1.1) \qquad \nu_j := \int_{-\infty}^{\infty} x^j W^2(x)dx, \; j = 0, 1, 2, \dots ,$$

converge. Consequently we can define orthonormal polynomials

$$(1.2) \qquad p_n(x) = p_n(W^2, x) = \gamma_n x^n + \cdots, \; \gamma_n = \gamma_n(W^2) > 0,$$

satisfying

$$\int_{-\infty}^{\infty} p_n(W^2, x)p_m(W^2, x)W^2(x)dx = \delta_{mn} \forall m, \; n.$$

We write the familiar three term recurrence relation in the form

$$(1.3) \qquad xp_n(x) = A_{n+1}p_{n+1}(x) + B_n p_n(x) + A_n p_{n-1}(x), \; A_n = \frac{\gamma_{n-1}}{\gamma_n}, \; n \geq 1.$$

We denote the polynomials of degree $\leq n$ by $\mathcal{P}_n$ and $C, C_1, C_2, \dots$ denote positive constants independent of n, x and $P \in \mathcal{P}_n$.

For those unfamiliar with the theory of orthogonal polynomials for weights on $(-\infty, \infty)$, the above formulations may seem strange, so let us try to motivate the notation: We write W^2 rather than W for the weight since this simplifies formulation

A. Cuyt (ed.), Nonlinear Numerical Methods and Rational Approximation II, 3–15.

of theorems (the awkward alternative would be to keep writing $W^{1/2}$). Moreover, the Q here may be thought of as generalizing the Hermite case where $Q(x) := \frac{1}{2}x^2$. Indeed our principal examples are the weights W_β^2, $\beta > 0$, defined by

$$(1.4) \qquad W_\beta^2(x) := \exp(-2Q_\beta(x)) = \exp(-|x|^\beta), \text{ where } Q_\beta(x) := \frac{1}{2}|x|^\beta.$$

The most spectacular recent developments in this topic have all involved potential theory, but we shall confine ourselves to one simple consequence thereof. G. Freud and P. Nevai in the 1970's realized the importance of *infinite–finite range inequalities.* It was H.N. Mhaskar and E.B. Saff that found the sharp form thereof in 1984 [18], [19] by proving that

$$\|PW\|_{L_\infty(\mathbb{R})} = \|PW\|_{L_\infty[-a_n,a_n]}, \; P \in \mathcal{P}_n,$$

where a_n is a number that depended only on n, not on the particular polynomial P of degree $\leq n$. More precisely, they proved that if $W = e^{-Q}$ and Q is even and convex, then the "best" choice of a_n is the positive root of the equation

$$(1.5) \qquad n = \frac{2}{\pi}\int_0^1 a_n t Q'(a_n t)\frac{dt}{\sqrt{1-t^2}}.$$

This number a_n, commonly called the *Mhaskar–Rahmanov–Saff number*, plays an important descriptive role in orthogonal polynomials. For example, if $Q = Q_\beta$, then

$$a_n = Cn^{1/\beta}, \; n \geq 1.$$

In Section 2 of this paper, we briefly survey results on zeros of orthogonal polynomials for weights on $\mathbb{R}$, such as largest zeros, zero distribution and spacing of zeros. In Section 3, we discuss the zero distribution of some special biorthogonal polynomials.

2. ZEROS OF ORTHOGONAL POLYNOMIALS

2.1. Largest Zeros

Since our weights $W^2 = e^{-2Q}$ have non–compact support, their largest zeros may be expected to approach infinity. How fast? It was G. Freud and P. Nevai who showed that x_{1n} grows roughly like $Q^{[-1]}(n)$ at least for smooth even Q, with $Q^{[-1]}$ denoting the inverse function of Q on $(0,\infty)$ [21]. Well in a fundamental 1984 paper [23], E.A. Rahmanov resolved one of Freud's conjectures, without knowing of its existence! He showed that if $W = e^{-Q}$, where for some $\beta > 1$,

$$(2.1) \qquad \lim_{|x|\to\infty} Q(x)/|x|^\beta = C_1 > 0,$$

then the largest zero x_{1n} of $p_n(W^2, x)$ satisfies

$$(2.2) \qquad \lim_{n\to\infty} x_{1n}/n^{1/\beta} = C_2 > 0.$$

Effectively, the constants C_1 and C_2 are related in such a way that if $a_n = a_n(Q)$, then

(2.3) $$\lim_{n\to\infty} x_{1n}/a_n = 1.$$

So a_n describes the asymptotic behaviour of the largest zero.

For the Hermite weight $W^2(x) = e^{-x^2}$, $a_n = \sqrt{2n}$, but much more precise asymptotics are known for x_{1n}. In fact for fixed $j \geq 1$, [27, p.132],

$$x_{jn} = \sqrt{2n}\Big(1 - 6^{-1/3}[2n]^{-2/3}\iota_j + o(n^{-2/3})\Big),\ n \to \infty,$$

where ι_j is the jth zero of Airy's function. (This is the solution of the differential equation $y'' - \frac{1}{3}xy = 0$ that remains bounded as $n \to \infty$). By placing hypotheses on the recurrence coefficients A_n rather than on the weight W^2, and by using extremal properties of x_{jn} as well as quadratic forms, Máté, Nevai and Totik [15] proved a certain invariance of this asymptotic behaviour: Assuming

(2.4) $$A_n = cn^\delta\left[1 + o(n^{-2/3})\right],\ n \to \infty,$$

for some c, $\delta > 0$, they proved that for each fixed j,

(2.5) $$x_{jn} = 2cn^\delta\Big[1 - 6^{-1/3}[n/\delta]^{-2/3}\iota_j + o(n^{-2/3})\Big],\ n \to \infty.$$

The asymptotic (2.4) is known to be true if for example $W^2 = W_\beta^2$, where $\beta = 2, 4, 6, \ldots$ [16]. Moreover the proof of Freud's conjecture on recurrence coefficients [11] under quite general conditions is evidence that (2.4) may well hold for arbitrary $\beta > 0$. Al. Magnus has further evidence in this direction [14].

An intermediate result between those of [15] and [23] is due to the author and A.L. Levin. Since we shall refer again to the hypotheses of [9], we record them:

Definition 2.1 Let $W := e^{-Q}$, where $Q : I\!R \to I\!R$ is even, continuous and Q'' is continuous in $(0,\infty)$, $Q' > 0$ in $(0,\infty)$, and for some A, $B > 1$,

(2.6) $$A \leq 1 + \frac{xQ''(x)}{Q'(x)} \leq B,\ x \in (0,\infty).$$

Then we write $W \in \mathcal{F}$.

For example, if $\beta > 1$, $W(x) = W_\beta(x) \in \mathcal{F}$, and here $A = B = \beta$. In [9, Cor. 1.2, p.466], it is shown (as an auxiliary result) that for $W \in \mathcal{F}$, there exists $C > 0$ such that

(2.7) $$\left|x_{1n}/a_n - 1\right| \leq Cn^{-2/3},\ n \geq 1.$$

This was recently extended to the weights W_β^2, $\beta \leq 1$ [10]. A reasonable conjecture, is that under the same, or even less, hypotheses,

(2.8) $$x_{1n}/a_n = 1 - k(Q)n^{-2/3} + o(n^{-2/3}),$$

where $k(Q)$ is independent of n. Obviously identifying $k(Q)$ in general is of some interest.

This merely scratches the surface. Goncar and Rahmanov [8] subsequently generalized Rahmanov's original work, as did others such as Mhaskar, Saff, Totik, and the author [12], [19], [20], [24]. What about non-even, or non-symmetric weights? Well there are two numbers a_n instead of just one! If $W = e^{-Q}$ is convex, where Q is not even, we define $a_n^{(\ell)} < 0$ and $a_n^{(r)} > 0$ by the equations

$$\frac{1}{\pi}\int_{a_n^{(\ell)}}^{a_n^{(r)}} \frac{Q'(t)\left(t - a_n^{(\ell)}\right)}{\sqrt{\left(a_n^{(r)} - t\right)\left(t - a_n^{(\ell)}\right)}}\,dt = n; \tag{2.9}$$

$$\frac{1}{\pi}\int_{a_n^{(\ell)}}^{a_n^{(r)}} \frac{Q'(t)\left(a_n^{(r)} - t\right)}{\sqrt{\left(a_n^{(r)} - t\right)\left(t - a_n^{(\ell)}\right)}}\,dt = -n. \tag{2.10}$$

These are easily seen to reduce to the defining equation for a_n in the case where Q is even. Under fairly general conditions on Q it is known that

$$\lim_{n\to\infty} x_{nn}/a_n^{(\ell)} = 1 = \lim_{n\to\infty} x_{1n}/a_n^{(r)}. \tag{2.11}$$

The first published result of this type appears in the work of Van Assche [31]. The recent work of Buyarov [1], [2] implicitly contains very general results of this type. Buyarov has not only generalized Rahmanov's results in their entirety, but also shown that some of the hypotheses are the minimum possible. We shall discuss this further later.

All this is essentially for Q of polynomial growth (or faster) at infinity. What about Q of slower than polynomial growth? Here the largest zero grows extremely rapidly [4], [30]. At the other end of the scale, for weights that decay very rapidly, so much so that Q grows faster than any polynomial at ∞, an old result of Erdös is surprisingly still the most general [5]. It asserts that if $W^2 = e^{-2Q}$, and Q grows so rapidly that $\forall r > 1$, $\exists C_r$ such that

$$\frac{y}{x} \geq r,\ |x| \geq C_r \Rightarrow Q(y) \geq 2Q(x), \tag{2.12}$$

then the smallest zero x_{nn} and largest zero x_{1n} satisfy

$$\lim_{n\to\infty} x_{nn}/u_n = 1;\ \lim_{n\to\infty} x_{1n}/\nu_n = 1, \tag{2.13}$$

where (u_n, ν_n) is the largest interval such that $Q(x) < n\log 2,\ x \in (u_n, \nu_n)$. Here $\log 2$ may be replaced by any fixed positive number.

2.2. Zero Distribution

Largest zeros are one thing. What about all the zeros, say *zero distribution?* Well what do we mean by this? If we are given a sequence of polynomials each with zeros in $(-1,1)$, then to each polynomial of degree n, we may assign a *counting measure*, that is, a measure assigning mass $\frac{1}{n}$ to each of the n zeros of the polynomial, with repetition according to multiplicity. This is a probability measure, that is, has total mass one. The behaviour of these counting measures as $n \to \infty$, gives an indication of how the zeros distribute themselves. Crucial in this is the fact that all these measures have support on a fixed compact set. This enables one to use Helly's Theorem or the Banach–Alaoglu Theorem to extract weakly convergent subsequences. (The reader unfamiliar with these tools should persevere, we shall explain below).

If we try to form counting measures for orthogonal polynomials on $\mathbb{R}$, we land up in the quandary that they have non–compact support. To circumvent this, several authors independently used *contracted counting measures*, that is at least for even weights, they divided all the zeros by the largest zero x_{1n} thereby reducing the contracted zeros to $[-1,1]$. Equivalently in many cases, we can divide by the Mhaskar–Rahmanov–Saff number a_n. For non–even weights, we contract $[x_{nn}, x_{1n}]$ to $[-1,1]$ by a linear transformation.

Walter Van Assche explored the alternative possibility of weighting the zeros, avoiding contraction [29]. Here we shall concentrate on the contractions, an idea apparently due to Erdös [5].

Let δ_z denote a Dirac delta or unit mass at $z \in \mathbb{C}$. We define the contracted zero counting measure μ_n for $p_n(W^2,x)$ corresponding to an even weight W^2 by

$$\mu_n := \frac{1}{n}\sum_{j=1}^{n} \delta_{x_{jn}/x_{1n}}. \tag{2.14}$$

Thus μ_n has total mass 1 and assigns mass $\frac{1}{n}$ to each contracted zero x_{jn}/x_{1n}. Note that if $\mathcal{L} \subset \mathbb{R}$, then

$$\begin{aligned}\mu_n(\mathcal{L}) &= \int_{\mathcal{L}} d\mu_n = \frac{1}{n}\cdot \text{ Number of zeros of } p_n(x\cdot x_{1n}) \in \mathcal{L}\\ &= \frac{1}{n}\cdot \text{ Number of contracted zeros } x_{jn}/x_{1n} \in \mathcal{L}\\ &= \text{ Proportion of contracted zeros } x_{jn}/x_{1n} \in \mathcal{L}.\end{aligned}$$

So this really is a counting measure. More generally for non–even weights, μ_n has the more complicated form

$$\mu_n := \frac{1}{n}\sum_{j=1}^{n} \delta_{-1+2\left(\frac{x_{jn}-x_{nn}}{x_{1n}-x_{nn}}\right)}. \tag{2.15}$$

The behaviour as $n \to \infty$ of μ_n describes the contracted zero distribution of the zeros of $p_n(W^2,x)$, $n \to \infty$. But how can we talk about limits of measures? We use the concept of *weak convergence*, which comes from functional analysis. We write

$$d\mu_n \xrightarrow{\star} d\mu, \; n \to \infty,$$

if for each continuous $f : [-1,1] \to \mathbb{R}$, we have

$$\lim_{n\to\infty} \int_{-1}^{1} f d\mu_n = \int_{-1}^{1} f d\mu.$$

Erdös showed in 1972 [5] that under (2.12), the contracted zero counting measures μ_n converge weakly to the arcsin distribution:

$$d\mu_n(x) \xrightarrow{\star} \frac{dx}{\pi\sqrt{1-x^2}}, \ n \to \infty. \tag{2.16}$$

He also noted that such a result could not be true for the Hermite weight and this began the race to determine the zero distribution for W_β^2, $\beta > 0$. In 1983–84 Mhaskar and Saff [18] and Rahmanov [23] independently proved the convergence to a contracted zero distribution for a class of weights W^2 including W_β^2. (Earlier J. Ullmann had determined what the limit distribution should be [28], with Nevai and Dehesa even earlier computing the moments of that distribution [22]). A sort of combination of the Mhaskar–Rahmanov–Saff results is that if for some $\beta > 0$, (2.1) is satisfied, then

$$d\mu_n \xrightarrow{\star} \nu_\beta(x)dx, \ n \to \infty, \tag{2.17}$$

where

$$\nu_\beta(x) = \frac{\beta}{\pi}\int_{|x|}^{1} \frac{y^{\beta-1}}{\sqrt{y^2-x^2}}dy, \ x \in (-1,1). \tag{2.18}$$

For those unfamiliar with weak convergence, an alternative formulation of (2.17) is easier to assimilate: $\forall -1 \leq a < b \leq 1$,

$$\begin{aligned} &\lim_{n\to\infty} \mu_n[a,b] \\ &= \lim_{n\to\infty} (\text{Proportion of zeros of } p_n \text{ in } [ax_{1n}, bx_{1n}]) = \int_a^b \nu_\beta(x)dx. \end{aligned} \tag{2.19}$$

Note that as $\beta \to \infty$, we have for each $x \in (-1,1)$,

$$\nu_\beta(x) \to \nu_\infty(x) = \frac{1}{\pi\sqrt{1-x^2}}.$$

So we may think of Erdös' result as the limiting case of the Mhaskar–Rahmanov–Saff result. In the other direction, as $\beta \to 0+$,

$$\nu_\beta(x)dx \xrightarrow{\star} d\delta_0(x),$$

that is we have weak convergence to a Dirac delta at 0. The latter zero distribution corresponds to weights that decrease extremely slowly, for example the Stieltjes–Wigert weight (or log–normal distribution)

$$W^2(x) = \exp(-k(\log x)^2), \ k > 0.$$

Some zero distribution results have been obtained in this context [30].

Generalizations of these results have been given in [8], [12], [20], [24], [31] and most recently by Buyarov [1], [2]. Moreover rates of weak convergence have been studied [17]. In dealing with general weights, we have to take account of two new factors: Firstly, for non-symmetric weights, we map $[x_{nn}, x_{1n}]$ linearly to $[-1, 1]$, rather than just divide all the zeros by x_{1n}. Secondly, the contracted measures need not converge weakly to a fixed measure, so the weak convergence of μ_n involves a sequence of explicitly defined measures rather than a single measure.

The complexities of the possible zero distributions probably require an innovative formulation, and Buyarov's results certainly do. The idea seems to be that to any sequence of orthonormal polynomials for a weight on the real line, there corresponds an increasing function $\rho : \mathbb{R} \to \mathbb{R}$ such that at least approximately, $\rho(-n) = x_{nn}$; $\rho(n) = x_{1n}$, $n \geq 1$. So we build our description around ρ.

First, we need some more notation. Recall that the Green's function $g_{[a,b]}(z)$ for the exterior of an interval $[a, b]$ with pole at ∞ is a function such that (i) $g_{[a,b]}$ is harmonic in $\mathbb{C} \setminus [a, b]$; (ii) $g_{[a,b]}$ has (boundary) value 0 on $[a, b]$; (iii) $g_{[a,b]}(z) - \log|z|$ is bounded as $|z| \to \infty$. An explicit formula is

$$g_{[a,b]}(z) := \log\left|\frac{2}{b-a}\left[z - \frac{a+b}{2} + \sqrt{(z-a)(z-b)}\right]\right| \tag{2.20}$$

where the branch of the square root is the principal one. Now we can give Buyarov's formulation:

Definition 2.2 Let $\rho : \mathbb{R} \to \mathbb{R}$ be a monotone increasing continuous function with $\rho(0) = 0$ and

$$\lim_{x \to -\infty} \rho(x) = -\infty; \ \lim_{x \to \infty} \rho(x) = \infty,$$

and

$$\lim_{x \to \infty} \rho(x)\rho(-x)/x^2 = 0. \tag{2.21}$$

Define $Q : \mathbb{R} \to \mathbb{R}$ by

$$Q(x) = \int_0^\infty g_{\left[\rho(-\tau),\rho(\tau)\right]}(x)d\tau. \tag{2.22}$$

Then we write $Q \in \mathcal{B}$. If in addition,

$$\lim_{y \to 1+} \sup_{x \in \mathbb{R}\setminus\{0\}} \frac{\rho(yx)}{\rho(x)} = 1, \tag{2.23}$$

then we write $Q \in \mathcal{B}+$.

Buyarov remarks that any function $Q : \mathbb{R} \to \mathbb{R}$ that is convex and whose derivatives increase (strictly?) belong to $\mathcal{B}$. The integral in (2.22) is trivially convergent: For large enough τ, $x \in \left[\rho(-\tau), \rho(\tau)\right]$ and then $g_{\left[\rho(-\tau),\rho(\tau)\right]}(x) = 0$. Buyarov's paper concentrates on logarithmic asymptotics for the orthonormal polynomials $p_n(W^2, x)$ corresponding to the weight $W^2 = e^{-2Q}$, but implicit in his paper are results on the contracted zero distribution.

For a start he implicitly shows that

$$\lim_{n\to\infty} x_{nn}/\rho(-n) = 1 = \lim_{n\to\infty} x_{1n}/\rho(n). \tag{2.24}$$

As a solution to a certain extremal problem he introduces a density function ν_n (it is the derivative of the equilibrium measure for the exterior field Q with total mass n): Let $\rho^{[-1]}$ denote the inverse function of ρ. Then set

$$\nu_n(x) := \int_{\left|\rho^{[-1]}(x)\right|}^{n} \frac{d\tau}{\pi\sqrt{\left(\rho(\tau)-x\right)\left(x-\rho(-\tau)\right)}},\ x \in \left[\rho(-n),\rho(n)\right].$$

The non-negative function $\nu_n(x)$ satisfies

$$\int_{\rho(-n)}^{\rho(n)} \nu_n(x)dx = n,$$

that is, is the density of a measure whose total mass over $\mathbb{R}$ is n. We need to shift this density function to $[-1,1]$ and reduce its mass to 1, so we define

$$\omega_n(t) := \frac{\rho(n)-\rho(-n)}{2n}\nu_n\left(\rho(-n)+\frac{t+1}{2}\left[\rho(n)-\rho(-n)\right]\right),\ t\in[-1,1]. \tag{2.25}$$

So,

$$\int_{-1}^{1} \omega_n(t)dt = 1.$$

Our contracted zero counting measure now has the form (2.15), and Buyarov implicitly shows that for $Q \in \mathcal{B}$,

$$d\mu_n(t) - \omega_n(t)dt \xrightarrow{\star} 0,\ n\to\infty. \tag{2.26}$$

If also $Q \in \mathcal{B}+$, then for any weight $\widehat{W}^2 = e^{-2\widehat{Q}}$, for which

$$\lim_{|x|\to\infty} \widehat{Q}(x)/Q(x) = 1, \tag{2.27}$$

the same zero distribution result holds. As a corollary, Buyarov treats weights of the form

$$Q(x) := \begin{cases} r_2|x|^\alpha, & x>0 \\ r_1|x|^\beta, & x<0 \end{cases},$$

assuming $\alpha > \beta > \frac{1}{2}$, $\alpha > 1$.

All this takes some getting used to, but the complicated formulations seem essential!

2.3. Uniform Spacing of the Zeros

It was G. Freud who proved the first results on spacing of successive zeros of orthogonal polynomials, with a view to applying these to weighted L_1 approximation. As a preliminary to proving bounds on orthogonal polynomials, A.L. Levin and the author obtained uniform estimates on the spacing of the zeros of $p_n(W^2, x)$ for $W^2 = e^{-2Q} \in \mathcal{F}$. In describing this spacing, we need a certain sequence of functions

$$\psi_n(x) := \max\left\{n^{-2/3},\, 1 - \frac{|x|}{a_n}\right\},\ x \in I\!R,\ n \geq 1. \tag{2.28}$$

These play much the same role for various weighted approximation problems on $I\!R$ as do the functions $\frac{1}{n}\max\left\{\frac{1}{n},\, \sqrt{1-x^2}\right\}$ in algebraic polynomial approximation problems on $[-1, 1]$. (Recall the phenomenon that the rate of algebraic polynomial approximation improves towards the endpoints of the interval). The result in [9, Cor. 1.2(b), p.467] states that for some C_1 and $C_2 > 0$ depending only on Q,

$$C_1 \leq \left(x_{j-1,n} - x_{j+1,n}\right) \Big/ \left(\frac{a_n}{n}\psi_n(x_{jn})^{-1/2}\right) \leq C_2 \tag{2.29}$$

uniformly for $n \geq 2$ and $2 \leq j \leq n-1$. This was slightly improved in [3, Lemma 4.4] where $x_{j-1,n} - x_{j+1,n}$ was replaced by $x_{jn} - x_{j+1,n}$. It is easier to assimilate this by restricting x_{jn} to lie in certain intervals. For example, if $\sigma < 1$ is fixed, then

$$C_3 \leq \left(x_{jn} - x_{j+1,n}\right) \Big/ \frac{a_n}{n} \leq C_4 \tag{2.30}$$

uniformly for n and j such that $|x_{jn}| \leq \sigma a_n/n$. Moreover, for each fixed $k \geq 1$, we have as $n \to \infty$,

$$C_5 \leq \left(x_{kn} - x_{k+1,n}\right) \Big/ \frac{a_n}{n^{2/3}} \leq C_6. \tag{2.31}$$

So the spacing between successive zeros seems to increase as we move to the largest zero. Accordingly, let us pose this as an open problem: Investigate, for $x_{jn} > 0$, the monotonicity of the spacing $x_{jn} - x_{j+1,n}$ as j increases.

It is noteworthy that all these spacing results apply to W_β^2 only when $\beta > 1$. For $\beta \leq 1$, the spacing is rather different. For $\beta < 1$, this is related to the indeterminacy of the corresponding moment problem, while $\beta = 1$ is on the boundary of determinacy. The interesting features of the spacing of zeros in the case $\beta = 1$ were already explored to some extent by Freud, Giroux and Rahman in the 1970's [7].

The author and A.L. Levin recently estimated the Christoffel functions for the weights $W_\beta^2(x)$, $\beta \leq 1$, and deduced spacing estimates for zeros. The description thereof requires the functions $\Lambda_n(x)$, which are given by different formulas for $|x| \leq a_n/2$ and for $|x| \geq a_n/2$. Set

$$\Lambda_n(x) := \begin{cases} (1+|x|)^{\beta-1}, & \beta < 1 \\ 1/\log\frac{2n}{1+|x|}, & \beta = 1 \end{cases},\quad |x| \leq a_n/2 \tag{2.32}$$

and

$$\Lambda_n(x) := \frac{a_n}{n}\psi_n(x)^{-1/2},\ |x| > a_n/2, \tag{2.33}$$

where $\psi_n(x)$ is defined by (2.28). It is shown in [10, Corollary 1.2] that there exist C_1 and C_2 such that uniformly for $n \geq 2$ and $2 \leq j \leq n$,

$$C_1 \leq (x_{j-1,n} - x_{j+1,n})/\Lambda_n(x_{jn}) \leq C_2.$$

In actual fact, we could define $\Lambda_n(x)$ by (2.33) for $|x| \geq \sigma a_n$ any fixed $\sigma > 0$, so for zeros restricted to that range, the spacing of the zeros is the same as that for W_β^2, $\beta > 1$. The difference lies closer to 0. We see that when $\beta = 1$, zeros near 0 are $(\log n)^{-1}$ apart, while for $\beta < 1$, such zeros are a constant distance apart.

Clearly there is ample room for extensions and generalizations here !

3. BIORTHOGONAL POLYNOMIALS

Many methods to accelerate the convergence of the sequence of partial sums $H_1(z)$, $H_2(z)$, $H_3(z), \ldots$ of a power series effectively generate weighted averages of the form

$$\sum_{j=0}^{n} c_{n,j} H_{n+j}(z) z^j \Big/ \sum_{j=0}^{n} c_{n,j} z^j .$$

For example, Padé approximants, the ϵ-algorithm, and the T-transformation generate weighted averages of this form.

The denominator polynomials in these weighted averages clearly determine the convergence acceleration method. In the modified T-transformation investigated by Sidi and Levin [25], the denominator polynomials have the form

$$D_n(z) := \sum_{j=0}^{n} \binom{n}{j} (j+1)^n (-z)^j . \tag{3.1}$$

I am inclined to call these *the Sidi polynomials*. To the orthogonal polynomial enthusiast, these polynomials have a particularly interesting biorthogonality property:

$$\int_0^1 (\log x)^j D_n(x) dx = 0,\ 0 \leq j \leq n-1. \tag{3.2}$$

In fact, it is easily seen that this relation determines $D_n(z) \in \mathcal{P}_n$, apart from multiplication by a constant factor. Moreover, (3.2) implies that the zeros of D_n are real, simple, and lie in $(0,1)$. The proof of (3.2) is easy: We set $z = e^t$ in (3.1) and multiply by e^t to obtain

$$\begin{aligned} e^t D_n(e^t) &= \sum_{j=0}^{n} \binom{n}{j} (-1)^j (j+1)^n e^{(j+1)t} \\ &= \sum_{j=0}^{n} \binom{n}{j} (-1)^j \left(\frac{d}{dt}\right)^n e^{(j+1)t} \\ &= \left(\frac{d}{dt}\right)^n \left[e^t (1-e^t)^n\right]. \end{aligned} \tag{3.3}$$

In the last line, we used the binomial theorem. Afficionados of the classical orthogonal polynomials will recognize this as *a Rodrigues type formula.* To prove (3.2) we set $x = e^t$ in the integral on the left–hand side there, and integrate successively by parts, using (3.3).

There is a vast literature on asymptotics of orthogonal polynomials for weights on finite intervals [21], [26] and precious little on asymptotics of such biorthogonal polynomials. So it seems reasonable to start with the D_n. Applying Cauchy's integral formula for derivatives to (3.3) gives a Schläfli type integral:

$$D_n(z) = \frac{n!}{2\pi i} \int_{\Gamma} \frac{e^{\nu}}{\nu} \left(\frac{1 - ze^{\nu}}{\nu} \right)^n d\nu, \tag{3.4}$$

where Γ is a positively oriented simple closed contour enclosing 0. This is a good (but difficult) candidate for steepest descent, and that was the method used in [13].

Here we shall restrict ourselves to discussing the zero distribution, which was an easy consequence of the asymptotics in [13]. Define

$$h(y) := \frac{\sin y}{y} e^{y \cot y - 1}, \; y \in [0, \pi]. \tag{3.5}$$

It is shown (by differentiation) in [13] that h is strictly decreasing in this interval from $h(0) = 1$ to $h(\pi) = 0$. Let us denote the inverse function of h by $h^{[-1]} : [0,1] \to [0, \pi]$. Furthermore, let y_{jn}, $1 \le j \le n$ denote the zeros of D_n and define the zero counting measure

$$\mu_n := \frac{1}{n} \sum_{j=1}^{n} \delta_{y_{jn}},$$

which has support on $(0, 1)$. It is shown in [13, Thm. 1.3] that

$$d\mu_n(x) \xrightarrow{\star} -\frac{dx}{\pi h'\left(h^{[-1]}(x)\right)}. \tag{3.6}$$

That is, $\forall\, 0 \le a < b \le 1$,

$$\lim_{n \to \infty} \mu_n[a, b] = \int_a^b -\frac{dx}{\pi h'\left(h^{[-1]}(x)\right)} = \frac{1}{\pi}\left(h^{[-1]}(a) - h^{[-1]}(b)\right). \tag{3.7}$$

This is quite different from the arcsin distribution that we obtain for wide classes of orthogonal polynomials associated with weights on finite intervals [26]. Clearly, this raises an interesting question: Suppose that $\psi : (0,1) \to I\!R$ is a strictly increasing function and suppose we determine monic polynomials

$$E_n(x) = x^n + \ldots \in \mathcal{P}_n$$

by the biorthogonality condition

$$\int_0^1 \psi(x)^j E_n(x) dx = 0, \; 0 \le j \le n - 1. \tag{3.8}$$

Determine the zero distribution of $\{E_n(x)\}_{n=0}^{\infty}$. Herbert Stahl has conjectured that the distribution of the zeros of E_n as $n \to \infty$ is determined by a certain integral equation, though the details are still being filled in. There seems to be some potential for research on various aspects of these polynomials!

Acknowledgements

All of the work reported on here is joint work with G. Criscuolo, B. Della Vecchia, A.L. Levin, G. Mastroianni, A. Sidi, and H. Stahl. The collaboration with them has been most stimulating.

References

1. Buyarov, V.S.: 1991, 'Logarithmic Asymptotics of Polynomials Orthogonal on the Real Axis', *Soviet Math. Dokl.*, Vol. no. **43**, pp. 743–746.
2. Buyarov, V.S.: 1992, 'Logarithmic Asymptotics of Polynomials Orthogonal on $\boldsymbol{R}$ with Asymmetric Weights', *Notes of the Academy of Sciences of the USSR*, Vol. no. **50**, pp. 789–795.
3. Criscuolo, G., Della Vecchia, B., Lubinsky, D.S. and Mastroianni, G.: 'Functions of the Second Kind for Freud Weights and Series Expansions of Hilbert Transforms', submitted.
4. Driver, K.A.: 1991, 'Nth Root Asymptotics for Extremal Errors Associated with Slowly Decreasing Weights', *Quaestiones Mathematicae*, Vol. no. **14**, pp. 337–354.
5. Erdős, P.: 1972, 'On the Distribution of the Roots of Orthogonal Polynomials', in: Proceedings of the Conference on the Constructive Theory of Functions (eds. G. Alexits, et. al.), Akademiai Kiado, Budapest, pp. 145–150.
6. Freud, G.: 1971, 'Orthogonal Polynomials', Pergamon Press/Akademiai Kiado, Budapest/Oxford.
7. Freud, G., Giroux, A. and Rahman, Q.I.: 1978, 'On Approximation by Polynomials with Weight $\exp(-|x|)$', *Canad. J. Math.*, Vol. no. **30**, pp. 358–372 (in French).
8. Goncar, A.A. and Rahmanov, E.A.: 1986, 'Equilibrium Measure and The Distribution of Zeros of Extremal Polynomials', *Math. USSR. Sbornik*, Vol. no. **53**, pp. 119–130.
9. Levin, A.L. and Lubinsky, D.S.: 1992, 'Christoffel Functions, Orthogonal Polynomials, and Nevai's Conjecture for Freud Weights', *Constructive Approximation*, Vol. no. **8**, pp. 463–535.
10. Levin, A.L. and Lubinsky, D.S.: 'Orthogonal Polynomials and Christoffel Functions for $\exp(-|x|^\alpha)$, $\alpha \leq 1$, *J. Approx. Theory*, to appear.
11. Lubinsky, D.S., Mhaskar, H.N. and Saff, E.B.: 1988, 'A Proof of Freud's Conjecture for Exponential Weights', *Constructive Approximation*, Vol. no. **4**, pp. 65–83.
12. Lubinsky, D.S., and Saff, E.B.: 1988, 'Strong Asymptotics for Extremal Polynomials Associated with Exponential Weights', *Springer Lecture Notes in Mathematics*, Vol. no. **1305**, Springer, Berlin.
13. Lubinsky, D.S. and Sidi, A.: 'Strong Asymptotics for Polynomials Biorthogonal to Powers of log x', submitted.
14. Magnus, Al.: 1993, 'Refined Asymptotics for Freud Recurrence Coefficients', in: Collection of Unsolved Problems of Linear and Complex Analysis (eds. V. Havin & N. Nikolskii), Springer Lecture Notes, Springer.
15. Máté, A., Nevai, P. and Totik, V.: 1986, 'Asymptotics for the Zeros of Orthogonal Polynomials Associated with Infinite Intervals', *J. London Math. Soc.*, Vol. no. **33**, pp. 303–310.
16. Máté, A., Nevai, P. and Zaslavsky, T.: 1985, 'Asymptotic Expansion of Ratios of Coefficients of Orthogonal Polynomials with Exponential Weights', *Trans. Amer. Math. Soc.*, Vol. no. **287**, pp. 495–505.
17. Mhaskar, H.N.: 1987, 'Some Discrepancy Theorems', in: Approximation Theory, Tampa (ed. E.B. Saff), Springer Lecture Notes, Vol. 1287, Springer, New York, pp. 117–131.
18. Mhaskar, H.N. and Saff, E.B.: 1984, 'Extremal Problems for Polynomials with Exponential Weights', *Trans. Amer. Math. Soc.*, Vol. no. **285**, pp. 203–234.

19. Mhaskar, H.N. and Saff, E.B.: 1985, 'Where Does the Sup-Norm of a Weighted Polynomial Live?', *Constructive Approximation*, Vol. no. **1**, pp. 71–91.
20. Mhaskar, H.N. and Saff, E.B.: 1987, 'Where Does the L_p-Norm of a Weighted Polynomial Live?', *Trans. of the Amer. Math. Soc.*, Vol. no. **303**, pp. 109–124.
21. Nevai, P.: 1986, 'Geza Freud, Orthogonal Polynomials and Christoffel Functions: A Case Study', *J. Approx. Theory*, Vol. no. **48**, pp. 3–167.
22. Nevai, P. and Dehesa, J.S.: 1979, 'On Asymptotic Average Properties of Zeros of Orthogonal Polynomials', *SIAM J. Math. Anal.*, Vol. no. **10**, pp. 1184–1192.
23. Rahmanov, E.A.: 1984, 'On Asymptotic Properties of Polynomials Orthogonal on the Real Axis', *Math. USSR. Sbornik*, Vol. no. **47**, pp. 155–193.
24. Saff, E.B. and Totik, V.: 'Logarithmic Potential with External Fields', Springer–Verlag, in press.
25. Sidi, A. and Levin, D.: 1983, 'Prediction Properties of the t-Transformation', *SIAM J. Num. Anal.*, Vol. no. **20**, pp. 589–598.
26. Stahl, H. and Totik, V.: 1992, 'General Orthogonal Polynomials, Encyclopaedia of Mathematics and its Applications', *Cambridge University Press*, Vol. no. **43**.
27. Szegö, G.: 1975, 'Orthogonal Polynomials', American Mathematical Society Colloquium Publications, Vol. no. 23, American Mathematical Society, Providence, Rhode Island, Fourth edition.
28. Ullman, J.L.: 1990, 'On Orthogonal Polynomials Associated with The Infinite Interval', in: Approximation Theory III, (ed. E.W. Cheney), Academic Press, New York, pp. 889–895.
29. Van Assche, W.: 1985, 'Weighted Zero Distribution for Polynomials Orthogonal on an Infinite Interval', *SIAM J. Math. Anal.*, Vol. no. **16**, pp. 1317–1334.
30. Van Assche, W.: 1987, 'The Ratio of q-like Orthogonal Polynomials', *J. Math. Anal. Applns.*, Vol. no. **128**, pp. 535–547.
31. Van Assche, W.: 1989, 'Norm Behaviour and Zero Distribution for Orthogonal Polynomials with Non-Symmetric Weights', *Constr. Approx.*, Vol. no. **5**, pp. 329–345.

SOME SEQUENCES ARISING IN THE CREATION OF NEW ORTHOGONAL POLYNOMIALS

JET WIMP and HARRY KIESEL
Drexel University, Philadelphia, PA 19104, USA

1. Introduction

Denote by $\widehat{U}_n(x)$ the monic Chebyshev polynomial of the second kind. These polynomials satisfy the recurrence relation

$$\widehat{U}_{n+1}(x) = x\widehat{U}_n(x) - \frac{1}{4}\widehat{U}_{n-1}(x), n = 0, 1, 2, \ldots, \widehat{U}_{-1}(x) = 0, \widehat{U}_0(x) = 1. \tag{1}$$

Two linearly independent solutions of (1) considered as a difference equation are

$$y_n^{(1)}(x) = \lambda_1^n, \lambda_1 = \frac{x + \sqrt{x^2-1}}{2}, y_n^{(2)}(x) = \lambda_2^n, \lambda_2 = \frac{x - \sqrt{x^2-1}}{2}, x \in \mathcal{C} - [-1, 1]. \tag{2}$$

Throughout this paper, we adopt the convention that $\sqrt{z^2-1}$ is analytic when $z \in \mathcal{C} - [-1, 1]$ and positive when z is real and >1. Note that this convention implies that the second solution above is subdominant to the first in the indicated region in the complex plane.

The family $\{\widehat{U}_n(x)\}$ is orthogonal with respect to the weight function

$$\omega(x) = \sqrt{1-x^2}, x \in [-1, 1], \tag{3}$$

whose Stieltjes transform is

$$F(z) = \int_{-1}^{1} \frac{\sqrt{1-x^2}}{z-x} dx = \pi \left[z - \sqrt{z^2-1} \right], z \in \mathcal{C} - [-1, 1]. \tag{4}$$

Now consider the sequence of polynomials $Q_n(x)$ defined by

$$Q_n(x) = (a_n x + b_n)\widehat{U}_{n-1}(x) + (1 - a_n)\widehat{U}_n(x), n = 0, 1, 2, \ldots, a_0 \neq 1. \tag{5}$$

With the exception of $Q_0(x)$ these polynomials are monic also. Working with monic polynomials makes the computations much less cumbersome.

We wish to determine conditions on the otherwise arbitrary sequences $\{a_n\}$, $\{b_n\}$ so that these new polynomials are also an orthogonal set. We note in passing that the same sort of computation could be done using general orthogonal polynomials rather than the special set $\widehat{U}_n(x)$. The difficulties in doing this are enormous, however. The

A. Cuyt (ed.), Nonlinear Numerical Methods and Rational Approximation II, 17–24.

relations involving the sequences $\{a_n\}, \{b_n\}$ are so complicated that they have so far resisted analysis.

Let $n \Rightarrow n+1$ in (5) and use (1) to get

$$Q_{n+1}(x) = (x + b_{n+1})\widehat{U}_n(x) + \frac{1}{4}(a_{n+1} - 1)\widehat{U}_{n-1}(x). \tag{6}$$

Let $n \Rightarrow n+1$ in this equation and again use (1) to get

$$Q_{n+2}(x) = \left\{\frac{1}{4}(a_{n+2} - 1) + x(x + b_{n+2})\right\}\widehat{U}_n(x) - \frac{1}{4}(x + b_{n+2})\widehat{U}_{n-1}(x). \tag{7}$$

Now consider (5), (6), (7) to be three equations in the two unknowns $\widehat{U}_n(x), \widehat{U}_{n-1}(x)$. They will have a solution if and only if the following determinant vanishes:

$$\begin{vmatrix} Q_n & a_n x + b_n & 1 - a_n \\ Q_{n+l} & \frac{1}{4}(a_{n+1} - 1) & x + b_{n+1} \\ Q_{n+2} & -\frac{1}{4}(x + b_{n+2}) & \{\frac{1}{4}(a_{n+2} - 1) + x(x + b_{n+2})\} \end{vmatrix} \tag{8}$$

Expanding this gives

$$\begin{aligned} &Q_n\left\{\frac{1}{4}(a_{n+1} - 1)\left[\frac{1}{4}(a_{n+2} - 1) + x(x + b_{n+2})\right] + \frac{1}{4}(x + b_{n+1})(x + b_{n+2})\right\} - \\ &Q_{n+1}\left\{(a_n x + b_n)\left[\frac{1}{4}(a_{n+2} - 1) + x(x + b_{n+2})\right] + \frac{1}{4}(1 - a_n)(x + b_{n+2})\right\} + \\ &Q_{n+2}\left\{(x + b_{n+1})(a_n x + b_n) - \frac{1}{4}(a_{n+1} - 1)(1 - a_n)\right\} = 0. \end{aligned} \tag{9}$$

If the family $\{Q_n\}$ are to be orthogonal, this must reduce to the standard three-term recurrence relation. We divide by the coefficient of Q_{n+2} to get

$$Q_{n+2} = \eta_n^{(2)} Q_{n+1} - \eta_n^{(1)} Q_n, \tag{10}$$

where

$$\begin{aligned} \eta_n^{(1)} &= \frac{1}{4}\frac{a_{n+1}}{a_n}\frac{\kappa_n^{(1)} x + \kappa_n^{(2)}}{W_n(x)}, \\ \eta_n^{(2)} &= x + (b_{n+2} - b_{n+1}) + \frac{\kappa_n^{(3)} x + \kappa_n^{(4)}}{W_n(x)}, \\ W_n(x) &= (x + b_{n+1})(a_n x + b_n) - \frac{1}{4}(a_{n+1} - 1)(1 - a_n), \end{aligned} \tag{11}$$

the $\kappa_n^{(i)}$ being independent of x,

$$\kappa_n^{(1)} = \frac{1}{4}(b_{n+2} a_{n+1} + b_{n+1}) - \frac{1}{4}\frac{a_{n+1}}{a_n}(b_{n+1} a_n + b_n),$$

$$\kappa_n^{(2)} = \frac{1}{4}\left[b_{n+2}b_{n+1} + \frac{1}{4}(a_{n+1}-1)(a_{n+2}-1)\right] - \frac{1}{4}\frac{a_{n+1}}{a_n}\left[b_{n+1}b_n + \frac{1}{4}(a_{n+1}-1)(a_n-1)\right], \tag{12}$$

$$\kappa_n^{(3)} = -b_{n+1}a_n(b_{n+2}-b_{n+1}) + \frac{1}{4}\left[a_n(a_{n+2}-1) - a_{n+1}(a_n-1)\right],$$

$$\kappa_n^{(4)} = -(b_{n+2}-b_{n+1})\left[b_{n+1}b_n + \frac{1}{4}(a_{n+1}-1)(a_n-1)\right] + \frac{1}{4}\left[b_n(a_{n+2}-1) - b_{n+2}(a_n-1)\right].$$

We must have $\kappa_n^{(i)} = 0$, and this gives the four equations,

$$(b_{n+2}a_{n+1} + b_{n+1}) - \frac{a_{n+1}}{a_n}(b_{n+1}a_n + b_n) = 0, \tag{13}$$

$$\left[b_{n+2}b_{n+1} + \frac{1}{4}(a_{n+1}-1)(a_{n+2}-1)\right] - \frac{a_{n+1}}{a_n}\left[b_{n+1}b_n + \frac{1}{4}(a_{n+1}-1)(a_n-1)\right] = 0, \tag{14}$$

$$b_{n+1}a_n(b_{n+2}-b_{n+1}) - \frac{1}{4}\left[a_n(a_{n+2}-1) - a_{n+1}(a_n-1)\right] = 0, \tag{15}$$

$$(b_{n+2}-b_{n+1})\left[b_{n+1}b_n + \frac{1}{4}(a_{n+1}-1)(a_n-1)\right] - \frac{1}{4}\left[b_n(a_{n+2}-1) - b_{n+2}(a_n-1)\right] = 0. \tag{16}$$

Note (13), (14) simplify. (13) can be written

$$\frac{(b_{n+2}a_{n+1} + b_{n+1})}{a_{n+1}} = \frac{(b_{n+1}a_n + b_n)}{a_n}, \tag{17}$$

which implies,

$$\frac{(b_{n+1}a_n + b_n)}{a_n} = C = \text{constant} = \frac{(b_1a_0 + b_0)}{a_0}, \tag{18}$$

or

$$b_{n+1} = C - \frac{b_n}{a_n}. \tag{19}$$

(14) can be written,

$$\frac{\left[b_{n+1}b_n + \frac{1}{4}(a_{n+1}-1)(a_n-1)\right]}{a_n} = D = \text{constant} = \left[\frac{b_1b_0 + \frac{1}{4}(a_1-1)(a_0-1)}{a_0}\right], \tag{20}$$

or

$$a_{n+1} = 1 + \frac{4Da_n - 4\left(C - \frac{b_n}{a_n}\right)b_n}{(a_n-1)}. \tag{21}$$

(19), (21), constitute a system of two first order nonlinear difference equations. It turns out the second pair of equations (15), (16) is redundant.

Theorem 1 *Equations (13) and (14) imply equations (15) and (16).*

Proof Using (19), (21), we express $a_{n+1}, a_{n+2}, b_{n+1}, b_{n+2}$ in terms of a_n, b_n. When these expressions are substituted into (15), (16), these equations are satisfied identically. ∎

In this paper we will consider only the special case $b_n \equiv 0$.

2. The case $b_n \equiv 0$

In this case (13), (14) collapse into a single second order non-linear difference equation,

$$a_{n+2} = 1 + a_{n+1}(1 - \frac{1}{a_n}), n = 0, 1, 2, \ldots, \tag{22}$$

and the recurrence relation for the polynomials $Q_n(x)$ becomes very simple:

$$Q_{n+2}(x) = xQ_{n+1}(x) - \frac{1}{4}\frac{a_{n+1}}{a_n}Q_n(x), n = 0, 1, 2, \ldots, Q_0(x) = 1 - a_0, Q_1(x) = x. \tag{23}$$

We can find a closed-form expression for a_n as follows. Make the substitution

$$c_n = a_n - 1. \tag{24}$$

Then (22) can be rewritten

$$\frac{c_{n+2}c_{n+1}}{1 + c_{n+1}} = \frac{c_{n+1}c_n}{1 + c_n}, \tag{25}$$

which implies

$$\frac{c_{n+1}c_n}{1 + c_n} = K = \text{constant}, \tag{26}$$

and K is given by

$$K = \frac{(a_0 - 1)(a_1 - 1)}{a_0}. \tag{27}$$

K is the salient quantity in everything that follows. It determines the support of the measure with respect to which the polynomials $Q_n(x)$ are orthogonal.

We will solve the recurrence (26) by assuming a solution of the form:

$$c_n = \frac{\alpha + \beta\lambda^n}{1 + \delta\lambda^n}. \tag{28}$$

Putting this into the recurrence, equating the coefficients of $\lambda^0, \lambda^n, \lambda^{2n}$, and utilizing an initial condition gives the four equations,

$$\alpha^2 = K(\alpha + 1), \tag{29}$$

$$\alpha\beta(\lambda + 1) = K\left[(\alpha + 1)\delta\lambda + (\beta + \delta)\right], \tag{30}$$

$$\beta^2 = K\delta(\beta + \delta), \tag{31}$$

$$c_0 = \frac{\alpha + \beta}{1 + \delta}. \tag{32}$$

The first equation gives,

$$\alpha = \frac{K + \sqrt{K^2 + 4K}}{2} \stackrel{\text{def}}{=} \Delta. \tag{33}$$

In equation (31), let $\beta/\delta = \tau$. We get $\tau^2 = K(\tau + 1)$, and this produces for τ the same value as α, leading to the constant solution, unless we take

$$\tau = \beta/\delta = \frac{K - \sqrt{K^2 + 4K}}{2} \stackrel{\text{def}}{=} \Delta^*. \tag{34}$$

We then have

$$\lambda = \frac{\Delta^*}{\Delta} = \frac{K - \sqrt{K^2 + 4K}}{K + \sqrt{K^2 + 4K}}, \delta = \frac{\Delta + 1 - a_0}{a_0 - 1 - \Delta^*}. \tag{35}$$

Note that $|\lambda| < 1, K \in \mathcal{C} - [-4, 0]$. Writing everything in terms of the sequence a_n gives

$$a_n = \frac{(a_0 - 1 - \Delta^*)(\Delta + 1) + (\Delta + 1 - a_0)(\Delta^* + 1)\lambda^n}{(a_0 - 1 - \Delta^*) + (\Delta + 1 - a_0)\lambda^n}, \tag{36}$$

$\lambda, \Delta, \Delta^*$ as above, and so

$$\lim_{n \to \infty} a_n = \Delta + 1 = \frac{K + 2 + \sqrt{K^2 + 4K}}{2}. \tag{37}$$

This is precisely what one gets by applying the theory of periodic continued fractions, see [3], Ch. 3.

One has to be careful about signs in (37). In accordance with the convention introduced after equation (2), we define

$$\sqrt{K^2 + 4K} = \begin{cases} |K^2 + 4K|^{1/2}, K > 0 \\ -|K^2 + 4K|^{1/2}, K < -4. \end{cases} \tag{38}$$

For $K \in [-4, 0]$ the behavior of a_n is very interesting.

Theorem 2 *Let a_0 be real and*

$$\theta = \pi - \arccos\left(\frac{\sqrt{|K|}}{2}\right), K \in [-4, 0]. \tag{39}$$

Then for

$$\theta = \frac{\nu\pi}{m}, \left[\frac{m}{2}\right] + 1 \le \nu \le m - 1, m = 1, 2, \ldots, \tag{40}$$

$\{a_n\}$ is an m-periodic sequence. For other values of θ, a_n is an unbounded sequence.

Proof The proof is a straightforward application of Klonecker's theorem and manipulation with complex exponentials, and is omitted. ∎

3. The distribution for the polynomials

The standard theory assures us that if a positivity condition is satisfied in a three-term recurrence for a set of polynomials, then there is a real distribution function with respect to which the polynomials are orthogonal. The recurrence (23) is not in the standard form because $Q_0 = 1 - a_0 \neq 1$, but this is easily repaired. We define a new recurrence relation for the sequence

$$\widehat{Q}_n(x) = \begin{cases} 1, n = 0 \\ Q_n(x), n > 0 \end{cases} \tag{41}$$

We have

$$\widehat{Q}_{n+2}(x) = x\widehat{Q}_{n+1}(x) - D_n\widehat{Q}_n(x), n = 0, 1, 2, \ldots, \tag{42}$$

with

$$\widehat{Q}_0 = 1, \widehat{Q}_1 = x, D_n = \begin{cases} \dfrac{1}{4}\dfrac{a_1(1-a_0)}{a_0}, n = 0, \\ \dfrac{1}{4}\dfrac{a_{n+1}}{a_n} = \dfrac{(1+\delta\lambda^n)(1+\delta\lambda^{n+3})}{4(1+\delta\lambda^{n+1})(1+\delta\lambda^{n+2})}, n > 0, \end{cases} \tag{43}$$

with δ, λ as in (35). We then have $\widehat{Q}_n(x) = Q_n(x), n \geq 1$.

If $D_n > 0$, there will exist a real distribution function with respect to which the polynomials $\widehat{Q}_n(x)$ are orthogonal. Even when the condition is violated, there will exist a (complex) measure for the polynomials, see the reference Askey and Ismail, [1][1]. It can be constructed by employing the minimal solution $\rho_n(x)$ of the recurrence (42).

Obviously, the minimal solution is

$$\rho_n(x) = a_n x \lambda_2^{n-1}(x) + (1 - a_n)\lambda_2^n(x), n = 1, 2, \ldots, \tag{44}$$

$$\lambda_2(x) = \frac{x - \sqrt{x^2 - 1}}{2}, x \in \mathcal{C} - [-1, 1]. \tag{45}$$

This gives $\rho_n(x)$ only for $n = 1, 2, \ldots$, but the value for $n = 0$ may be constructed by using the recurrence in the backward direction. The Stieltjes transform of the measure will be given by

$$F(z) = \frac{\rho_0(z)}{z\rho_0(z) - \rho_1(z)}. \tag{46}$$

The Askey and Ismail reference provides a contour integral representation of the linear functional, and the path of integration may be deformed to give the continuous component of the measure plus the contributions due to mass points. It turns out that in all cases the measure has a continuous component as well as mass points. The results of this computation are given below.

Note that since the Carleman condition is satished, $\sum D_n^{-1/2} = \infty$, the measure will be unique in any event.

[1] We thank David Masson for suggesting this alternative approach.

It is interesting that a direct approach is possible, namely, one based on a generating function for the $Q_n(x)$. To be implemented, however, this approach requires a computer algebra, such as MAPLE. The computations could not conceivably be done by hand.

Define

$$\nu_1 = \Delta + 1, \nu_2 = \frac{(\Delta + 1 - a_0)(\Delta^* + 1)}{(a_0 - 1 - \Delta^*)}, \nu_3 = \frac{(\Delta + 1 - a_0)}{(a_0 - 1 - \Delta^*)}, \lambda = \frac{\Delta^*}{\Delta}, \tag{47}$$

$$g(x,t) = \sum_{n=0}^{\infty} A_n t^n Q_n(x), A_n = 1 + \nu_3 \lambda^n, \tag{48}$$

$$h(x,t) = \sum_{n=0}^{\infty} t^n \widehat{U}_n(x) = \frac{4}{t^2 - 4xt + 4}. \tag{49}$$

Theorem 3 *We have*

$$g(x,t) = (1 - \nu_1 + \nu_1 xt)h(x,t) + (\nu_3 - \nu_2 + \nu_2 xt)h(x, \lambda t). \tag{50}$$

Proof Elementary ■

Let $\mathcal{L}$ denote a linear functional on the space of all polynomials, and

$$\mathcal{L}\{g(x,t)g(x,u)\} = \sum_{m,n=0}^{\infty} A_n A_m t^n u^m \mathcal{L}\{Q_n(x)Q_m(x)\} \stackrel{\text{def}}{=} V(t,u). \tag{51}$$

Clearly, $\{Q_n(x)\}$ will be an orthogonal family with respect to the functional $\mathcal{L}$ iff $V(t,u)$ is a function of tu alone. A necessary and sufficient condition for this is

$$V(t,u) = V(1,tu) \tag{52}$$

We now define two linear functionals. For $K > 0, a_0 \neq 1$, let

$$\mathcal{L}\{f\} = \frac{1}{2\sqrt{K}}\left(\frac{(1+a_0)\sqrt{K}}{(1-a_0)} - \sqrt{K+4}\right)\left[f\left(\frac{i\sqrt{K}}{2}\right) + f\left(-\frac{i\sqrt{K}}{2}\right)\right] + \frac{1}{\pi}\int_{-1}^{1} \frac{f(t)\sqrt{1-t^2}}{t^2 + \frac{K}{4}} dt. \tag{53}$$

For $K < -4, a_0 \neq 1$, let

$$\mathcal{L}^*\{f\} = \frac{-1}{2\sqrt{|K|}}\left(\frac{(1+a_0)\sqrt{|K|}}{(1-a_0)} - \sqrt{|K|-4}\right)\left[f\left(\frac{\sqrt{|K|}}{2}\right) + f\left(-\frac{\sqrt{|K|}}{2}\right)\right] + \frac{1}{\pi}\int_{-1}^{1} \frac{f(t)\sqrt{1-t^2}}{\frac{|K|}{4} - t^2} dt. \tag{54}$$

The application of MAPLE confirms that for either of the above functionals, $V(t,u) - V(1,tu) = 0$, so

$$\mathcal{L}\{\widehat{Q}_n \widehat{Q}_m\} = \mathcal{L}^*\{\widehat{Q}_n \widehat{Q}_m\} = 0, m \neq n. \tag{55}$$

This, of course, is precisely what the application of the Stieltjes inversion formula to the expression (46) gives. The computer algebra computations are truly horrendous: the expression for $V(t,u)$ is a linear combination of four rational functions whose numerators and denominators are multinomials of degree 9 in u and t, and it is unlikely that hand computations could ever reproduce this result. Even with MAPLE, we are not able to show from our expression for $V(t,u)$ that $\mathcal{L}\{\widehat{Q}_n(x)^2\} \neq 0$.[2] This would involve proving that the Taylor series for the expression for $V(t,t)$ has non-vanishing coefficients. Fortunately, closed form expressions for the values

$$\mathcal{L}\{\widehat{Q}_n(x)^2\} \stackrel{\text{def}}{=} h_n, \mathcal{L}^*\{\widehat{Q}_n(x)^2\} \stackrel{\text{def}}{=} h_n^* \tag{56}$$

may be found from the formula in Erdélyi, [2], 10.3(8). We have

$$h_n = \frac{a_n(1-a_0)}{4^n a_0}\mathcal{L}\{1\} = \frac{2a_n}{4^n}, \tag{57}$$

while $h_n^* = -h_n$.

It is easily verified that the linear functional $\mathcal{L}^*$ is positive definite in the two cases

$$i)\ a_0 > 1, \text{or} \tag{58}$$

$$ii)\ a_0 < -\frac{1}{4}(\sqrt{|K|} - \sqrt{|K|-4})^2. \tag{59}$$

Finally, we point out that the case $K = -4$ in (54), yields a special case of some polynomials studied by Koornwinder, [4].

References

1. Askey, R., and Ismail, M. E., *Recurrence Relations, Continued Fractions, and Othogonal Polynomials*, Memoirs AMS, v 49, Providence, RI, (1984).
2. Erdélyi, A, et al, *Higher Transcendental Functions*, 3 v., McGraw-Hill, NY (1953).
3. Jones, William B., and Thron, W. J., *Continued Fractions: Analytic Theory and Applications*, Encyclopedia of Mathematics and its Applications, v. 11, Addison-Wesley, Reading, Mass. (1980).
4. Koornwinder, Tom H., Orthogonal polynomials with weight function $(1-x)^\alpha(1+x)^\beta + M\delta(x+1) + N\delta(x-1)$, Canad. Math. Bull. 27 (2),205-214 (1984).

[2] We thank the referee for observing that the computer algebra approach suffers from this drawback.

CONVERGENCE OF LAGRANGE INTERPOLATION FOR FREUD WEIGHTS IN WEIGHTED $L_P(\mathbb{R})$, $0 < P \leq 1$

D.M. MATJILA
Department of Mathematics, University of the North, Private Bag X1106, Sovenga 0727, Rep. of South Africa.

Abstract. We determine the necessary and sufficient conditions for convergence in weighted $L_p(\mathbf{R})$, $0 < p \leq 1$ of Lagrange Interpolation to a certain class of continuous functions taken at the zeros of orthonormal polynomials associated with Freud Weights of type $W_\beta(x) := \exp\bigl(-|x|^\beta/2\bigr)$, $x \in \mathbf{R}$, and $\beta > 1$.

1. INTRODUCTION AND RESULTS

The result of this paper closes the gap $0 < p \leq 1$ in the work of D.S. Lubinsky and the author in [3], wherein are given for $1 < p < \infty$, the necessary and sufficient conditions for the convergence of a Lagrange Interpolation process (for a class of continuous functions characterized by some decay behaviour) associated with the Freud Weights of A.L. Levin and D.S. Lubinsky in their long paper [2].

Consequently, the results of this paper and those of [3] generalizes all results on mean convergence associated with Freud Weights including the work on the Hermite weight $W_2^2(x) := \exp(-x^2)$, $x \in \mathbb{R}$ done by P. Nevai in [7] and subsequently the work on the generalized Hermite weight $W_\beta^2(x) := |x|^\beta \exp(-x^2)$, $x \in \mathbb{R}$, $\beta > -1$ by S. Bonan [1].

Let $W := \exp(-Q)$, where $Q : \mathbb{R} \to \mathbb{R}$ is even and continuous in $\mathbb{R}$, Q'' is continuous in $(0, \infty)$ and $Q' > 0$ in $(0, \infty)$, while for some A, $B > 1$,

$$A \leq \left[\frac{d}{dx}\{xQ'(x)\}\right] \Big/ Q'(x) \leq B, \; x \in (0, \infty).$$

Then we call such a W a Freud Weight and denote the class of all such weights by $\mathcal{F}L^2$. For example $W(x) = W_\beta(x) := \exp\bigl(-|x|^\beta/2\bigr)$, $x \in \mathbb{R}$, $\beta > 1$ is in $\mathcal{F}L^2$. The orthonormal polynomials associated with the Freud Weights $W^2 = \exp(-2Q)$ are denoted by $p_n(W^2, x)$ or $p_n(x)$ for simplicity. These polynomials satisfy

$$\int_{\mathbb{R}} p_n(x)p_m(x)W^2(x)dx = \delta_{mn}, \quad m, n = 0, 1, 2, \ldots .$$

We denote the zeros of $p_n(x)$ by

$$-\infty < x_{n,n} < x_{n-1,n} < \cdots < x_{2,n} < x_{1,n} < +\infty.$$

A. Cuyt (ed.), *Nonlinear Numerical Methods and Rational Approximation II*, 25–35.

For a function $f : \mathbb{R} \to \mathbb{R}$, we denote the Lagrange Interpolation polynomial of degree at most $n-1$ to f at the zeros of $p_n(x)$ by $L_n[f](x)$, so that

$$L_n[f](x) = \sum_{j=1}^{n} \ell_{jn}(x) f(x_{jn}), \tag{1.1}$$

where

$$\ell_{jn}(x) := \frac{p_n(x)}{p_n'(x_{jn})(x - x_{jn})}, \quad 1 \le j \le n,$$

are the nth fundamental polynomials for W^2.

Let $\mathcal{X}$ be a linear space of continuous functions $f : \mathbb{R} \to \mathbb{R}$. Define a functional $\|.\|_{L_p(\mathbb{R})} : \mathcal{X} \to \mathbb{R}$ by

$$\|f\|_{L_p(\mathbb{R})} := \left(\int_{\mathbb{R}} |f(x)|^p \, dx \right)^{1/p}, \tag{1.2}$$

for every $f \in \mathcal{X}$. Then our result is:

Theorem 1.1. Let $W \in \mathcal{F}L^2$. Fix $0 < p \le 1$ and choose $\alpha > 0$. Set

$$\hat{\alpha} := \min \left\{ \frac{2}{p} - \frac{1}{p^2}, \alpha \right\}. \tag{1.3}$$

Choose $\Delta \in \mathbb{R}$. Then for

$$\lim_{n \to \infty} \left\| (f - L_n[f])(x) W(x) (1 + |x|)^{-\Delta} \right\|_{L_p(\mathbb{R})} = 0 \tag{1.4}$$

to hold for every function $f : \mathbb{R} \to \mathbb{R}$ satisfying

$$\lim_{|x| \to \infty} |fW|(x) (1 + |x|)^{\alpha} = 0 \tag{1.5}$$

it is sufficient that

$$\Delta > \frac{1}{p} - \hat{\alpha} \tag{1.6A}$$

and necessary that

$$\Delta > \frac{1}{p} - \min\{1, \alpha\}. \tag{1.6B}$$

Remark. For $\frac{1}{2} < p \le 1$ and $0 < \alpha \le \frac{2}{p} - \frac{1}{p^2}$, the necessary and sufficient conditions became identical.

2. THE SUFFICIENT CONDITIONS

We establish the sufficient conditions by following the method of Bonan [1] which entails the use of results on convergence in weighted $L_p(\mathbb{R})$, $p > 1$, and a careful application of Hölder's inequality. The following is the result for $1 < p < 4$.

Theorem 2.1. Let $W \in \mathcal{F}L^2$ and $1 < p < 4$. Suppose $\alpha > 0$. Then for

$$\lim_{n\to\infty} \left\| (f - L_n[f])(x) W(x) (1+|x|)^{-\Delta} \right\|_{L_p(\mathbb{R})} = 0$$

to hold for every continuous function $f : \mathbb{R} \to \mathbb{R}$ satisfying (1.5), it is necessary and sufficient that

$$\Delta > \frac{1}{p} - \min\{1, \alpha\}.$$

Proof. This is Theorem 1.3 in [3]. ∎

Proof of the sufficient conditions. Let $1 < q,\ q' < \infty$, such that $\frac{1}{q} + \frac{1}{q'} = 1$. By Hölder's inequality,

$$\begin{aligned} &\left\| (f - L_n[f])(x) W(x) (1+|x|)^{-\Delta} \right\|^p_{L_p(\mathbb{R})} \\ &= \int_{\mathbb{R}} \left| (f - L_n[f])(x) W(x) (1+|x|)^{-\Delta/q'} (1+|x|)^{-\Delta/q} \right|^p dx \\ &\le \left(\int_{\mathbb{R}} \left| (f - L_n[f])(x) W(x) (1+|x|)^{-\Delta/q'} \right|^{pq} dx \right)^{1/q} \\ &\quad \times \left(\int_{\mathbb{R}} (1+|x|)^{-\Delta p q'/q} dx \right)^{1/q'} . \end{aligned} \tag{2.1}$$

Using Theorem 2.1 we infer that the right–hand side of (2.1) tends to zero as n tends to ∞, provided that,

$$-\Delta p q'/q < -1 \tag{2.2}$$

and $\Delta/q' > \frac{1}{pq} - \min\{1, \alpha\}$. Since $\frac{2}{p} - p^2 \le 1$, we can replace this by the requirement that

$$\Delta/q' > \frac{1}{pq} - \hat{\alpha}, \quad \text{where} \quad 1 < pq < 4. \tag{2.3}$$

From (2.2) we obtain

$$\Leftrightarrow q < 1 + \Delta p. \tag{2.4}$$

Conditions in (2.3) yield,

$$\Leftrightarrow q > (1 + \Delta p)/p(\hat{\alpha} + \Delta) \tag{2.5}$$

and

$$1 < pq < 4 \Leftrightarrow 1/p < q < 4/p. \tag{2.6}$$

We then consolidate inequalities in (2.4–6) into

$$\max\Big\{1/p,\ (1 + \Delta p)/p(\hat{\alpha} + \Delta)\Big\} < q < \min\{4/p,\ 1 + \Delta p\}.$$

Finally, the existence of q will be ascertained if we can show that

$$\max\Big\{1/p,\ (1 + \Delta p)/p(\hat{\alpha} + \Delta)\Big\} < \min\{4/p,\ 1 + \Delta p\}.$$

(i) $1/p < 1 + \Delta p$:
This follows from the fact that $\hat{\alpha} \leq \frac{2}{p} - \frac{1}{p^2}$ in view of (1.3). Thus

$$1 + \Delta p > 1 + p\left(\frac{1}{p} - \hat{\alpha}\right) > 1 + p\left(\frac{1}{p} - \frac{2}{p} + \frac{1}{p^2}\right) = 1/p,$$

as $\Delta > \frac{1}{p} - \hat{\alpha}$.

(ii) $(1 + \Delta p)/p(\hat{\alpha} + \Delta) < 1 + \Delta p$.
From the fact that $\Delta > \frac{1}{p} - \hat{\alpha}$ follows that

$$(1 + \Delta p)/p(\hat{\alpha} + \Delta) < (1 + \Delta p)p(1/p) = 1 + \Delta p.$$

(iii) $(1 + \Delta p)/p(\hat{\alpha} + \Delta) < 4/p$.
Note that (iii) is equivalent to

$$\Delta > \frac{4\hat{\alpha} - 1}{p - 4} = \frac{1}{p} - \left(\frac{1}{p} - \frac{4\hat{\alpha} - 1}{p - 4}\right),$$

which will be implied by (1.6), provided we can show that

$$\begin{aligned} &\hat{\alpha} < \frac{1}{p} - \frac{4\hat{\alpha} - 1}{p - 4} \\ \Leftrightarrow\ &\hat{\alpha} > \frac{2}{p} - \frac{4}{p^2} \\ \Leftrightarrow\ &\min\left\{\frac{2}{p} - \frac{1}{p^2}, \alpha\right\} > \frac{2}{p} - \frac{4}{p^2}. \end{aligned}$$

Since $\alpha > 0$ and the right–hand side is negative, (iii) follows.

This completes the proof of the sufficient conditions. ∎

3. THE NECESSARY CONDITIONS

The establishment of the necessary conditions requires among other things, the *Uniform Boundedness Principle.* However, as a result of the fact that we have in (1.2) a metric and not a norm (see Rao [8, pp.189–190]), we are then compelled to make use of the general form of the Uniform Boundedness Principle.

Theorem 3.1. Let X be a linear metric space of second category equipped with an invariant metric. Suppose $\{T_n : n \in A\}$ is a family of continuous mappings of X into a similar linear metric space Y such that ($\|.\|$ denoting the invariant metric of Y)

$$\|T_n(f+g)\| \le \|T_n(f)\| + \|T_n(g)f\| \tag{3.1a}$$

$$\|T_n(af)\| = \|aT_n(f)\| \tag{3.1b}$$

for all $f,\ g \in X,\ a \in \mathbb{R}$ (or $\mathbb{C}$). Then if $\{T_n(f) : n \in A\}$ is bounded in Y for each $f \in X$, we have

$$\lim_{f\to 0} T_n(f) = 0, \tag{3.2}$$

uniformly for $n \in A$.

Proof. This is Proposition 11, p.292 in Rao[8]. ■

We also need the *Mhaskar–Rahmanov–Saff* number a_u associated with $W := \exp(-Q)$ in $\mathcal{F}L^2$ (see Mhaskar and Saff [5], [6]). For any $u > 0$, a_u is the positive root of the equation

$$u = \frac{2}{\pi}\int_0^1 a_u t Q'(a_u t)\Big/\sqrt{1-t^2}\,dt.$$

Under the conditions on Q, which guarantee that $Q(s)$ and $Q'(s)$ increase strictly in $(0,\infty)$, a_u is uniquely defined and increases with increasing u. It grows like $Q^{[-1]}(u)$, where $Q^{[-1]}$ denotes the inverse of Q in $(0,\infty)$. For example if $Q(x) = |x|^\beta$, $\beta > 1$, then $a_n = Cn^{1/\beta}$, $n \ge 1$, where C depends on β only.

Throughout, $C, C_1, C_2, \ldots$, will denote positive constants independent of n or $x \in \mathbb{R}$ or a polynomial P, which are not necessarily the same from line to line. The symbol $\sim$ will be used in the following sense: If $\{c_n\}$ and $\{d_n\}$ are real sequences, then we write $c_n \sim d_n$ if there exist C_1 and C_2 such that

$$C_1 \le c_n/d_n \le C_2, \qquad \text{for } n \text{ large enough}.$$

A similar notation is used for functions and sequences of functions. We also need the following result:

Theorem 3.2. Let $W \in \mathcal{F}L^2$. Define

$$\psi_n(x) := \max\left\{n^{-2/3},\ 1-\frac{|x|}{a_n}\right\}, \qquad \text{for } n \ge 1 \text{ and } x \in \mathbb{R}.$$

Then

(a) uniformly for $2 \le j \le n-1$,

(3.3) $$x_{j-1,n} - x_{j+1,n} \sim \frac{a_n}{n} \psi_n(x_{jn})^{-1/2},$$

(b)

(3.4) $$\sup_{x\in\mathbb{R}} |p_n(x)| W(x) \left|1 - \frac{|x|}{a_n}\right|^{1/4} \sim a_n^{-1/2},$$

(c) let $0 < p < \infty$. Then there exists C such that for $n \ge 1$, and any polynomial P of degree at most n,

(3.5) $$\|PW\|_{L_p(\mathbb{R})} \le C\|PW\|_{L_p[-a_n,a_n]},$$

(d) uniformly for $n \ge 1$, $1 \le j \le n$, and $x \in \mathbb{R}$,

(3.6) $$|\ell_{jn}(x)| \sim (a_n^{3/2}/n) W(x_{jn}) \psi_n(x_{jn})^{-1/4} \frac{|p_n(x)|}{|x - x_{jn}|},$$

(e) there exists C_1, such that uniformly for $n \ge 1$, $1 \le j \le n$, and

(3.7) $$|x - x_{jn}| \le C_1(a_n/n)\psi_n(x_{jn})^{-1/2},$$

we have

(3.8) $$|p_n(x)| W(x) \sim (n/a_n^{3/2}) \psi_n(x_{jn})^{1/4} |x - x_{jn}|,$$

(f) for any $0 < p < 4$,

(3.9) $$\|p_n W\|_{L_p(\mathbb{R})} \sim a_n^{\frac{1}{p}-\frac{1}{2}}.$$

Proof. (a), (b), (c) are respectively Corollary 1.2(b), Corollary 1.4, Theorem 1.8 in [2]. (d), (e), (f) are respectively Lemma 2.6(a), (c) and Theorem 1 in [4]. ∎

Proof of the necessary conditions. Here we follow the method of Lubinsky and Matjila in [3]. Let $\eta : \mathbb{R} \to (0, \infty)$ be an even continuous function that is decreasing in $[0, \infty)$ with

(3.10) $$\log(2 + |x|)^{-p/2} \le \eta(x), \ x \in [0, \infty)$$

but

(3.11) $$\lim_{x\to\infty} \eta(x) = 0.$$

Let X be the space of continuous functions $f : \mathbb{R} \to \mathbb{R}$ such that

$$\|f\|_X := \max_{x\in\mathbb{R}} |fW|(x)\big(1+|x|\big)^{\alpha}\eta(x)^{-1} < \infty. \tag{3.12}$$

Furthermore, let Y be the space of continuous functions $f : \mathbb{R} \to \mathbb{R}$ with

$$\|f\|_Y := \left\| f(x)W(x)\big(1+|x|\big)^{-\Delta} \right\|_{L_p(\mathbb{R})}^p < \infty, \tag{3.13}$$

for some $\Delta \in \mathbb{R}$. We then note that Y is not a normed space with respect to (3.13) for $0 < p < 1$, but however it is a metric space with a metric

$$d(f,g) := \left\| (f-g)(x)W(x)\big(1+|x|\big)^{-\Delta} \right\|_{L_p(\mathbb{R})}^p. \tag{3.14}$$

Observe that each $f \in X$ satisfies (1.5) and so our hypothesis ensures that

$$\lim_{n\to\infty} \big\| f - L_n[f] \big\|_Y = 0,$$

in other words, for every $\delta > 0$, there exists n_0 such that $n \geq n_0$ implies that

$$\big\| f - L_n[f] \big\|_Y \leq \delta, \qquad \text{for each } f \in X. \tag{3.15}$$

Now setting $T_n[.] := (I - L_n)[.]$, where I is the identity operator and $L_n[.]$ is the Lagrange Interpolation polynomial, we see that T_n, $n \geq 1$, satisfies (3.1 a–b), and from (3.2) and (3.15) we can conclude that there exists C_1 such that

$$\|f\|_X \leq 1 \Rightarrow \big\| f - L_n[f] \big\|_Y \leq C_1, \quad \text{for} n \geq n_0.$$

In particular, for $\hat{f} := f/\|f\|_X$, $f \in X$ and $f \neq 0$, we obtain

$$\|\hat{f}\|_X = 1 \Rightarrow \big\| \hat{f} - L_n[\hat{f}] \big\|_Y \leq C_2,$$

uniformly for $n \geq 1$, in other words,

$$\big\| f - L_n[f] \big\|_Y \leq C_2 \|f\|_X^p, \tag{3.16}$$

uniformly for $n \geq 1$ and $f \in X$ such that $\|f\|_X \leq 1$.

In particular, as $L_1[f] = f(0)$ (recall $p_1(x) = x$) we obtain for every continuous function $f : \mathbb{R} \to \mathbb{R}$ with $f(0) = 0$, that

$$\|f\|_Y \leq C_2 \|f\|_X^p, \tag{3.17}$$

provided that the right–hand side is finite. Hence (3.16) and (3.17) yield

$$\big\| L_n[f] \big\|_Y \leq C_3 \|f\|_X^p, \tag{3.18}$$

for every $n \geq 1$ and every continuous f with $f(0) = 0$ for which the right–hand side is finite. Now choose g_n, $n \geq 1$, such that g_n is continuous in $\mathbb{R}$, $g_n = 0$ in $[0, \infty) \cup (-\infty, a_{n/2})$,

$$\|g_n\|_X = 1, \tag{3.19}$$

and for $x_{jn} \in [-a_{n/2}, 0)$,

$$(g_n W)(x_{jn})\big(1+|x_{jn}|\big)^{\alpha}\eta(x_{jn})^{-1} \text{ sign } \ (p_n'(x_{jn})) = 1.$$

Then for $x > 0$, our choice of g_n and (3.6) we obtain, on following the steps of [3],

$$\begin{aligned}
\big|L_n[g_n](x)\big| &= \left| \sum_{x_{jn}\in[-a_{n/2},0)} g_n(x_{jn}) \frac{p_n(x)}{p_n'(x_{jn})(x-x_{jn})} \right| \\
&\geq C_4\big(a_n^{3/2}/n\big) \sum_{x_{jn}\in[-a_{n/2},0)} \frac{|p_n(x)|}{x+|x_{jn}|}\big(1+|x_{jn}|\big)^{-\alpha}\eta(x_{jn}) \\
&\geq C_5\eta(a_n)a_n^{1/2}|p_n(x)| \sum_{x_{jn}\in[-a_{n/2},0)} \frac{x_{j-1,n}-x_{j+1,n}}{x+|x_{jn}|}\big(1+|x_{jn}|\big)^{-\alpha},
\end{aligned}$$

after using (3.3) in the last inequality.

At least for $x \geq 1$, we have

$$x+|x_{jn}| \sim x+|t|, \quad \text{for } \ t \in \big(x_{j+1,n}, x_{j-1,n}\big).$$

Thus for $x \geq 1$,

$$\begin{aligned}
\big|L_n[g_n](x)\big| &\geq C_6\eta(a_n)a_n^{1/2}|p_n(x)| \int_0^{a_{n/4}} \frac{(1+t)^{-\alpha}}{x+t}dt \\
&\geq C_7\eta(a_n)a_n^{1/2}|p_n(x)| \int_0^{\min\{a_{n/4},x\}} \frac{(1+t)^{-\alpha}}{2x}dt \\
&\geq C_8\eta(a_n)a_n^{1/2}\big(|p_n(x)|/x\big) \begin{cases} 1, & \alpha>1 \\ \log(1+\min\{a_{n/4},x\}), & \alpha=1 \\ \big(\min\{a_{n/4},x\}\big)^{1-\alpha}, & \alpha<1 \end{cases}.
\end{aligned}$$

Let us set

$$(\log x)^\star := \begin{cases} \log(1+x), & \alpha=1 \\ 1, & \alpha\neq 1 \end{cases}.$$

By considering $x \in [1, a_{n/4}]$ and $x \in [a_{n/4}, 2a_n]$ separately, we can rewrite the above inequality as

$$\big|L_n[g_n]\big|(x) \geq C_9\eta(a_n)a_n^{1/2}|p_n(x)|x^{-\alpha^\star}(\log x)^\star \tag{3.20}$$

for $x \in [1, 2a_n]$, where we set $\alpha^\star = \min\{1,\alpha\}$.

Now using the inequality (3.10) in (3.20) we obtain

$$\big|L_n[g_n]\big|(x) \geq C_{10}(\log n)^{-p/2}a_n^{1/2}|p_n(x)|x^{-\alpha^\star},$$

and (3.19) and (3.20) yield as in [3],

$$\begin{aligned}C_4 &= C_4\|g_n\|_X^p \geq \left\|L_n[g_n]\right\|_Y \geq \left\|L_n[g_n](x)W(x)(1+|x|)^{-\Delta}\right\|_{L_p[1,2a_n]}^p \\ &= \int_1^{2a_n} \left|L_n[g_n](x)W(x)(1+|x|)^{-\Delta}\right|^p dx \\ &\geq C_{11}(\log n)^{-p^2/2} a_n^{p/2} \int_1^{2a_n} \left[\left|p_n(x)\right|W(x)(1+|x|)^{-(\alpha^*+\Delta)}\right]^p dx.\end{aligned}$$

Thus

$$\begin{aligned}C_{12}(\log n)^{p^2} &\geq a_n^{p/2} \int_1^{2a_n} \left[|p_n W|(x)(1+|x|)^{-(\alpha^*+\Delta)}\right]^p dx \\ &= a_n^{p/2} \left(\int_1^{\sigma a_n} + \int_{\sigma a_n}^{2a_n}\right) \left[|pW|(x)(1+|x|)^{-(\alpha^*+\Delta)}\right]^p dx\end{aligned} \tag{3.21}$$

where $\sigma \in (0,1)$ is sufficiently small. It is an easy consequence of (3.4–5) that

$$\int_{\sigma a_n}^{2a_n} |p_n W|^p(x)dx \sim \int_0^{a_n} |p_n W|^p(x)dx \sim \int_{\mathbb{R}} |p_n W|^p(x)dx,$$

so we can replace (3.21) by

(3.22)

$$\begin{aligned}C_{13}(\log n)^{p^2/2} &\geq a_n^{p/2}\left[\left(\int_1^{\sigma a_n} + \int_{\sigma a_n}^{2a_n}\right)|p_n W|^p(x)(1+|x|)^{-(\alpha^*+\Delta)p} dx\right] \\ &\geq a_n^{p/2}\left[\int_1^{\sigma a_n} |p_n W|^p(x)(1+|x|)^{-(\alpha^*+\Delta)p} dx + C_{14} a_n^{-(\alpha^*+\Delta)p} \int_{\mathbb{R}} |p_n W|^p(x)dx\right] \\ &\geq a_n^{p/2} \int_1^{\sigma a_n} |p_n W|^p(x)(1+|x|)^{-(\alpha^*+\Delta)p} dx + C_{15} a_n^{-(\alpha^*+\Delta)p+1}\end{aligned}$$

where we have applied (3.9) to the second integral in the penultimate step. Now we determine a suitable lower bound for the integral in (3.22). From (3.3) we have

$$x_{j,n} - x_{j+2,n} \sim (a_n/n), \quad \text{for } |x_{jn}| \leq 2\sigma a_n \tag{3.23}$$

and by Theorem 3.2 (e),

$$|p_n W|(x) \sim \left(n/a_n^{3/2}\right)|x - x_{jn}|, \quad \text{for } |x - x_{jn}| \leq \delta a_n/n,$$

where $\delta \in (0,1)$ is small enough. Also

$$|p_n W|(x) \sim a_n^{-1/2}, \quad \text{for } (\delta a_n/2n) \leq |x - x_{jn}| \leq (\delta a_n/n).$$

Let $\mathcal{S} := \Big\{ j : |x_{jn}| \le \sigma a_n/2 \text{ and } j \text{ even} \Big\}$. Then

$$\begin{aligned}
&\int_0^{\sigma a_n} |p_n W|^p(x)\big(1+|x|\big)^{-(\alpha^\star+\Delta)p}\,dx \\
&\quad\le \frac{1}{2}\sum_{j\in\mathcal{S}} \int_{x_{j+2,n}}^{x_{j,n}} |p_n W|^p(x)\big(1+|x|\big)^{-(\alpha^\star+\Delta)p}\,dx \\
&\quad\le C_{16} a_n^{-p/2} \sum_{j\in\mathcal{S}} \int_{x_{j,n}-(\delta a_n/n)}^{x_{j,n}-(\delta a_n/2n)} \big(1+|x|\big)^{-(\alpha^\star+\Delta)p}\,dx
\end{aligned}$$

as $\Big[x_{jn}-(\delta a_n/n),\, x_{jn}-(\delta a_n/2n)\Big] \subset \big[x_{j+2,n},\, x_{jn}\big]$ is true for δ small enough. Thus

(3.24)

$$\begin{aligned}
&\int_0^{\sigma a_n} |p_n W|^p(x)\big(1+|x|\big)^{-(\alpha^\star+\Delta)p}\,dx \\
&\quad\ge C_{17} a_n^{-p/2} \sum_{j\in\mathcal{S}} \frac{a_n}{n}\big(1+|x_{jn}|\big)^{-(\alpha^\star+\Delta)p} \\
&\quad\ge C_{18} a_n^{-p/2} \sum_{j\in\mathcal{S}} \big(x_{j,n}-x_{j+2,n}\big)\big(1+|x_{jn}|\big)^{-(\alpha^\star+\Delta)p} \quad \text{(after using (3.23))} \\
&\quad\ge C_{19} a_n^{-p/2} \int_0^{\delta a_n/2} \big(1+|t|\big)^{-(\alpha^\star+\Delta)p}\,dt \\
&\quad= C_{20} a_n^{-p/2} \begin{cases} 1, & (\alpha^\star+\Delta)p>1 \\ \log n, & (\alpha^\star+\Delta)p=1 \\ a_n^{-(\alpha^\star+\Delta)p+1}, & (\alpha^\star+\Delta)p<1 \end{cases}.
\end{aligned}$$

Combining (3.22) and (3.24) we obtain

$$C_{21}(\log n)^{p^2/2} \ge C_{22} \begin{cases} 1, & (\alpha^\star+\Delta)p>1 \\ \log n, & (\alpha^\star+\Delta)p=1 \\ a_n^{-(\alpha^\star+\Delta)p+1}, & (\alpha^\star+\Delta)p<1 \end{cases} + C_{15} a_n^{-(\alpha^\star+\Delta)p+1}$$

which is possible for large n, only provided that

$$\Delta > \frac{1}{p} - \alpha^\star.$$

This proves the necessary conditions. ■

Acknowledgements

I am indebted to my mentor D.S. Lubinsky for guidance in the preparation of this paper.

References

1. Bonan S.: 1982, 'Weighted Mean Convergence of Lagrange Interpolation', Ph.D. Thesis, Columbus, Ohio.
2. Levin, A.L. and Lubinsky, D.S.: 1992, Christoffel Functions, 'Orthogonal Polynomials and Nevai's Conjecture for Freud Weights', *Constr. Approx.*, Vol. no. **8**, pp. 463–535.
3. Lubinsky, D.S. and Matjila, D.M.: 'Necessary and Sufficient Conditions for Mean Convergence of Lagrange Interpolation for Freud Weights', to appear in SIAM J. Math. Anal.
4. Lubinsky, D.S. and Moricz, F.: 'The Weighted L_p–Norms of Orthonormal Polynomials for Freud Weights', to appear in J. Approx. Theory.
5. Mhaskar, H.N. and Saff, E.B.: 1984, 'Extremal Problems for Polynomials with Exponential Weights', *Trans. Amer. Math. Soc.*, Vol. no. **285**, pp. 203–234.
6. Mhaskar, H.N. and Saff, E.B.: 1985, 'Where Does the Sup–Norm of a Weighted Polynomial Live?', *Constr. Approx.*, Vol. no. **1**, pp. 71–91.
7. Nevai, P.: 1980, 'Mean Convergence of Lagrange Interpolation, II', *J. Approx. Theory*, Vol. no. **30**, pp. 263–276.
8. Rao, M.M.: 1987, 'Measure Theory and Integration', John–Wiley & Sons, Inc., New York.

ORTHOGONALITY AND BOUNDARY INTERPOLATION

ADHEMAR BULTHEEL
Department of Computer Science
K.U. Leuven
Belgium

PABLO GONZALEZ-VERA
Department of Mathematical Analysis
University of La Laguna
Tenerife, Spain

ERIK HENDRIKSEN
Department of Mathematics
University of Amsterdam
The Netherlands

and

OLAV NJÅSTAD
Department of Mathematical Sciences
University of Trondheim - NTH
Norway

Abstract. Let $\{\alpha_n : n = 1, 2, ...\}$ be a sequence of (not necessarily distinct) points on the unit circle. Set $\omega_0(z) \equiv 1$, $\omega_n(z) = (z - \alpha_1)(z - \alpha_2)...(z - \alpha_n)$ for $n = 1, 2, ...,$ and let $\mathcal{L}$ denote the linear space spanned by the functions $\{1/\omega_n : n = 0, 1, 2, ...\}$. Let M be a quasi-definite linear functional on $\mathcal{L} \cdot \mathcal{L}$, and define the inner product $<,>$ by $< f, g >= M(fg_*)$, where $g_*(z) = \overline{g(1/\overline{z})}$. (In particular M may be a positive definite functional given by $M(f) = \int_{-\pi}^{\pi} f(e^{it})d\mu(t)$, where μ is a measure such that all functions in $\mathcal{L} \cdot \mathcal{L}$ are μ-integrable). Let $\{\varphi_n\}$ be an orthogonal system obtained from the basis $\{1/\omega_n\}$ by the Gram-Schmidt method and define the associated functions $\{\psi_n\}$ by $\psi_n(z) = M(\frac{t+z}{t-z}[\varphi_n(t) - \varphi_n(z)])$. Interpolation of ψ_n/φ_n to $M(\frac{t+z}{t-z})$ (or to the formal series $\mu_0 + 2\sum_{m=1}^{\infty} \mu_m z \omega_{m-1}(z)$, $\mu_m = M(1/\omega_m)$) at interpolation points α_k is discussed.

1. Introduction

Let $\{\alpha_n\}$ be a sequence of (not necessarily distinct) points on the unit circle $\boldsymbol{T} = \{z \in \boldsymbol{C} : |z| = 1\}$. We define the polynomials ω_n by

$$\omega_0 = 1, \quad \omega_n(z) = \prod_{m=1}^{n} (z - \alpha_m), \quad n = 1, 2, \ldots. \tag{1.1}$$

The linear space spanned by the functions $\{\frac{1}{\omega_0}, \frac{1}{\omega_1}, ..., \frac{1}{\omega_n}\}$ shall be denoted by $\mathcal{L}_n$, i.e.

$$\mathcal{L}_n = Span\{\frac{1}{\omega_0}, \frac{1}{\omega_1}, ..., \frac{1}{\omega_n}\}. \tag{1.2}$$

A. Cuyt (ed.), *Nonlinear Numerical Methods and Rational Approximation II*, 37–47.

Further we define

$$\mathcal{L} = \cup_{n=0}^{\infty} \mathcal{L}_n. \tag{1.3}$$

The functions in $\mathcal{L}_n$ are exactly the functions that may be written in the form

$$R(z) = \frac{p(z)}{\omega_n(z)}, \quad p \in \Pi_n, \tag{1.4}$$

where Π_n denotes the space of polynomials of degree at most n. (This follows from the basic facts about partial fraction decomposition.)

The space $\mathcal{L} \cdot \mathcal{L}$ is defined by $\mathcal{L} \cdot \mathcal{L} = \{f \cdot g : f \in \mathcal{L}, g \in \mathcal{L}\}$, and similarly for other products of this form.

Let M be a given linear functional defined at least on $\mathcal{L} \cdot \mathcal{L}$. (Note that M is then defined on $\mathcal{L}$, since $\mathcal{L} = \mathcal{L} \cdot 1 \subset \mathcal{L} \cdot \mathcal{L}$.) The functional M may in particular be defined on $\mathcal{L} \cdot \mathcal{L}$ by

$$M[R] = \int_{-\pi}^{\pi} R(e^{i\theta})\, d\mu(\theta), \tag{1.5}$$

where μ is a finite Borel measure on $[-\pi, \pi]$ with the property that all the integrals $\int_{-\pi}^{\pi} [\omega_m(e^{i\theta})\omega_n(e^{i\theta})]^{-1}\, d\mu(\theta)$, $m, n = 0, 1, 2, \ldots$, exist.

For every function f on $\hat{C}$ we define its substar transform f_* by

$$f_*(z) = \overline{f(1/\overline{z})}. \tag{1.6}$$

We note that if $R \in \mathcal{L}_n$, then also $R_* \in \mathcal{L}_n$, since

$$[\frac{1}{\omega_m(z)}]_* = (-1)^m \alpha_1 \ldots\ldots \alpha_m \frac{z^m}{\omega_m(z)}. \tag{1.7}$$

We define the Hermitian form $<,>$ on $\mathcal{L} \otimes \mathcal{L}$ by

$$< P, Q >= M[P \cdot Q_*]. \tag{1.8}$$

We shall assume that this form is quasi-definite, i.e.

$$< P, P > \neq 0, \quad \text{for} \quad P \in \mathcal{L},\ P \not\equiv 0. \tag{1.9}$$

This is in particular the case for the form associated by (1.5), which is positive definite.

Let $\{\varphi_n\}$ be the orthonormal system (unique except for a unimodular constant factor) associated with the sequence $\{\frac{1}{\omega_n}\}$ such that $Span\,\{\varphi_0, \ldots, \varphi_n\} = Span\,\{\frac{1}{\omega_0}, \ldots, \frac{1}{\omega_n}\}$ for all n. We may write

$$\varphi_n(z) = \frac{p_n(z)}{\omega_n(z)}, \quad p_n \in \Pi_n. \tag{1.10}$$

The functions of the second kind $\{\psi_n\}$ (associated with $\{\varphi_n\}$) are defined by

$$\psi_0 = 1, \ \psi_n(z) = M[\frac{t+z}{t-z}[\varphi_n(t) - \varphi_n(z)]], \ n = 1,2,\dots . \tag{1.11}$$

It is easily seen that the expression in the outer brackets [ ] belongs to $\mathcal{L}_n$ both as a function of z and as a function of t, and M is here meant to operate on its argument considered as a function of t. The function ψ_n then belongs to $\mathcal{L}_n$, and it may be written as

$$\psi_n(z) = \frac{q_n(z)}{\omega_n(z)}, \ q_n \in \Pi_n. \tag{1.12}$$

Let us now assume that the functional M is defined also for the functions $\eta_z^{(k,j)}(t) = \dfrac{z^j t^k}{t-z}$ for $j,k = 0,1$, $z \in \hat{C} - T$. (By M being defined for the function $f(z,t)$ for $z = \infty$ we mean that M is defined for the function $f(\zeta^{-1},t)$ for $\zeta = 0$). Also assume that M is defined on the extended space $\tilde{\mathcal{L}} \cdot \tilde{\mathcal{L}}$, where

$$\tilde{\mathcal{L}} = Span\{\frac{z^j t^k}{\omega_n(t)(t-z)} : \tag{1.13}$$

$$n = 0,1,2,\dots, \ j,k = 0,1, \ z \in (\hat{C} - T) \cup \bigcup_{n=1}^{\infty}\{\alpha\}\}.$$

We note that M is then in particular defined for t and t^2.

We may then define the Riesz-Herglotz transform Ω_M of M by

$$\Omega_M(z) = M[\frac{t+z}{t-z}]. \tag{1.14}$$

(Here again M operates on its argument considered as a function of t.) The functional M defined by (1.5) is defined on $\tilde{\mathcal{L}} \cdot \tilde{\mathcal{L}}$. For the transform Ω_M of this functional we shall also write Ω_μ, thus

$$\Omega_\mu(z) = \int_{-\pi}^{\pi} \frac{e^{i\theta} + z}{e^{i\theta} - z} \, d\mu(\theta). \tag{1.15}$$

The function $g_t(z) = \dfrac{t+z}{t-z}$ has formal Newton series expansions

$$\frac{t+z}{t-z} \sim 1 + 2\sum_{k=1}^{\infty} \frac{z\omega_{k-1}(z)}{\omega_k(t)} \tag{1.16}$$

and

$$\frac{t+z}{t-z} \sim -1 + 2\sum_{k=1}^{\infty} \frac{\alpha_k t^k \omega_{k-1}(z)}{z^k \omega_k(t)}. \tag{1.17}$$

(Cf. Section 2. For Newton series expansions, see also [12].) By operating formally by M on (1.16) and (1.17) we get the Newton series expansions

$$\Omega_M(z) \sim \mu_0 + 2\sum_{k=1}^{\infty} \mu_k\, z\omega_{k-1}(z) \tag{1.18}$$

and

$$\Omega_M(z) \sim \nu_0 + 2\sum_{k=1}^{\infty} \nu_k \frac{\omega_{k-1}(z)}{z^k}, \tag{1.19}$$

where the moments $\{\mu_n\}$, $\{\nu_n\}$ are defined by

$$\mu_n = M[\frac{1}{\omega_n(t)}], \quad n = 0, 1, 2... \tag{1.20}$$

and

$$\nu_0 = -\mu_0, \quad \nu_n = \alpha_n\, M[\frac{t^n}{\omega_n(t)}], \quad n = 1, 2, \ldots . \tag{1.21}$$

We define the polynomials P_n and Q_n by

$$P_n(z) = \mu_0 + 2\sum_{k=1}^{n} \mu_k\, z\omega_{k-1}(z), \tag{1.22}$$

$$Q_n(z) = \nu_0 + 2\sum_{k=1}^{n} \nu_k \frac{\omega_{k-1}(z)}{z^k}. \tag{1.23}$$

For basic information concerning orthogonal functions associated with spaces $\mathcal{L}_n$ as defined above, see [8], cf. also [7].

We shall in this paper discuss some interpolation properties connecting the rational functions $\{\frac{\psi_n(z)}{\varphi_n(z)}\}$, the Riesz-Herglotz transform $\Omega_M(z)$ and the Newton series (1.18)-(1.19). It will be pointed out that $\{\frac{\psi_n(z)}{\varphi_n(z)}\}$ is an (n, n) multipoint Padé approximant for $\Omega_M(z)$ at the interpolation table

$$\{0, \infty, \alpha_1, \alpha_1, \alpha_2, \alpha_2, ..., \alpha_{n-1}, \alpha_{n-1}, \alpha_n\}.$$

- Related connections where all the interpolation points are in the interior of the unit circle are discussed e.g. in [4,5,6,7,9,16] and in references found there. See also [11].

A special case of the situation treated in this paper is when the sequence $\{\alpha_n\}$ consists of a finite number of points cyclically repeated. The closely related (through the Cayley transform) cyclic situation when the points $\{\alpha_n\}$ lie on the real axis has been investigated e.g. in [14,15,16] and in the references found there. When in particular two points are cyclically repeated (for convenience ∞ and 0), the orthogonal functions are orthogonal Laurent polynomials. (See e.g. [13,17]. Cf. also [3].) When all the points $\{\alpha_n\}$ coalesce to one point (for convenience ∞), the orthogonal functions are orthogonal polynomials. (See e.g. [1,2,10].)

2. Interpolation results

In addition to the given interpolation points $\{\alpha_n\}$ on $\boldsymbol{T}$, the points at the origin and at infinity will play a role in our discussion. We shall use the notation

$$\alpha_0 = 0, \quad \alpha_\infty = \infty. \tag{2.1}$$

Let A_n denote the interpolation table $\{\alpha_0, \alpha_1, ..., \alpha_n\}$, and $\tilde{A}_n$ the table $\{\alpha_\infty, \alpha_1, ..., \alpha_n\}$.

We note that the space $\mathcal{L}_n$ is independent of the order of the points $\{\alpha_1, ..., \alpha_n\}$ and the space $\mathcal{L}$ is independent of the order of the points $\{\alpha_n : n = 1, 2, ...\}$. We also note that for each n the functions $\{\varphi_0, \varphi_1, ..., \varphi_n\}$ and $\{\psi_0, \psi_1, ..., \psi_n\}$ are determined solely by the table A_n (or $\tilde{A}_n$) and the values of the functional M on $\mathcal{L}_n \cdot \mathcal{L}_n$.

Proposition 2.1. *The polynomial P_n interpolates Ω_M at the table A_n in the sense that*

$$\Omega_M(z) - P_n(z) = z\,\omega_n(z)\,F_n(z) \tag{2.2}$$

and the polynomial Q_n interpolates Ω_M at the table $\tilde{A}_n$ in the sense that

$$\Omega_M(z) - Q_n(z) = \frac{\omega_n(z)}{z^{n+1}}\,G_n(z). \tag{2.3}$$

Here F_n and G_n are functions which are defined at least for $\{\alpha_0, \alpha_1, ..., \alpha_n\}$ and $\{\alpha_\infty, \alpha_1, ..., \alpha_n\}$, respectively.

Proof: It can easily be verified by induction that

$$1 + 2\sum_{k=1}^{n} \frac{z\omega_{k-1}(z)}{\omega_k(t)} = \frac{t+z}{t-z}\left[1 - \frac{z\omega_n(z)}{t\omega_n(t)}\right] - \frac{z\omega_n(z)}{t\omega_n(t)} \tag{2.4}$$

and

$$\frac{t+z}{t-z} - \left[1 + 2\sum_{k=1}^{n} \frac{z\omega_{k-1}(z)}{\omega_k(t)}\right] = \frac{2z\omega_n(z)}{(t-z)\omega_n(t)}. \tag{2.5}$$

Similarly (e.g. by applying the substar transform) we get

$$1 - 2\sum_{k=1}^{n} \alpha_k \frac{t^k\omega_{k-1}(z)}{z^k\omega_k(t)} = -\frac{t+z}{t-z}\left[1 - \frac{t^{n+1}\omega_n(z)}{z^{n+1}\omega_n(t)}\right] - \frac{t^{n+1}\omega_n(z)}{z^{n+1}\omega_n(t)} \tag{2.6}$$

and

$$\frac{t+z}{t-z} + \left[1 - 2\sum_{k=1}^{n} \alpha_k \frac{t^k\omega_{k-1}(z)}{z^k\omega_k(t)}\right] = \frac{2t^{n+1}\omega_n(z)}{z^n(t-z)\omega_n(t)}. \tag{2.7}$$

Operating on Equations (2.5) and (2.7) by M we get

$$\Omega_M(z) - [\mu_0 + 2\sum_{k=1}^{n} \mu_k z\omega_{k-1}(z)] = 2z\omega_n(z)\, M[\frac{1}{(t-z)\omega_n(t)}] \tag{2.8}$$

and

$$\Omega_M(z) - [\nu_0 + 2\sum_{k=1}^{n} \nu_k \frac{\omega_{k-1}(z)}{z^k}] = 2\frac{\omega_n(z)}{z^{n+1}}\, M[\frac{zt^{n+1}}{(t-z)\omega_n(t)}]. \tag{2.9}$$

The left sides of (2.8) and (2.9) are the left sides of (2.2) and (2.3). We introduce

$$F_n(z) = 2M[\frac{1}{(t-z)\omega_n(t)}], G_n(z) = 2M[\frac{zt^{n+1}}{(t-z)\omega_n(t)}]. \tag{2.10}$$

It follows from the assumption of M being defined on $\tilde{\mathcal{L}} \cdot \tilde{\mathcal{L}}$ (see Section 1) that $F_n(z)$ and $G_n(z)$ are defined (as finite numbers) for $z = \alpha_1, ..., \alpha_n$. Also both $F_n(0) = 2M[\frac{1}{t\omega_n(t)}]$ and $G_n(\infty) = 2M[(\frac{t^{n+1}}{(\zeta t - 1)\omega_n(t)})_{\zeta=0}]$ are defined by assumption.

□

We shall make use of alternative expressions for $\psi_n(z)$ to the one given in the definition (1.11).

Lemma 2.2. *Let $m = 0, 1, ..., n-1$. Then we may write*

$$\psi_n(z) = M[\frac{t+z}{t-z}(\frac{\omega_m(z)}{\omega_m(t)}\varphi_n(t) - \varphi_n(z))]. \tag{2.11}$$

Proof: We may write

$$\psi_n(z) - M[\frac{t+z}{t-z}(\frac{\omega_m(z)}{\omega_m(t)}\varphi_n(t) - \varphi_n(z))] = M[\frac{(t+z)(\omega_m(t) - \omega_m(z))\varphi_n(t)}{(t-z)\omega_m(t)}]. \tag{2.12}$$

The argument of M on the right side is of the form $\frac{R_m(t)\varphi_n(t)}{\omega_m(t)}$, $R_m \in \Pi_m$, hence the functional value is zero by orthogonality.

□

We shall use the following notation:

$$\Delta_{n,s} = \psi_n(z) + \varphi_n(z)\, P_s(z), \tag{2.13}$$

$$\tilde{\Delta}_{n,s} = \psi_n(z) + \varphi_n(z)\, Q_s(z). \tag{2.14}$$

We recall that p_n and q_n denote the numerator polynomials of φ_n and ψ_n, cf. (1.10) and (1.12).

Proposition 2.3. *For $m < n$ and arbitrary s the following formula is valid:*

$$q_n(z) + p_n(z)P_s(z)\omega_s(z) = \\ 2zM[\frac{\omega_s(t)\omega_n(z)\omega_m(z)p_n(t) - \omega_s(z)\omega_n(t)\omega_m(t)p_n(z)}{(t-z)\omega_s(t)\omega_n(t)\omega_m(t)}]. \tag{2.15}$$

Proof: Taking into account (1.20), (1.22) and (2.11) we may write for $m < n$:

$$\Delta_{n,s} = M[\frac{t+z}{t-z}(\frac{\omega_m(z)}{\omega_m(t)}\varphi_n(t) - \varphi_n(z)) + \varphi_n(z)(1 + 2\sum_{k=1}^{s}\frac{z\omega_{k-1}(z)}{\omega_k(t)})] \tag{2.16}$$

and by using (2.4) with n replaced by s in the summation we get

$$\Delta_{n,s} = M[\frac{(t+z)\omega_m(z)\varphi_n(t)}{(t-z)\omega_m(t)} - \frac{2z\omega_s(z)\varphi_n(z)}{(t-z)\omega_s(t)}]. \tag{2.17}$$

Substitution of the expansion (1.10) for φ_n then yields

$$\Delta_{n,s} = M[\frac{(t+z)\omega_s(t)\omega_n(z)\omega_m(z)p_n(t) - 2z\omega_s(z)\omega_n(t)\omega_m(t)p_n(z)}{(t-z)\omega_s(t)\omega_n(t)\omega_m(t)\omega_n(z)}]. \tag{2.18}$$

This may again be written as

$$\Delta_{n,s} = M[\frac{2z[\omega_s(t)\omega_n(z)\omega_m(z)p_n(t) - \omega_s(z)\omega_n(t)\omega_m(t)p_n(z)]}{(t-z)\omega_s(t)\omega_n(t)\omega_m(t)\omega_n(z)}] \\ + \omega_m(z)M[\frac{p_n(t)}{\omega_n(t)\omega_m(t)}]. \tag{2.19}$$

The argument of M in the last term may be written as $\varphi_n(t)\dfrac{1}{\omega_m(t)}$, hence the last term is zero since $m < n$. (Note that the argument of the first term to the right in (2.19) belongs to the domain of definition of M, since the argument in the last term in (2.19) and the argument in (2.17) (hence also in (2.16)) does. Thus

$$\Delta_{n,s} = \frac{2z}{\omega_n(z)}M[\frac{\omega_s(t)\omega_n(z)\omega_m(z)p_n(t) - \omega_s(z)\omega_n(t)\omega_m(t)p_n(z)}{(t-z)\omega_s(t)\omega_n(t)\omega_m(t)}]. \tag{2.20}$$

Multiplication by $\omega_n(z)$ gives (2.15).

□

Proposition 2.4. *For $m < n$ and arbitrary s the following formula is valid:*

$$q_n(z) + p_n(z)Q_s(z) \\ = \frac{2}{z^s}M[\frac{tz^s\omega_s(t)\omega_n(z)\omega_m(z)p_n(t) - t^{s+1}\omega_s(z)\omega_n(t)\omega_m(t)p_n(z)}{(t-z)\omega_s(t)\omega_n(t)\omega_m(t)}]. \tag{2.21}$$

Proof: Using (1.21) and (1.23) we get for $m < n$:

$$\tilde{\Delta}_{n,s} = M[\frac{t+z}{t-z}(\frac{\omega_m(z)}{\omega_m(t)}\varphi_n(t) - \varphi_n(z)) + \varphi_n(z)(-1 + 2\sum_{k=1}^{s}\alpha_k\frac{t^k\omega_{k-1}(z)}{z^k\omega_k(t)})] \tag{2.22}$$

and by (2.6) with n replaced by s in the summation this yields

$$\tilde{\Delta}_{n,s} = M[\frac{(t+z)\omega_m(z)\varphi_n(t)}{(t-z)\omega_m(t)} - \frac{2t^{s+1}\omega_s(z)\varphi_n(z)}{z^s(t-z)\omega_s(t)}]. \tag{2.23}$$

This may be written as

$$\tilde{\Delta}_{n,s} = M[\frac{(t+z)z^s\omega_s(t)\omega_n(z)\omega_m(z)p_n(t) - 2t^{s+1}\omega_s(z)\omega_n(t)\omega_m(t)p_n(z)}{z^s(t-z)\omega_s(t)\omega_n(t)\omega_m(t)\omega_n(z)}]. \tag{2.24}$$

Rearranging we get

$$\tilde{\Delta}_{n,s} = M[\frac{2[tz^s\omega_s(t)\omega_n(z)\omega_m(z)p_n(t) - t^{s+1}\omega_s(z)\omega_n(t)\omega_m(t)p_n(z)]}{z^s(t-z)\omega_s(t)\omega_n(t)\omega_m(t)\omega_n(z)}] \tag{2.25}$$

$$-\omega_m(z)\,M[\frac{p_n(t)}{\omega_n(t)\omega_m(t)}].$$

The last term is zero (cf. the proof of Proposition 2.3), and so (2.21) follows by multiplication by $\omega_n(z)$.

□

For each n we define interpolation tables $B_{2n-1}^{(n)}$ and $\tilde{B}_{2n-1}^{(n)}$ by

$$B_{2n-1}^{(n)} = \{\alpha_0, \beta_1^{(n)}, \beta_2^{(n)}, ..., \beta_{2n-1}^{(n)}\} \tag{2.26}$$

$$\tilde{B}_{2n-1}^{(n)} = \{\alpha_\infty, \beta_1^{(n)}, \beta_2^{(n)}, ..., \beta_{2n-1}^{(n)}\}, \tag{2.27}$$

where

$$\beta_k^{(n)} = \alpha_k \text{ for } 1 \le k \le n, \;\; \beta_k^{(n)} = \alpha_{k-n} \text{ for } n < k \le 2n-1. \tag{2.28}$$

We further define

$$\omega_0^{(n)} = 1, \;\; \omega_k^{(n)}(z) = \prod_{j=1}^{k}(z - \beta_j^{(n)}) \text{ for } k = 1, 2, ..., 2n-1. \tag{2.29}$$

Note that

$$\omega_k^{(n)}(z) = \omega_n(z)\omega_{k-n}(z) \text{ for } n < k \le 2n-1. \tag{2.30}$$

We also define moments $\mu_k^{(n)}$, $\nu_k^{(n)}$ by

$$\mu_k^{(n)} = M[\frac{1}{\omega_k^{(n)}(t)}], \;\; k = 0, 1, 2, ..., 2n-1, \tag{2.31}$$

$$\nu_0^{(n)} = -\mu_0^{(n)}, \;\; \nu_k^{(n)} = M[\frac{\alpha_k t^k}{\omega_k^{(n)}(t)}], \;\; k = 1, 2, ..., 2n-1. \tag{2.32}$$

Then $\mu_k^{(n)} = \mu_k$, $\nu_k^{(n)} = \nu_k$ for $k = 0, 1, ..., n$. Finally we introduce polynomials $P_{2n-1}^{(n)}$ and $Q_{2n-1}^{(n)}$ defined by

$$P_{2n-1}^{(n)}(z) = \mu_0^{(n)} + 2 \sum_{k=1}^{2n-1} \mu_k^{(n)} z \omega_{k-1}^{(n)}(z), \tag{2.33}$$

$$Q_{2n-1}^{(n)}(z) = \nu_0^{(n)} + 2 \sum_{k=1}^{2n-1} \nu_k^{(n)} \frac{\omega_{k-1}^{(n)}(z)}{z^k}. \tag{2.34}$$

Proposition 2.5. *The polynomial $P_{2n-1}^{(n)}(z)$ interpolates $\Omega_M(z)$ in the sense that*

$$\Omega_M(z) - P_{2n-1}^{(n)}(z) = z\omega_n(z)\omega_{n-1}(z) f_{2n-1}(z) \tag{2.35}$$

and the polynomial $Q_{2n-1}^{(n)}(z)$ interpolates $\Omega_M(z)$ in the sense that

$$\Omega_M(z) - Q_{2n-1}^{(n)}(z) = \frac{\omega_n(z)\omega_{n-1}(z)}{z^{2n}} g_{2n-1}(z). \tag{2.36}$$

Here $f_{2n-1}(z)$ has finite values for $z = \alpha_0, \alpha_1, ..., \alpha_n$, and $g_{2n-1}(z)$ has finite values for $z = \alpha_\infty, \alpha_1, ..., \alpha_n$.

Proof: This follows by applying Proposition 2.1 to the tables $B_{2n-1}^{(n)}$ and $\tilde{B}_{2n-1}^{(n)}$, taking (2.30) into account.

□

Proposition 2.6. *The following formula is valid:*

$$q_n(z) + p_n(z) P_{2n-1}^{(n)}(z) = z\omega_n(z)\omega_{n-1}(z) F_{2n-1}(z), \tag{2.37}$$

where $F_{2n-1}(z)$ has finite values for $z = \alpha_0, \alpha_1, ..., \alpha_n$.

Proof: It follows by applying Proposition 2.3 to the table $B_{2n-1}^{(n)}$, setting $m = n-1$, $s = 2n-1$ and taking into account (2.30), that

$$q_n(z) + p_n(z) P_{2n-1}^{(n)}(z) = 2z\omega_n(z)\omega_{n-1}(z) M\left[\frac{p_n(t) - p_n(z)}{(t-z)\omega_n(t)\omega_{n-1}(t)}\right]. \tag{2.38}$$

The function $F_{2n-1}(z) = 2M\left[\frac{p_n(t) - p_n(z)}{(t-z)\omega_n(t)\omega_{n-1}(t)}\right]$ is defined and finite at $z = \alpha_0, \alpha_1, ..., \alpha_n$ by assumption, the expression in brackets belonging to the domain of definition of M.

□

Proposition 2.7. *The following formula is valid:*

$$q_n(z) + p_n(z) Q_{2n-1}^{(n)}(z) = \frac{\omega_n(z)}{z} G_{2n-1}(z), \tag{2.39}$$

where $G_{2n-1}(z)$ has finite values for $z = \alpha_\infty, \alpha_1, ..., \alpha_n$.

Proof: It follows by applying Proposition 2.4 to the table $\tilde{B}^{(n)}_{2n-1}$, setting $m = 0$, $s = 2n-1$ and taking into account (2.30), that

$$q_n(z) + p_n(z)Q^{(n)}_{2n-1} = 2\frac{\omega_n(z)}{z} M[\frac{tz^{2n-1}\omega_{n-1}(t)p_n(t) - t^{2n}\omega_{n-1}(z)p_n(z)}{(t-z)\omega_n(t)\omega_{n-1}(t)z^{2n-2}}] \quad (2.40)$$

The function $G_{2n-1}(z) = 2M[\frac{tz^{2n-1}\omega_{n-1}(t)p_n(t) - t^n\omega_{n-1}(z)p_n(z)}{(t-z)\omega_n(t)\omega_{n-1}(t)z^{2n-2}}]$ is defined and finite for $z = \alpha_1, ..., \alpha_n$. Furthermore $H_{2n-1}(\zeta) = G_{2n-1}(\zeta^{-1}) = 2M[\frac{t\omega_{n-1}(t)p_n(t) - \zeta^{2n-1}t^{2n}\omega_{n-1}(\zeta^{-1})p_n(\zeta^{-1})}{(t\zeta-1)\omega_n(t)\omega_{n-1}(t)}]$ is finite for $\zeta = 0$, since the argument for M in this expression belongs to the domain of definition of M for $\zeta = 0$.

□

Theorem 2.8. *The rational function $\frac{q_n(z)}{p_n(z)} = \frac{\psi_n(z)}{\varphi_n(z)}$ interpolates $-\Omega_M(z)$ linearly at the table $\{\alpha_0, \alpha_\infty, \alpha_1, \alpha_1, \alpha_2, \alpha_2, ..., \alpha_{n-1}, \alpha_{n-1}, \alpha_n\}$ in the sense that*

$$q_n(z) + p_n(z)\Omega_M(z) = A_n(z)\, z\omega_n(z)\omega_{n-1}(z) \quad (2.41)$$

and

$$q_n(z) + p_n(z)\Omega_M(z) = B_n(z) \cdot \frac{\omega_n(z)}{z} \quad (2.42)$$

where $A_n(z)$ and $B_n(z)$ are functions such that $A_n(z)$ is finite for $z = \alpha_0, \alpha_1, ..., \alpha_n$ and $B_n(z)$ is finite for $z = \alpha_\infty, \alpha_1, ..., \alpha_n$.

Proof: We may write

$$q_n(z) + p_n(z)\Omega_M(z) = q_n(z) + p_n(z)P^{(n)}_{2n-1}(z) + p_n(z)[\Omega_M(z) - P^{(n)}_{2n-1}(z)] \quad (2.43)$$

Formula (2.41) now follows from (2.35) and (2.37). Similarly we may write

$$q_n(z) + p_n(z)\Omega_M(z) = q_n(z) + p_n(z)Q^{(n)}_{2n-1}(z) + p_n(z)[\Omega_M(z) - Q^{(n)}_{2n-1}(z)] \quad (2.44)$$

and so Formula (2.42) follows from (2.36) and (2.39).

□

Remark. The rational function $\frac{p_n(z)}{q_n(z)}$ satisfies $2n+1$ interpolation conditions for $-\Omega_M(z)$, while the degrees of the polynomials $p_n(z)$ and $q_n(z)$ are both at most n. Thus $\frac{p_n(z)}{q_n(z)}$ is a multipoint Padé approximant to $-\Omega_M(z)$. (For a more detailed

discussion of multipoint Pad approximants, see e.g. [15]. The essential requirement is that the rational function satisfies a number of interpolation conditions at least equal to one plus the sum of the degrees of numerator and denominator.)
Instead of introducing the new tables $B_{2n-1}^{(n)}$ and $\tilde{B}_{2n-1}^{(n)}$ for each n we might introduce one new table $B = \{\alpha_0, \alpha_\infty, \alpha_1, \alpha_1, \alpha_2, \alpha_2, ..., \alpha_n, \alpha_n, ...\}$. Orthogonality would be defined in terms of the table $A = \{\alpha_0, \alpha_1, \alpha_2, ..., \alpha_n, ...\}$, interpolation would be at points of B. Moments, however, would have to be defined in terms of the tables $B_{2n-1}^{(n)}$, $\tilde{B}_{2n-1}^{(n)}$.

References

1. N.I. Akhiezer, *The classical moment problem and some related questions in analysis*, Hafner, New York, 1965.
2. C. Brezinski, *Padé-Type Approximants and Orthogonal Polynomials*, Birkhäuser, Basel/Boston/Stuttgart, 1980.
3. A. Bultheel, *Laurent series and their Padé approximations*, Operator Theory: Advances and Applications, vol. 27, Birkhäuser Verlag, Basel, 1987.
4. A. Bultheel and P. Dewilde, Orthogonal functions related to the Nevanlinna-Pick problem, *Mathematical Theory of Networks and Systems*, Proceedings MTNS Conference, Delft, The Netherlands (Ed. P. Dewilde) Western Periodicals, North Hollywood, 1979, 207-211.
5. A. Bultheel, P. Gonzalez-Vera, E. Hendriksen and O. Njåstad, A Szegö theory for rational functions, Technical Report TW-131, K.U. Leuven, Dept. of Computer Science, May 1990.
6. A. Bultheel, P. Gonzalez-Vera, E. Hendriksen and O. Njåstad, The computation of orthogonal rational functions and their interpolating properties, *Numerical Algorithms* 2 (1992) 85-114.
7. A. Bultheel, P. Gonzalez-Vera, E. Hendriksen and O. Njåstad, Moment problems and orthogonal functions, *J. Comp. Appl. Math.*, 48 (1993) 49-68.
8. A. Bultheel, P. Gonzalez-Vera, E. Hendriksen and O. Njåstad, Orthogonal rational functions with poles on the unit circle, *J. Math. Anal. Appl.*, to appear.
9. A. Bultheel, P. Gonzalez-Vera, E. Hendriksen and O. Njåstad, Orthogonal rational functions on the unit circle and the real line, Koninklijke Akademie van Wetenshappen Verhandelingen, Afd. Natuurkunde, Erste Reeks.
10. T.S. Chihara, *Introduction to Orthogonal Polynomials*, Mathematics and Its Applications Series, Gordon & Breach, New York/London/Paris, 1978.
11. M.M Djrbashian, A survey on the theory of orthogonal systems and some open problems, *Orthogonal Polynomials: Theory and Practice* (Ed. P. Nevai, Kluwer, Dordrecht/Boston/London 1990) 135-146.
12. M.A. Gallucci and W.B. Jones, Rational Approximations corresponding to Newton series (Newton-Padé approximants), *J. Approximation Theory* 17 (1976) 366-392.
13. W.B. Jones, O. Njåstad and W.J. Thron, Orthogonal Laurent polynomials and the strong Hamburger moment problem, *J. Math. Anal. Appl.* 98 (1984) 528-554.
14. O. Njåstad, An extended Hamburger moment problem, *Proc. Edinburgh Math. Soc.* Ser. II, 28 (1985) 167-183.
15. O. Njåstad, Multipoint Padé approximation and orthogonal rational functions, *Non-linear Numerical Methods and Rational Approximation* (ed.: A. Cuyt, D. Reidel Publ. Comp., Dordrecht 1988) 259-270.
16. O. Njåstad, A modified Schur algorithm and an extended Hamburger moment problem, *Trans. Amer. Math. Soc.* 327 (1991) 283-311.
17. O. Njåstad and W.J. Thron, The theory of sequences of orthogonal L-polynomials, in *Padé Approximants and Continued Fractions* (H. Waadeland and H. Wallin, eds.), Det Kongelige Norske Videnskabers Selskab, Skrifter, 1983), 54-91.

BREAKDOWNS IN THE COMPUTATION OF ORTHOGONAL POLYNOMIALS

C. BREZINSKI
Laboratoire d'Analyse Numérique et d'Optimisation
UFR IEEA - M3
Université des Sciences et Technologies de Lille
F-59655 Villeneuve d'Ascq Cedex, France
Email: brezinsk@omega.univ-lille1.fr

and

M. REDIVO-ZAGLIA
Dipartimento di Elettronica e Informatica
Università degli Studi di Padova
Via Gradenigo 6/a
I-35131 Padova, Italy
Email: elen@elett1.dei.unipd.it

Abstract. When computing recursively the members of a family of formal orthogonal polynomials, a division by zero can occur, thus producing a breakdown in the algorithm which has to be stopped. In this paper, such breakdowns are analyzed in detail and classified. It is also showed how to avoid them in some particular cases. Applications to Padé approximation, extrapolation methods and Lanczos method for solving systems of linear equations are discussed.

Key words: Orthogonal polynomials, Lanczos method, Padé approximation, extrapolation methods.

1. Introduction

Let c be the linear functional on the space of complex polynomials defined by

$$c(x^i) = c_i \quad \text{for} \quad i = 0, 1, \ldots,$$

where the c_i's are given complex numbers.

The family of orthogonal polynomials $\{P_k\}$ with respect to c is defined by

$$c\left(x^i P_k(x)\right) = 0 \quad \text{for} \quad i = 0, \ldots, k-1.$$

These polynomials are given by the determinantal formula

$$P_k(x) = d_k \begin{vmatrix} c_0 & \cdots & c_k \\ \vdots & & \vdots \\ c_{k-1} & \cdots & c_{2k-1} \\ 1 & \cdots & x^k \end{vmatrix},$$

where d_k is an arbitrary nonzero constant.

A. Cuyt (ed.), *Nonlinear Numerical Methods and Rational Approximation II*, 49–59.

When the linear functional c can be represented as the integral on the real line of some positive measure, then the usual orthogonal polynomials are recovered. Otherwise, such polynomials are called formal orthogonal polynomials with respect to c.

Orthogonal polynomials (usual and formal) have many applications in numerical analysis and in applied mathematics. The role of the usual ones in Gaussian quadrature methods is, for example, well known. Formal orthogonal polynomials have applications in Padé approximation [3], Lanczos methods for solving systems of linear equations [16] and in extrapolation methods which are used for accelerating the convergence of sequences [8].
In these three applications, the normalizing factor d_k is chosen differently. However, they can all be written under the form

$$P_k(x) = \frac{\begin{vmatrix} c_0 & \cdots & c_k \\ \vdots & & \vdots \\ c_{k-1} & \cdots & c_{2k-1} \\ 1 & \cdots & x^k \end{vmatrix}}{\begin{vmatrix} c_0 & \cdots & c_k \\ \vdots & & \vdots \\ c_{k-1} & \cdots & c_{2k-1} \\ a_0 & \cdots & a_k \end{vmatrix}}, \tag{1}$$

where the a_i's are numbers which can eventually depend on k.

We shall consider three different cases, corresponding to three different normalizations

- case 1: In Padé approximation, the choice $a_0 = \cdots = a_{k-1} = 0$ and $a_k = 1$ is made, which corresponds to monic polynomials.
- case 2: In Lanczos methods, the choice $a_0 = 1$ and $a_1 = \cdots = a_k = 0$ is considered. It corresponds to $P_k(0) = 1$.
- case 3: In extrapolation methods, we take $a_0 = \cdots = a_k = 1$ which corresponds to $P_k(1) = 1$.

Let us remark that, in the cases 2 and 3, the a_i's do not depend on k.

Thus, in each case, P_k exists and is unique if and only the determinant

$$H_k = \begin{vmatrix} c_0 & \cdots & c_k \\ \vdots & & \vdots \\ c_{k-1} & \cdots & c_{2k-1} \\ a_0 & \cdots & a_k \end{vmatrix} \tag{2}$$

is different from zero.

We shall now see that, when computing recursively some family of formal orthogonal polynomials, a division by zero can occur in the relation used. We shall examine in detail the meaning of such a division by zero (called a *breakdown*) and classify such breakdowns. We shall also see how to avoid them.

2. True and ghost breakdown

The computation of the polynomials $\{P_k\}$ is usually performed via some recurrence relationship.

It is well known that a family of (usual or formal) orthogonal polynomials satisfies a three-term recurrence relationship of the form

$$P_{k+1}(x) = (A_{k+1}x + B_{k+1})P_k(x) - C_{k+1}P_{k-1}(x) \tag{3}$$

for $k = 0, 1, \ldots$ with $P_{-1}(x) = 0$ and P_0 equals to an arbitrary nonzero constant. Imposing the orthogonality conditions, we obtain

$$A_{k+1}\, c\left(x^k P_k\right) - C_{k+1}\, c\left(x^{k-1}P_{k-1}\right) = 0 \tag{4}$$

$$A_{k+1}\, c\left(x^{k+1} P_k\right) + B_{k+1}\, c\left(x^k P_k\right) - C_{k+1}\, c\left(x^k P_{k-1}\right) = 0. \tag{5}$$

The normalization condition provides a third equation and thus we obtain three equations for the three unknowns $A_{k+1}, B_{k+1}, C_{k+1}$.

Let us examine the three preceding cases in detail.

- case 1: The polynomials P_k are monic and thus we must have $A_{k+1} = 1$. It follows from (4) that

$$C_{k+1} = c\left(x^k P_k\right)/c\left(x^{k-1}P_{k-1}\right).$$

Then, from (5), we obtain

$$B_{k+1} = \left[C_{k+1}\, c\left(x^k P_{k-1}\right) - c\left(x^{k+1}P_k\right)\right]/c\left(x^k P_k\right).$$

Thus, in order to be able to use (3), $c(x^{k-1}P_{k-1})$ and $c(x^k P_k)$ must be different from zero. If one of these quantities is zero, then a *breakdown* occurs in the algorithm which has to be stopped. Since P_k exists, then $c(x^{k-1}P_{k-1})$ is different from zero. From (1), we see that $c(x^k P_k) \neq 0$ if and only if the determinant H_{k+1} given by (2) is different from zero, that is, since

$$H_{k+1} = \begin{vmatrix} c_0 & \cdots & c_k & c_{k+1} \\ \vdots & & \vdots & \vdots \\ c_k & \cdots & c_{2k} & c_{2k+1} \\ 0 & \cdots & 0 & 1 \end{vmatrix} = \begin{vmatrix} c_0 & \cdots & c_k \\ \vdots & & \vdots \\ c_k & \cdots & c_{2k} \end{vmatrix},$$

if and only if P_{k+1} exists.

A breakdown which occurs if and only if the next polynomial P_{k+1} does not exist, will be called a *true breakdown*.

Let us recall that, in the usual case described above (polynomials orthogonal with respect to a positive measure), all the quantities $c(x^k P_k)$ are strictly positive and thus all the polynomials P_k exist and no breakdown can occur in the three-term recurrence relationship.

- case 2: the normalization condition $P_{k+1}(0) = 1$ gives, as the third equation

$$B_{k+1} - C_{k+1} = 1.$$

The determinant D_k of the 3×3 system giving A_{k+1}, B_{k+1} and C_{k+1} is equal to

$$D_k = -c\left(x^k P_k\right)\left[c\left(x^k P_k\right) - c\left(x^k P_{k-1}\right)\right] - c\left(x^{k-1} P_{k-1}\right) c\left(x^{k+1} P_k\right).$$

When $k = 1$, we have

$$D_1 = -\frac{c_0^2}{c_1^2}\left(c_2^2 - c_1 c_3\right).$$

On the other hand, P_2 exists and is unique if and only if

$$H_2 = \begin{vmatrix} c_0 & c_1 & c_2 \\ c_1 & c_2 & c_3 \\ 1 & 0 & 0 \end{vmatrix} = c_1 c_3 - c_2^2 \neq 0.$$

Thus, we have $D_1 = H_2 c_0^2 / c_1^2$.
It follows that, if P_2 does not exist, a breakdown will occur in the three-term recurrence relationship. But, on the other hand, even if $H_2 \neq 0$, D_1 can be zero (if $c_0 = 0$) and a breakdown can occur in the recurrence relationship. Such a breakdown is not due to the non-existence of the polynomial P_2, but to the relation used for its computation. This kind of breakdown will be called a *ghost breakdown.*

– case 3: the normalization condition $P_{k+1}(1) = 1$ gives, as the third equation

$$A_{k+1} + B_{k+1} - C_{k+1} = 1.$$

The determinant D_k of the 3×3 system giving A_{k+1}, B_{k+1} and C_{k+1} is equal to

$$\begin{aligned} D_k &= -c\left(x^k P_k\right)\left[c\left(x^k P_k\right) - c\left(x^k P_{k-1}\right) - c\left(x^{k-1} P_{k-1}\right)\right] \\ &\quad - c\left(x^{k-1} P_{k-1}\right) c\left(x^{k+1} P_k\right). \end{aligned}$$

When $k = 1$, we have

$$D_1 = -\frac{c_0^2}{(c_0 - c_1)^2}\left(c_0 c_2 - c_1^2 - c_2^2 - c_0 c_3 + c_1 c_2 + c_1 c_3\right).$$

On the other hand, P_2 exists and is unique if and only if

$$H_2 = \begin{vmatrix} c_0 & c_1 & c_2 \\ c_1 & c_2 & c_3 \\ 1 & 1 & 1 \end{vmatrix} = c_0 c_2 - c_1^2 - c_2^2 - c_0 c_3 + c_1 c_2 + c_1 c_3 \neq 0.$$

Thus, we have $D_1 = H_2 c_0^2 / (c_0 - c_1)^2$.
It follows that, if P_2 does not exist, a breakdown will occur. But, on the other hand, even if $H_2 \neq 0$, D_1 can be zero (if $c_0 = 0$) and a ghost breakdown can occur in the recurrence relationship.

There are other possible procedures for computing recursively the orthogonal polynomials P_k. One of them consists in computing simultaneously the family $\left\{P_k^{(1)}\right\}$ of polynomials orthogonal with respect to the linear functional $c^{(1)}$ defined by

$$c^{(1)}\left(x^i\right) = c\left(x^{i+1}\right) = c_{i+1} \qquad \text{for } i = 0, 1, \ldots.$$

These polynomials, which can be normalized according to choices similar to those made for the P_k's, are given by the determinantal formula

$$P_k^{(1)}(x) = \frac{\begin{vmatrix} c_1 & \cdots & c_{k+1} \\ \vdots & & \vdots \\ c_k & \cdots & c_{2k} \\ 1 & \cdots & x^k \end{vmatrix}}{\begin{vmatrix} c_1 & \cdots & c_{k+1} \\ \vdots & & \vdots \\ c_k & \cdots & c_{2k} \\ b_0 & \cdots & b_k \end{vmatrix}}, \tag{6}$$

where the b_i's are numbers which can eventually depend on k. They are such that

$$c^{(1)}\left(x^i P_k^{(1)}\right) = 0 \qquad \text{for } i = 0, \ldots, k-1.$$

There are various relations between the polynomials of the two families $\{P_k\}$ and $\left\{P_k^{(1)}\right\}$ (see [3, 9]). For example we have

$$P_k(x) = \alpha_k P_{k-1}(x) + \beta_k x P_{k-1}^{(1)}(x)$$

and

$$P_k^{(1)}(x) = \alpha_k' P_{k-1}^{(1)}(x) + \beta_k' P_k(x),$$

where the coefficients appearing in these relations are obtained by imposing the orthogonality conditions to both sides and the normalizing conditions (which are not necessarily the same for both families).

Obviously, a true or a ghost breakdown can arise in these relations due to the nonexistence of either some polynomial P_k or some polynomial $P_k^{(1)}$ (true breakdown) or due to a division by zero in the expressions of $\alpha_k, \beta_k, \alpha_k'$ or β_k' (ghost breakdown).

We shall now explain how to avoid breakdowns in some particular cases.

3. Avoiding true breakdowns

The case of the true breakdown was completely solved by Draux in his thesis [18] when the polynomials are monic. In our knowledge, it has not been solved in another case. Before explaining his results, let us slightly change our notations. We shall now give a name only to the existing polynomials, that is P_k will be the monic polynomial of degree n_k of the family. The next polynomial will be denoted by P_{k+1}

and its degree will be $n_{k+1} = n_k + m_k$ with $m_k \geq 1$. All the polynomials of the degrees $n_k+1, \ldots, n_k+m_k-1$ do not exist and thus they were not given a name. It was proved by Draux that m_k is given by the conditions

$$c(x^i P_k) = 0 \qquad \text{for } i = 0, \ldots, n_k + m_k - 2 \tag{7}$$

$$\neq 0 \qquad \text{for } i = n_k + m_k - 1. \tag{8}$$

These polynomials can be recursively computed by the relationship

$$P_{k+1}(x) = \left(\alpha_0 + \cdots + \alpha_{m_k-1} x^{m_k-1} + x^{m_k}\right) P_k(x) - C_{k+1} P_{k-1}(x) \tag{9}$$

for $k = 0, 1, \ldots$, with $P_{-1}(x) = 0, P_0(x) = 1, C_1 = 0$ and

$$C_{k+1} = c\left(x^{n_k+m_k-1} P_k\right) / c\left(x^{n_k-1} P_{k-1}\right)$$

$$\alpha_{m_k-1}\, c\left(x^{n_k+m_k-1} P_k\right) + c\left(x^{n_k+m_k} P_k\right) = C_{k+1}\, c\left(x^{n_k} P_{k-1}\right)$$

$$\cdots\cdots\cdots\cdots\cdots\cdots\cdots\cdots\cdots\cdots$$

$$\alpha_0\, c\left(x^{n_k+m_k-1} P_k\right) + \cdots + \alpha_{m_k-1}\, c\left(x^{n_k+2m_k-2} P_k\right) + c\left(x^{n_k+2m_k-1} P_k\right) = C_{k+1}\, c\left(x^{n_k+m_k-1} P_{k-1}\right).$$

Thus, since $c(x^{n_k+m_k-1} P_k)$ is different from zero, this system is never singular, which shows that no breakdown (true or ghost) can occur in the relation (9). In fact, the relation (9) consists in jumping over the polynomials which does not exist and computing only the existing ones (which are called *regular*).

We shall now consider the case where the polynomials P_k are normalized by the condition $P_k(0) = 1$ and the polynomials $P_k^{(1)}$ are monic. It can be seen from (1) and (6) that the condition for the existence and uniqueness of P_k and $P_k^{(1)}$ is the same, namely that

$$\begin{vmatrix} c_1 & \cdots & c_k \\ \vdots & & \vdots \\ c_k & \cdots & c_{2k-1} \end{vmatrix} \neq 0.$$

Such normalizations are those made in some of the algorithms used for implementing Lanczos method for solving a system of linear equations. This method will be discussed in detail in section 5.

It can be proved [12] that the following relation holds

$$P_{k+1}(x) = P_k(x) - x\left(\beta_0 + \cdots + \beta_{m_k-1} x^{m_k-1}\right) P_k^{(1)}(x)$$

where n_k is the degree of $P_k^{(1)}$ and m_k is defined by (7) and (8) with $c^{(1)}$ instead of c and $P_k^{(1)}$ instead of P_k. The β_i's are given as the solution of

$$\beta_{m_k-1}\, c^{(1)}\left(x^{n_k+m_k-1} P_k^{(1)}\right) = c\left(x^{n_k} P_k\right)$$

. . . .

$$\beta_0\, c^{(1)}\left(x^{n_k+m_k-1}P_k^{(1)}\right)+\cdots+\beta_{m_k-1}\, c^{(1)}\left(x^{n_k+2m_k-2}P_k^{(1)}\right)=c\left(x^{n_k+m_k-1}P_k\right).$$

The polynomials $P_k^{(1)}$ are computed by the three–term recurrence relationship (9) as given above, with $P_k^{(1)}$ instead of P_k and using $c^{(1)}$ instead of c for computing its unknown coefficients.

Thus, in this algorithm, no breakdown (true or ghost) can occur which provides a breakdown–free algorithm for implementing Lanczos method [12]. The only possible breakdown is the incurable hard one (see [21] for a complete explanation of the meaning of this) which arises when $n_k = n$, the dimension of the space, and the corresponding polynomial does not exist.

4. Avoiding ghost breakdowns

We shall now consider some other recurrence relations which will also give rise to other algorithms for implementing Lanczos method (see section 5).

Let us, for example, consider the recurrence relationship

$$P_{k+1}^{(1)}(x) \;=\; t_k(x)\, P_k^{(1)}(x) - \gamma_k P_k(x) \tag{10}$$

with the same normalizations conditions as above, that is $P_k(0) = 1$ and $P_k^{(1)}$ monic. t_k is a monic polynomial of degree m_k and γ_k a constant. Imposing the orthogonality conditions, we obtain γ_k and the coefficients of t_k as the solution of a system of linear equations which is not singular if and only if $c(x^{n_k}P_k) \neq 0$, that is, in other words, if and only if P_k has exactly the degree n_k. Thus this relation can be subject of a ghost breakdown.

If we consider the recurrence relation

$$P_{k+1}^{(1)}(x) \;=\; e_{k+1}P_{k+1}(x) + e'_{k+1}P_k^{(1)}(x) \tag{11}$$

then a ghost breakdown also occurs if $c(x^{n_k}P_k) = 0$.

It is possible to avoid such ghost breakdowns by jumping farther than necessary *at the preceding step*, that is jumping over polynomials which exist. When, at the iteration k, we have $c\,(x^{n_k}P_k) = 0$ the process has to be stopped due to a ghost breakdown, even if the polynomials P_{k+1} and $P_{k+1}^{(1)}$ exist. The only possibility for avoiding such a ghost breakdown is not to compute the regular polynomials P_k and $P_k^{(1)}$ but, starting again from the polynomials P_{k-1} and $P_{k-1}^{(1)}$, to compute the first regular polynomials P_k and $P_k^{(1)}$ with a degree n'_k greater than n_k and such that $c\left(x^{n'_k}P_k\right) \neq 0$.

Let us now describe in more detail how this can be performed and how the polynomials P_{k+1} can be simultaneously computed.

We assume that the polynomials P_{k-1} (of the degree n_{k-1} at most) and $P_{k-1}^{(1)}$ (of exactly the degree n_{k-1}) have been computed. First, for avoiding a true breakdown, we look for the next regular polynomials P_k (of the degree $n_k = n_{k-1} + m_{k-1}$ at most) and $P_k^{(1)}$ (of exactly the degree $n_k = n_{k-1} + m_{k-1}$). After having determined these polynomials, we compute the quantity $c\,(x^{n_k} P_k)$ which will be needed in the next iteration for computing the polynomial $P_{k+1}^{(1)}$. If this quantity is zero, we shall try to compute the polynomials $\overline{P}_k$ of the degree $n_k' = n_k + m_k'$ at most and $\overline{P}_k^{(1)}$ of exactly the degree $n_k' = n_k + m_k'$, with $m_k' = 1$. If these polynomials do not exist or if $c\left(x^{n_k'} \overline{P}_k\right) = 0$, we increase m_k' by one and try to compute again the new polynomials $\overline{P}_k$ and $\overline{P}_k^{(1)}$. This process is continued until these polynomials can be computed and the condition is satisfied. Thus, we shall finally obtain two regular polynomials $\overline{P}_k$ and $\overline{P}_k^{(1)}$ (with respectively the degrees n_k' at most and n_k') such that $c\left(x^{n_k'} \overline{P}_k\right) \neq 0$. Taking these polynomials as P_k and $P_k^{(1)}$ respectively, it is now possible to avoid the ghost breakdown in the computation of P_{k+1} and $P_{k+1}^{(1)}$.

Now, the problem is to compute the polynomials $\overline{P}_k$ of the degree $n_k' = n_{k-1} + m_{k-1} + m_k'$ at most and $\overline{P}_k^{(1)}$ of exactly the degree $n_k' = n_{k-1} + m_{k-1} + m_k'$. It is no more possible to use the formulae given above for avoiding a true breakdown because these formulae are only valid when all the polynomials of the degrees between $n_{k-1} + 1$ and $n_k' - 1$ do not exist. But, in our case, we know that there exists at least one polynomial $P_k^{(1)}$ of the degree $n_k < n_k'$. Thus other relations have to be used. They can be found in [13]. For example, we have

$$P_{k+1}(x) = (1 - x\,v_k(x))\,P_k(x) - x\,w_k(x) P_k^{(1)}(x)$$

where w_k is a polynomial of the degree $m_k - 1$ at most and v_k a polynomial of the degree $m_k - 2$ at most. The coefficients of these polynomials are obtained by imposing the orthogonality conditions. This recurrence relationship is now free from true and ghost breakdowns.

When using other recurrence relations or when a different normalization is chosen for our polynomials, a ghost breakdown can be produced by a condition different from $c\,(x^{n_k} P_k) = 0$. For curing such ghost breakdowns, it is possible to proceed in a similar way since the quantity which can produce a breakdown at the iteration k can always be evaluated at the step $k - 1$.

When the absolute value of quantity responsible for a true or a ghost breakdown is different from zero but small, then rounding errors can affect seriously the algorithm. This situation is known under the name of *near–breakdown*. The remedy is again to jump over the quantities that are badly evaluated and to compute only those which exist and can be safely computed. Then, in this situation, it is also necessary to jump over some regular polynomials and thus, the same type of relations than above have to be used.

In the next section, we shall discuss in detail how the previous procedures apply to Lanczos method. Let us mention that similar techniques have been used for avoiding breakdowns and near–breakdowns in the computation of Padé approximants [7], in extrapolation algorithms [6] (see [5] for the vector case) and in the Mühlbach–Neville–Aitken algorithm for interpolation by a linear combination of functions [4]. Breakdowns can be similarly avoided in the computation of multivariate rational interpolants [1].

5. Lanczos method and around

Lanczos method, proposed in [22] for transforming a matrix into a similar tridiagonal one, can also be used as an iterative method for solving a system of linear equations [23].

We consider in $\mathbf{C}^n$ the system of linear equations

$$Av = b.$$

Let x_0 be an arbitrary vector and let $r_0 = b - Ax_0$ be the corresponding residual vector. Let $E_k = \text{span}\left(r_0, Ar_0, \ldots, A^{k-1}r_0\right)$ and $F_k = \text{span}\left(y, A^*y, \ldots, A^{*\,k-1}y\right)$ where y is an arbitrary non-zero vector and where A^* is the conjugate transpose of A.

For solving this system, Lanczos method consists in constructing a sequence of vectors (x_k) defined by

$$\begin{aligned} (x_k - x_0) &\in E_k \\ r_k = b - Ax_k &\perp F_k. \end{aligned}$$

An important property of Lanczos method is its finite convergence, namely that $\exists k \leq n$ such that $r_k = 0$ and $x_k = v = A^{-1}b$.

It is easy to see that the residual vector r_k satisfies

$$r_k = P_k(A)r_0$$

where P_k is the polynomial of degree at most k, normalized by the condition $P_k(0) = 1$, belonging to the family of formal orthogonal polynomials with respect to the linear functional c defined by

$$c\left(x^i\right) = c_i = \left(y, A^i r_0\right), \qquad i = 0, 1, \ldots.$$

Thus Lanczos method consists in computing P_k, then r_k and finally x_k such that $r_k = b - Ax_k$ without, of course, inverting A, that is

$$x_k = x_0 - R_{k-1}(A)r_0$$

where $P_k(x) = 1 + xR_{k-1}(x)$.

The polynomials P_k are computed recursively and each particular recurrence relationship gives rise to a different Lanczos algorithm; see [16] for a review based on orthogonal polynomials and [20] for the classical approach using linear algebra techniques (let us mention that this reference is almost never quoted in the papers dealing with the subject, although it contains much material).

These recurrence relationships can be subject of true and ghost breakdowns and near–breakdowns as well. These breakdowns can be avoided by using the techniques described in the previous sections.

In [12, 13, 14], three algorithms for avoiding the true breakdown were proposed. They are called MRZ, SMRZ and BMRZ. In these three algorithms, P_{k+1} is always computed from P_k and $P_k^{(1)}$ as indicated above.

In the MRZ, $P_{k+1}^{(1)}$ is computed from $P_{k-1}^{(1)}$ and $P_k^{(1)}$ by the usual three-term recurrence relationship. In the SMRZ, $P_{k+1}^{(1)}$ is obtained by (10) from $P_k^{(1)}$ and P_k. The third algorithm, called BMRZ, is a generalization of the second relation between adjacent families of orthogonal polynomials (see Brezinski [3], relation 2.9, p.92 or p.95), and it was also proposed by Draux in [18] (pp. 397–398). In this algorithm, $P_{k+1}^{(1)}$ is expressed in terms of P_{k+1} and $P_k^{(1)}$ by means of (11).

Other breakdown-free algorithms, quite close to ours, have also been obtained by Gutknecht [19]. The same problem has been treated by Baheux [2] for other recurrence relations.

Not all the recurrence relations for avoiding a true-breakdown can be generalized into relations for avoiding a near-breakdown. For instance, it was shown in [13] that the MRZ can be generalized into the algorithm GMRZ, the SMRZ can be generalized into the algorithm BSMRZ, but that the BMRZ cannot be generalized.

Similar techniques have been used in [10, 11, 17] for avoiding a breakdown and a near-breakdown in the CGS of Sonneveld [24] and in the Bi-CGSTAB of Van der Vorst [25]. In particular, the subroutines given in [10] and [11] make use of the procedure described in section 4 for avoiding ghost breakdowns due to $c(x^{n_k}P_k) = 0$. All these questions are discussed in more detail in [15].

References

1. H. Allouche, A. Cuyt, Well-defined determinant representations for non-normal multivariate rational interpolants, *Numerical Algorithms*, to appear.
2. C. Baheux, *Algorithmes d'Implémentation de la Méthode de Lanczos*, Thèse, Université des Sciences et Technologies de Lille, 1994.
3. C. Brezinski, *Padé-type Approximation and General Orthogonal Polynomials.*, ISNM vol. 50, Birkhäuser, Basel, 1980.
4. C. Brezinski, The Mühlbach-Neville-Aitken algorithm and some extensions, *BIT*, 20 (1980) 444–451.
5. C. Brezinski, Other manifestations of the Schur complement, *Linear Algebra Appl.*, 111 (1988) 231–247.
6. C. Brezinski, Algebraic properties of the E-transformation, in *Numerical Analysis and Mathematical Modelling*, Banach Center Publications, vol. 24, PWN, Warsaw, 1990, pp. 85–90.
7. C. Brezinski, M. Morandi Cecchi, M. Redivo-Zaglia, The reverse bordering method, *SIAM J. Matrix Anal. Appl.*, to appear.
8. C. Brezinski, M. Redivo-Zaglia, *Extrapolation Methods. Theory and Practice*, North-Holland, Amsterdam, 1991.
9. C. Brezinski, M. Redivo-Zaglia, A new presentation of orthogonal polynomials with applications to their computation, *Numerical Algorithms*, 1 (1991) 207–222.
10. C. Brezinski, M. Redivo-Zaglia, Treatment of near-breakdown in the CGS algorithm, *Numerical Algorithms*, to appear.
11. C. Brezinski, M. Redivo-Zaglia, Look-ahead in Bi-CGSTAB and other methods for linear systems, to appear.
12. C. Brezinski, M. Redivo-Zaglia, H. Sadok, A breakdown-free Lanczos type algorithm for solving linear systems, *Numer. Math.*, 63 (1992) 29–38.
13. C. Brezinski, M. Redivo-Zaglia, H. Sadok, Avoiding breakdown and near-breakdown in Lanczos type algorithms, *Numerical Algorithms*, 1 (1991) 261–284.
14. C. Brezinski, M. Redivo-Zaglia, H. Sadok, Addendum to "Avoiding breakdown and near-breakdown in Lanczos type algorithms", *Numerical Algorithms*, 2 (1992) 133–136.
15. C. Brezinski, M. Redivo-Zaglia, H. Sadok, Breakdowns in the implementation of the Lánczos method for solving linear systems, *Inter. J. Comp. Math. with Applics.*, to appear.
16. C. Brezinski, H. Sadok, Lanczos type algorithms for solving systems of linear equations,

Appl. Numer. Math., 11 (1993) 443–473.

17. C. BREZINSKI, H. SADOK, Avoiding breakdown in the CGS algorithm, *Numerical Algorithms*, 1 (1991) 199–206.
18. A. DRAUX, *Polynômes Orthogonaux Formels. Applications*, LNM 974, Springer–Verlag, Berlin, 1983.
19. M.H. GUTKNECHT, A completed theory of the unsymmetric Lanczos process and related algorithms, Part I, *SIAM J. Matrix Anal. Appl.*, 13 (1992) 594–639.
20. M.R. HESTENES, The conjugate–gradient method for solving linear systems, in *Proceedings of the Sixth Symposium on Applied Mathematics*, J. Curtiss ed., Amer. Math. Soc., Providence, 1956, pp.83–102.
21. W.D. JOUBERT *Generalized Conjugate Gradient and Lanczos Methods for the Solution of Nonsymmetric Systems of Linear Equations*, Ph.D. Thesis, The University of Texas at Austin, 1983.
22. C. LANCZOS, An iteration method for the solution of the eigenvalue problem of linear differential and integral operators, *J. Res. Natl. Bur. Stand.*, 45 (1950) 255–282.
23. C. LANCZOS, Solution of systems of linear equations by minimized iterations, *J. Res. Natl. Bur. Stand.*, 49 (1952) 33–53.
24. P. SONNEVELD, CGS, a fast Lanczos–type solver for nonsymmetric linear systems, *SIAM J. Sci. Stat. Comp.*, 10 (1989) 36–52.
25. H.A. VAN DER VORST, Bi-CGSTAB: a fast and smoothly converging variant of Bi-CG for the solution of nonsymmetric linear systems, *SIAM J. Sci. Stat. Comp.*, 13 (1992) 631–644.

RECURRENCE RELATIONS IN THE TABLE OF VECTOR ORTHOGONAL POLYNOMIALS

J. VAN ISEGHEM
U.F.R de Mathématiques
Université des Sciences et Technologies de Lille
F-59655 Villeneuve d'Ascq Cedex, France
Email: jvaniseg@ano.univ-lille1.fr

Abstract. Vector orthogonal polynomials appeared as the denominators of vector approximants ([2]). To compute these last ones or to study the orthogonality itself, it is useful to be able to move in the table of the polynomials. It is obviously the first step before studying non regular cases of vector-orthogonality.

1. Introduction

If $f_1, \ldots, f_d$ are d functions, holomorphic in a neighbourhood of zero, a vector linear functional Γ, i.e. defined from $\mathbb{C}[[X]]$ to $\mathbb{C}^d$, can be defined

$$f_\alpha(t) = \sum_{i \geq 0} c_i^\alpha t^i, \quad \Gamma_i = (c_i^1, \ldots, c_i^d)^t$$

$$\mathbb{F}(t) = \sum_{i \geq 0} \Gamma_i t^i, \quad \Gamma(x^i) = \Gamma_i, \quad \Gamma = (c^1, \ldots, c^d).$$

With these notations, it follows, as in the scalar case $\Gamma\left(\frac{1}{1-xt}\right) = \mathbb{F}(t)$. The vector orthogonal polynomials (or of dimension d) are then defined by the following relations (R), for $r = nd + k, 0 \leq k < d$, s integer

$$(R) \begin{cases} \Gamma(x^i P_r^s(x)) = 0 & i = s, \ldots, s+n-1 \\ c^\alpha(x^{s+n} P_r^s(x)) = 0 & \alpha = 1, \ldots, k \end{cases}$$

The condition $c^{k+1}(x^{s+n} P_r^s(x)) \neq 0$ for all n, s, r, is equivalent to the fact that all the polynomials P_r^s exist, are unique and of degree equal to their lower index. This case is called non degenerated and is the only one to be considered here.

The polynomial $\tilde{P}_r^s(x) = x^r P_r^s(x^{-1})$ appears as the denominator of the vector Padé approximant $\mathbb{N}/D = [s+r-1/r]$ of degree (s+r-1,r), totally defined by the accuracy through order principle ($[q/d]$ is the integer part of the fraction)

$$\mathbb{F}(t) - [p/q](t) = O(t^{p+[q/d]+1}).$$

If simultaneous Padé approximants ([1]) are considered, for some choices of the parameters the denominators are actually vector orthogonal polynomials.

A. Cuyt (ed.), Nonlinear Numerical Methods and Rational Approximation II, 61–69.

So, the search of recurrence relations in the table of the P_r^s is of interest, on one hand for the study of vector orthogonality, on the other hand for the computation of vector Padé approximants. Moreover it is a necessary step for the study of non normal tables, that is when some polynomials do not exist or are not of maximum degree.

In that field, the diagonal relation, the Q.D algorithm and the cross rule have already been extended to the vector case ([3]). In the scalar case, all the algorithms can be proved through the Q.D algorithm, so here the vector Q.D algorithm will be used in a similar way. All the relations (except the three terms one (q)) will have d+2 terms.

2. Basic relations

The polynomials $(P_r^s)_{s,r\geq 0}$ as defined in [2], are monic, and satisfy a recurrence relation of order $d+1$, i.e. with $(d+2)$ terms for all $s \geq 0$. Two adjacent families $(P_r^s)_r$ and $(P_r^{s+1})_r$ are linked by the QD vector algorithm ([3]).

They are placed in an array P, where s is constant on a diagonal, and r is constant on a column.

The basic relations, already known, are now recalled. The two first ones, (q) and (e), define the quantities q and e appearing in the QD algorithm

$$(q) \qquad P_{r+1}^s = xP_r^{s+1} - q_{r+1}^s P_r^s, s \geq 0, r \geq 0$$

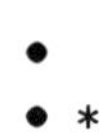

$$(e) \qquad P_r^{s+1} = P_r^s - \sum_{r-d}^{r-1} e_{r,i}^s P_i^{s+1}, \quad s \geq 0, r \geq 0$$

This formula is valid for $r \geq 0$ if $e_{r,i}^s = 0$, $i < 0$, the index i being contained between $r-d$ and $r-1$. The table is,here, always assumed to be normal, i.e. all polynomials P_r^s are exactly of degree r. This is equivalent ([3]) to the fact that all generalized Hankel determinants are non zero. Because $q_{r+1}^s = \dfrac{H_r^s H_{r+1}^{s+1}}{H_r^{s+1} H_{r+1}^s}$, if the table is normal, then all the quantities q_r^s are non zero.

The diagonal relation (Δ) can be deduced from the two preceding ones by two ways, which gives rise to the following QD algorithm, linking the quantities $q_r^s, e_{r,i}^s$ for two consecutives diagonal indices s and s+1.

$$(\Delta)\ P_{r+1}^s = (x - q_{r+1}^s - e_{r,r-1}^s)P_r^s - \sum_{r+1-d}^{r-1} (e_{r,i}^s q_{i+1}^s + e_{r,i-1}^s) P_i^s - q_{r+1-d}^s e_{r,r-d}^s P_{r-d}^s$$

$$(QD) \begin{cases} q_{r+1}^{s+1} + e_{r,r-1}^{s+1} = q_{r+1}^s + e_{r+1,r}^s \\ e_{r,i}^{s+1} q_{i+1}^{s+1} + e_{r,i-1}^{s+1} = e_{r,i}^s q_{r+1}^s + e_{r+1,i}^s, \quad i = r-d+1, \cdots, r-1 \\ e_{r,r-d}^{s+1} q_{r-d+1}^{s+1} = e_{r,r-d}^s q_{r+1}^s \end{cases}$$

It is known ([3]) that any two relations taken from (q),(e), (Δ) imply the third one. We will see that using (Δ) and one relation ((q) in general) is not less general than using the three relations as is typically done for the scalar case.

3. Horizontal and diagonal-horizontal relations

It is now possible to prove horizontal relations, and (ΔH) relations (partly diagonal, partly horizontal relations), which are

(Δ) $\quad\quad$ $\Delta_0 H$ $\quad\quad$ $\Delta_k H$

Taking (Δ) and (e), the following relation is obtained from Δ, using the QD formulas; it contains two terms of indices (r,s),(r+1,s-1) and the d others on the diagonal s

$$P^s_{r+1} = P^{s-1}_{r+1} - \sum_{r-d+1}^{r} e^{s-1}_{r+1,i} P^s_i, \quad s \geq 0, r \geq 0$$

$$P^s_{r+1} = (x - q^{s-1}_{r+1} - e^{s-1}_{r+1,r}) P^s_r - \sum_{r+1-d}^{r-1} (e^{s-1}_{r,i} q^{s-1}_{r+1} + e^{s-1}_{r+1,i}) P^s_i - q^{s-1}_{r+1} e^{s-1}_{r,r-d} P^s_{r-d}$$

$$(\Delta_0 H) \quad P^{s-1}_{r+1} = (x - q^{s-1}_{r+1}) P^s_r - \sum_{i=r-d}^{r-1} e^{s-1}_{r,i} q^{s-1}_{r+1} P^s_i$$

Now the terms on the diagonal s can be expressed in terms of the diagonal $s+1$, using (q) in reverse order, that is $P^s_i = (xP^{s+1}_i - P^s_{i+1})/q^s_{i+1}$, of course if the $q^s_{i+1} \neq 0$. If the table is normal, that is if all the polynomials are assumed to exist and to be of maximum degree, then all the q^s_i are non zero, as it has been recalled in part 2. Then recursively, a diagonal-horizontal relation (Δ_kH) is obtained with k+1 terms on the horizontal and d-k on the diagonal of index s+k.
The proof is done by induction on k. Let us suppose that the relation (Δ_kH) is the following

$$(\Delta_k H) \qquad P^{s-1}_{r+1} = (x - q^{s-1}_{r+1} + a^s_r) P^s_r - \sum_{i=1}^{k-1} a^s_{r-i} x^i P^{s+i}_{r-i} + x^k \sum_{i=r-d}^{r-k} b^{s+k}_i P^{s+k}_i.$$

Going from (Δ_kH) to (Δ_{k+1}H) is done by induction on h. The first step is (Δ_kH) itself. The last sum of the last formula contains the diagonal terms, the formula (q) is used. Let us suppose for some h between 0 and d-k

$$\begin{aligned}
x^k \sum_{r-d}^{r-k} b_i^{s+k} P_i^{s+k} &= x^{k+1} \sum_{i=r-d}^{r-d+h-1} \beta_i^h P_i^{s+k+1} + x^k \sum_{i=r-d+h}^{r-k} \alpha_i^h P_i^{s+k} \\
&= x^{k+1} \sum_{i=r-d}^{r-d+h-1} \beta_i^h P_i^{s+k+1} + x^k \frac{\alpha_{r-d+h}^h}{q_{r-d+h+1}^{s+k}} (x P_{r-d+h}^{s+k+1} - P_{r-d+h+1}^{s+k}) \\
&\quad + x^k \sum_{i=r-d+h+1}^{r-k} \alpha_i^h P_i^{s+k} \\
&= x^{k+1} \sum_{i=r-d}^{r-d+h} \beta_i^{h+1} P_i^{s+k+1} + x^k \sum_{i=r-d+h+1}^{r-k} \alpha_i^{h+1} P_i^{s+k}
\end{aligned}$$

So, $\alpha_i^{h+1}, \beta_i^{h+1}$ are defined recursively by

$$\begin{aligned}
\alpha_i^{h+1} &= \alpha_i^h \; i = r-d+h+2, r-k, \quad & \alpha_{r-d+h+1}^{h+1} &= \alpha_{r-d+h+1}^h - \frac{\alpha_{r-d+h}^h}{q_{r-d+h+1}^{s+k}} \\
\beta_i^{h+1} &= \beta_i^h \; i = r-d, r-d+h-1, \quad & \beta_{r-d+h}^{h+1} &= \frac{\alpha_{r-d+h}^h}{q_{r-d+h+1}^{s+k}}
\end{aligned}$$

For h=d-k, there are no more terms P_i^{s+k} except P_{r-k}^{s+k}, and so we get the formula (Δ_{k+1}H) with $a_{r-k}^s = -\alpha_{r-k}^{d-k}$ and $b_i^{s+k+1} = b_i^{d-k}, i = r-d, r-k-1$

$$(\Delta_{k+1} H) \;\; P_{r+1}^{s-1} = (x - q_{r+1}^{s-1} + a_r^s) P_r^s - \sum_{i=1}^{k} a_{r-i}^s x^i P_{r-i}^{s+i} + x^{k+1} \sum_{i=r-d}^{r-k-1} b_i^{s+k+1} P_i^{s+k+1}.$$

Finally for k=d, all the terms are on the horizontal line

$$(H) \qquad P_{r+1}^{s-1} = (x - q_{r+1}^{s-1} + a_r^s) P_r^s - \sum_{i=1}^{d} a_{r-i}^s x^i P_{r-i}^{s+i}.$$

Looking only at the last term P_{r-d}^{s+k} of (Δ_kH), $b_{r-d}^{s+k+1} = \frac{1}{q_{r-d+1}^{s+k} b_{r-d}^{s+k}}$, so it can be seen that $a_{r-d}^s = \dfrac{e_{r,r-d}^s}{\prod_1^{d-1} q_{r-d+1}^{s+i}}$.

4. Vertical and vertical-diagonal relations

Starting from the diagonal relation, the terms are recursively moved back on the vertical line containing the polynomials of degree r-d, by a direct use of the (q) relation (without division).

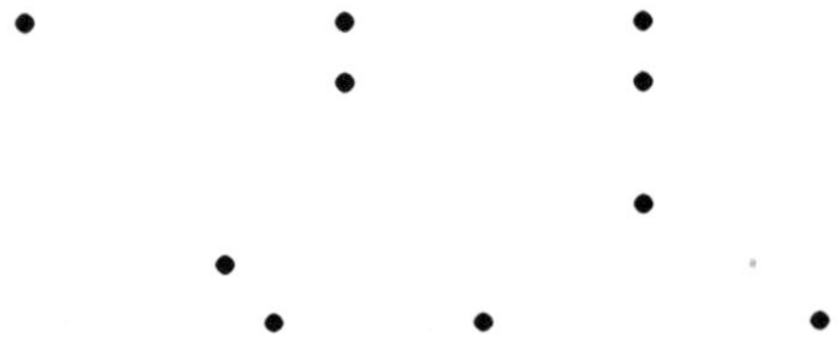

(Δ) $\qquad V\Delta_1 \qquad V\Delta_k$

The first step is to obtain (VΔ_1), containing the polynomials P^s_{r-d}, P^{s+1}_{r-d} and the d other ones on the diagonal of index s+1. It is a particular case because a simplification by x occurs.

$$(\Delta)\ P^s_{r+1} = (x - q^s_{r+1} - e^s_{r,r-1})P^s_r - \sum_{r+1-d}^{r-1} (e^s_{r,i}q^s_{i+1} + e^s_{r,i-1})P^s_i - q^s_{r+1-d}e^s_{r,r-d}P^s_{r-d}$$

Replacing P^s_{r+1}, P^s_r using relation (q), it follows

$$xP^{s+1}_r = (x - e^s_{r,r-1})xP^{s+1}_{r-1} - \Big(q^s_r(x - e^s_{r,r-1}) + (e^s_{r,r-1}q^s_r + e^s_{r,r-2})\Big)P^s_{r-1} + \ldots$$

The term $q^s_{i+1}e^s_{r,r-i}P^s_i$ will cancel for each i, and finally, after simplification by x, it follows $(V\Delta_1)$

$$P^{s+1}_r = (x-e^s_{r,r-1})P^{s+1}_{r-1} - \sum_{i=r-d}^{r-2} \Big((-1)^i q^s_{i+2}\cdots q^s_r x + e^s_{r,i}\Big)P^{s+1}_i + (-1)^d q^s_{r-d+1}\cdots q^s_r P^s_{r-d}$$

Now let us suppose that it is possible to write the following formula (VΔ_k), the formula (VΔ_{k+1}) is then proved

$$(V\Delta_k) \qquad x^{k-1}P^{s+k}_{r-k+1} = \sum_{i=r-d}^{r-k} B^{s+k}_i x^{k-1}P^{s+k}_i + V_k$$

$$V_k = \sum_{i=1}^{k-1} A_i x^{i-1}P^{s+i}_{r-d} + (-1)^d q^s_{r-d+1}\cdots q^s_r P^s_{r-d}$$

where $\deg A_i = 1$ and $\deg B_i = 1$.

Using the relation (q) for the left hand side, we obtain from $(V\Delta_k)$

$$x^kP^{s+k+1}_{r-k} = (q^{s+k}_{r-k+1} + B^{s+k}_{r-k})x^{k-1}P^{s+k}_{r-k} \sum_{i=r-d}^{r-k-1} B^{s+k}_i x^{k-1}P^{s+k}_i + V_k.$$

The formula (VΔ_{k+1}) is then proved by induction on j, $j = k$ and the previous relation being the first step of the proof. Let us write for $k+1 \le j \le d$

$$\begin{aligned}
x^kP^{s+k+1}_{r-k} &= \sum_{r-j}^{r-k-1} C^j_i x^k P^{s+k+1}_i + \sum_{r-d}^{r-j} D^j_i x^{k-1}P^{s+k}_i + V_k \\
&= \sum_{r-j}^{r-k-1} C^j_i x^k P^{s+k+1}_i + D^j_{r-j}x^kP^{s+k+1}_{r-j-1} \\
&\quad +(D^j_{r-j+1} - q^{s+k}_{r-j}D^j_{r-j})x^{k-1}P^{s+k}_{r-j-1} + \sum_{r-d}^{r-j-1} D^j_i x^{k-1}P^{s+k}_i + V_k \\
&= \sum_{r-j-1}^{r-k-1} C^{j+1}_i x^kP^{s+k+1}_i + \sum_{r-d}^{r-j-1} D^{j+1}_i x^{k-1}P^{s+k}_i + V_k
\end{aligned}$$

where the C_i^j, D_i^j are defined by

$$
\begin{aligned}
C_i^{j+1} &= C_i^j, i = r-j, r-k-1, \qquad C_{r-j-1}^{j+1} = D_{r-j}^j \\
D_i^{j+1} &= D_i^j, i = r-d, r-j-2, \qquad D_{r-j-1}^{j+1} = D_{r-j-1}^j - q_{r-j}^{s+k} D_{r-j}^j
\end{aligned}
$$

When P_{r-d}^{s+k} is the only polynomial remaining on the diagonal s+k, the relation (VΔ_{k+1}) is obtained

$$
\begin{aligned}
(V\Delta_{k+1}) \qquad x^k P_{r-k}^{s+k+1} &= \sum_{r-d}^{r-k-1} B_i^{s+k+1} x^k P_i^{s+k+1} + V_{k+1} \\
B_i^{s+k+1} &= C_i^d,\ i = r-d, r-k+1 \\
A_k = D_{r-d}^d,\quad V_{k+1} &= A_k x^{k-1} P_{r-d}^{s+k} + V_k
\end{aligned}
$$

Because of the initial formula, all the C_i^j, D_i^j are of degree 1, and so are the A_i and B_i^{s+k+1}. Now for $k = d$, this last formula is the vertical relation (V) (the first term is easy to find since $B_{r-k-1}^{s+k+1} = C_{r-k-1}^d = C_{r-k-1}^{k+1} = q_{r-k-1}^{s+k} + B_{r-k}^{s+k}$)

$$
\begin{aligned}
(V) \qquad & x^d P_{r-d}^{s+d+1} = (x+\alpha_d) x^{d-1} P_{r-d}^{s+d} + \sum_{i=1}^{d-1} A_i x^{i-1} P_{r-d}^{s+i} + (-1)^d q_{r-d+1}^s \cdots q_r^s P_{r-d}^s \\
& \deg A_i = 1, \quad \alpha_d = q_{r-d+1}^{s+d} + q_{r-d+2}^{s+d-1} + \ldots + q_r^{s+1} - e_{r,r-1}^s
\end{aligned}
$$

5. Ascending staircase and antidiagonal

As it has been seen, the form of the tail of the relation can be modified by use of relation (q) of the Q-D algorithm and so only the form of the right part of the relation is of importance from the point of view of the existence of the relation.

The first relation (AS1) is obtained by eliminating the term P_{r+1}^s between relations (q) and (e)

$$
\begin{aligned}
P_{r+1}^s &= P_{r+1}^{s-1} - \sum_{r-d+1}^{r} e_{r+1,i}^{s-1} P_i^s \\
&= x P_r^{s+1} - q_{r+1}^s P_r^s \\
(AS1)\ P_{r+1}^{s-1} &= x P_r^{s+1} + (e_{r+1,r}^{s-1} - q_{r+1}^s) P_r^s + \sum_{r-d+1}^{r-1} e_{r+1,i}^{s-1} P_i^s .
\end{aligned}
$$

Similarly, the second (AS2) is obtained from (Δ_1H) and (q)

$$
\begin{aligned}
P_{r+1}^{s-1} &= (x - q_{r+1}^{s-1} + a_r^s) P_r^s + x \sum_{r-d}^{r-1} b_i^{s+1} P_i^{s+1} \\
&= x P_r^s - q_{r+1}^{s-1} P_r^{s-1} \\
(AS2)\ q_{r+1}^{s-1} P_r^{s-1} &= (q_{r+1}^{s-1} - a_r^s) P_r^s - x \sum_{r-d}^{r-1} b_i^{s+1} P_i^{s+1}
\end{aligned}
$$

$(AS1)$ $(AS2)$

An anti-diagonal relation can also be obtained, antidiagonal is to be understood for the three last terms in the following relation.
Starting from (AS1) with the polynomials $P^s_r, P^{s-1}_{r+1}, P^{s+1}_r$, then P^{s-1}_{r+1}, P^{s+1}_r are replaced through (AS2) with respectively $(r+1, s-1)$ and $(r, s+1)$ in south-east position and the 'antidiagonal' is obtained linking $P^{s-2}_{r+1}, P^s_r, P^{s+2}_{r-1}$ and polynomials of the diagonals of index s and s+2.

$$(AS1)\ \ P^{s-1}_{r+1} = xP^{s+1}_r + (e^{s-1}_{r+1,r} - q^s_{r+1})P^s_r + \sum_{r-d+1}^{r-1} e^{s-1}_{r+1,i}P^s_i .$$

$$(q^{s-2}_{r+2} - a^{s-1}_{r+1})P^{s-1}_{r+1} = q^{s-2}_{r+2}P^{s-2}_{r+1} + x\sum_{i=r-d+1}^{r} b^s_i P^s_i$$

$$(q^s_{r+1} - a^{s+1}_r)P^{s+1}_r = q^s_{r+1}P^s_r + x\sum_{i=r-d}^{r-1} b^{s+2}_i P^{s+2}_i$$

The antidiagonal $(\mathbf{a\Delta})$ is as follows

$$\begin{aligned}\frac{q^{s-2}_{r+2}}{q^{s-2}_{r+2} - a^{s-1}_{r+1}}P^{s-2}_{r+1} &= \Big(x\big(\frac{q^s_{r+1}}{q^s_{r+1} - a^{s+1}_r} - \frac{b^s_r}{q^{s-2}_{r+2} - a^{s-1}_{r+1}}\big) + e^{s-1}_{r+1,r} - q^s_{r+1}\Big)P^s_r \\ &- \sum_{i=r-d+1}^{r-1}\big(xb^s_i \frac{xb^s_i}{q^{s-2}_{r+2} - a^{s-1}_{r+1}} - e^{s-1}_{r+1,i}\big)P^s_i \\ &+x^2\frac{b^{s+2}_{r-1}}{q^s_{r+1} - a^{s+1}_r}P^{s+2}_{r-1} + x^2\sum_{i=r-d}^{r-2}\frac{b^{s+2}_i}{q^s_{r+1} - a^{s+1}_r}P^{s+2}_i\end{aligned}$$

The relation involves the polynomials of the following scheme

$(\mathbf{a\Delta})$

From the computational point of view, there are large differences, as in the scalar case, between ascending and descending relations. If the polynomial P^s_r is known, computing P^{s-1}_{r+1} on the horizontal, P^s_{r+1} on the diagonal or P^{s+1}_r on the vertical needs new data from the initial series and new terms of the sequences (q^j_i) and $(e^k_{i,j})$ computed by the Q-D algorithm, using the first column and going from diagonal s to s+1. Ascending in the north-east corner does not need new data, as in the Baker algorithm, and the coefficients are computed by the Q-D algorithm going from column r to r+1.

The point is to know what must be known in one place (r,s) to compute all the polynomials surrounding this place, and to be able to start again from any surrounding point. We suppose that what is known is

$$P^s_{r-d}, \dots P^s_r,\ q^{s-1}_i, i = r-d+1, \dots r+1,\ e^{s-1}_{i,j}, j = i, \dots i-d,\ i = r-d, \dots, r+1.$$

The data is just what is needed to compute P^s_{r+1} by the diagonal relation. Then (q) is used twice, first with $x = 0$ to have the value of $q^s_i, i = r-d+1, r+1$, second to compute $P^{s+1}_i, i = r-d, r$, and so the diagonal of index s+1 is obtained up to the index r. P^{s+2}_{r-1} is obtained similarly by (q), having P^{s+2}_r and P^{s+1}_{r-1}. P^{s-1}_{r+1} is obtained by $(\Delta_0 H)$, then P^{s-2}_{r+1}, P^{s+1}_r are obtained by (AS1). So all the polynomials surrounding P^s_r are obtained. The diagonals below or above the diagonal s are obtained through the relation (q) used forward or backward.

The only thing that would really improve these algorithms would be computations with large steps, i.e. going from a column of index kd to the next of index (k+1)d. The first kind of step is

$$\begin{array}{lll} \bullet & & \\ \vdots & & \\ \bullet & \cdots & * \end{array}$$

The algorithm exists and is the iteration of relation (q)

$$P^s_r = x^k P^{s+k}_{r-k} + \sum_0^{k-1} \lambda_i x^i P^{s+i}_{r-k}$$

The coefficients are computed recursively

$$\lambda^{k+1}_i = \lambda^k_{i-1} - \lambda^k_i q^{s+i}_{r-k}\ i = 1, k, \lambda^k_0 = (-1)^k \prod_0^k q^s_{r-i}$$

Another proof was given in [2], using another normalisation of the polynomials and not involving the Q-D algorithm.

Using this relation for k=d, and the vertical relation it is possible to prove the existence of a second kind of large step relation

$$\begin{array}{lll} \bullet & \cdots & * \\ \vdots & & \\ \bullet & & \end{array}$$

This proof does not give rise to a simple computation of the coefficients, and another one has to be found.

For the two other kinds of large step relation that can exist, i.e. with the column of polynomials on the right, it is not clear if they can be useful for computational needs and they have not been searched.

References

1. M. de Bruin, *Simultaneous Padé Approximants and Orthogonality*, 74-84 LN 1171 Polynmes Orthogonaux et Appl. eds C.Brezinski et coll.
2. J. Van Iseghem, *Vector Padé Approximants*, IMACS congress 1985, Num. Math and Appl. R. Vichnevestsky and J. Vignes eds. Elsevier Sci. Pub. North-Holland 73-77 (1986).
3. J. Van Iseghem, *Vector Orthogonal Relations, Vector QD Algorithm*,JCAM 19,141-150, 1987.

GEGENBAUER–SOBOLEV ORTHOGONAL POLYNOMIALS

FRANCISCO MARCELLÁN*
Dto. de Ingeniería. Escuela Politécnica Superior. Universidad Carlos III. 28913 Leganés. Madrid. Spain

and

TERESA E. PÉREZ† and MIGUEL A. PIÑAR†
Dto. de Matemática Aplicada. Facultad de Ciencias. Universidad de Granada. 18071 Granada. Spain

Abstract. In this paper, orthogonal polynomials in the Sobolev space $W^{1,2}([-1,1],\rho^{(\alpha)},\lambda\rho^{(\alpha)})$, where $\rho^{(\alpha)}=(1-x^2)^{\alpha-\frac{1}{2}}, \alpha>-\frac{1}{2}$ and $\lambda\geq 0$, are studied. For these non-standard orthogonal polynomials algebraic and differential properties are obtained, as well as the relation with the classical Gegenbauer polynomials. Finally, some properties concerning the localization and separation of the zeros of these polynomials are deduced.

Key words: Orthogonal polynomials, Sobolev spaces, Gegenbauer polynomials

1. Introduction

The aim of this paper is the study of the *non-standard* sequence of orthogonal polynomials associated to the inner product

$$(f,g)_S=\int_{-1}^{1}f(x)g(x)(1-x^2)^{\alpha-\frac{1}{2}}dx+\lambda\int_{-1}^{1}f'(x)g'(x)(1-x^2)^{\alpha-\frac{1}{2}}dx, \qquad (1.1)$$

where $\alpha>-\frac{1}{2}$, and $\lambda\geq 0$. These polynomials will be called *Gegenbauer–Sobolev orthogonal polynomials*. They can be considered a generalization of the classical Gegenbauer polynomials, since, for $\lambda=0$, the previous inner product is the usual one for the Gegenbauer polynomials. Moreover, for $\alpha=\frac{1}{2}$, (1.1) is the simplest case of an inner product in a Sobolev space.

The particular case $\alpha=\frac{1}{2}$ has been widely studied by several authors. Althammer, in 1962 ([1]), uses these polynomials to obtain simultaneously an approximation for a function and its derivative. He obtains several algebraic properties, a relation between these orthogonal polynomials and Legendre polynomials, a Rodrigues-type formula, and finally, he shows that the polynomials have all their zeros real, different and contained in the interval $[-1,1]$. Later, Gröbner, in [3], by using a variational technique deduces again the Rodrigues-type formula, and Schäfke ([6]) extends the algebraic study of the polynomials. The relative position of the zeros with respect to the roots of the Legendre polynomials has been studied by Cohen in [2]. Further-

* This research was partially supported by CICYT, contract PB89-0181-C02-01
† This research was partially supported by Junta de Andalucía, Grupo de Investigación 1107.

A. Cuyt (ed.), Nonlinear Numerical Methods and Rational Approximation II, 71–82.

more, Gegenbauer–Sobolev orthogonal polynomials appear as an interesting example of the so called *symmetrically coherent pairs* by Iserles et al. ([4]).

In this work, we will generalize these properties to the more general case $\alpha > -\frac{1}{2}$.

The next section contains the notations and basic algebraic properties of the Gegenbauer–Sobolev polynomials.

In section 3 we define a differential operator on the linear space of the real polynomials. This operator allows us to relate the Sobolev inner product with the usual one. From the self-adjoint character for the differential operator, we will deduce the existence of a difference-differential relation and a Rodrigues-type formula.

Finally, we will obtain some properties about the zeros of the polynomials. For every value of α and λ, we will show that the n-th degree Gegenbauer–Sobolev polynomial has exactly n real and different zeros, which interlace the zeros of the Gegenbauer polynomial with the same degree. In the case $\alpha \geq \frac{1}{2}$, all the zeros belong to the interval $[-1,1]$, and, for $-\frac{1}{2} < \alpha < \frac{1}{2}$, at most two zeros are outside the interval.

2. Gegenbauer–Sobolev Orthogonal Polynomials

From now on, we will denote by $(.,.)_S$ the Sobolev inner product defined by (1.1), where $\alpha > -\frac{1}{2}, \lambda \geq 0$, and by $(.,.)$ the usual Gegenbauer inner product. Let us denote by $\{Q_n^{(\alpha)}\}_n$ the monic orthogonal polynomial sequence associated to the inner product (1.1); it will be called *Gegenbauer–Sobolev Monic Orthogonal Polynomial Sequence (MOPS)*. Also, we will denote by $\{C_n^{(\alpha)}\}_n$ the sequence of the classical monic Gegenbauer polynomials (see Szegö [7, sec. 4.7]).

From the symmetry of the inner product $(.,.)_S$ we can easily deduce the symmetrical character of the Gegenbauer–Sobolev polynomials, $Q_n^{(\alpha)}(x)$, that is

$$Q_n^{(\alpha)}(-x) = (-1)^n Q_n^{(\alpha)}(x).$$

First, we are going to relate the Gegenbauer–Sobolev polynomials with the classical Gegenbauer polynomials.

Proposition 2.1 *If we denote by*

$$\xi_n^{(\alpha)} = \frac{(n+2)(n+1)}{4(n+\alpha+1)(n+\alpha)},$$

then we have

$$C_n^{(\alpha)}(x) - \xi_{n-2}^{(\alpha)} C_{n-2}^{(\alpha)}(x) = Q_n^{(\alpha)}(x) - d_{n-2}(\lambda) Q_{n-2}^{(\alpha)}(x), \tag{2.1}$$

where

$$d_n(\lambda) = \xi_n^{(\alpha)} \frac{(C_n^{(\alpha)}, C_n^{(\alpha)})}{(Q_n^{(\alpha)}, Q_n^{(\alpha)})_S} = \xi_n^{(\alpha)} \frac{k_n^{(\alpha)}}{\tilde{k}_n^{(\alpha)}}, \tag{2.2}$$

and $k_n^{(\alpha)} = (C_n^{(\alpha)}, C_n^{(\alpha)})$, $\quad \tilde{k}_n^{(\alpha)} = (Q_n^{(\alpha)}, Q_n^{(\alpha)})_S$, $\quad n \geq 0$.

Proof. Let $R_n^{(\alpha)}(x)$ be the polynomial

$$R_n^{(\alpha)}(x) = C_n^{(\alpha)}(x) - \xi_{n-2}^{(\alpha)} C_{n-2}^{(\alpha)}(x).$$

Expanding this polynomial in terms of the Gegenbauer–Sobolev polynomials we get

$$R_n^{(\alpha)}(x) = Q_n^{(\alpha)}(x) - \sum_{m=1}^{[n/2]} d_{n-2m}(\lambda) Q_{n-2m}^{(\alpha)}(x).$$

By using some well known relations (see Szegö [7, rel. 4.7.14, 4.7.29]):

$$C_n^{(\alpha-1)}(x) = C_n^{(\alpha)}(x) - \xi_{n-2}^{(\alpha)} C_{n-2}^{(\alpha)}(x), \quad (C_n^{(\alpha)})'(x) = n C_{n-1}^{(\alpha+1)}(x), \tag{2.3}$$

we obtain

$$d_{n-2m}(\lambda) = -\frac{(R_n^{(\alpha)}, Q_{n-2m}^{(\alpha)})_S}{(Q_{n-2m}^{(\alpha)}, Q_{n-2m}^{(\alpha)})_S} = -\frac{(C_n^{(\alpha)} - \xi_{n-2}^{(\alpha)} C_{n-2}^{(\alpha)}, Q_{n-2m}^{(\alpha)})}{\tilde{k}_{n-2m}^{(\alpha)}}.$$

Therefore $d_{n-2m}(\lambda) = 0, \quad 2 \le m \le [n/2]$. Finally

$$d_{n-2}(\lambda) = -\frac{(C_n^{(\alpha)} - \xi_{n-2}^{(\alpha)} C_{n-2}^{(\alpha)}, Q_{n-2}^{(\alpha)})}{\tilde{k}_{n-2}^{(\alpha)}} = \xi_{n-2}^{(\alpha)} \frac{k_{n-2}^{(\alpha)}}{\tilde{k}_{n-2}^{(\alpha)}}. \qquad \square$$

Next, we are going to obtain an useful relation for the computation of the coefficients $d_n(\lambda)$

Proposition 2.2 *The sequence $d_n(\lambda)$ satisfy the following recurrence relation*

$$d_n(\lambda) = \frac{\xi_n^{(\alpha)} \gamma_n^{(\alpha)} \gamma_{n-1}^{(\alpha)}}{\gamma_{n-1}^{(\alpha)}(\lambda n^2 + \gamma_n^{(\alpha)}) + \xi_{n-2}^{(\alpha)}(\xi_{n-2}^{(\alpha)} - d_{n-2}(\lambda))}, \tag{2.4}$$

with the initial conditions

$$\begin{aligned} d_1(\lambda) &= \frac{3}{2(\alpha+2)(\alpha+1)[1+2(\alpha+1)\lambda]}, \\ d_2(\lambda) &= \frac{3}{(\alpha+3)(\alpha+2)\left[1 + \dfrac{8(\alpha+2)(\alpha+1)}{2\alpha+1}\lambda\right]}. \end{aligned}$$

Here, $\gamma_n^{(\alpha)}$ denotes the coefficient in the three term recurrence relation for the Gegenbauer polynomials:

$$C_{n+1}^{(\alpha)}(x) = x C_n^{(\alpha)}(x) - \gamma_n^{(\alpha)} C_{n-1}^{(\alpha)}(x), \quad \gamma_n^{(\alpha)} = \frac{n(n+2\alpha-1)}{4(n+\alpha)(n+\alpha-1)}. \tag{2.5}$$

Proof. It suffices to consider $(Q_{n+2}^{(\alpha)}, Q_n^{(\alpha)})_S = 0$, then replacing (2.1), we get

$$d_n(\lambda) = \frac{-(R_{n+2}^{(\alpha)}, R_n^{(\alpha)})_S}{(R_n^{(\alpha)}, R_n^{(\alpha)})_S + d_{n-2}(\lambda)(R_n^{(\alpha)}, R_{n-2}^{(\alpha)})_S}. \qquad \square$$

Remark. Equation (2.4) gives a Jacobi continued fraction for $d_n(\lambda)$ which should therefore be the ratio of the values of some orthogonal polynomials in λ. These polynomials can be used in the computation of the Gegenbauer–Sobolev orthogonal polynomials.

The next proposition gives an expression for the norm of the Gegenbauer–Sobolev polynomials.

Proposition 2.3 *For every value of* $n \geq 0$ *we have*

$$\tilde{k}_n^{(\alpha)} = (Q_n^{(\alpha)}, Q_n^{(\alpha)})_S = k_n^{(\alpha)}(1 + 2(n+\alpha)\lambda A_n),$$

where $\{A_n\}_n$ *denotes the sequence defined by* $A_n = \dfrac{(Q_n^{(\alpha)})'(1)}{C_n^{(\alpha)}(1)}$.

Proof. Expanding $C_{n-1}^{(\alpha+1)}$ in terms of the kernels, using the reproducing property of the kernels and the differential relation (2.3), we deduce

$$\begin{aligned}(Q_n^{(\alpha)}, Q_n^{(\alpha)})_S &= (Q_n^{(\alpha)}, C_n^{(\alpha)})_S = k_n^{(\alpha)} + n\lambda((Q_n^{(\alpha)})', C_{n-1}^{(\alpha+1)}) = \\ &= k_n^{(\alpha)} + (n+\alpha)\lambda((Q_n^{(\alpha)})', \frac{k_n^{(\alpha)}}{C_n^{(\alpha)}(1)}K_n(x,1) - \frac{k_n^{(\alpha)}}{C_n^{(\alpha)}(-1)}K_n(x,-1)) = \\ &= k_n^{(\alpha)}\left[1 + 2(n+\alpha)\lambda\frac{(Q_n^{(\alpha)})'(1)}{C_n^{(\alpha)}(1)}\right]. \qquad \square\end{aligned}$$

3. The linear operator $\mathcal{F}$

We define a linear operator $\mathcal{F}$ on the linear space of the the real polynomials $\mathbb{P}$, by means of the expression

$$\mathcal{F} = (1-x^2)\mathcal{I} + \lambda(2\alpha - 1)xD - \lambda(1-x^2)D^2, \tag{3.1}$$

where $\mathcal{I}$ denotes the identity operator and D is the usual derivative.

Under the appropriate conditions, this operator allows us to relate the Sobolev inner product with the usual Gegenbauer product.

Proposition 3.1 *Let* $p(x), q(x)$ *be arbitrary real polynomials, then*

$$((1-x^2)p(x), q(x))_S = \int_{-1}^{1} p(x)\mathcal{F}q(x)\rho^{(\alpha)}(x)dx = (p(x), \mathcal{F}q(x)).$$

Proof. Integration by parts in the second term in the Sobolev inner product leads to

$$((1-x^2)p(x), q(x))_S = \int_{-1}^{1}(1-x^2)p(x)q(x)\rho^{(\alpha)}(x)dx -$$

$$-\lambda\int_{-1}^{1}(1-x^2)p(x)[q'(x)\rho^{(\alpha)}(x)]'dx = \int_{-1}^{1}p(x)\mathcal{F}q(x)\rho^{(\alpha)}(x)dx. \quad \square$$

Theorem 3.2 *The differential operator $\mathcal{F}$, defined by (3.1), is self-adjoint with respect to the inner product (1.1), that is*

$$(\mathcal{F}p, q)_S = (p, \mathcal{F}q)_S, \quad \forall p, q \in \mathbb{P}.$$

Proof. It suffices to show this result for the elements of the basis

$$\mathcal{B} = \{r_n(x) = (1-x)^{[\frac{n+1}{2}]}(1+x)^{[\frac{n}{2}]}, n \geq 0\},$$

of the linear space $\mathbb{P}$, taking integration by parts. □

The two previous results are essential in the rest of the paper. As a direct consequence of the proposition 3.1, we can establish the relation between the Gegenbauer–Sobolev orthogonal polynomials and the classical Gegenbauer polynomials.

Proposition 3.3 *For every value of $n \geq 2$, we have*

$$(x^2-1)C_n^{(\alpha)}(x) = Q_{n+2}^{(\alpha)}(x) + a_n^{(n)}Q_n^{(\alpha)}(x) + a_{n-2}^{(n)}Q_{n-2}^{(\alpha)}(x), \tag{3.2}$$

where $a_n^{(n)} = -\dfrac{\gamma_n^{(\alpha)}\gamma_{n-1}^{(\alpha)}}{\xi_{n-2}^{(\alpha)}} - d_n(\lambda), \quad a_{n-2}^{(n)} = \dfrac{\gamma_n^{(\alpha)}\gamma_{n-1}^{(\alpha)}}{\xi_{n-2}^{(\alpha)}}d_{n-2}(\lambda).$

Proof. From (2.5) we get

$$(x^2-1)C_n^{(\alpha)}(x) = C_{n+2}^{(\alpha)}(x) + (\gamma_{n+1}^{(\alpha)} + \gamma_n^{(\alpha)} - 1)C_n^{(\alpha)}(x) + \gamma_n^{(\alpha)}\gamma_{n-1}^{(\alpha)}C_{n-2}^{(\alpha)}(x).$$

In the other hand,

$$1 - \gamma_n^{(\alpha)} - \gamma_{n+1}^{(\alpha)} - \xi_n^{(\alpha)} = \frac{(n+2\alpha-2)(n+2\alpha-1)}{4(n+\alpha)(n+\alpha-1)} = \frac{\gamma_n^{(\alpha)}\gamma_{n-1}^{(\alpha)}}{\xi_{n-2}^{(\alpha)}},$$

therefore, using (2.1), we conclude

$$(x^2-1)C_n^{(\alpha)}(x) = C_{n+2}^{(\alpha)}(x) - \xi_n^{(\alpha)}C_n^{(\alpha)}(x) - \frac{\gamma_n^{(\alpha)}\gamma_{n-1}^{(\alpha)}}{\xi_{n-2}^{(\alpha)}}\left[C_n^{(\alpha)}(x) - \xi_{n-2}^{(\alpha)}C_{n-2}^{(\alpha)}(x)\right] =$$

$$= Q_{n+2}^{(\alpha)}(x) - d_n(\lambda)Q_n^{(\alpha)}(x) - \frac{\gamma_n^{(\alpha)}\gamma_{n-1}^{(\alpha)}}{\xi_{n-2}^{(\alpha)}}\left[Q_n^{(\alpha)}(x) - d_{n-2}(\lambda)Q_{n-2}^{(\alpha)}(x)\right]. \quad \square$$

Proposition 3.4 *The Gegenbauer–Sobolev polynomial verify the following relation*

$$-\mathcal{F}Q_n^{(\alpha)}(x) = C_{n+2}^{(\alpha)}(x) + b_n^{(n)}C_n^{(\alpha)}(x) + b_{n-2}^{(n)}C_{n-2}^{(\alpha)}(x), \tag{3.3}$$

where $b_n^{(n)} = -\left\{\xi_n^{(\alpha)} + \dfrac{\xi_n^{(\alpha)}}{\xi_{n-2}^{(\alpha)}}\dfrac{\gamma_n^{(\alpha)}\gamma_{n-1}^{(\alpha)}}{d_n(\lambda)}\right\}, \quad b_{n-2}^{(n)} = \xi_n^{(\alpha)}\dfrac{\gamma_n^{(\alpha)}\gamma_{n-1}^{(\alpha)}}{d_n(\lambda)}.$

Proof. If we expand $-\mathcal{F}Q_n^{(\alpha)}$ in terms of the Gegenbauer polynomials

$$-\mathcal{F}Q_n^{(\alpha)}(x) = C_{n+2}^{(\alpha)}(x) + \sum_{m=0}^{[n/2]} b_{n-2m}^{(n)} C_{n-2m}^{(\alpha)}(x),$$

the coefficients can be computed from proposition 3.1

$$b_{n-2m}^{(n)} = -\frac{(\mathcal{F}Q_n^{(\alpha)}(x), C_{n-2m}^{(\alpha)}(x))}{k_{n-2m}^{(\alpha)}} = -\frac{(Q_n^{(\alpha)}(x), (1-x^2)C_{n-2m}^{(\alpha)}(x))_S}{k_{n-2m}^{(\alpha)}},$$

and consequently, since $Q_n^{(\alpha)}(x)$ is an orthogonal polynomial, we get $b_{n-2m}^{(n)} = 0$, for $2 \leq m \leq [n/2]$. The other coefficients can be computed in the following way

$$b_{n-2}^{(n)} = -\frac{(Q_n^{(\alpha)}, (1-x^2)C_{n-2}^{(\alpha)})_S}{k_{n-2}^{(\alpha)}} = \frac{\tilde{k}_n^{(\alpha)}}{k_{n-2}^{(\alpha)}} = \gamma_n^{(\alpha)}\gamma_{n-1}^{(\alpha)}\frac{\tilde{k}_n^{(\alpha)}}{k_n^{(\alpha)}} = \xi_n^{(\alpha)}\frac{\gamma_n^{(\alpha)}\gamma_{n-1}^{(\alpha)}}{d_n(\lambda)}.$$

Finally, using the previous proposition we obtain

$$b_n^{(n)} = -\frac{(\mathcal{F}Q_n^{(\alpha)}(x), C_n^{(\alpha)}(x))}{k_n^{(\alpha)}} = -\left\{\xi_n^{(\alpha)} + \frac{\xi_n^{(\alpha)}}{\xi_{n-2}^{(\alpha)}}\frac{\gamma_n^{(\alpha)}\gamma_{n-1}^{(\alpha)}}{d_n(\lambda)}\right\}. \quad \square$$

Remark. In the particular case $\alpha = 1/2$, the above expression coincides with the relation obtained by Althammer in [1, exp. 3.6].

4. Differential properties

Now, we are going to obtain a difference-differential relation satisfied by the polynomials, which coefficients are deduced from the propositions in the previous section.

Proposition 4.1 (Difference–Differential Relation) *For every $n \geq 2$, the following relation holds*

$$-\mathcal{F}Q_n^{(\alpha)}(x) = Q_{n+2}^{(\alpha)}(x) + c_n^{(n)}Q_n^{(\alpha)}(x) + c_{n-2}^{(n)}Q_{n-2}^{(\alpha)}(x), \tag{4.1}$$

where $c_n^{(n)} = -d_n(\lambda) - \dfrac{\xi_n^{(\alpha)}}{\xi_{n-2}^{(\alpha)}}\dfrac{\gamma_n^{(\alpha)}\gamma_{n-1}^{(\alpha)}}{d_n(\lambda)}, \quad c_{n-2}^{(n)} = \gamma_n^{(\alpha)}\gamma_{n-1}^{(\alpha)}\dfrac{\xi_n^{(\alpha)}}{\xi_{n-2}^{(\alpha)}}\dfrac{d_{n-2}(\lambda)}{d_n(\lambda)}.$

Proof. Expand $-\mathcal{F}Q_n^{(\alpha)}$ in terms of the polynomials $\{Q_n^{(\alpha)}\}$

$$-\mathcal{F}Q_n^{(\alpha)}(x) = Q_{n+2}^{(\alpha)}(x) + \sum_{i=0}^{[n/2]} c_{n-2m}^{(n)} Q_{n-2m}^{(\alpha)}(x),$$

where, from the self-adjoint character of $\mathcal{F}$, we obtain

$$c_{n-2m}^{(n)} = -\frac{(\mathcal{F}Q_n^{(\alpha)}, Q_{n-2m}^{(\alpha)})_S}{(Q_{n-2m}^{(\alpha)}, Q_{n-2m}^{(\alpha)})_S} = -\frac{(Q_n^{(\alpha)}, \mathcal{F}Q_{n-2m}^{(\alpha)})_S}{\tilde{k}_{n-2m}^{(\alpha)}},$$

thus $c_{n-2m}^{(n)} = 0$, for $2 \leq m \leq [n/2]$.

The non–null coefficients are deduced in the following way

$$c_{n-2}^{(n)} = -\frac{(Q_n^{(\alpha)}, \mathcal{F}Q_{n-2}^{(\alpha)})_S}{\tilde{k}_{n-2}^{(\alpha)}} = \frac{\tilde{k}_n^{(\alpha)}}{\tilde{k}_{n-2}^{(\alpha)}} = \gamma_n^{(\alpha)}\gamma_{n-1}^{(\alpha)}\frac{\xi_n^{(\alpha)}}{\xi_{n-2}^{(\alpha)}}\frac{d_{n-2}(\lambda)}{d_n(\lambda)},$$

and, from propositions 2.1 and 3.4, we conclude

$$\begin{aligned} c_n^{(n)} &= -\frac{(\mathcal{F}Q_n^{(\alpha)}, Q_n^{(\alpha)})_S}{\tilde{k}_n^{(\alpha)}} = \frac{(C_{n+2}^{(\alpha)} + b_n^{(n)}C_n^{(\alpha)} + b_{n-2}^{(n)}C_{n-2}^{(\alpha)}, Q_n^{(\alpha)})_S}{\tilde{k}_n^{(\alpha)}} = \\ &= \frac{(C_{n+2}^{(\alpha)}, Q_n^{(\alpha)})_S}{\tilde{k}_n^{(\alpha)}} + b_n^{(n)} = \frac{(Q_{n+2}^{(\alpha)} - d_n(\lambda)Q_n^{(\alpha)} + \xi_n^{(\alpha)}C_n^{(\alpha)}, Q_n^{(\alpha)})_S}{\tilde{k}_n^{(\alpha)}} + b_n^{(n)} = \\ &= -d_n(\lambda) - \frac{\xi_n^{(\alpha)}}{\xi_{n-2}^{(\alpha)}}\frac{\gamma_n^{(\alpha)}\gamma_{n-1}^{(\alpha)}}{d_n(\lambda)}. \qquad \square \end{aligned}$$

Next, we will obtain a *Rodrigues-type formula* for the Gegenbauer–Sobolev polynomials. This result generalizes the usual Rodrigues formula, since when $\lambda = 0$ both expressions are the same. In the same way, that formula generalizes the expression obtained by Gröbner in [3], for $\alpha = 1/2$, using a variational technique.

Proposition 4.2 (Rodrigues–type formula) *The following relation holds*

$$\begin{aligned} (1-x^2)^{\alpha-\frac{1}{2}}\mathcal{F}Q_n^{(\alpha)}(x) &= (-1)^n\frac{\Gamma(n+2\alpha)}{\Gamma(2n+2\alpha)}(1-x^2) \\ D^n&\left\{\left[(1-x^2)^n + 2\lambda(n+\alpha)A_n(1-x^2)^{n-1}\right](1-x^2)^{\alpha-\frac{1}{2}}\right\}. \qquad (4.2) \end{aligned}$$

Proof. Using proposition 3.4 and the three term recurrence relation for the Gegenbauer polynomials (2.5), we get

$$\begin{aligned} \mathcal{F}Q_n^{(\alpha)}(x) &= (1-x^2)C_n^{(\alpha)}(x) + \frac{\gamma_n^{(\alpha)}\gamma_{n-1}^{(\alpha)}}{\xi_{n-2}^{(\alpha)}}2\lambda(n+\alpha)A_n\left[C_n^{(\alpha)}(x) - \xi_{n-2}^{(\alpha)}C_{n-2}^{(\alpha)}(x)\right] = \\ &= (1-x^2)C_n^{(\alpha)}(x) - \frac{n+2\alpha-1}{2(n+\alpha-1)}\lambda A_n\left[-2(n+\alpha-1)C_n^{(\alpha)}(x) + nxC_{n-1}^{(\alpha)}(x)\right]. \end{aligned}$$

And, by substitution of the structure relation for the Gegenbauer polynomials (see Szegö [7, rel. 4.7.27]), we deduce

$$\begin{aligned} \mathcal{F}Q_n^{(\alpha)}(x) &= (1-x^2)C_n^{(\alpha)}(x) - \\ &-\frac{n+2\alpha-1}{2(n+\alpha-1)}\lambda A_n\left[(1-x^2)(C_{n-1}^{(\alpha)})'(x) - (2\alpha-1)xC_{n-1}^{(\alpha)}(x)\right]. \end{aligned}$$

Now, we can introduce the weight function $\rho^{(\alpha)}(x) = (1-x^2)^{\alpha-\frac{1}{2}}$, and using the usual Rodrigues formula for the Gegenbauer polynomials, we conclude

$$\mathcal{F}Q_n^{(\alpha)}(x) = (-1)^n\frac{\Gamma(n+2\alpha)}{\Gamma(2n+2\alpha)}\frac{1-x^2}{(1-x^2)^{\alpha-\frac{1}{2}}}D^n\left[(1-x^2)^{n+\alpha-\frac{1}{2}}\right] -$$

$$-\frac{n+2\alpha-1}{2(n+\alpha-1)}\lambda A_n(-1)^{n-1}\frac{\Gamma(n+2\alpha-1)}{\Gamma(2n+2\alpha-2)}\frac{1-x^2}{(1-x^2)^{\alpha-\frac{1}{2}}}D^n\left[(1-x^2)^{n-1+\alpha-\frac{1}{2}}\right]=$$

$$=\frac{(1-x^2)(-1)^n}{(1-x^2)^{\alpha-\frac{1}{2}}}\frac{\Gamma(n+2\alpha)}{\Gamma(2n+2\alpha)}D^n\left\{\left[(1-x^2)+2\lambda(n+\alpha)A_n\right](1-x^2)^{n+\alpha-\frac{3}{2}}\right\}.\square$$

5. Zeros

The behavior of the zeros for the Gegenbauer–Sobolev polynomials has been studied in the particular case $\alpha = 1/2$. In this case, Althammer, in [1], shows that the orthogonal polynomials of degree n has exactly n different and real zeros contained in the interval $[-1,1]$. Besides, Cohen, in [2], shows that these zeros separate the roots of the Legendre polynomial of degree $n-1$, for $\lambda \geq 2/n$. Recently, Meijer, in [5], has obtained analogous results for the Gegenbauer–Sobolev polynomials when λ is *sufficiently large* and $\alpha \geq 1/2$.

In this section, from the properties of the linear operator $\mathcal{F}$, we will show that for every value of $\lambda \geq 0$ and $\alpha > -1/2$, the polynomial $Q_n^{(\alpha)}(x)$ has n different and real zeros, at least $n-2$ of them are contained in the interval $[-1,1]$, moreover, these zeros separate those of $C_n^{(\alpha)}(x)$. For $\alpha \geq 1/2$, all the zeros are contained in $[-1,1]$. First, we need some technical results.

Lemma 5.1 *For $\alpha \geq 1/2$, we have $Q_n^{(\alpha)}(1) > 0$.*

Proof. In order to obtain the sign of $Q_n^{(\alpha)}(1)$, we can divide expression (2.1) by $C_n^{(\alpha)}(1)$ and if we denote by $B_n = \dfrac{Q_n^{(\alpha)}(1)}{C_n^{(\alpha)}(1)}$, we can deduce

$$B_n=\frac{2(2\alpha-1)(n+\alpha-1)}{(n+2\alpha-1)(n+2\alpha-2)}+\frac{n(n-1)}{(n+2\alpha-1)(n+2\alpha-2)}\frac{k_{n-2}^{(\alpha)}}{\tilde{k}_{n-2}^{(\alpha)}}B_{n-2}.$$

Finally, since $B_1 = B_2 = 1$, if $\alpha \geq 1/2$, we have $B_n > 0, \forall n \in \mathbb{N}$, and therefore $Q_n^{(\alpha)}(1) > 0$. $\square$

Remark. If $-1/2 < \alpha < 1/2$, no global properties about the sign of B_n can be deduced, but if there exists $n_0 \in \mathbb{N}$ such that $B_{n_0} \leq 0$, then $B_n < 0, \forall n > n_0$. In this way, for $-1/2 < \alpha < 1/2$, if some polynomial $Q_n^{(\alpha)}(x)$ has a pair of symmetrical roots outside the interval $[-1,1]$, the higher degree polynomials verify the same property. Analyzing the sign of $(Q_n^{(\alpha)})'(1)$, we immediately deduce that these roots have odd multiplicity. From the expressions

$$B_3=\frac{(2\alpha-1)(\alpha+2)}{(\alpha+1)(2\alpha+1)}+\frac{3}{(\alpha+1)(2\alpha+1)[1+2\lambda(\alpha+1)]},$$

$$B_4=\frac{(2\alpha-1)(\alpha+3)}{(\alpha+1)(2\alpha+3)}+\frac{6}{(\alpha+1)(2\alpha+3)\left[1+8\lambda\dfrac{(\alpha+1)(\alpha+2)}{(2\alpha+1)}\right]},$$

we deduce that $B_3, B_4 < 0$, for λ sufficiently large and $-1/2 < \alpha < 1/2$, and therefore, we can always obtain inner products like (1.1) such that $B_n < 0, \forall n \geq 3$.

In order to study the behavior of the roots of the Gegenbauer–Sobolev polynomials we need to consider two different cases, according to the parity of the polynomial. Both cases are similar, however a few technical differences forces to consider them separately. But, for sake of brevity, we shall consider only the case when the degree of the polynomial is even.

Lemma 5.2 *Let $p(x)$ be a polynomial with degree $2k$ containing only powers of even degree. Then*
i) If $\lambda = 0$ or $\alpha = 1/2$, there exists an unique polynomial $\hat{p}(x)$ with degree $2k$ containing only powers of even degree, such that

$$\mathcal{F}\hat{p}(x) = (1 - x^2)p(x),$$

ii) If $\lambda \neq 0$ and $\alpha \neq 1/2$, there exists an unique polynomial $\hat{p}(x)$ with degree $2k$ containing only powers of even degree, and an unique constant $\hat{c}(p)$, such that

$$\mathcal{F}\hat{p}(x) = (1 - x^2)p(x) + (1 - x^2)^2\hat{c}(p).$$

Moreover, $\hat{p}'(1) = \hat{p}'(-1) = 0$.

Proof. i) If $\lambda = 0$, the linear operator $\mathcal{F}$ is the operator associated to the multiplication by $(1 - x^2)$ and then $\hat{p}(x) = p(x)$. In the case $\alpha = 1/2$, we have $\mathcal{F} = (1 - x^2)[I - \lambda D^2]$ and it suffices to take $\hat{p}(x) = [I - \lambda D^2]^{-1}p(x)$.
ii) Expand the polynomials in terms of the base of the even polynomials given by $\{(1 - x^2)^i, i \geq 0\}$, thus, we get

$$(1 - x^2)p(x) = \sum_{i=1}^{k+1} a_i(1 - x^2)^i, \quad \hat{p}(x) = \sum_{i=0}^{k} b_i(1 - x^2)^i.$$

We are going to compute the coefficients $b_0, b_1, \ldots, b_k$, and the constant $\hat{c}(p)$, from the coefficients $a_1, a_2, \ldots, a_{k+1}$. First, we must remark that the image of $(1 - x^2)^i$ by means of the operator $\mathcal{F}$ can be expressed by

$$\mathcal{F}(1 - x^2)^i = (1 - x^2)^{i+1} + 4\lambda i(\alpha + i - 1)(1 - x^2)^i - 2\lambda i(2\alpha + 2i - 3)(1 - x^2)^{i-1}.$$

In this way, we obtain the following linear system

$$\begin{aligned} b_k &= a_{k+1}, \\ b_{k-1} + 4\lambda k(\alpha + k - 1)b_k &= a_k, \\ \cdots \quad \cdots &\quad \cdots, \\ b_{i-1} + 4\lambda i(\alpha + i - 1)b_i - 2\lambda(i + 1)(2\alpha + 2i - 1)b_{i+1} &= a_i, \quad 3 \leq i \leq k - 1, \quad (5.1) \\ \cdots \quad \cdots &\quad \cdots, \\ b_1 + 8\lambda(\alpha + 1)b_2 - 6\lambda(2\alpha + 3)b_3 &= a_2 + \hat{c}(p), \\ b_0 + 4\lambda\alpha b_1 - 4\lambda(2\alpha + 1)b_2 &= a_1, \\ -2\lambda(2\alpha - 1)b_1 &= 0. \end{aligned}$$

The unique solution of the preceding system of linear equations can be obtained in a recursive way, extracting each coefficient b_i from the corresponding equation. Moreover, since $\lambda \neq 0$ and $\alpha \neq 1/2$, the last equation gives $b_1 = 0$, and therefore $\hat{p}'(-1) = \hat{p}'(1) = 0$. □

Corollary 5.3 *In the conditions of the previous lemma, if* $\lambda \neq 0$ *and* $\alpha \neq 1/2$, *for every polynomial* $q(x)$ *we have*

$$(q, \hat{p})_S = \int_{-1}^{1} p(x)q(x)\rho^{(\alpha)}(x)dx + \hat{c}(p)\int_{-1}^{1}(1-x^2)q(x)\rho^{(\alpha)}(x)dx.$$

Proof. It suffices to integrate by parts. □

Lemma 5.4 *Let* $0 < x_1 < x_2 < \ldots < x_k < 1$, k *real numbers, and consider the polynomial*

$$p(x) = \prod_{i=1}^{k}(x^2 - x_i^2),$$

then, if $\lambda \neq 0$ *and* $\alpha \neq 1/2$, *we have* $\hat{c}(p) > 0$.

Proof. The polynomial $p(x)$ can be expressed in the following way

$$p(x) = (-1)^k \prod_{i=1}^{k}[(1-x^2) - (1-x_i^2)].$$

If we apply lemma 5.2, with the same notations we get $b_1 = 0$, and the solution of the linear system (5.1) can be computed in a recursive way.

From the expansion of $p(x)$ in the basis $\{(1-x^2)^i, i \geq 0\}$, we have $a_{k+1} = (-1)^k$, and using the Cardano–Vieta formulae, we get

$$(-1)^{i-1}a_i > 0, \quad i = 2, 3, \ldots, k+1, \quad a_1 = (1-x_1^2)(1-x_2^2)\ldots(1-x_k^2) > 0.$$

Thus, we obtain $\hat{c}(p) > 0$. □

Lemma 5.5 *For* $n \geq 4$, *we have*

$$\int_{-1}^{1}(1-x^2)Q_{2n}^{(\alpha)}(x)\rho^{(\alpha)}(x)dx = \frac{16(\alpha+2)(\alpha+1)}{2\alpha+1}\lambda k_2^{(\alpha)}\prod_{i=1}^{n-1} d_{2n-2i}(\lambda).$$

Proof. It suffices to apply repeatedly equality (2.1), since $Q_2^{(\alpha)}(x) = C_2^{(\alpha)}(x)$. □

The next theorem shows the real and simple character for the positive zeros of the Gegenbauer–Sobolev polynomials of even degree. The corresponding result for the negative zeros comes from the symmetrical character of the polynomial.

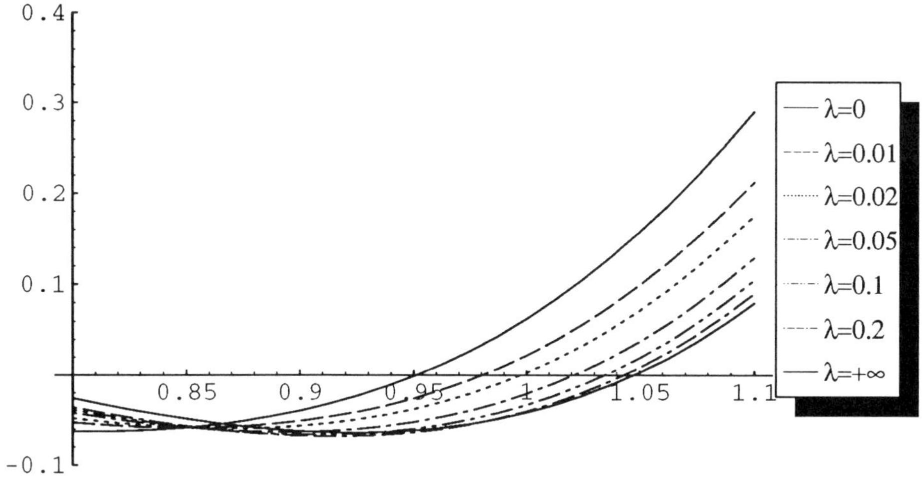

Fig. 1. Graphs of $Q_5^{(\alpha)}(x)$ in the proximity of $x = 1$, for $\alpha = 0$ and different values of λ.

Theorem 5.6 *Let $\lambda > 0$, then the polynomial $Q_{2n}^{(\alpha)}(x), n \geq 2$, has exactly $2n$ different real roots, and at least $2n-2$ of them are contained in the interval $[-1,1]$. For $\alpha \geq 1/2$, all the roots are contained in the interval $[-1,1]$.*

Moreover, the positive roots of $Q_{2n}^{(\alpha)}(x)$ separate those of $C_{2n}^{(\alpha)}(x)$. If we denote by $0 < x_{n,1} < x_{n,2} < \ldots < x_{n,n}$ the positive roots of $C_{2n}^{(\alpha)}(x)$, and by $y_{n,1} < y_{n,2} < \ldots < y_{n,n}$ those of $Q_{2n}^{(\alpha)}(x)$, then

$$0 < x_{n,1} < y_{n,1} < x_{n,2} < y_{n,2} < \ldots < x_{n,n} < y_{n,n}.$$

Proof. Consider the polynomial

$$\omega_i(x) = \prod_{\substack{j=1 \\ j \neq i}}^{n} (x^2 - x_{n,j}^2).$$

Then, if $\alpha \neq 1/2$, by lemma 5.2, there exists a polynomial $\hat{\omega}_i(x)$ with degree $2n-2$, and a constant $\hat{c}(\omega_i)$ such that

$$\mathcal{F}\hat{\omega}_i(x) = (1-x^2)\omega_i(x) + (1-x^2)^2\hat{c}(\omega_i).$$

From corollary 5.3, we get

$$0 = (Q_{2n}^{(\alpha)}, \hat{\omega}_i)_S = \int_{-1}^{1} \omega_i(x) Q_{2n}^{(\alpha)}(x)\rho^{(\alpha)}(x)dx + \hat{c}(\omega_i)\int_{-1}^{1}(1-x^2)Q_{2n}^{(\alpha)}(x)\rho^{(\alpha)}(x)dx.$$

The *Gaussian quadrature formula* based on the zeros of $C_{2n}^{(\alpha)}(x)$ leads to

$$0 = (Q_{2n}^{(\alpha)}, \hat{\omega}_i)_S = \lambda_i \omega_i(x_{n,i}) Q_{2n}^{(\alpha)}(x_{n,i}) + \hat{c}(\omega_i)\int_{-1}^{1}(1-x^2)Q_{2n}^{(\alpha)}(x)\rho^{(\alpha)}(x)dx,$$

where $\lambda_i > 0$ denotes the corresponding Christoffel numbers, and from lemmas 5.4 and 5.5, we deduce that the second term is positive. Thus

$$\lambda_i \omega_i(x_{n,i}) Q_{2n}^{(\alpha)}(x_{n,i}) = \lambda_i \frac{(C_{2n}^{(\alpha)})'(x_{n,i})}{2x_{n,i}} Q_{2n}^{(\alpha)}(x_{n,i}) < 0,$$

then $Q_{2n}^{(\alpha)}(x_{n,i}) \neq 0$ and therefore $sgn(Q_{2n}^{(\alpha)}(x_{n,i})) = (-1)^{n-i+1}$.

In this way, $Q_{2n}^{(\alpha)}(x)$ changes sign between any two consecutive zeros of $C_{2n}^{(\alpha)}(x)$, and since $sgn(Q_{2n}^{(\alpha)}(x_{n,n})) = -1$, we deduce that $Q_{2n}^{(\alpha)}(x)$ has n different real roots which separate those of $C_{2n}^{(\alpha)}(x)$, in the announced position.

If $\alpha = 1/2$, taking integration by parts in the second term of the inner product we get

$$\begin{aligned} 0 &= (Q_{2n}^{(1/2)}, \hat{\omega}_i)_S = \int_{-1}^{1} \hat{\omega}_i(x) Q_{2n}^{(1/2)}(x)dx + \lambda \int_{-1}^{1} \hat{\omega}_i'(x)(Q_{2n}^{(1/2)})'(x)dx = \\ &= \int_{-1}^{1} \omega_i(x) Q_{2n}^{(1/2)}(x)dx + \lambda \left[Q_{2n}^{(1/2)}(1)\hat{\omega}_i'(1) - Q_{2n}^{(1/2)}(-1)\hat{\omega}_i'(-1) \right] = \\ &= \lambda_i \omega_i(x_{n,i}) Q_{2n}^{(1/2)}(x_{n,i}) + \lambda \left[Q_{2n}^{(1/2)}(1)\hat{\omega}_i'(1) - Q_{2n}^{(1/2)}(-1)\hat{\omega}_i'(-1) \right]. \end{aligned}$$

Using the same reasoning as in lemma 5.4, we obtain

$$\hat{\omega}_i'(1) = (-2)b_1 > 0, \quad \hat{\omega}_i'(-1) = 2b_1 < 0,$$

and we get $\lambda_i \omega_i(x_{n,i}) Q_{2n}^{(1/2)}(x_{n,i}) < 0$, therefore $sgn(Q_{2n}^{(1/2)}(x_{n,i})) = (-1)^{n-i+1}$, and the conclusion follows in the same way.

Finally, we must remember that, from lemma 5.1, for $\alpha \geq 1/2$, we have $Q_{2n}^{(\alpha)}(1) > 0$, and the position of the last root $y_{n,n}$ is $y_{n,n} < 1$. □

Acknowledgements

The authors wish to thank the referes for their constructive comments.

References

1. Althammer, P. *Eine Erweiterung des Orthogonalitätsbegriffes bei Polynomen und deren Anwendung auf die beste Approximation*, J. Reine Angew. Math. 211 (1962), 192-204.
2. Cohen, E.A. *Zero distribution and behavior of orthogonal polynomials in the Sobolev space* $W^{1,2}[-1,1]$, SIAM J. Math. Anal. 6 (1975), 105-116.
3. Gröbner, W. *Orthogonale Polynomsysteme, die gleichzeitig mit $f(x)$ auch deren Ableitung $f'(x)$ aproximieren*, in Funktionalanalysis, Approximationstheorie, Numerische Mathematik, ISNM 7, Birkhäuser, Basel 1967, 24-32.
4. Iserles, A., Koch, P. E., Nørsett, S. P., Sanz-Serna, J.M. *On polynomials orthogonal with respect to certain Sobolev Inner Products*, J. Approx. Th. 65 (1991). 151-175.
5. Meijer, H.G. *Coherent pairs and zeros of Sobolev-type orthogonal polynomials.* Indag. Math., N. S., 4(2), (1993), 163-176.
6. Schäfke, F.W. *Zu den Orthogonalpolynomen von Althammer*, J. Reine Angew. Math. 252 (1972). 195-199.
7. Szegö, G., Orthogonal Polynomials, Amer. Math. Soc. Colloq. Publ. 23, Providence, RI, 1975 (4th edition).

INVERSE PROBLEMS : RATIONAL MODIFICATIONS

A. CACHAFEIRO and C. SUÁREZ
Departamento de Matemática Aplicada
E.T.S. de Ingenieros Industriales
Universidad de Vigo
36.280 Vigo, España

Abstract. In the present paper we consider a regular and hermitian functional u defined in the space of Laurent polynomials and we study the solutions T of the equation $(z-\alpha)(z^{-1}-\overline{\alpha})T = u$ for $\alpha \in \mathbf{C} - \{0\}$. If we fix a solution T we obtain the corresponding sequence of moments, characterize the regularity and we determine the expression for the sequence of monic orthogonal polynomials related to T. Finally we study the positive definite case, we obtain the relation between the associated measures and we apply these results to the functional u induced by the Lebesgue normalized measure on the unit circle.

Key words: orthogonal polynomials, measures on the unit circle, regular functionals, rational modifications.

1. Introduction

On the real line it has been studied the connection between the systems of orthogonal polynomials with respect to a weight function and its modified form by a rational function. This relation is established by the generalized Christoffel's formula, which is an elegant expression although it is hardly very useful from the computational point of view ([16]). It includes, as a particular case, the Christoffel's formula related to weight functions modified by polynomials. In a more general situation, this study has been developed for polynomial and inverse polynomial modifications of regular functionals ([2] and [10]).

Since in the unit circle the only positive trigonometric polynomials are of the form $| h(z) |^2$ with $z = e^{i\theta}$ and $h(z)$ an algebraic polynomial with complex coefficients which has no zeros in the open unit disk, it has been investigated polynomial and rational modifications of measures of the form $| h(z) |^2 d\mu$ and $\dfrac{d\mu}{| h(z) |^2}$. Following the scheme designed by Gautschi for the real line ([4]), it is possible to do this study by a sequence of elementary steps in such a way that the polynomial and the rational cases can be reduced to the study of $| z-\alpha |^2 d\mu$ and $\dfrac{d\mu}{| z-\alpha |^2}$ respectively.

For the polynomial case it has been obtained in [6] and [7] similar formulas to those of Paszkowski ([11]) related to the real case. In [3] the direct problem has been studied , that is, the modification of a regular functional by a positive trigonometric polynomial and very similar results to those corresponding to the positive definite case have been obtained.

With respect to the rational modifications of measures on the unit circle, Szegö

A. Cuyt (ed.), *Nonlinear Numerical Methods and Rational Approximation II*, 83–96.

proved in [15] that the orthonormal polynomials related to the measure μ such that $d\mu(\theta) = \dfrac{d\theta}{2\pi| h(e^{i\theta}) |^2}$ are given by $\varphi_n(z) = z^n\overline{h}(\frac{1}{z})$ for $n \geq k$ with k the degree of the polynomial $h(z)$, as above.
In [7] a representation is established for the orthogonal polynomials associated to rational modifications of measures of the type $\dfrac{d\mu}{| z-\alpha |^2}$ and the expression for the polynomials is given in terms of the initial ones and the Walsh functions.
The study of rational modifications of linear functionals appears for the first time in [14]. In that work the problem is posed as the inverse problem to that corresponding to polynomial modifications of functionals and the results derived are more general to those related to the positive definite case.
In this paper we deal with the consideration of rational modifications of linear functionals following the direction pointed out in [14].

2. Hermitian Solutions of the Functional Equation $(z-\alpha)(z^{-1}-\overline{\alpha})T = u$

Let $\Delta = span\{z^k\}_{k\in\mathbf{Z}}$ be the space of Laurent polynomials and $u : \Delta \longrightarrow \mathbf{C}$ a linear and hermitian functional, that is, $u(z^n) = \overline{u(z^{-n})}$ for all $n \in \mathbf{N}$. Given $\alpha \in \mathbf{C}-\{0\}$ we study the linear and hermitian functionals $T : \Delta \longrightarrow \mathbf{C}$ which are solutions of the equation $(z-\alpha)(z^{-1}-\overline{\alpha})T = u$. If we denote by $u_n = u(z^n)$ and $c_n = T(z^n)$ for $n \in \mathbf{Z}$ it holds that

2.1. THEOREM

Given $c_0 \in \mathbf{R}-\{0\}$ and $c_{-1} = \overline{c_1} \in \mathbf{C}$ with c_1 verifying $-\overline{\alpha}c_1+(1+| \alpha |^2)c_0-\alpha\overline{c_1} = u_0$ then

(i) For $| \alpha | \neq 1$

$$c_n = -\frac{1}{\overline{\alpha}(1-| \alpha |^2)}\sum_{k=1}^{n}\frac{1-| \alpha |^{2k}}{\overline{\alpha}^{k-1}}u_{n-k} + \frac{1-| \alpha |^{2(n+1)}}{\overline{\alpha}^n(1-| \alpha |^2)}c_0 - \frac{\alpha(1-| \alpha |^{2n})}{\overline{\alpha}^n(1-| \alpha |^2)}c_{-1} \quad (1)$$

and $c_{-n} = \overline{c_n}$ for all $n \geq 1$

(ii) For $| \alpha | = 1$

$$c_n = -\sum_{k=1}^{n} k\alpha^k u_{n-k} + (n+1)\alpha^n c_0 - n\alpha^{n+1}c_{-1} \quad (2)$$

and $c_{-n} = \overline{c_n}$ for all $n \geq 1$.
(The above expressions are valid for $n = 0$ if we agree that the sum in c_n vanishes for $n = 0$).

Proof. Since $\forall n \in \mathbf{Z} \quad (z-\alpha)(z^{-1}-\overline{\alpha})T(z^n) = u(z^n)$ we get for $n \in \mathbf{Z}$ the difference equation $-\overline{\alpha}c_{n+1} + (1+| \alpha |^2)c_n - \alpha c_{n-1} = u_n$.
By hypothesis we deduce that $c_{-n} = \overline{c_n}$. Then it suffices to solve the equation for $n \geq 1$. It is clear that the difference equation can be written in the following form :

$$C_{n+1} = AC_n + B_n$$

with

$$A = \begin{pmatrix} \bar{\alpha}^{-1}(1+|\alpha|^2) & -\bar{\alpha}^{-1}\alpha \\ 1 & 0 \end{pmatrix}, \quad C_{n+1} = \begin{pmatrix} c_{n+1} \\ c_n \end{pmatrix}$$

and

$$B_n = \begin{pmatrix} -\bar{\alpha}^{-1}u_n \\ 0 \end{pmatrix}.$$

We distinguish between two cases depending on the eigenvalues of A.
(i) If $|\alpha| \neq 1$ then the eigenvalues are $\lambda_1 = \bar{\alpha}^{-1}$ and $\lambda_2 = \alpha$ and therefore $A = PDP^{-1}$ with

$$D = \begin{pmatrix} \bar{\alpha}^{-1} & 0 \\ 0 & \alpha \end{pmatrix}.$$

and

$$P = \begin{pmatrix} \bar{\alpha}^{-1} & \alpha \\ 1 & 1 \end{pmatrix}.$$

By successive iterations

$$C_{n+1} = A^{n+1}C_0 + \sum_{k=0}^{n-1} A^{n-k}B_k + B_n = PD^{n+1}P^{-1}C_0 + P\left(\sum_{k=0}^{n-1} D^{n-k}P^{-1}B_k\right) + B_n$$

By operating we get (1).
(ii) If $|\alpha| = 1$ then $\lambda_1 = \lambda_2 = \alpha$. By proceeding in the same way we get (2). □

In the following we always assume that T is a hermitian solution of the functional equation $(z-\alpha)(z^{-1}-\overline{\alpha})T = u$ with initial conditions c_0 and c_1.
If we denote by δ_α and δ'_α the Dirac distribution and the derivative of the Dirac distribution respectively, we have:

2.2. THEOREM

For $|\alpha| \neq 1$, T has one of the two non equivalent following descompositions
(i)

$$T = T_0 + \frac{c_0 - \alpha\overline{c_1}}{1-|\alpha|^2}\delta_{\bar{\alpha}^{-1}} + \frac{c_0 - \overline{\alpha}c_1}{1-|\alpha|^2}\delta_\alpha \tag{3}$$

with T_0 the hermitian functional defined by

$$\forall n \geq 0 \quad T_0(z^n) = -\frac{1}{1-|\alpha|^2}\left(\sum_{k=1}^{n} \frac{1-|\alpha|^{2k}}{\overline{\alpha}^k}u_{n-k} + u_0\delta_\alpha(z^n)\right)$$

and $T_0(z^{-n}) = \overline{T_0(z^n)}$
(ii)

$$T = T_1 + \frac{\overline{\alpha}c_1 - |\alpha|^2 c_0}{1-|\alpha|^2}\delta_{\bar{\alpha}^{-1}} + \frac{\alpha\overline{c_1} - |\alpha|^2 c_0}{1-|\alpha|^2}\delta_\alpha \tag{4}$$

with T_1 the hermitian functional defined by

$$\forall n \geq 0 \quad T_1(z^n) = -\frac{1}{1-|\alpha|^2}\left(\sum_{k=1}^{n} \frac{1-|\alpha|^{2k}}{\overline{\alpha}^k} u_{n-k} - u_0\delta_{\bar{\alpha}^{-1}}(z^n)\right)$$

and $T_1(z^{-n}) = \overline{T_1(z^n)}$.

Proof. We try to write T in the following form $T = L + M\delta_{\bar{\alpha}^{-1}} + N\delta_\alpha$ with L a hermitian functional and M and N constants. Since $T(z^{-n}) = \overline{T(z^n)}$ and $L(z^{-n}) = \overline{L(z^n)}$ for all $n \geq 1$, we have $(M-\overline{N})\bar{\alpha}^{-n} = (\overline{M}-N)\alpha^n$ for all $n \geq 1$ and then $(M-\overline{N})\bar{\alpha}^{-(n+1)} = (M-\overline{N})\alpha\bar{\alpha}^{-n}$. Therefore $(M-\overline{N})\bar{\alpha}^{-1} = (M-\overline{N})\alpha$ which implies $M = \overline{N}$. Next we try to break down T in this way.
From (1) we get

$$c_n = -\frac{1}{\overline{\alpha}(1-|\alpha|^2)}\sum_{k=1}^{n}\frac{1-|\alpha|^{2k}}{\overline{\alpha}^{k-1}}u_{n-k} + \frac{c_0-\alpha\overline{c_1}}{1-|\alpha|^2}\frac{1}{\overline{\alpha}^n} + \frac{\alpha\overline{c_1}-|\alpha|^2 c_0}{1-|\alpha|^2}\alpha^n \qquad (5)$$

Since T is hermitian we have

$$\alpha\overline{c_1} - |\alpha|^2 c_0 = -\overline{\alpha}c_1 + c_0 - u_0 \qquad (6)$$

(i) If we substitute (6) in (5) we get (3).
(ii) In the same way, if we substitute $c_0 - \alpha\overline{c_1}$ in (5) we obtain (4). □
(The terms with α^n or $\bar{\alpha}^{-n}$ that remains in T_0 and T_1 prevent their first moment is zero.)

2.3. THEOREM

For $|\alpha| = 1$, T has the following descompositions:

$$T = T_2 + c_0\delta_\alpha + i\alpha(Im\alpha\, Rec_1 - Re\alpha\, Imc_1)\delta'_\alpha \qquad (7)$$

with T_2 the hermitian functional defined by $T_2(z^n) = -\sum_{k=1}^{n-1} k\alpha^k u_{n-k} - \frac{1}{2}\alpha u_0 \delta'_\alpha(z^n)$

for $n \geq 0$ and $T_2(z^{-n}) = \overline{T_2(z^n)}$.
(We again understand that for $n = 0, 1$ the first term in $T_2(z^n)$ is zero).

Proof. First we see that if we write $T = L + M\delta_\alpha + N\delta'_\alpha$ with L a hermitian functional then it holds that $M \in \mathbf{R}$ and $N = \pm i\alpha|N|$. Indeed, from $T(1) = L(1)+M$ we conclude $M \in \mathbf{R}$ and from $T(z^{-n}) = \overline{T(z^n)}$ for $n \geq 0$, we obtain $-N\alpha^{n-1} = \overline{N}\alpha^{n+1}$ and therefore $-N = \overline{N}\alpha^2$. In order to study the solutions of this equation write $N = |N|e^{i\Psi}$ and $\alpha = e^{i\theta}$. Then $|N|e^{i(\Psi+\Pi)} = |N|e^{-i\Psi}e^{2i\theta}$ and thus $\Psi = \theta + (2k-1)\frac{\Pi}{2}$ with $k \in \mathbf{Z}$. By taking $k = 0, 1$ we deduce the result. We note that $N \in \mathbf{R} - \{0\}$ for $\alpha = \pm i$.
From $-\overline{\alpha}c_1 - \alpha\overline{c_1} = u_0 - 2c_0$ and $-\overline{\alpha}c_1 = -(Re\alpha\, Rec_1 + Im\alpha\, Imc_1) - i(Re\alpha\, Imc_1 - Im\alpha\, Rec_1)$ it follows that

$$\alpha\overline{c_1} = c_0 - \frac{1}{2}u_0 - i(Re\alpha\, Imc_1 - Im\alpha\, Rec_1) \qquad (8)$$

By taking into account (8) we obtain from (2)

$$c_n = -\sum_{k=1}^{n-1} k\alpha^k u_{n-k} - \frac{1}{2}n\alpha^n u_0 + \alpha^n c_0 - in\alpha^n(Im\alpha\, Rec_1 - Re\alpha\, Imc_1)$$

which implies (7).
In order to show that T_2 is hermitian, consider

$$c_{-n} = T_2(z^{-n}) + c_0\overline{\alpha}^n + i\alpha(Im\alpha\, Rec_1 - Re\alpha\, Imc_1)n\overline{\alpha}^{n+1} \tag{9}$$

By taking conjugates in (8) and by substituting in (9) we get $c_{-n} = T_2(z^{-n}) + (n+1)c_0\overline{\alpha}^n - nc_1\overline{\alpha}^{n+1} - \frac{1}{2}n\overline{\alpha}^n u_0$. On the other hand, by identifying the preceding expression with the conjugate of (2) we obtain the result. □

Remark. If we denote the sequences of moments related to T_i by $c_{ni} = T_i(z^n)$ it holds that T_0, T_1 and T_2 are hermitian solutions of the equation $(z-\alpha)(z^{-1} - \overline{\alpha})T = u$ verifying the initial conditions $c_{00} = \dfrac{u_0}{|\alpha|^2 - 1}, c_{10} = \bar{\alpha}^{-1}\dfrac{u_0}{|\alpha|^2-1}, c_{01} = -\dfrac{u_0}{|\alpha|^2-1}, c_{11} = -\alpha\dfrac{u_0}{|\alpha|^2-1}, c_{02} = 0$ and $c_{12} = -\frac{1}{2}u_0\alpha$ respectively.

3. The Regular Case

We recall the following definitions :
A linear and hermitian functional $L : \Delta \longrightarrow \mathbf{C}$ is called regular if the principal submatrices of the moment matrix are nonsingular, that is,

$$\forall n \geq 0 \qquad \Delta_n = det\left(L(z^{i-j})\right)_{\substack{i=0,..,n\\ j=0,..,n}} \neq 0.$$

In this case $\quad \forall n \geq 0 \quad \mathrm{e_n} = \dfrac{\Delta_\mathrm{n}}{\Delta_{\mathrm{n}-1}} \neq 0$ with $\Delta_{-1} = 1$.
A linear and hermitian functional $L : \Delta \longrightarrow \mathbf{C}$ is called positive definite if $\Delta_n > 0$ for all $n \in \mathbf{N}$. In this case $\mathrm{e_n} > 0$ for all n.
The regularity of a functional L is equivalent to the existence of a monic orthogonal polynomial sequence $\{L_n(z)\}_{n\in\mathbf{N}}$ related to L. In that case we write $M.O.P.S.(L)$. and it holds that $L(L_n(z)\overline{L_n}(z^{-1})) = \mathrm{e_n}$.

First of all we recall a result concerning the direct problem ([3]) :

3.1. LEMMA

Let T be a linear hermitian and regular functional, $\alpha \in \mathbf{C} - \{0\}$, $\{T_n(z)\}_{n\in\mathbf{N}}$ the $M.O.P.S.(T)$ and $\{H_n(z,y)\}_{n\in\mathbf{N}}$ the sequence of kernels related to T defined by $H_n(z,y) = \sum_{k=0}^{n} E_k{}^{-1}T_k(z)\overline{T_k(y)}$ with $E_k = T(T_k(z)\overline{T_k}(z^{-1}))$.
If $u = (z-\alpha)(z^{-1} - \overline{\alpha})T$ then u is regular if and only if

$$\forall n \geq 0 \qquad H_n(\alpha,\alpha) \neq 0$$

and the $M.O.P.S.(u)$ $\{P_n(z)\}_{n\in\mathbf{N}}$ is

$$\forall n \geq 1 \qquad (z-\alpha)P_{n-1}(z) = T_n(z) - \frac{T_n(\alpha)}{H_{n-1}(\alpha,\alpha)}H_{n-1}(z,\alpha) \tag{10}$$

Moreover

$$\forall n \geq 1 \qquad E_n {e_{n-1}}^{-1} - 1 = -\frac{|\, T_n(\alpha)\,|^2}{H_{n-1}(\alpha,\alpha)}{e_{n-1}}^{-1} \tag{11}$$

and

$$\forall n \geq 1 \qquad E_n {e_{n-1}}^{-1} = \frac{H_{n-1}(\alpha,\alpha)}{H_n(\alpha,\alpha)} \tag{12}$$

where $e_n = u(P_n(z)\overline{P_n}(z^{-1}))$.
Proof. See [3]. □

The next subject is to study the regularity of the hermitian solutions T of the functional equation $(z-\alpha)(z^{-1}-\overline{\alpha})T = u$.
Henceforth we will assume that u is regular and $\{P_n(z)\}_{n\in\mathbf{N}}$ is $M.O.P.S.(u)$. Also we will denote by $\{K_n(z,y)\}_{n\in\mathbf{N}}$ the sequence of kernels related to u defined by $K_n(z,y) = \sum_{k=0}^{n} \frac{1}{e_k} P_k(z)\overline{P_k(y)}$.

3.2. THEOREM

T is regular if and only if

$$\forall n \geq 1 \qquad c_0 - \sum_{k=0}^{n-1} {e_k}^{-1} |\, u_k(\alpha) + \overline{P_k}(\alpha^{-1})(\overline{c_1} - \overline{\alpha} c_0)\,|^2 \neq 0$$

where for $\alpha \neq 0$ and

$$\forall k \geq 0, \qquad u_k(\alpha) = u\left(\frac{\overline{P_k}(z^{-1}) - \overline{P_k}(\alpha^{-1})}{z-\alpha}\right).$$

The $M.O.P.S.(T)$ $\{T_n(z)\}_{n\in\mathbf{N}}$ is given by :

$$\forall n \geq 2 \qquad \frac{T_n(z) - T_n(\alpha)}{z-\alpha} = P_{n-1}(z) -$$

$$-T_n(\alpha)\sum_{k=0}^{n-2} {e_k}^{-1} u_k(\alpha) P_k(z) - T_n(\alpha)(\overline{c_1} - \overline{\alpha} c_0) K_{n-2}(z, \frac{1}{\overline{\alpha}}) \tag{13}$$

with:

$$\forall n \geq 2 \qquad T_n(\alpha) = -\left(\overline{u_{n-1}(\alpha)} + P_{n-1}(\bar{\alpha}^{-1})(c_1 - \alpha c_0)\right) \times$$

$$\times \left(c_0 - \sum_{k=0}^{n-2} e_k{}^{-1} | \, u_k(\alpha) + \overline{P_k}(\alpha^{-1})(\overline{c_1} - \overline{\alpha} c_0) \, |^2 \right)^{-1} \tag{14}$$

and $T_1(z) = z - c_1 c_0{}^{-1}$

Proof. " $\Rightarrow$ " If we assume that T is regular then

$$\forall n \geq 2 \qquad \frac{T_n(z) - T_n(\alpha)}{(z-\alpha)} = P_{n-1}(z) + \sum_{k=0}^{n-2} \lambda_{nk} P_k(z) \tag{15}$$

with $\quad \lambda_{nj} e_j = u \left(\frac{T_n(z) - T_n(\alpha)}{(z-\alpha)} \overline{P_j}(z^{-1}) \right)$ for $0 \leq j \leq n-2$.

By definition of T $\quad \lambda_{nj} e_j = (z-\alpha)(z^{-1} - \overline{\alpha}) T \left(\frac{T_n(z) - T_n(\alpha)}{(z-\alpha)} \overline{P_j}(z^{-1}) \right)$ from which, by applying the orthogonality with respect to T we get

$$\lambda_{nj} e_j = -T_n(\alpha) T[(1 - z\overline{\alpha}) z^{-1} \overline{P_j}(z^{-1})] \tag{16}$$

with

$$T[(1 - z\overline{\alpha}) z^{-1} \overline{P_j}(z^{-1})] = T\left((1 - z\overline{\alpha}) z^{-1} [\overline{P_j}(z^{-1}) - \overline{P_j}(\alpha^{-1}) + \overline{P_j}(\alpha^{-1})]\right) =$$

$$= T\left((1 - z\alpha) z^{-1} (z-\alpha) \frac{\overline{P_j}(z^{-1}) - \overline{P_j}(\alpha^{-1})}{z-\alpha}\right) + \overline{P_j}(\alpha^{-1})(\overline{c_1} - \overline{\alpha} c_0) =$$

$$= u_j(\alpha) + \overline{P_j}(\alpha^{-1})(\overline{c_1} - \overline{\alpha} c_0) \tag{17}$$

Then by taking into account (16) and (17) and by substituting in (15) we obtain (13).

The sequence $\{T_n(z)\}_{n \geq 2}$ will be completely determined if we give the sequence of parameters $\{T_n(\alpha)\}_{n \geq 2}$. In order to do that, we use Lemma 3.1.

If we apply the functional T to both members of expression (10) and we calculate $T[(z-\alpha)P_{n-1}(z)]$ like in (17) for $j = n-1$ it follows

$$T_n(\alpha) = -\overline{[u_{n-1}(\alpha) + \overline{P_{n-1}}(\alpha^{-1})(\overline{c_1} - \overline{\alpha} c_0)]} H_{n-1}(\alpha, \alpha) \tag{18}$$

Now, in order to obtain $H_{n-1}(\alpha, \alpha)$ in terms of u we use expressions (11) and (12) from which we conclude

$$(H_n(\alpha, \alpha))^{-1} = (H_{n-1}(\alpha, \alpha))^{-1} - e_{n-1}{}^{-1} | \, u_{n-1}(\alpha) + \overline{P_{n-1}}(\alpha^{-1})(\overline{c_1} - \overline{\alpha} c_0) \, |^2 \tag{19}$$

If we apply repeatedly the preceding formula :

$$\forall n \geq 1 \qquad (H_n(\alpha, \alpha))^{-1} = c_0 - \sum_{k=0}^{n-1} e_k{}^{-1} | \, u_k(\alpha) + \overline{P_k}(\alpha^{-1})(\overline{c_1} - \overline{\alpha} c_0) \, |^2 \tag{20}$$

Then we obtain the condition of regularity and by substituting (20) in (18) we get (14).

" $\Leftarrow$ " If we write

$$S_{n-2}(z) = \sum_{k=0}^{n-2} e_k{}^{-1}[u_k(\alpha) + \overline{P_k}(\alpha^{-1})(\overline{c_1} - \overline{\alpha}c_0)]P_k(z)$$

then (13) turns into

$$\forall n \geq 2 \qquad T_n(z) = (z-\alpha)P_{n-1}(z) - T_n(\alpha)(z-\alpha)S_{n-2}(z) + T_n(\alpha) \tag{21}$$

Then it suffices to prove that the sequence defined by (21) satisfies the following orthogonality conditions :
$T(T_n(z)\overline{T_j}(z^{-1})) = 0$ for $0 \leq j \leq n-1$ and $T(T_n(z)\overline{T_n}(z^{-1})) \neq 0$ for all n.
(i) Indeed, for $2 \leq j \leq n$

$$T(T_n(z)\overline{T_j}(z^{-1})) = T[(z-\alpha)P_{n-1}(z)\overline{T_j}(z^{-1})] -$$

$$T_n(\alpha)T[(z-\alpha)S_{n-2}(z)\overline{T_j}(z^{-1})] + T_n(\alpha)T(\overline{T_j}(z^{-1})) \tag{22}$$

We obtain each term in (22) by using (21) in order to evaluate $\overline{T_j}(z^{-1})$. Then

$$T[(z-\alpha)P_{n-1}(z)\overline{T_j}(z^{-1})] = T[(z-\alpha)P_{n-1}(z)(z^{-1}-\overline{\alpha})\overline{P_{j-1}}(z^{-1})] -$$

$$\overline{T_j(\alpha)}T[(z-\alpha)P_{n-1}(z)(z^{-1}-\overline{\alpha})\overline{S_{j-2}}(z^{-1}) + \overline{T_j(\alpha)}T((z-\alpha)P_{n-1}(z)] =$$

$$\delta_{n-1,j-1}e_{n-1} + \overline{T_j(\alpha)}T((z-\alpha)P_{n-1}(z)) \tag{23}$$

with:

$$T[(z-\alpha)P_{n-1}(z)] =$$

$$T\left((z-\alpha)\frac{P_{n-1}(z) - P_{n-1}(\bar{\alpha}^{-1})}{z-\bar{\alpha}^{-1}}(z-\bar{\alpha}^{-1})\right) + T[(z-\alpha)P_{n-1}(\bar{\alpha}^{-1})] =$$

$$-\bar{\alpha}^{-1}u\left(z\frac{P_{n-1}(z) - P_{n-1}(\bar{\alpha}^{-1})}{z-\bar{\alpha}^{-1}}\right) + (c_1 - \alpha c_0)P_{n-1}(\bar{\alpha}^{-1}) =$$

$$-\bar{\alpha}^{-1}u\left((z-\bar{\alpha}^{-1}+\bar{\alpha}^{-1})\frac{P_{n-1}(z) - P_{n-1}(\bar{\alpha}^{-1})}{z-\bar{\alpha}^{-1}}\right) + (c_1 - \alpha c_0)P_{n-1}(\bar{\alpha}^{-1}) =$$

$$\bar{\alpha}^{-1}P_{n-1}(\bar{\alpha}^{-1})u_0 - \bar{\alpha}^{-2}\overline{u\left(\frac{\overline{P_{n-1}}(z^{-1}) - \overline{P_{n-1}}(\alpha^{-1})}{z^{-1}-\alpha^{-1}}\right)} + (c_1 - \alpha c_0)P_{n-1}(\bar{\alpha}^{-1}) =$$

$$\overline{(c_1 - \alpha c_0 + \bar{\alpha}^{-1}u_0)P_{n-1}(\bar{\alpha}^{-1}) + \bar{\alpha}^{-1}u\left((z-\alpha+\alpha)\frac{\overline{P_{n-1}}(z^{-1}) - \overline{P_{n-1}}(\alpha^{-1})}{z-\alpha}\right)} =$$

$$(c_1 - \alpha c_0)P_{n-1}(\bar{\alpha}^{-1}) + \overline{u_{n-1}(\alpha)} \tag{24}$$

By substituting (24) in (23) and by taking into account the hypothesis we get

$$T[(z-\alpha)P_{n-1}(z)\overline{T_j}(z^{-1})] =$$

$$\delta_{n-1,j-1}e_{n-1} - \overline{T_j(\alpha)}T_n(\alpha)\left(c_0 - \sum_{k=0}^{n-2} e_k{}^{-1}\mid u_k(\alpha) + \overline{P_k}(\alpha^{-1})(\overline{c_1} - \overline{\alpha}c_0)\mid^2\right) \tag{25}$$

In the same way, if we calculate the second and the third sum in (22) we obtain:

$$\forall j < n \qquad T[(z-\alpha)S_{n-2}(z)\overline{T_j}(z^{-1})] = u_{j-1}(\alpha) + \overline{P_{j-1}}(\alpha^{-1})(\overline{c_1} - \overline{\alpha}c_0) +$$

$$\overline{T_j(\alpha)} \sum_{k=j-1}^{n-2} e_k{}^{-1}\mid u_k(\alpha) + \overline{P_k}(\alpha^{-1})(\overline{c_1} - \overline{\alpha}c_0)\mid^2$$

and

$$T[(z-\alpha)S_{n-2}(z)\overline{T_n}(z^{-1})] = 0.$$

Similarly for the third term in the right member of (22) we have

$$T(\overline{T_j}(z^{-1})) = \overline{T[(z-\alpha)P_{j-1}(z)]} - \overline{T_j(\alpha)T[(z-\alpha)S_{j-2}(z)]} + \overline{T_j(\alpha)}c_0 = u_{j-1}(\alpha) +$$

$$\overline{P_{j-1}}(\alpha^{-1})(\overline{c_1} - \overline{\alpha}c_0) - \overline{T_j(\alpha)}\sum_{k=0}^{j-2} e_k{}^{-1}\mid u_k(\alpha) + \overline{P_k}(\alpha^{-1})(\overline{c_1} - \overline{\alpha}c_0)\mid^2 + \overline{T_j(\alpha)}c_0$$

Finally by substituting in (22) we get for $2 \le j < n$ $T(T_n(z)\overline{T_j}(z^{-1})) = 0$ and $T(T_n(z)\overline{T_n}(z^{-1})) \neq 0$.
(ii) For $j = 1$ and $n \ge 2$

$$T(T_n(z)\overline{T_1}(z^{-1})) = T[(z-\alpha)P_{n-1}(z)\overline{T_1}(z^{-1})] -$$

$$T_n(\alpha)T[(z-\alpha)S_{n-2}(z)\overline{T_1}(z^{-1})] + T_n(\alpha)T(\overline{T_1}(z^{-1})) \tag{26}$$

We obtain the first term in the right member of (26) by substituting $T_1(z)$ and by taking into account (24):

$$T[(z-\alpha)P_{n-1}(z)\overline{T_1}(z^{-1})] =$$

$$-(\overline{\alpha} - \overline{c_1}c_0{}^{-1})T_n(\alpha)\left(c_0 - \sum_{k=0}^{n-2} e_k{}^{-1}\mid u_k(\alpha) + \overline{P_k}(\alpha^{-1})(\overline{c_1} - \overline{\alpha}c_0)\mid^2\right)$$

In the same way we calculate the second term in the right member of (26)

$$T[(z-\alpha)S_{n-2}(z)\overline{T_1}(z^{-1})] = (\overline{c_1}-\overline{\alpha}c_0)+$$

$$(\overline{\alpha}-\overline{c_1}c_0{}^{-1})\left(\sum_{k=0}^{n-2} e_k{}^{-1}\mid u_k(\alpha)+\overline{P_k}(\alpha^{-1})(\overline{c_1}-\overline{\alpha}c_0)\mid^2\right)$$

and $T[\overline{T_1}(z^{-1})]=0$ by construction. Then

$$T[T_n(z)\overline{T_1}(z^{-1})]=0$$

(iii) Since $T_1(z)=z-c_1c_0{}^{-1}$ it follows that

$$T[T_1(z)\overline{T_1}(z^{-1})] = \frac{c_0^2-\mid c_1\mid^2}{c_0} \neq 0$$

and finally
(iv) For $n\geq 2$

$$T[T_n(z)] = T[(z-\alpha)P_{n-1}(z)] - T_n(\alpha)T[(z-\alpha)S_{n-2}(z)] + T_n(\alpha)c_0 = 0.$$

Therefore, by taking into account (i), (ii), (iii) and (iv) we conclude T is regular. □

Remarks.
1) If T is a regular solution of $(z-\alpha)(z^{-1}-\overline{\alpha})T=u$ then for $n\geq 1$, it holds that $T_n(\alpha)=0$ if and only if $u_{n-1}(\alpha)+\overline{P_{n-1}}(\alpha^{-1})(\overline{c_1}-\overline{\alpha}c_0)=0$. In this case $T_n(z)=(z-\alpha)P_{n-1}(z)$.
2) The expression

$$-[u_k(\alpha)+\overline{P_k}(\alpha^{-1})(\overline{c_1}-\overline{\alpha}c_0)]$$

plays the same role as the Walsh functions of second kind in the positive definite case ([7]).
3) Equivalent characterizations for the regularity of T are given in [14] in terms of the associated polynomials of the first kind and by using the point α^{-1}.

3.3. EXAMPLES

Let u be the associated functional to the Lebesgue normalized measure on the unit circle, defined by:

$$u_n = \begin{cases} 0, & \text{if } n\neq 0; \\ 1, & \text{if } n=0 \end{cases}$$

By applying Theorem 3.2 and by taking into account that $P_n(z)=z^n$ for all n we have:
(1) For $\mid\alpha\mid\neq 1$ the condition of regularity is

$$\forall n\geq 1 \qquad c_0-\mid\overline{c_1}-\overline{\alpha}c_0\mid^2\frac{1-\mid\alpha\mid^{2n}}{(1-\mid\alpha\mid^2)\mid\alpha\mid^{2n-2}}\neq 0$$

and the $M.O.P.S.(T)$ is given by

$$\forall n \geq 2 \qquad \frac{T_n(z) - T_n(\alpha)}{(z-\alpha)} = z^{n-1} - T_n(\alpha)(\overline{c_1} - \overline{\alpha} c_0) \sum_{k=0}^{n-2} \frac{z^k}{\alpha^k}$$

with

$$T_n(\alpha) = \overline{\alpha}^{1-n}(\alpha c_0 - c_1)\left(c_0 - |\, \overline{c_1} - \overline{\alpha} c_0 \,|^2 \frac{1 - |\,\alpha\,|^{2n-2}}{(1 - |\,\alpha\,|^2)|\,\alpha\,|^{2n-4}}\right)^{-1}$$

and $T_1(z) = z - c_1 c_0{}^{-1}$.
Next we analyse some particular cases:
i) $T = T_0$ is regular and $T_n(z) = z^{n-1}(z - \bar{\alpha}^{-1})$ for all $n \geq 1$.
ii) $T = T_1$ is regular and $T_n(z) = z^{n-1}(z - \alpha)$ for all $n \geq 1$.
iii) If we take T such that $c_1 = 0$ which implies $c_0 = \dfrac{1}{1 + |\,\alpha\,|^2}$ we have that T is regular and the $M.O.P.S.(T)$ is

$$T_n(z) = z^n + \alpha\left(\frac{1 - |\,\alpha\,|^{2(n-2)}}{|\,\alpha\,|^{2(n-2)} - |\,\alpha\,|^2}\right) z^{n-1} + \alpha\overline{\alpha}^{1-n}\left(\frac{|\,\alpha\,|^{2(n-2)}(1 - |\,\alpha\,|^2)}{|\,\alpha\,|^{2(n-2)} - |\,\alpha\,|^2}\right)$$

and $T_1(z) = z$.
(2) For $|\,\alpha\,| = 1$ the condition of regularity is $c_0 - n|\,\overline{c_1} - \overline{\alpha} c_0\,|^2 \neq 0$ for all $n \geq 1$ and the $M.O.P.S.(T)$ is

$$T_n(z) = z^n - \left(\alpha + \overline{\alpha}^{n-2} T_n(\alpha)(\overline{c_1} - \overline{\alpha} c_0)\right) z^{n-1} + T_n(\alpha)(\overline{c_1}\alpha - |\,\alpha\,|^2 c_0 + 1)$$

with

$$\forall n \geq 2 \qquad T_n(\alpha) = \alpha^{n-1}(\alpha c_0 - c_1)\left(c_0 - (n-1)|\,\overline{c_1} - \overline{\alpha} c_0\,|^2\right)^{-1}$$

and $T_1(z) = z - c_1 c_0{}^{-1}$.
i) In particular, the functional verifying the initial conditions $c_0 = \frac{1}{2}$ and $c_1 = \alpha i$ is regular and the $M.O.P.S.(T)$ is

$$\forall n \geq 2 \qquad T_n(z) = z^n - \alpha \frac{5n-2}{5n-7} z^{n-1} + \alpha^n \frac{3+4i}{5n-7}$$

$$T_1(z) = z - 2\alpha i.$$

4. The Positive Definite Case

Next we analyse the situation when T is a positive definite linear functional. We recall that in this case there exists a finite positive measure on $[0, 2\pi]$ such that $c_n = \int_0^{2\pi} e^{in\theta} d\mu(\theta)$ ([5]).

4.1. THEOREM

Let u be a positive definite linear functional and $c_0 > 0$.
(i) If $c_0 >| c_1 |$ then T is positive definite if and only if

$$\forall n \geq 2 \qquad c_0 - \sum_{k=0}^{n-1} e_k{}^{-1} | u_k(\alpha) + \overline{P_k}(\alpha^{-1})(\overline{c_1} - \overline{\alpha} c_0) |^2 > 0$$

In this case, if we denote by μ and ν the measures associated with the sequences $\{T_n(z)\}_{n\in\mathbf{N}}$ and $\{P_n(z)\}_{n\in\mathbf{N}}$ respectively, then the absolutely continuous parts verify the following relation:

$$\mu'(\theta) = \frac{\nu'(\theta)}{(1+|\alpha|^2) - 2Re(\overline{\alpha}e^{i\theta})} \tag{27}$$

(ii) If $c_0 \leq| c_1 |$ then T is not a positive definite functional.

Proof. Since $E_n = e_{n-1}\dfrac{H_{n-1}(\alpha,\alpha)}{H_n(\alpha,\alpha)}$ for all $n \geq 1$ and $H_1(\alpha,\alpha) = \dfrac{c_0}{c_0{}^2 - | c_1 |^2}$, it is necessary that $c_0 >| c_1 |$ in order that T is positive definite. The preceding condition forces to $H_n(\alpha,\alpha) > 0$ for all $n \geq 2$ and then by taking into account (20) we get the result.
Next we obtain the relations between the measures associated with both sequences for which we use the technique described in [1]. We assume u is normalized, that is, $u_0 = 1$ and we denote by $F(z) = c_0 + 2\sum_{k=1}^{\infty} c_{-n}z^n$ and by $U(z) = 1 + 2\sum_{k=1}^{\infty} u_{-n}z^n$. From the difference equation verified by $\{c_n\}_{n\in\mathbf{N}}$, given in Theorem 2.1, it follows that

$$F(z)[(1+|\alpha|^2) - \alpha z^{-1} - \overline{\alpha}z] = U(z) - \alpha c_{-1} + \overline{\alpha}c_1 + \overline{\alpha}c_0 z - \alpha c_0 z^{-1} \tag{28}$$

Then

$$Re(F(e^{i\theta}) = \frac{Re(U(e^{i\theta})}{(1+|\alpha|^2) - 2Re(\overline{\alpha}e^{i\theta})}$$

If we use that $\mu'(\theta) = Re(F(e^{i\theta}))$ and $\nu'(\theta) = Re(U(e^{i\theta}))$ we get the result. □

4.2. THEOREM

Let u be a positive definite linear functional and assume that the associated measure ν is absolutely continuous. Let T be a positive definite solution of the equation $(z-\alpha)(z^{-1} - \overline{\alpha})T = u$ and let μ be the measure associated with T.
If we denote by $c_{0_s} = c_0 - \int_0^{2\pi} d\mu'(\theta)$ and by $c_{1_s} = c_1 - \int_0^{2\pi} e^{i\theta} d\mu'(\theta)$ with $\mu'(\theta)$ given by (28) then:
(i) For $| \alpha |\neq 1$

$$\forall n \in \mathbf{Z} \qquad T(z^n) = \int_0^{2\pi} e^{in\theta} d\mu'(\theta) + \frac{1}{1-|\alpha|^2}(\overline{\alpha}c_{1_s} - |\alpha|^2 c_{0_s})\delta_{\overline{\alpha}^{-1}}(z^n) +$$

$$\frac{1}{1-|\alpha|^2}(\alpha\overline{c_{1_s}} - |\alpha|^2 c_{0_s})\delta_\alpha(z^n) \qquad \text{with } z = e^{i\theta} \tag{29}$$

(ii) For $|\alpha| = 1$

$$\forall n \in \mathbf{Z} \qquad T(z^n) = \int_0^{2\pi} e^{in\theta} d\mu'(\theta) + c_{0s}\delta_\alpha(z^n) +$$

$$i\alpha(Im\alpha\, Rec_{1s} - Re\alpha\, Imc_{1s})\delta'_\alpha(z^n) \qquad \text{with } z = e^{i\theta} \tag{30}$$

Proof. (i) From (3)

$$c_n = -\frac{1}{1-|\alpha|^2}\int_0^{2\pi}\left[e^{in\theta}\left(\frac{e^{-in\theta}-\overline{\alpha}^n}{e^{i\theta}\overline{\alpha}^n(e^{-i\theta}-\overline{\alpha})}\right)-\left(\frac{\alpha e^{in\theta}-\alpha^{n+1}}{e^{i\theta}-\alpha}\right)\right]d\nu'(\theta)+$$

$$\frac{c_0-\alpha\overline{c_1}}{1-|\alpha|^2}\delta_{\overline{\alpha}^{-1}}(z^n)+\frac{c_0-\overline{\alpha}c_1-u_0}{1-|\alpha|^2}\delta_\alpha(z^n)$$

If we take into account the connection between μ' and ν' given by (27) together with $u_0 = 1$ then we obtain

$$c_n = \int_0^{2\pi} e^{in\theta}d\mu'(\theta) + \frac{\alpha}{1-|\alpha|^2}\left(\frac{1}{\overline{\alpha}^n}-\alpha^n\right)\int_0^{2\pi}e^{-i\theta}d\mu'(\theta)+$$

$$\frac{1}{1-|\alpha|^2}\left(\overline{\alpha}\alpha^{n+1}-\frac{1}{\overline{\alpha}^n}\right)\int_0^{2\pi}d\mu'(\theta)+\frac{c_0-\alpha\overline{c_1}}{1-|\alpha|^2}\delta_{\overline{\alpha}^{-1}}(z^n)+\frac{c_0-\overline{\alpha}c_1-1}{1-|\alpha|^2}\delta_\alpha(z^n)$$

that is, by using (6)

$$c_n = \int_0^{2\pi} e^{in\theta}d\mu'(\theta) + \frac{1}{1-|\alpha|^2}(-\alpha\overline{c_{1s}}+c_{0s})\delta_{\overline{\alpha}^{-1}}(z^n)+$$

$$\frac{1}{1-|\alpha|^2}(\alpha\overline{c_{1s}}-|\alpha|^2c_{0s})\delta_\alpha(z^n).$$

As may be easily seen $-\alpha\overline{c_{1s}}+c_{0s} = \overline{\alpha}c_{1s}-|\alpha|^2c_{0s}$ from which the result follows for $n \in \mathbf{N}$. By taking into account $c_{-n} = \overline{c_n}$ we get (29).
(ii) By proceeding in a similar way we get (30). □

4.3. EXAMPLES

Again let u be the Lebesgue functional on the unit circle. By applying the preceding results we deduce:
For $|\alpha| = 1$ there are not solutions T positive definite.
For $|\alpha| < 1$ T is positive if and only if $T = T_1$.
For $|\alpha| > 1$ T is positive definite if and only if $c_0 > c_1$ and $\dfrac{c_0}{c_0{}^2-|c_1|^2} \le |\alpha|^2$.
In particular T_0 is positive definite.
Moreover for T_0 and T_1 we get that $c_{os} = c_{1s} = 0$ which implies that μ is absolutely continuous with respect to the Lebesgue measure and

$$d\mu(\theta) = \frac{d\theta}{(1+|\alpha|^2)-2Re(\overline{\alpha}e^{i\theta})}. \qquad \square$$

References

1. Alfaro, M., Alfaro, M. P., Guadalupe, J., Vigil, L. *Correspondance entre suites de polynômes orthogonaux et functions de la boule unitè de* $H^\infty(D)$, Lecture Notes in Math. 1.171 (1985), 158-163.
2. Chihara, T. *An introduction to Orthogonal Polynomials*, Gordon and Breach, New York (1978).
3. Garcia Lazaro, P. *Distribuciones y polinomios Ortogonales*, Doctoral Dissertation, Pub. Seminarion Mat. García de Galdeano, Serie II, Sec.2 32 (1991).
4. Gautschi, W. *An algorithmic implementation of the generalized Christoffel theorem*, Numerical Integration, G. Hammerlin Editor, Birkhauser, Basel (1982), 89-106.
5. Geronimus, Y. *Polynomials orthogonal on a circle and their applications*, Transl. Amer. Math. Soc., serie 1, vol. 3 (1962), 1-78.
6. Godoy, E., Marcellan, F. *An analog of the Christoffel formula for polynomial modification of a measure on the unit circle*, Bull. Un. Mat. Ital. 7, 5-A (1991), 1-12.
7. Godoy, E., Marcellan, F. *Orthogonal Polynomials and Rational modifications of measures*, Canad. Math. Bull. (to appear).
8. Henrici, P. *Elements of Numerical Analysis*, John Wiley and Sons, Inc., New York (1964).
9. Ismail, M. H., Ruedemann, R. W. *Relation between Polynomials Orthogonal on the unit circle with respect to different weights*, J. Approx. Theory 71 (1) (1992), 39-60.
10. Maroni, P. *Sur la suite de polynômes orthogonaux associée a la forme* $u = \delta_c + \lambda(x-c)^{-1}L$, Periodica Mathematica Hungarica 21 (3) (1990), 223-248.
11. Paszkowski, S. *Sur des transformations de polynômes orthogonaux multiplication et division de function de poids par un polynôme*, Lecture Notes in Math. 1.171 (1985), 239-246.
12. Rudin, W. *Functional Analysis*, McGraw-Hill B. Co. (1973).
13. Rudin, W. *Real and Complex Analysis*, McGraw-Hill, London (1970).
14. Suarez Rodriguez, C. *Polinomios Ortogonales relativos a modificaciones de funcionales regulares y hermitianos*, Doctoral Dissertation, Univ. de Vigo (1993).
15. Szego, G. *Orthogonal Polynomials*, Amer. Math. Soc. Colloq. Publ. 23, Providence, R. I. (1975, 4th edition).
16. Uvarov, V. *The connection between systems of polynomials orthogonal with respect to different distribution functions*, URSS. Comput. Math. and Phys. 9, n 6 (1969), 25-36.

NORMALITY AND ERROR FORMULAE FOR SIMULTANEOUS RATIONAL APPROXIMANTS TO NIKISHIN SYSTEMS

K. DRIVER
Mathematics Department, University of Witwatersrand, P.O. WITS 2050, Johannesburg, South Africa.

and

H. STAHL*
Technische Fachhochschule, Berlin / FB 2, Luxemburger Straße 10, D-13353 Berlin 65, Germany.

Abstract. Normality and error formulae for simultaneous rational approximants will be discussed. Simultaneous rational approximants generalize Padé approximants and continued fractions in a very natural way. As in the case of Padé approximants so also here Markov functions are especially interesting and important. The common denominator of the simultaneous approximants satisfies a multiple orthogonality relation, which in the case of Markov functions is defined by m positive Borel measures supported on the real line. Up to now only two types of systems of Markov functions are known for which more than trivial results can be proved. These are the Angelesco and the Nikishin systems. The present paper concentrates on Nikishin systems. For them, new results about normality, and new error formulae are reviewed, and the proofs sketched.

Key words: Nikishin Systems, simultaneous rational approximants, Hermite-Padé polynomials, multiple orthogonality, normality, orthogonal polynomials.
AMS (MOS) Subject classification: 41A21, 30A10

1. Simultaneous Rational Approximants to Markov Functions

Let m positive Borel measures $\mu_1, \ldots, \mu_m$ be given with compact support $\text{supp}(\mu_j) \subseteq \mathbb{R}$ and define

$$f_j(z) := \int \frac{d\mu_j(x)}{x - z}, \quad j = 1, \ldots, m. \tag{1}$$

Let

$$n = (n_1, \ldots, n_m) \in \mathbb{N}^m, \tag{2}$$

be a multi-index and define

$$|n| := n_1 + \cdots + n_m. \tag{3}$$

By $\mathcal{P}_n$ we denote the set of polynomials of degree at most n.

Definition 1.1 For each multiindex $n \in \mathbb{N}^m$ there exists a vector of polynomials $(Q_0, \ldots, Q_m) \in \mathcal{P}_{|n|} \times \mathcal{P}_{|n|-1} \times \cdots \times \mathcal{P}_{|n|-1} \setminus \{(0, \ldots, 0)\}$ satisfying

$$Q_0(z) f_j(z) - Q_j(z) = \mathcal{O}(z^{-n_j - 1}) \quad \text{as } z \to \infty,\ j = 1, \ldots, m, \tag{4}$$

* Research supported by the Deutsche Forschungsgemeinschaft (AZ: Sta 299/4–2).

A. Cuyt (ed.), Nonlinear Numerical Methods and Rational Approximation II, 97–105.

where $\mathcal{O}(\cdot)$ denotes Landau's capital oh. The vector $(Q_1/Q_0, \ldots, Q_m/Q_0)$ is called a *simultaneous rational approximant* to the vector of Markov functions $(f_1, \ldots, f_m)$.

Remark It is clear from (4) that in the vector $(Q_0, \ldots, Q_m)$, the polynomial Q_0 plays an exceptional role. If Q_0 is determined, then $Q_1, \ldots, Q_m$ simply follow as principal parts of $Q_0 f_j$, $j = 1, \ldots, m$.

1.1. Normality

Because of the prominent role that is played by the denominator polynomial Q_0 in determining the vector $(Q_0, \ldots, Q_m)$, the definition of normality can be based solely on Q_0.

Definition 1.2 A multiindex $n = (n_1, \ldots, n_m) \in \mathbb{N}^m$ is called *normal* if Q_0 is uniquely determined by (4) up to a non-zero constant factor and if $\deg(Q_0) = |n|$.

1.2. Multiple Orthogonality

As an analogue of the orthogonality relation satisfied by the denominator polynomial of a Padé approximant, we have in the case of simultaneous rational approximants the following result.

Theorem 1.1 *For each multiindex $n = (n_1, \ldots, n_m) \in \mathbb{N}^m$, a polynomial $Q_0 \in \mathcal{P}_{|n|}$, $Q_0 \not\equiv 0$, satisfies relation (4) (with appropriate polynomials $Q_1, \ldots, Q_m \in \mathcal{P}_{|n|-1}$) if and only if*

$$\int x^k Q_0(x) d\mu_j(x) = 0, \quad k = 0, \ldots, n_j - 1, \quad j = 1, \ldots, m. \tag{5}$$

Definition 1.3 Relation (5) defines what is called a *multiple orthogonality* of Q_0 with respect to the measures $\mu_1, \ldots, \mu_m$ and the multidegrees $n_1, \ldots, n_m$.

2. Nikishin Systems

Unlike ordinary orthogonality, multiple orthogonality (5) does not in general determine the polynomial Q_0 in $\mathcal{P}_{|n|}$ up to a constant factor. Likewise, there is in general no known result that can be considered as the analogue of Markov's theorem. Up to now there are only two types of systems of Markov functions for which positive results have been proved. The first one is an Angelesco System (cf. [Angelesco A.,1918], [Gonchar A.A. and Rakhmanov E.A.,1981]) and the second one a Nikishin System introduced by E.M. Nikishin in [Nikishin E.M.,1979] and [Nikishin E.M.,1980].

2.1. Definition of a Nikishin System

Let m positive Borel measures

$$\mu_1, \ldots, \mu_m \geq 0, \qquad \text{with} \qquad \operatorname{supp}(\mu_j) \subseteq \mathbb{R} \text{ compact}, \ j = 1, \ldots, m, \tag{6}$$

be given. We assume that two consecutive measures are supported by two intervals that have at most a common end point, i.e.

$$\mathrm{Int}(I(\mu_j)) \cap I(\mu_{j+1}) = \emptyset \quad \text{for } j = 1, \ldots, m-1, \tag{7}$$

where $I(\mu_i)$ denotes the convex hull of $\mathrm{supp}(\mu_i)$.

Definition 2.1 Assume that (7) holds and let the family of functions $\{f_{jk} \mid k = 1, \ldots, j+1,\ j = 0, \ldots, m\}$ be recursively defined by

$$f_{j,j+1}(z) :\equiv 1, \qquad j = 0, \ldots, m, \tag{8}$$

$$f_{jk}(z) := \int \frac{f_{j-1,k}(x) d\mu_j(x)}{x - z}, \quad k = 1, \ldots, j,\ j = 1, \ldots, m. \tag{9}$$

Then the vector of functions

$$(f_1, \ldots, f_m) := (f_{m1}, \ldots, f_{mm}) \tag{10}$$

forms the *Nikishin system* generated by the measures $\{\mu_j\}_{j=1}^m$.

Remarks (1) Each f_j, $j = 1, \ldots, m$, is a Markov function or its negative. Their generating measures $f_{m-1,j}(x)d\mu_m(x)$, $j = 1, \ldots, m$, have identical support, and are all modifications of the last measure μ_m. In the case when $I(\mu_j)$ and $I(\mu_{j+1})$ have a common endpoint, we assume that all the involved integrals exist.

(2) Each row of the table $\{f_{jk}\}$ again forms a Nikishin system. Note that the measures $\mu_1, \ldots, \mu_m$ play a different role in the definition of a Nikishin system from that played by the measures in (1).

2.2. Multiple Orthogonality

The multiple orthogonality relation (5) has in the case of a Nikishin system the form

$$\int x^k Q_0(x) f_{m-1,j}(x) d\mu_m(x) = 0, \quad k = 0, \ldots, n_j - 1,\ j = 1, \ldots, m, \tag{11}$$

(cf. [DRIVER K. and STAHL H. 1], Proof of Theorem 1.1). If we define the function

$$F(z) = F(p_1, \ldots, p_m; z) := \sum_{j=1}^{m} f_{m-1,j}(z) p_j(z) \tag{12}$$

with $(p_1, \ldots, p_m) \in \mathcal{P}_{n_1-1} \times \cdots \times \mathcal{P}_{n_m-1}$, then it follows from (11) that

$$\int F(x) Q_0(x) d\mu_m(x) = 0. \tag{13}$$

This orthogonality will turn out to be a basic tool in the investigation of properties of Nikishin systems. Note that $\mathcal{P}_{n_1-1} \times \cdots \times \mathcal{P}_{n_m-1}$ is a linear space of dimension $|n|$. We start with the study of normality.

3. Normality of Nikishin Systems

In [Driver K. and Stahl H. 1], it has been shown that practically half the multi-indices of the table associated with a Nikishin system are normal. This investigation extends earlier results by Nikishin in [Nikishin E.M.,1980], where he has shown that all multi-indices $n = (n_1, \ldots, n_m) \in \mathbb{N}^m$ of the form

$$n_j = \begin{cases} k+1, & j \leq q, \quad k \in \mathbb{N},\ 0 \leq q \leq m, \\ k, & j > q, \end{cases}$$

are normal.

3.1. The New Normality Results

Let N_m^* be the subset of $\mathbb{N}^m$ defined by

$$N_m^* := \{n = (n_1, \ldots, n_m) \in \mathbb{N}^m \mid n_j \geq \max(n_1, \ldots, n_{j-1}) - 1,\ j = 2, \ldots, m\}. \quad (14)$$

Theorem 3.1 (cf. [Driver K. and Stahl H. 1], Theorem 4.1).
(i) *All multiindices $n \in N_m^*$ are normal.*
(ii) *For $n \in N_m^*$ the common denominator polynomial Q_0 has exactly $|n|$ simple zeros in the interior of $I(\mu_m)$.*

Remark So far we do not know whether there exist non-normal indices in $\mathbb{N}^m \setminus N_m^*$. At least for the special case of $m = 2$ this possibility can be excluded.

Theorem 3.2 (cf. [Driver K. and Stahl H. 1], Theorem 4.3). *If $m = 2$, then all multi-indices $n = (n_1, n_2) \in \mathbb{N}^2$ are normal, and the polynomial Q_0 has exactly $|n|$ simple zeros in* $\mathrm{Int}(I(\mu_2))$.

3.2. Some Elements of the Proof of Theorem 3.1

The main tool in Nikishin's investigation [Nikishin E.M.,1979] and [Nikishin E.M.,1979] was the theory of algebraic Chebychev systems (AT-systems). Apparently, the use of AT-systems was also the motivation for the recursive definition (9). In the proof of Theorem 4.1 in [Driver K. and Stahl H. 1], properties of the multipoint version of Hermite-Padé polynomials of type I play a decisive role.

Let $n \in \mathbb{N}^m$ be a multiindex and we assume that $|n| - 1$ points $x_1, \ldots, x_{|n|-1} \in \mathbb{R} \setminus I(\mu_m)$ are given. Set

$$w(z) := \prod_{k=1}^{|n|-1} (z - x_k). \quad (15)$$

Then there exists a vector of polynomials $(p_1, \ldots, p_m) \in \mathcal{P}_{n_1-1} \times \cdots \times \mathcal{P}_{n_m-1} \setminus \{(0, \ldots, 0)\}$ satisfying

$$\sum_{j=1}^{m} f_{m-1,j}(z) p_j(z) = \mathcal{O}(w(z)) \quad \text{as } z \to Z(w). \quad (16)$$

By $Z(w)$ we denote the set of all zeros of the polynomial w. In the diagonal case $n_1 = \cdots = n_m$ the elements of the vector $(p_1, \ldots, p_m)$ are known as Hermite-Padé polynomials of type I (in the nondiagonal case there are slight differences to the usual definition (cf. [Aptekarev A.I., Stahl H.,1992])).

Under certain restrictions for the multiindex it is possible to prove an integral representation for the left-hand side of (16).

Theorem 3.3 (cf. [Driver K. and Stahl H. 1], Theorem 4.2(c)). *If* $n_m \geq \max(n_1, \dots, n_{m-1}) - 1$, *then*

$$\sum_{j=1}^{m} (f_{m-1,j}p_j)(z) = \frac{w(z)}{g(z)} \int \sum_{j=1}^{m-1} (f_{m-2,j}p_j)(x) \frac{g(x)d\mu_m(x)}{w(x)(x-z)} \tag{17}$$

with an arbitrary polynomial $g \in \mathcal{P}_{|n|-n_m} \setminus \{0\}$.

Remark The condition in Theorem 3.3 is satisfied for all $n \in N_m^*$.

Using Theorem 3.3 recursively yields

Corollary 3.4 We have

$$\sum_{j=1}^{m} f_{m-1,j}(z)p_j(z) \neq 0 \quad \text{for all} \quad z \in \mathbb{C} \setminus (I(\mu_{m-1}) \cup Z(w)). \tag{18}$$

Theorem 3.1 can be proved by Corollary 3.4 in the following way: (i) Assume that Q_0 has only $N < |n|$ odd zeros in $\text{Int}(I(\mu_m))$. (ii) In accordance with (16) and Corollary 3.4 it is possible to find a vector of polynomials $(p_1, \dots, p_m) \in \mathcal{P}_{n_1-1} \times \dots \times \mathcal{P}_{n_m-1}$ such that

$$F(z) := \sum_{j=1}^{m} f_{m-1,j}(z)p_j(z) \tag{19}$$

has simple zeros where Q_0 has an odd zero and $F(z) \neq 0$ elsewhere in $\text{Int}(I(\mu_m))$. Hence, FQ has no sign change on $I(\mu_m)$, and consequently $\int FQ_0 d\mu_m \neq 0$. (iii) The last inequality contradicts (13) and part (ii) of Theorem 3.1 is proved. Part (i) of the theorem then follows rather immediately. For more details see [Driver K. and Stahl H. 1], Proof of Theorem 4.2.

4. The Investigation of the Remainder Functions

For the investigation of the convergence of the simultaneous rational approximants it is necessary to study the remainder functions

$$R_j(z) := Q_0(z)f_j(z) - Q_j(z), \quad j = 1, \dots, m. \tag{20}$$

4.1. The Particular Case of R_m

It turns out that an investigation of R_m is much easier than that of $R_1, \dots, R_{m-1}$. We have

Theorem 4.1 (cf. [Driver K. and Stahl H. 2], Theorem 2.3.2(a)). *If* $n_m \geq \max(n_1, \ldots, n_{m-1}) - 1$, *then*

$$\int x^k R_m(x) f_{m-2,j}(x) d\mu_{m-1}(x) = 0, \quad k = 0, \ldots, n_j - 1, \; j = 1, \ldots, m-1. \quad (21)$$

Remark The condition in Theorem 4.1 is satisfied for all multiindices $n \in N_m^*$.

Theorem 4.1 shows that R_m satisfies a multiple orthogonality relation. Since the functions $\{f_{m-2,1}, \ldots, f_{m-2,m-2}\}$ again form a Nikishin system, the next theorem follows practically as a corollary of Theorem 4.1.

Theorem 4.2 (cf. [Driver K. and Stahl H. 2], Theorem 2.3.2(b)). *For* $n \in N_m^*$ *we have*
(i) R_m *has exactly* $|n| - n_m$ *simple zeros* $y_1, \ldots, y_{|n|-m_n}$ *in* $\mathrm{Int}(I(\mu_{m-1}))$.
(ii) R_m *has a zero of order* $n_m + 1$ *at infinity.*
(iii) $R_m(z) \neq 0$ *for* $z \in \mathbb{C} \setminus (I(\mu_m) \cup \{y_k\})$.

Using the $|n| - n_m$ zeros of R_m in $\mathrm{Int}(I(\mu_{m-1}))$, we define the polynomial

$$w_{m-1}(z) := \prod_{k=1}^{|n|-n_m} (z - y_k). \quad (22)$$

The next theorem can in some sense be considered as the analogue of Markov's Theorem for the m-th component of the simultaneous rational approximants.

Theorem 4.3 (cf. [Driver K. and Stahl H. 2], Theorem 2.3.3). *Let* $n \in N_m^*$. *Then*
(i) we have

$$\int x^k Q_0(x) \frac{d\mu_m(x)}{w_{m-1}(x)} = 0, \quad k = 0, \ldots, |n| - 1, \quad (23)$$

(ii) and for the approximation error we obtain the representation

$$f_m(z) - \frac{Q_m(z)}{Q_0(z)} = \frac{w_{m-1}(z)}{Q_0(z) g(z)} \int \frac{Q_0(x) g(x) d\mu_m(x)}{w_{m-1}(x)(x - z)} \quad (24)$$

with an arbitrary $g \in \mathcal{P}_{|n|} \setminus \{0\}$.

Remarks (1) Part (i) defines an ordinary orthogonality for Q_0. Since $d\mu/w_m$ does not change sign on $I(\mu_m)$, this orthogonality has rather strong consequences; one gets, for instance, Theorem 4.1 from (23) as a byproduct.

(2) Formula (24) shows that the last component Q_m/Q_0 is actually a multi-point Padé approximant to f_m with a total number of $2|n| + 1$ interpolation points. Choosing $g = Q_0$ in (24) yields a useful representation for the approximation error.

4.2. The Remainder Functions $R_1, \ldots, R_{m-1}$

The nice results in Theorem 4.3 make it very desirable to have similar results for the other remainder functions $R_1, \ldots, R_{m-1}$. However, it turns out that this is much more difficult if $m > 2$. In order to obtain orthogonality relations for R_j, $j = 1, \ldots, m-1$, that are in some sense analogous to the multiple orthogonality satisfied by R_m in (21), we introduce auxiliary Nikishin systems $\{\hat{f}_{mk}\}$ which are closely related to $\{f_{mk}\}$. With respect to the new systems, we can prove a set of orthogonality relations that will enable us to establish the existence of a number of zeros of R_j in $I(\mu_{m-1})$, however, the set of multiindices for which these relations hold is restricted to the near diagonal case.

We recall that if $f_k(z)$ is a Markov function

$$f_k(z) = \int \frac{d\mu_k(x)}{x - z}, \tag{25}$$

then the reciprocal defines again a Markov function up to a transformation. We have (cf. [Stahl H. and Totik,V.,1992], Lemma 6.3.5)

$$\frac{1}{f_k(z)} = a + bz - \int \frac{d\tilde{\mu}_k(x)}{x - z}, \quad a, b \in \mathbb{R}, \tag{26}$$

where $\tilde{\mu}_k$ is a positive measure with $I(\tilde{\mu}_k) \subseteq I(\mu_k)$ and the set $[\mathrm{supp}(\mu_k) \setminus \mathrm{supp}(\tilde{\mu}_k)] \cup [\mathrm{supp}(\tilde{\mu}_k) \setminus \mathrm{supp}(\mu_k)]$ consists of isolated points.

Definition 4.1 *Let j be fixed, $1 \le j \le m-1$. We define*

$$\{\hat{\mu}_k\}_{k=1}^{m} := \{\mu_1, \ldots, \mu_{j-2}, f_{jj}\mu_{j-1}, f_{j+1,j}\tilde{\mu}_j, \ldots, f_{mj}\tilde{\mu}_{m-1}, \tilde{\mu}_m\}, \tag{27}$$

where $\tilde{\mu}_k$ is the measure associated with $1/f_{kj}$ as in (26).

Let $\{\hat{f}_{ki}\}_{k=0,\ldots,m; i=1,\ldots,k+1}$, be the family of functions in the Nikishin system defined by the measures $\{\hat{\mu}_k\}_{k=1}^{m}$, as in (9). Note that there are $m-1$ different auxiliary systems, and for each system j is fixed, $1 \le j \le m-1$. The following theorem corresponds to Theorem 4.1 in the case of R_m.

Theorem 4.4 (cf. [Driver K. and Stahl H. 3], Section 2). *Let j be fixed, $1 \le j \le m-1$. Assume that*

$$n_k \le n_j, \quad k = 1, \ldots, m-1, \quad n_m < n_j. \tag{28}$$

Then there exist constants $c_{j+1}, \ldots, c_{m-1} \in \mathbb{R}$ such that

$$\int t^l R_j(t) \hat{f}_{m-2,k}(t)\, d\tilde{\mu}_{m-1}(t) = 0, \qquad l = 0, \ldots, n_k - 1, \tag{29}$$

$$k = 1, \ldots, j-1 \tag{30}$$

$$\int t^l R_j(t) \left[\hat{f}_{m-2,k-1}(t) + c_k \hat{f}_{m-2,k}(t)\right] d\tilde{\mu}_{m-1}(t) = 0, \quad l = 0, \ldots, n_k - 1, \quad k = j+1, \ldots, m-1, \tag{31}$$

$$\int t^l R_j(t)\, d\tilde{\mu}_{m-1}(t) = 0, \qquad l = 0, \ldots, n_m - 1. \tag{32}$$

We observe that the orthogonality relations for R_j are different for the indices preceding n_j and those that follow the j-th component of the multi-index. In the case of $j = m$, this problem does not arise since all the indices precede n_m and this highlights the fact that we have more complete results for the m-th remainder function R_m, than we do for $R_1, \ldots, R_{m-1}$.

Let the sequence N_d of multiindices be defined as

$$N_d := \{n = (n_1, \ldots, n_m) \in \mathbb{N}^m \mid n_1 = \cdots = n_{m-1} = n_m + 1 = s \in \mathbb{N}\}. \tag{33}$$

i.e. N_d is the diagonal in $\mathbf{N}^m$ minus one in the last component. It turns out that for this sequence we are able to achieve the most complete results in our investigation. The best we can prove with respect to the zeros of $R_1, \ldots, R_{m-1}$ is

Theorem 4.5 (cf. [Driver K. and Stahl H. 3], Section 2). *We assume that N_d is defined by (33) and $n = (n_1, \ldots, n_m) \in N_d$.*

(i) Each function R_j, $j = 1, \ldots, m-1$, has at least $|n| - n_j - \epsilon_j$ simple zeros in the interval $\text{Int}(I(\mu_{m-1}))$, *where $\epsilon_{m-1} = 0$, $\epsilon_j = 1$ for $j = 1, \ldots, m-2$.*

(ii) Each R_j, $j = 1, \ldots, m-1$, has a zero of order at least $n_j + 1$ at infinity.

(iii) There is an additional zero of order one for each R_j, $j = 1, \ldots, m-2$, in $\overline{\mathbb{R}} \setminus I(\mu_m)$. If this zero lies at infinity, then R_j has a zero of order $n_j + 2$ at infinity, otherwise R_j has $|n| - n_j$ simple zeros in $\mathbb{R} \setminus I(\mu_m)$.

Remark We do not know whether the additional zero mentioned in part (iii) can really lie outside of $I(\mu_{m-1})$. Note that this question is the difference between Theorem 4.1 and 4.5.

In connection with the zeros of the remainder functions $R_1, \ldots, R_m$ we make the following conjecture:

Conjecture 4.6 *Let $n = (n_1, \ldots, n_m) \in N_d$, then the $|n| - n_j$ zeros of R_j, $j = 1, \ldots, m-1$, in $\mathbb{R} \setminus I(\mu_m)$ interlace with the $|n| - n_m$ zeros of R_m on $Int(I(\mu_{m-1}))$.*

At least in the case $m = 2$ the conjecture is true (cf. [Driver K. and Stahl H. 2], Section 4).

The proof of Theorem 4.5 relies heavily on Theorem 4.4 but both proofs are too long to be given here. We refer the reader to [Driver K. and Stahl H. 3]. Our final theorem is the analogue of Theorem 4.3 for all $j = 1, \ldots, m$ but now restricted to $n \in \mathbf{N}_d$.

Let the zeros of R_j be denoted by y_{ji}, $j = 1, \ldots, m$, $i = 1, \ldots, |n| - n_j - \delta_j$, $\delta_m = \delta_{m-1} = 0$, $\delta_j = 1$ or 0 for $j = 1, \ldots, m-2$, depending on whether the additional zero of Theorem 4.5 (iii) lies at infinity or in $\mathbb{R} \setminus I(\mu_m)$. Define

$$w_{m-1,j}(z) := \prod_{i=1}^{s_j} (z - y_{ji}), \quad s_j := |n| - n_j - \delta_j, \quad j = 1, \ldots, m. \tag{34}$$

Comparing (34) with (22) shows that $w_{m-1,m} = w_{m-1}$.

Theorem 4.6 (cf. [Driver K. and Stahl H. 3], Section 2). *Let* $n = (n_1, \ldots, n_m) \in N_d$ *and* N_d *be defined by (33). Then*
(i) we have

$$\int x^l Q_0(x) \frac{f_{m-1,j}(x)}{w_{m-1,j}(x)} d\mu_m(x) = 0, \quad l = 0, \ldots, |n| - 1, \; j = 1, \ldots, m, \tag{35}$$

(ii) and for the approximation error in the jth *component, we obtain the representation*

$$\left(f_j - \frac{Q_j}{Q_0}\right)(z) = \frac{w_{m-1,j(z)}}{(gQ_0)(z)} \int \frac{(gQ_0)(x) f_{m-1,j}(x)}{w_{m-1,j}(x)(x-z)} d\mu_m(x), \; j = 1, \ldots, m, \tag{36}$$

with an arbitrary $g \in \mathcal{P}_{|n|} \setminus \{0\}$.

Theorem 4.6 will be the basis for the investigation of the convergence of simultaneous rational approximants to Nikishin Systems. In [Bustamente J., Lopez G.,] it has been proved that diagonal sequences of simultaneous rational approximants $(Q_1/Q_0, \ldots, Q_m/Q_0)$ converge locally uniformly in $\overline{\mathbb{C}} \setminus I(\mu_m)$ in each component to $(f_1, \ldots, f_m)$. However, error formula (36) will allow us to determine the exact rate of convergence. Its determination involves a potential-theoretic equilibrium problem related to vector potentials, which were introduced and used in [Gonchar A.A. and Rakhmanov E.A.,1981] and [Gonchar A.A. and Rakhmanov E.A.,1985]. The investigation will be the subject of a forthcoming paper. A non-optimal estimate of the rate of convergence can be found in [Driver K. and Stahl H. 2], Section 2.4.

References

Angelesco A.: 1918, Sur l'approximation simultanée de plusieurs *C.R. Acad, Sci.* Paris 167 (1918), 629–631.

Aptekarev A.I., Stahl H.: 1992, Asymptotics of Hermite-Padé polynomials. *Progress in Approximation Theory*, eds. A.A. Gonchar and E.B. Saff, Springer Verl. (1992), 127–167.

Bustamente J., Lopez G.: Hermite-Padé approximation to a Nikishin type system of analytic functions, will appear in *Mat. Sbornik.*

Driver K. and Stahl H. 1: Normality in Nikishin systems. Will appear in *Indagationes Mathematicae.*

Driver K. and Stahl H. 2: Simultaneous rational approximants to Nikishin systems I. *Manuscript.*

Driver K. and Stahl H. 3: Simultaneous rational approximants to Nikishin systems II. *Manuscript.*

Gonchar A.A. and Rakhmanov E.A.: 1981, On the convergence of simultaneous Padé approximants for systems of Markov type functions. *Trudy Mat. Inst. Steklov.* **157** (1981), 31–48; Engl. transl.: *Proc. Stekl. Math. Inst.* Issue **3** (1983), 31–50 .

Gonchar A.A. and Rakhmanov E.A.: 1985, The equilibrium problem for vector potentials. *Uspekhi Mat. Nauk* **40** (1985), 155–156 (Russian).

Nikishin E.M.: 1979, A system of Markov functions. *Vestnik Moskov. Univ. Ser. I Mat. Mekh.* 1979, 60–63; Engl. transl.: *Moskov Univ. Math. Bull.* **34** (1979).

Nikishin E.M.: 1980, On simultaneous Padé approximants. *Mat.Sb.* **113 (155)** (1980), 499–519; Engl. transl.: *Math. USSR Sb.* **41 (4)** (1982), 409–425.

Stahl H. and Totik,V.: 1992, General Orthogonal Polynomials, Vol. **43**, *Encyclopedia of Math. and Its Appl.*, Cambridge University Press 1992.

RATIONAL

INTERPOLATION

THE MULTIPOINT PADÉ TABLE AND GENERAL RECURRENCES FOR RATIONAL INTERPOLATION

MARTIN H. GUTKNECHT
*Interdisciplinary Project Center for Supercomputing, ETH Zurich, ETH-Zentrum, CH-8092 Zurich, Switzerland**

Abstract. We first review briefly the Newton-Padé approximation problem and the analogous problem with additional interpolation conditions at infinity, which we call multipoint Padé approximation problem. General recurrence formulas for the Newton-Padé table combine either two pairs of Newton-Padé forms or one such pair and a pair of multipoint Padé forms. We show that, likewise, certain general recurrences for the multipoint Padé table compose two pairs of multipoint Padé forms to get a new pair of multipoint Padé forms. We also discuss the possibility of superfast, *i.e.*, $O(n \log^2 n)$ algorithms for certain rational interpolation problems.

Key words: Newton-Padé approximation, multipoint Padé approximation, rational interpolation, recurrences, fast algorithms

0. Introduction

The rational interpolation problem, which can be formulated in various ways, is readily reduced to a particularly structured linear system of equations. For a rational interpolant with denominator degree n, one can derive, for example, a homogeneous $n \times (n+1)$ system with a divided-difference matrix (see [38], [45], [23]), which contains higher divided differences in each column, or a homogeneous system with a Löwner matrix [13], which contains first divided differences, or yet other structured systems. But, in any case, while the application of the Gauss algorithm would require $O(n^3)$ operations, there are recursive $O(n^2)$ algorithms for solving these structured systems. Most of the classical methods (see [36], [45], [22] for surveys), such as the determination of a Thiele continued fraction, only work under genericness assumptions, whereas the Kronecker algorithm [33], [45], [2] and the generalizations of the Thacher-Tukey algorithm and the Thiele fraction proposed by Graves-Morris and Hopkins [22], [21] and by Werner [46], [47] are generally applicable in exact arithmetic, *i.e.*, are so-called reliable. The latter is also true for the more recent recurrences of Antoulas and Anderson [3] and the algorithms of Gutknecht [25], Van Barel and Bultheel [42], [44], and Beckermann [4], which are all closely related to Werner's algorithm. However, if subject to roundoff, all these methods are numerically unstable in cases of near-singularity of some linear systems that belong to rational interpolants of lower degrees constructed during the recursive process.

Therefore, in [23] we described new, more general recurrences that allow us to avoid ill-conditioned intermediate interpolants. These recurrences are modeled after those of the Cabay-Meleshko algorithm [10], [37] for computing Padé approximants,

* Email: `mhg@ips.ethz.ch`

A. Cuyt (ed.), Nonlinear Numerical Methods and Rational Approximation II, 109–135.

which was shown to be (weakly) stable. But whereas the recurrences of Cabay and Meleshko follow a diagonal of the Padé table, those in [23] are for the Newton-Padé table and allow us to move in any direction in which $m + n$ does not decrease. In particular, we can follow a pair of adjacent rows or antidiagonals of the table. While the general recurrence for a diagonal can be understood as combining two pairs of Newton-Padé forms, other paths require to combine a pair Newton-Padé forms with a pair of multipoint Padé forms. The latter represent multipoint Padé approximants, by which we mean rational interpolants with some interpolation data prescribed at infinity. (We need to point out that, in contrast to our terminology introduced in [26] and [23], the notion multipoint Padé approximant is sometimes used as equivalent to Newton-Padé approximant.) Here, we review in condensed form the background of this result, present general recurrences for both Newton-Padé forms and multipoint Padé forms, draw a number of conclusions, and discuss the merits and pitfalls of this approach, including the possibility of superfast $O(n \log^2 n)$ algorithms. For more information on the background we refer the reader to [23] and the many references cited there.

Such general recurrences are the basis of look-ahead algorithms for rational and Padé approximation, and for the solution of the corresponding structured linear systems. The notion of a look-ahead procedure came up in connection with the unsymmetric Lanczos algorithm [39], [27], [24], [14], which is closely related to a diagonal recurrence in the Padé table and the fast solution of a Hankel system. A look-ahead algorithm requires, in addition to a general recurrence, a look-ahead strategy, *i.e.*, a rule for choosing the step size such that the recurrence remains stable. Although our theoretical results give some hints on how to choose such a strategy here, this topic is not yet discussed in detail. For the diagonal Padé recursion, a complete analysis was given in [10]. For various forms of Padé row recursions, a partial analysis is given in [28], [30], [29]. These papers also reference several articles on fast Hankel and Toeplitz solvers that make use of the look-ahead philosophy in a somewhat different way. The case of rational interpolation still needs further investigation. However, some first numerical results are presented in [7]. There are extensions of this approach to vector-valued and matrix-valued Padé approximation and rational interpolation; see, *e.g.*, [5], [8], [9], [15], [41], [43], [40].

As in [25] and [23] we assume here for notational convenience that the interpolation data are given in Newton series form. This dispenses us of the need of using much more complicated notation for multiple interpolation points and the corresponding derivatives. However, we need to point out that our formulas, in particular the recurrences, are readily translated into formulas for different representations of the data. We do not consider the formal Newton series as a prime computational tool. In particular, if all interpolation points are distinct and well apart, it is much easier to represent the data as a set of function values than to transform these into divided differences. Only if some of the points are clustered, it is preferable to represent the data corresponding to these points by a Newton series, *i.e.*, by a Newton interpolation polynomial if there are only finitely many data, as one can assume in practice.

When using directly the function values one is free to choose the order in which the data are taken into consideration. With an appropriate strategy, related to pivoting in Gaussian elimination, Graves-Morris improved the numerical stability of the Thiele-Thacher-Tukey-Werner interpolation process considerably; see [21], [20], [47]. This idea could be combined with our approach, but, for simplicity, is here left out of consideration.

1. Newton-Padé Approximation

In some applications of rational interpolation the data are given as quotients of values. Treating the numerators and denominators of these quotients separately does not make the problem much more difficult. In fact, then it becomes more symmetric. It is also no extra effort to allow both the data and the points to be complex. For notational convenience we assume that the data for the interpolation points z_j $(j = 0, 1, \ldots)$ are given as a quotient of two (finite or infinite) formal Newton series:

$$h(z) := -\frac{f(z)}{g(z)}, \quad f(z) := \sum_{k=0}^{\infty} \phi_k t_k(z), \quad g(z) := \sum_{k=0}^{\infty} \gamma_k t_k(z), \tag{1.1}$$

where

$$t_0(z) :\equiv 1, \quad t_k(z) := \prod_{j=0}^{k-1} (z - z_j), \quad k = 1, 2, \ldots. \tag{1.2}$$

The two formal Newton series must be relatively prime in the sense that they do not vanish simultaneously at any z_k. While a single formal Newton series can be used to represent a series of finite complex function values and derivatives at arbitrary points in the complex plane, we can represent also poles of h with the representation (1.1). It is well known that the coefficients ϕ_k and γ_k in (1.1) are the kth *divided differences* at $z_0, \ldots, z_k$ of f and g, respectively.

We let $\mathcal{N}_Z$ be the set of all formal Newton series for the fixed sequence $Z := \{z_j\}_{j=0}^{\infty}$ of interpolation points. $\mathcal{N}_Z$ becomes a commutative algebra if addition, scalar multiplication, and multiplication are defined pointwise (using the Leibniz rule for the derivatives of the product) [45]. By $O(t_l)$ we denote any element of $\mathcal{N}_Z$ that has zeros at $z_0, \ldots z_{l-1}$, multiple ones being understood as zeros of formal derivatives. In other words, $g = O(t_l)$ if and only if the Newton series of g starts with the t_l-term. If $l \le 0$, the symbol $O(t_l)$ is understood as a void condition.

$\mathcal{P}_m$ is the space of complex *polynomials* p of exact degree $\partial p \le m$, and $\mathcal{R}_{m,n}$ is the set of *rational functions* $r = p/q$ with $p \in \mathcal{P}_m$, $q \in \mathcal{P}_n$, $q \ne 0$. If $r = p/q$ with relatively prime polynomials p and q of exact degrees ∂p and ∂q, r is said to be of *exact type* $(\partial p, \partial q)$ and to have the *defect*

$$\delta := \delta_{m,n} := \min\{m - \partial p, n - \partial q\} \tag{1.3}$$

in $\mathcal{R}_{m,n}$. If $\delta > 0$, r is called *degenerate* in $\mathcal{R}_{m,n}$. In particular, if $p = 0$, $\partial p := -\infty$, and hence $r = 0$ has exact type $(-\infty, 0)$ and defect n in $\mathcal{R}_{m,n}$.

We augment $\mathcal{R}_{m,n}$ by the constant ∞, which can be represented as $1/0$ and has exact type $(0, -\infty)$ and defect m in $\mathcal{R}_{m,n}$. Let $\overline{\mathcal{R}}_{m,n} := \mathcal{R}_{m,n} \cup \{\infty\}$. Note that then $r \in \overline{\mathcal{R}}_{m,n}$ if and only if $1/r \in \overline{\mathcal{R}}_{n,m}$.

Given a relatively prime pair $(f,g) \in \mathcal{N}_Z \times \mathcal{N}_Z$, the double sequence of *linearized rational interpolation problems* or *Newton-Padé approximation problems* for (f,g) consists in finding, for each pair $(m,n) \in \mathbf{N} \times \mathbf{N}$ the pairs $(p,q) \in \mathcal{P}_m \times \mathcal{P}_n$ for which

$$g(z)p(z) + f(z)q(z) = O(t_{m+n+1}(z)). \tag{1.4}$$

Any such pair $(p,q) \in \mathcal{P}_m \times \mathcal{P}_n$ is called an (m,n) *Newton-Padé form* (NPF) of (f,g). The associated rational function $r_{m,n} := p/q$ is an (m,n) *Newton-Padé approximant* (NPA) of (f,g) [17]. When writing p/q we do not assume that p and q are relatively prime, but we think of p/q as the rational function obtained after cancellation of common factors of p and q. Therefore p/q has to be distinguished from (p,q). Actually, there may be no relatively prime solution pair.

The pair (p,q), which only depends on the $(m+n)$th partial sums of the Newton series f and g, is clearly never uniquely determined. But it is easy to prove (see, *e.g.*, [23]) that for fixed m and n all solutions (p,q) of (1.4) yield the same rational function p/q, *i.e.*, the (m,n) NPA $r_{m,n}$ is well defined. We call it a *true* rational interpolant if $r_{m,n}$ interpolates $h = -f/g$ at the first $m+n+1$ data points, which is the case unless there exists no relatively prime pair (p,q).

The set of all NPFs for a particular interpolation problem is characterized in the following fundamental theorem, which is essentially due to Maehly and Witzgall [34]. A simple proof is given in [23].

Theorem 1. *The general solution $(p,q) \in \mathcal{P}_m \times \mathcal{P}_n$ of (1.4) is*

$$(p,q) = (\hat{p}_{m,n}\, s_{m,n}\, w,\ \hat{q}_{m,n}\, s_{m,n}\, w), \tag{1.5}$$

where $\hat{p}_{m,n}$, $\hat{q}_{m,n}$, and $s_{m,n}$ are polynomials that are uniquely determined up to a common scaling factor of $\hat{p}_{m,n}$ and $\hat{q}_{m,n}$, while $w \in \mathcal{P}_{\delta_{m,n}-\partial s_{m,n}}$ is arbitrary. $\hat{p}_{m,n}$ and $\hat{q}_{m,n}$ are relatively prime, $s_{m,n}$ is a monic divisor of t_{m+n+1} of degree $\partial s_{m,n} \le \delta_{m,n}$, and $\delta_{m,n}$ is the defect of $r_{m,n} = \hat{p}_{m,n}/\hat{q}_{m,n}$ in $\overline{\mathcal{R}}_{m,n}$. The zeros of $s_{m,n}$ are the points z_k where the pair $(\hat{p}_{m,n}, \hat{q}_{m,n})$ does not satisfy the interpolation conditions.

From this theorem we learn in particular that if $\partial s_{m,n} > 0$, then $(\hat{p}_{m,n}, \hat{q}_{m,n})$ is not a solution of the linearized problem, and, hence, $r_{m,n} = \hat{p}_{m,n}/\hat{q}_{m,n}$ is not a true interpolant of h. The points where $r_{m,n}$ does not interpolate up to the required order are called *unattainable*. These points are exactly the zeros of $s_{m,n}$. For this reason $s_{m,n}$ is called the *deficiency polynomial*. For a summary of the connections between the nonlinear rational interpolation problem and the Newton-Padé approximation problem, see [23] and the references cited there.

Among the other easy consequences of Theorem 1 there is the following Characterization Theorem.

Theorem 2. *(NPA Characterization Theorem) The function $r \in \overline{\mathcal{R}}_{m,n}$ with defect δ is the (m,n) NPA of $h = -f/g$ if and only if r interpolates h in at least $m+n+1-\delta$ data points.*

Associated with a double sequence of Newton-Padé problems defined by (1.4) there is a *Newton-Padé table* which covers a quarter of the (m,n)-plane and contains as its (m,n) entry the NPA $r_{m,n}$. For the recursive computation of rational

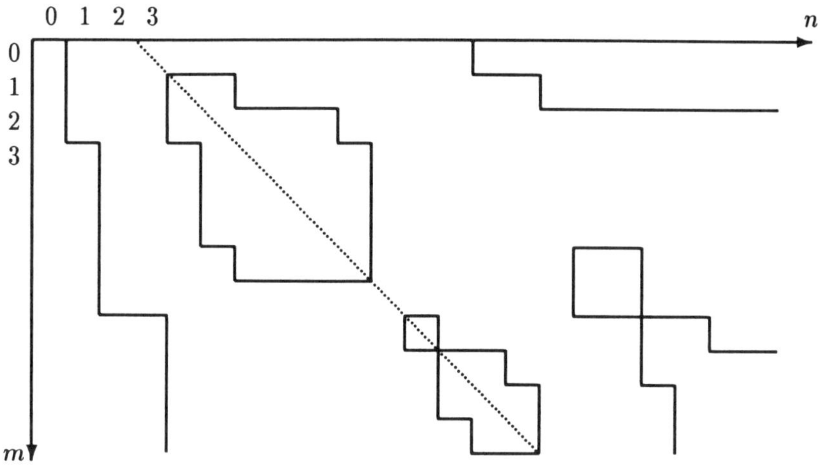

Fig. 1. A Newton-Padé table with four singular blocks: at left a block of the constant ∞, at the top right a block of the constant 0, in the middle a singular block consisting of three components (with the common diagonal dotted), and at the bottom right the infinite block of h (assumed to be rational).

interpolants one often follows some path in the Newton-Padé table. For example, this path may be a diagonal, an antidiagonal, or a staircase. (By a 'diagonal' we mean the main diagonal or any of its infinitely many upper or lower codiagonals.) However, in such algorithms special measures need to be taken if two interpolants are identical on the path. Hence, it is important to understand in which situations identical entries can occur. Moreover, for floating-point computations it is equally important to deal appropriately with nearly identical interpolants.

Sets of more than one equal entry are called *singular blocks* of the Newton-Padé table. The block structure of the Newton-Padé table was first investigated by Claessens [11]. The following formulation of the Block Structure Theorem was given in [25]. Additional results from [25] omitted here concern the characterization of the blocks of the zero and the infinity function, and of the infinite block occurring when h is rational; see Fig. 1.

Theorem 3. *(NP Block Structure Theorem) Let* $r = \hat{p}/\hat{q} \neq 0, \infty$ *be an NPA of* $h = -f/g$ *of exact type* $(\partial\hat{p}, \partial\hat{q})$. *Then the block of* r *in the Newton-Padé table of* h *is a (finite or infinite) union of squares whose upper left corners lie at or below the location* $(\partial\hat{p}, \partial\hat{q})$ *on the diagonal passing through this location.*

It was also shown in [25] that this statement is sharp, *i.e.*, any block satisfying the above description can actually occur. In general, a singular block in the Newton-Padé table needs not even be connected. It may consist of a finite or even infinite number of disjoint components, see Fig. 1. There is an easy interpretation for the form of these blocks. Every antidiagonal of the table is associated with an interpolation condition: *e.g.*, the one through $m + n = k =$ const is associated with the data $(z_k, h(z_k))$ if $z_k \neq z_i$ $(\forall i < k)$. If r is an NPA whose block contains at least one point of the above antidiagonal, then this block becomes broader or narrower at this antidiagonal depending on whether the new linearized interpolation condition is satisfied by r or not. For a more detailed discussion of this see [25]. Once the block gets narrower, r is certainly no longer a true interpolant. But it may even happen that an NPA r is for no index pair (m, n) a true interpolant. In fact this is true if and only if the $(\partial\hat{p}, \partial\hat{q})$ entry does not belong to the block of r. For a particular $h = -f/g$ and r, this situation can be avoided by reordering the data. However, this means considering another formal Newton series representation and, hence, another Newton-Padé table.

2. Multipoint Padé Approximation

While the rational interpolation problem of the previous section includes the case of prescribed poles at finite interpolation points, it does not allow any interpolation condition at infinity, except the prescription of the degrees m and n, which determine whether the NPA r has a pole, a zero, or a finite nonzero value at infinity. In this section we include interpolation conditions at infinity.

Again, we want to consider a double sequence of approximants. However, since conditions at infinity fix the difference between numerator and denominator degree, one of the two parameters has now a new meaning: the first, μ, determines the bias between using information at infinity and at finite interpolation points; the second, n, still denotes the nominal denominator degree of the rational function, *i.e.*, the maximum number of finite poles (with account of their multiplicity).

The data at the finite interpolation points are again assumed to be given in the Newton series form (1.1), although, as mentioned before, we suggest to use in practice the direct representation by function values and derivatives unless the interpolation points are clustered. Assuming that the function does not have a multiple zero at ∞, we prescribe its behavior there by a quotient of two formal power series in z^{-1},

$$\hat{h}(z) := -\frac{\hat{f}(z)}{\hat{g}(z)}, \quad \hat{f}(z) := \sum_{k=-\infty}^{\iota} \hat{\phi}_k z^k, \quad \hat{g}(z) := \sum_{k=-\infty}^{0} \hat{\gamma}_k z^k, \tag{2.1}$$

with $\iota \geq -1$ and $\hat{\phi}_\iota \neq 0$ or $\hat{\gamma}_0 \neq 0$. Of course, we could choose $\iota = 0$ or $\iota = -1$ without restricting generality, but we prefer to keep the parameter ι (iota) since we will encounter data of the above form for different values of ι, and, a priori, we will only know that $\hat{\phi}_\iota$ or $\hat{\gamma}_0$ does not vanish, but not which one.

It has been shown in [23] and will be further discussed below that for those general recurrences for the Newton-Padé table that do not just follow a diagonal, one has to solve rational interpolation problems whose data are of the form (1.1), (2.1).

We let $\mathcal{L}$ denote the linear space of *formal Laurent series* $y(z) = \sum_{k=-\infty}^{\infty} \eta_k z^k$ with complex coefficients. We need also subspaces of $\mathcal{L}$ that are of the form

$$\begin{aligned}
\mathcal{L}_{l:m} &:= \{y \in \mathcal{L};\ \eta_k = 0 \text{ if } k < l \text{ or } k > m\}, \\
\mathcal{L}_l &:= \mathcal{L}_{l:\infty} = \{y \in \mathcal{L};\ \eta_k = 0 \text{ if } k < l\}, \\
\mathcal{L}_m^\star &:= \mathcal{L}_{-\infty:m} = \{y \in \mathcal{L};\ \eta_k = 0 \text{ if } k > m\}.
\end{aligned}$$

In particular, $\mathcal{L}_0$ and $\mathcal{L}_0^\star$ are the sets of *formal power series* in z and $1/z$, respectively. When $y \in \mathcal{L}_m^\star$, we write $y(z) = O_-(z^m)$, and if additionally $y \notin \mathcal{L}_{m-1}^\star$, we may express this as $\partial y = m$. The set of all polynomials (*i.e.*, of formal power series with finitely many terms) is denoted by $\mathcal{P}$.

As in [23] we consider now the following double sequence of *multipoint Padé approximation problems*: given $(f, g) \in \mathcal{N}_Z \times \mathcal{N}_Z$, as in (1.1), and $(\hat{f}, \hat{g}) \in \mathcal{L}_\iota^\star \times \mathcal{L}_0^\star$ as in (2.1), let, for each index pair $[\mu; n] \in \mathbb{Z} \times \mathbb{N}$, the nonnegative integer $m := m(\mu; n)$ be defined by

$$m := \max\{\iota - \mu - 1, \iota + n, \mu\}, \tag{2.2}$$

and determine the pairs

$$(u, v) \in \begin{cases} \mathcal{L}_{\mu+n+1:\iota+n} \times \mathcal{P}_n = \mathcal{L}_{\iota+n-m:\iota+n} \times \mathcal{P}_n & \text{if } \mu \leq -n-1, \\ \mathcal{P}_{\iota+n} \times \mathcal{P}_n = \mathcal{P}_m \times \mathcal{P}_n & \text{if } -n-1 \leq \mu \leq \iota + n, \\ \mathcal{P}_\mu \times \mathcal{P}_n = \mathcal{P}_m \times \mathcal{P}_n & \text{if } \mu \geq \iota + n, \end{cases} \tag{2.3}$$

which satisfy the conditions $(u, v) \neq (0, 0)$ and

$$\hat{g}(z)u(z) + \hat{f}(z)v(z) = O_-(z^\mu), \tag{2.4a}$$

$$g(z)u(z) + f(z)v(z) = O(t_{\mu+n+1}(z)), \tag{2.4b}$$

where $t_j(z) \equiv 1$ if $j \leq 0$. Any such pair (u, v) is here called a $[\mu; n]$ *multipoint Padé form (MPF)* of $(\hat{f}, \hat{g}; f, g)$, and

$$r_{\mu;n}(z) := \frac{u(z)}{v(z)} \tag{2.5}$$

is referred to as the corresponding $[\mu; n]$ *multipoint Padé approximant (MPA)*.

It was shown in [23] that in analogy to the situation in Padé and Newton-Padé approximation, for fixed μ and n, the $[\mu; n]$ MPFs (u, v) are equivalent in the sense that they yield a uniquely determined rational function u/v. Hence, also the MPA $r_{\mu;n}$ is well defined.

Let us discuss the three cases in (2.3). In the first one, where $n \leq -\mu - 1 = m - \iota$, (2.4b) is void and (2.4a) implies $\iota + n - \mu = m + n + 1$ conditions at $z = \infty$ for the pair (u, v) with a total of $m + n + 2$ parameters. Hence, we have a one-point Padé approximation problem at infinity. In the third case, $\iota + n \leq \mu = m$, (2.4a) is void, while (2.4b) is identical to the condition (1.4) for an (m, n) NPA. Finally, in the second case, both (2.4a) and (2.4b) are in effect, and we have a *proper* multipoint Padé problem with $\iota + n - \mu$ conditions at $z = \infty$ and $\mu + n + 1$ conditions at finite interpolation points. Hence, there is a total of $\iota + 2n + 1$ conditions for $(u, v) \in \mathcal{P}_{\iota+n} \times \mathcal{P}_n$.

In summary, the three cases in (2.3) have the following interpretation:

$\mu \leq -n-1$:	Padé problem at $z = \infty$,	$m = \iota - \mu - 1$,
$-n-1 < \mu < \iota + n$:	proper multipoint Padé problem,	$m = \iota + n$,
$\iota + n \leq \mu$:	Newton-Padé problem,	$m = \mu$.

The MPAs for a data set $(\hat{f}, \hat{g}; f, g)$ can be collected in a *multipoint Padé table* $\{r_{\mu;n}\ ;\ [\mu; n] \in \mathbb{Z} \times \mathbb{N}\}$ that covers a half-plane. We let the n-axis point to the right and the μ-axis point to the bottom. If $\iota = 0$, the proper MPAs lie in a 90° sector with horizontal axis of symmetry between $\mu = -1$ and $\mu = 0$; cf. Fig. 2. Above this sector the table contains Padé approximants at infinity, below there are NPAs. If $\iota > 0$, the lower border of this sector lies farther down. If $\iota = -1$, it is farther up by one entry. In this case $r_{-1;0}(z) \equiv 0$, as is seen from (2.3)–(2.4b).

Note that n is the nominal denominator degree of $r_{\mu;n}$, but μ is *not* the numerator degree, except in the NPA sector of the table, where $\mu = m$. The number of parameters in the numerator is always $m + 1$.

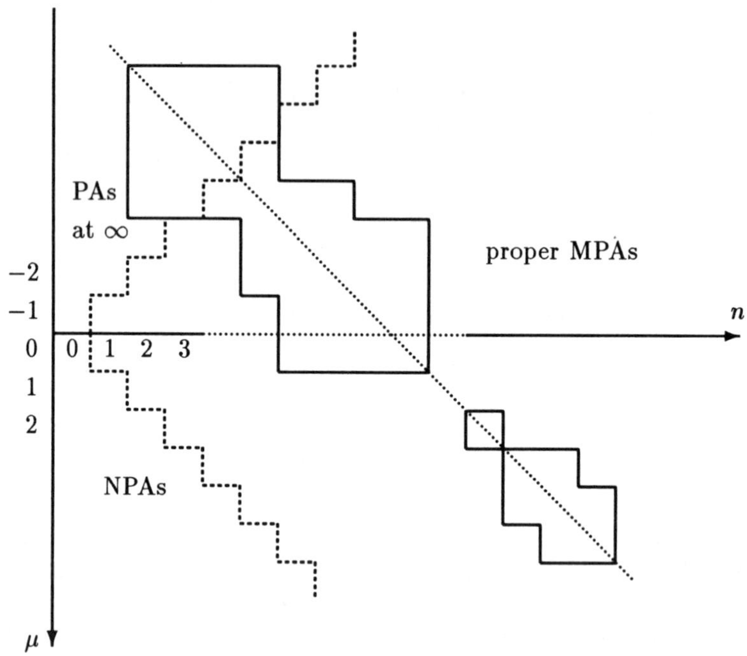

Fig. 2. The case $\iota = 0$ of a multipoint Padé table with its 90° sector of proper multipoint Padé approximants and the two adjoining sectors of Padé approximants at infinity and Newton-Padé approximants, which are separated by dashed staircases. Also shown is a typical singular block with its dotted diagonal.

Again, one can describe the set of solutions of the defining equations (2.2)–(2.4b). This result was given in [26]. We cite it from [23], where it was reformulated.

Theorem 1'. *In the case $-n - 1 < \mu < \iota + n$ of proper multipoint Padé approximation the general form of the MPFs (u, v) defined by (2.2)–(2.4b) is*

$$(u, v) = (\hat{u}_{\mu;n}\, s_{\mu;n}\, w,\ \hat{v}_{\mu;n}\, s_{\mu;n}\, w), \tag{2.6}$$

where $\hat{u}_{\mu;n}, \hat{v}_{\mu;n}$ are up to scaling uniquely determined relatively prime polynomials, and $s_{\mu;n}$ is a monic divisor of $t_{\mu+n+1}$ of degree $\partial s_{\mu;n} \leq \min\{\delta, \nu\}$, with $\delta := \delta_{\mu;n}$ (≥ 0) being the defect of $r_{\mu;n} := \hat{u}_{\mu;n}/\hat{v}_{\mu;n}$ in $\overline{\mathcal{R}}_{\iota+n,n}$, and

$$\nu := \nu_{\mu;m} := \mu - \partial(\hat{g}\hat{u}_{\mu;n} + \hat{f}\hat{v}_{\mu;n}) \ (\geq 0) \tag{2.7}$$

being the number of extra (linearized) interpolation conditions fulfilled at ∞ *by* $(\hat{u}_{\mu;n}, \hat{v}_{\mu;n})$. *Finally,* w *is an arbitrary nonzero polynomial of degree*

$$\partial w \le \min\{\delta, \nu\} - \partial s_{\mu;n}. \tag{2.8}$$

The zeros of the deficiency polynomial $s_{\mu;n}$ *are the finite interpolation points* z_k *(with appropriate multiplicity) at which the pair* $(\hat{u}_{\mu;n}, \hat{v}_{\mu;n})$ *does not satisfy the interpolation conditions. The deficiency at* ∞,

$$\hat{\sigma} := \hat{\sigma}_{\mu;n} := \begin{cases} 0 & \text{if } \nu \ge \delta, \\ \delta - \nu & \text{if } \nu < \delta, \end{cases} \tag{2.9}$$

indicates the number of (linearized) interpolation conditions at ∞ *that are not fulfilled by a pair* (u, v) *of the form* (2.6) *with* $\partial s_{\mu;n} + \partial w = \delta$. *It satisfies*

$$0 \le \hat{\sigma} \le \min\{n - \mu, \delta - \partial s_{\mu;n}\}. \tag{2.10}$$

To avoid the treatment of several cases we restricted Theorem 1' to proper MPFs, *i.e.*, to $-n-1 < \mu < \iota + n$. However, when the appropriate interpretations of void conditions are made, the theorem remains true for $\mu \le -n-1$ and $\mu \ge \iota + n$. In fact, then one has a one-point Padé approximation problem at ∞ or a Newton-Padé approximation problem, respectively, and one can therefore just refer to Theorem 1.

In view of a general MPA characterization theorem, it is useful to define the defect δ for all three types of MPAs. Assume r is any rational function that can be written as $r = \hat{u}/\hat{v}$ with $(\hat{u}, \hat{v})$ satisfying (2.3) and $\hat{u}$, $\hat{v}$ being relatively prime. In the case $\mu \le -n-1$, where $m = \iota - \mu - 1$ and thus $\iota + n - m \le 0$, we call $(\hat{u}, \hat{v})$ relatively prime if the polynomials $z^{\iota+m-n}\hat{u}(z)$ and $\hat{v}(z)$ are relatively prime. We define the $[\mu; n]$-*defect* of r by

$$\delta := \delta_{\mu;n} := \max\{\mu - \partial\hat{u}, \iota + n - \partial\hat{u}, n - \partial\hat{v}\}. \tag{2.11}$$

For proper MPAs (*i.e.*, if $-n-1 \le \mu \le \iota + n$), this is just the defect of r in $\overline{\mathcal{R}}_{\iota+n,n}$. For NPAs (*i.e.*, for $\mu \ge \iota + n$, where $m = \mu$), it is the defect of r in $\overline{\mathcal{R}}_{m,n}$. Finally, for PAs at ∞ (*i.e.*, for $\mu \le -n-1$, where $m = \iota - \mu - 1$), $\delta_{\mu;n}$ is equal to the defect of the function $z \mapsto r(1/z)$ in $\overline{\mathcal{R}}_{m,n}$.

With this definition, the following analogue of Theorem 2 holds [23].

Theorem 2'. *(MPA Characterization Theorem) Let* $[\mu; n] \in \mathbb{Z} \times \mathbb{N}$, *let* (u, v) *satisfy* (2.3), *and let* δ *be the* $[\mu; n]$-*defect of* $r = u/v$. *Then* r *is the* $[\mu; n]$ *MPA if it satisfies at least* $\iota + 2n + 1 - \delta$ *of the* $\iota + 2n + 1$ *nonlinear interpolation conditions implied by* (2.4a)–(2.4b).

If the finite interpolation points z_k all coalesce at $z = 0$, our proper multipoint Padé approximation problem becomes a two-point Padé approximation problem. Then g and f are formal power series, and the term $O(t_{\mu+n+1})$ in (2.4b) becomes $O(z^{\mu+n+1})$. Such two-point Padé approximation problems for two formal power series $\hat{h} := -\hat{f}/\hat{g}$ and $h := -f/g$ in $1/z$ and z, respectively, have been investigated by several authors; see [23] for references and comments. The two-point analog of our multipoint Padé table is called *M-table*. It was shown by Cooper, Magnus, and McCabe [12] that in the M-table all singular blocks are either square or infinite, exactly as in the case of the classical Padé table. The only new feature is the possibility of a new type of an infinite block, a half-plane $\{(\mu, n) \in \mathbb{Z} \times \mathbb{N};\ n \geq n_0\}$. We showed in [23] that also in the multipoint Padé table the typical singular blocks have the same form as in the Newton-Padé table, except that the portion in the Padé approximation sector must be square; see Fig. 2. For the formulation of this Block Structure Theorem it is important to have the following characterization for the diagonal of a singular block [23].

Lemma. *Let $r = \hat{u}/\hat{v}$ be an MPA. Assume that $\hat{u} \neq 0$ and $\hat{v} \neq 0$ are relatively prime. Then those entries in the block of r where the $[\mu; n]$-defect δ satisfies*

$$\hat{g}(z)\hat{u}(z) + \hat{f}(z)\hat{v}(z) \equiv O_{-}(z^{\mu-\delta}) \tag{2.12}$$

lie on a particular diagonal of the multipoint Padé table. The case $\mu - \delta = -\infty$ is permitted and means that the upper edge of the square block and hence also its diagonal are at infinity.

We call the constant difference $\mu - \delta$ in (2.12) the *order at* ∞ of $(\hat{u}, \hat{v})$ and denote it by μ_∞. With this definition, the following theorem holds [23].

Theorem 3'. *(MP Block Structure Theorem) Let the data $(\hat{f}, \hat{g}; f, g)$ be of the form (1.1) and (2.1), with $\iota = 0$. Let $r = \hat{u}/\hat{v}$ be an MPA of $(\hat{f}, \hat{g}; f, g)$, and let μ_∞ be the order at ∞ of $(\hat{u}, \hat{v})$. Assume that $\hat{u} \neq 0$ and $\hat{v} \neq 0$ are relatively prime.*

Then the block of r in the multipoint Padé table of $(\hat{f}, \hat{g}; f, g)$ is a finite or infinite union of squares whose upper left corners lie at or below the location $[\mu_\infty, \partial\hat{v}]$ on the diagonal passing through this location. If not empty, the intersection of the block with the sector $\mu < -n - 1$ is equal to the intersection of a square with this sector; and the upper left corner of this square is then at $[\mu_\infty, \partial\hat{v}]$.

As for Theorem 3 the strength of this result is that it is best possible: again, one can prove that any block of the described form can appear; see [23].

3. Recursive Computation of NPAs and MPAs: Basic Ideas

The general recurrence relations for Newton-Padé approximants (NPAs) and multipoint Padé approximants (MPAs) that we want to discuss in the rest of this paper are modeled after the recurrences applied by Cabay and Meleshko [37], [10] for the stable computation of a diagonal sequence of Padé approximants. Under more restrictive assumptions these recurrences were used before by Gragg, Gustavson, Warner, and Yun [18], who modified the recurrences of Brent, Gustavson, and Yun [6]. Here we consider NPAs and MPAs instead of Padé approximants, and a more general type of recurrence that allows us to proceed in other directions too. Whereas some of the other algorithms that have been proposed for singular tables take special small steps to follow the border of the block when encountering a singular one, here the idea is to take a large step crossing the block (as in [25]) or even several blocks. The option of jumping over several blocks allows us to avoid ill-conditioned interpolants that could occur as intermediate results. This is a necessity for a numerically stable algorithm. Moreover, this option allows us to apply recursive doubling and opens up the possibility of superfast $O(n \log^2 n)$ algorithms. Such algorithms were suggested in [6] for staircase Padé sequences, in [8] for diagonal matrix Padé sequences, and in [28] for row Padé sequences. Here we will point out that for a certain class of rational interpolation problems we can still attain the same low complexity.

A crucial tool is the usage of *basic pairs* of NPFs and MPFs. According to Theorems 1 and 1', NPFs and MPFs are determined uniquely up to scaling if and only if in (1.5) and (2.6), respectively, the polynomial w is of degree 0, *i.e.*, a constant. It can be seen that w is a constant if and only if, in the respective table, we are either at a normal entry or at a position at the border of a singular block. In view of the possible form of singular blocks, the latter conditions are fulfilled if the previous or the following entry on the same diagonal is different. Hence, we are going to use pairs of NPFs (and MPFs, respectively) with the property that they are upper left and lower right neighbors, respectively, of each other, but belong to different blocks. This then implies that these NPFs are unique up to scaling, and that the denominators (and also the numerators) are relatively prime except for the possibility of a common zero at an interpolation point, in which case either the corresponding NPAs are not true interpolants or the data function g (or f, respectively) has a zero there also. The second member of these pairs will be called *weakly regular.* (We would call it regular if we knew additionally that the NPA is a true interpolant, but we will not have this knowledge in general.) Instead of basic pairs consisting of a regular NPA and its upper left neighbor, one could also use basic pairs consisting of an NPA (then called weakly column-regular) and its upper neighbor, or an NPA (then called weakly row-regular) and its left neighbor, or an NPA and its lower left neighbor. The last type of basic pairs appears implicitly in the classical Kronecker algorithm [33], [45], [2].

Basic pairs appear at many places in the literature, at least implicitly; see, *e.g.*, [1], [3] (Assumption (3.8)), [4], [42], [44]. In the Padé case they can be traced back to the Euclidean algorithm [6], [18] and are related to Bézoutiants, the Christoffel-Darboux formula, and inversion formulas for Hankel and Toeplitz matrices; see, *e.g.*, [16], [19], [31], [32]. Heinig and Rost [31] call the denominators of the basic pairs *fundamental solutions.* Implicitly, the basic pairs considered here also come up in one version of the methods of Werner [46], [47] and Gutknecht [25] that produces a generalization of the Thiele fraction called a diagonal G-fraction in [25], because its convergents are the distinct entries on a particular diagonal of the Newton-Padé table. In all these situations, one proceeds from one basic pair to the next one, so that the two pairs have a common member. However, the full algorithmic power of basic pairs becomes apparent in conjunction with the mentioned recurrences that allow us to jump over arbitrary many blocks, like in [37], [10], [28], [23].

Antoulas and Anderson, who consider rational interpolants with $\max\{\partial p, \partial q\}$ chosen minimal, give in Lemma 3.2 of [3] a recurrence that is said to allow one to go from any pair of such interpolants to any new interpolant satisfying the union of the previously fulfilled interpolation conditions, plus, possibly, some additional ones. However, the proof of the formula is not given, and, in any case, its application becomes complicated if many new data are added at the same interpolation point, as one then needs to compute higher derivatives of a quotient. In their Theorem 3.9, whose proof is said to serve as a guideline for the one of their Lemma 3.2, the authors turn then to what we call a basic pair and add only one new interpolation condition at a time.

Briefly, the general recurrence from [23] (stated as Theorem 9 below) says that a basic pair of NPFs of types $(m-1, n-1)$ and (m, n) can be updated to a basic pair with types $(m+\kappa-1, n+k-1)$ and $(m+\kappa, n+k)$ satisfying additional $\kappa+k$ interpolation conditions. For such a step of length $\kappa+k$ one needs to compute the so-called residuals of the previous pair, and a pair of $[\kappa; k-1]$ and $[\kappa+1; k]$ MPFs. Here, κ is restricted to $-k \leq \kappa \leq k$. Hence, viewed from the weakly regular (m, n) NPF (p_n, q_n) the newly constructed weakly regular $(m+\kappa, n+k)$ NPF (p_{n+k}, q_{n+k}) lies in the Newton-Padé table in a 90° sector with horizontal axis, see Fig. 3. If we want to move instead in a 90° sector with vertical axis we can apply the same theorem, but with (f, g), (p_n, q_n), (m, n) replaced by (g, f), (q_n, p_n), (n, m). In the case where $\kappa = k$ (and also when $\kappa = k-1$), the two MPFs are NPFs. If, moreover, all interpolation points coalesce at 0, the MPFs are just Padé forms, and the recurrence can be seen to be equivalent to the one applied by Cabay and Meleshko [10].

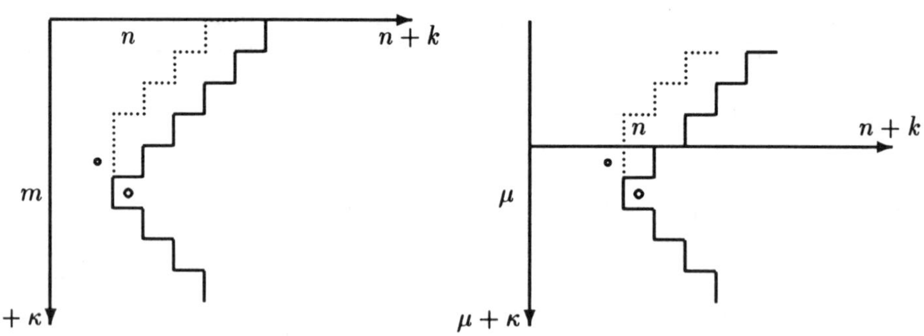

Fig. 3. At left: the index range of the new $(m+\kappa, n+k)$ NPA (solid line) and its upper left neighbor (dotted line) that can be constructed according to Theorem 9 below. At right: the index range of the new $[\mu+\kappa; n+k]$ MPA (solid line) and its upper left neighbor (dotted line) that can be constructed according to Theorem 9'.

For the multipoint Padé table we present in Theorem 9' below a completely analogous recurrence: starting from a basic pair consisting of a $[\mu-1, n-1]$ and a $[\mu, n]$ MPF for $(\hat{f}, \hat{g}; f, g)$, we can construct a basic pair consisting of a $[\mu+\kappa-1, n+k-1]$ and a $[\mu+\kappa, n+k]$ MPF for $(\hat{f}, \hat{g}; f, g)$ by computing a basic pair consisting of a $[\kappa, k-1]$ and a $[\kappa+1, k]$ MPF for another multipoint Padé problem that is made up of the residuals of the first pair. Again, κ is restricted to $-k \leq \kappa \leq k$, so that we can move in the multipoint Padé table in a 90° sector with horizontal axis whose cusp is at $[\mu; n]$; see Fig. 3.

The natural way to apply these recurrences consists in computing NPFs (or MPFs) with slowly increasing n, leaving out values of n where the NPF (or MPF) and its upper left neighbor are not numerically well conditioned or not sufficiently independent. However, there is the other option of applying *recursive doubling.* Assume we want to compute some (n, n) NPA, where n is large. For simplicity, let n be a power of 2 and assume that all the entries on the main diagonal of the Newton-

Padé table are distinct, and, hence, regular. According to the NPA recurrence, we can split up the problem and first compute a pair consisting of a $(n/2-1, n/2-1)$ and a $(n/2, n/2)$ NPF and then, for a multipoint Padé problem that depends on these solutions, compute a pair consisting of a $[n/2; n/2-1]$ MPF and a $[n/2+1; n/2]$ MPF. (Actually, here these MPFs can be understood as NPFs.) To fulfill these two different tasks we can again apply, for each one, the same kind of splitting, using the NPA recurrence for the first task and the MPA recurrence for the second one. This can be continued until we get to problems where $n = 1$, and thus the solution becomes trivial. Unfortunately, this application of the *divide-and-conquer* strategy has a weak point: since the data for each subproblem depend on the solution of the previous problem, the subproblems cannot be solved in parallel. Hence, there is no gain from the point of view of parallel computing. The divide-and-conquer strategy is here not applied in its form amenable to parallel computation, but in its original sense, where an enemy is split up and its sections are conquered one after another. Nevertheless there can be some gain here too, since the original problem is split up repeatedly until each subproblem becomes trivial. If the work for the reduction is sufficiently small, the computational cost for solving the original problem may be smaller than when applying the straightforward recurrence. We will return to this question in §6.

4. General Recurrence Relations for NPAs

Assuming that, for the sequence of NPFs we are going to construct, the numerator degree $m = m(n)$ is a function of the denominator degree n, we write $\{(p_n, q_n)\}$ instead of $\{(p_{m(n),n}, q_{m(n),n})\}$. For example, $m(n) = n + l$, with fixed l, for a diagonal sequence, and $m(l) = l$, with fixed l, for a row sequence in the Newton-Padé table. (Note that m has here another meaning than in §2.) Actually, there is normally no point in computing the full sequence $\{(p_n, q_n)\}$; only the subsequence of weakly regular NPFs, or rather, the subsequence of well-conditioned weakly regular NPFs will be constructed, together with their upper left neighbors

$$(\grave{p}_n, \grave{q}_n) := (p_{m(n)-1,n-1}, q_{m(n)-1,n-1}). \tag{4.1}$$

(p_n, q_n) and $(\grave{p}_n, \grave{q}_n)$ can be normalized by letting the sum of the squares of the coefficients of each pair be 1. Other options for normalization follow from Lemma 4 below.

When $n = 0$ or $m(n) = 0$, the upper left neighbor $(\grave{p}_n, \grave{q}_n)$ lies outside the Newton-Padé table, but the following definitions prove to be useful:

$$\begin{array}{llll} \grave{p}_0(z) := t_{m-1}(z), & \grave{q}_0(z) :\equiv 0, & \text{if } n = 0, m > 0; \\ \grave{p}_0(z) :\equiv 0, & \grave{q}_0(z) := t_{n-1}(z), & \text{if } n > 0, m = 0; \\ \grave{p}_0(z) :\equiv 1, & \grave{q}_0(z) :\equiv 0, & \text{if } n = m = 0 \text{ and } q_0 \neq 0; \\ \grave{p}_0(z) :\equiv 0, & \grave{q}_0(z) :\equiv 1, & \text{if } n = m = 0 \text{ and } q_0 = 0; \end{array} \tag{4.2}$$

We could use the fourth definition whenever $p_0 \neq 0$ (and, actually, we should when $|q_0|$ is small). That would leave a free choice between the third and the fourth definition when $p_0 \neq 0$ and $q_0 \neq 0$.

These special definitions allow the following interpretation: we can extend the Newton-Padé table to the left by adding an infinite block of the constant ∞, and we can extend it above by adding an infinite block of the constant 0, as suggested in [19] for the Padé table. The quadrant $m < 0$, $n < 0$ can be filled with either 0 or ∞, but if $r_{0,0} = 0$ (or ∞), it is preferable to choose ∞ (or 0, respectively).

We call the NPF $(p_n, q_n) := (p_{m(n),n}, q_{m(n),n})$ *weakly regular* if

$$\frac{p_n}{q_n} \neq \frac{\grave{p}_n}{\grave{q}_n}, \quad i.e., \quad \grave{p}_n q_n - p_n \grave{q}_n \neq 0 \in \mathcal{P}. \tag{4.3}$$

Likewise, we call (p_n, q_n) *weakly row-regular* if the $(m(n), n-1)$ and the $(m(n), n)$ NPA differ, and we call it *weakly column-regular* if the $(m(n)-1, n)$ and the $(m(n), n)$ NPA differ. According to the Block Structure Theorem, (p_n, q_n) is weakly regular if and only if it is weakly row-regular or weakly column-regular.

The *residual* e_n of the NPF (p_n, q_n) is defined by

$$g(z)p_n(z) + f(z)q_n(z) = t_{m(n)+n+1}(z)e_n(z). \tag{4.4}$$

This residual can be written as a formal Newton series for the points $z_{m(n)+n+i}$, $i = 1, 2, \ldots$, *i.e.*, as a series of the polynomials

$$t_0^{(m+n)}(z) :\equiv 1, \quad t_k^{(m+n)}(z) := \prod_{i=1}^{k}(z - z_{m(n)+n+i}), \quad k = 1, 2, \ldots. \tag{4.5}$$

In practice, this Newton series can be replaced by some other representation of the data determining the residual. The definitions of the residual e_n of (p_n, q_n) and the residual $\grave{e}_n$ of $(\grave{p}_n, \grave{q}_n)$ can be summarized in

$$[g \;\; f]\begin{bmatrix} \grave{p}_n & p_n \\ \grave{q}_n & q_n \end{bmatrix} = t_{m(n)+n-1}[\grave{e}_n \;\; \tau_n e_n], \tag{4.6}$$

where

$$\tau_n(z) := (z - z_{m(n)+n-1})(z - z_{m(n)+n}) = t_2^{(m+n-2)}(z). \tag{4.7}$$

When $n = 0$ and $m > 0$, (4.6) still holds with

$$\grave{e}_0 = g, \quad e_0 = (gp_0 + fq_0)/t_{m+1}, \tag{4.8}$$

where q_0 is a nonzero constant that can be normalized to 1. Moreover, (4.3) holds when $q_0 \neq 0$. Analogous statements can be made for the other special cases from (4.2).

The following three lemmas were proved in [28] for $m, n > 0$. Here, we cover also the cases $n = 0$, $m > 0$ and $n > 0$, $m = 0$. The latter always follows from the former by symmetry. We leave out the case $m = n = 0$ because it requires special treatment.

Lemma 4. *Let (p_n, q_n) be an $(m(n), n)$ NPF (with $n > 0$ or $m(n) > 0$), and let $\grave{e}_n$ be the residual of an $(m(n)-1, n-1)$ NPF $(\grave{p}_n, \grave{q}_n)$. Then the following statements are equivalent:*

(i) (p_n, q_n) is weakly regular, i.e., *(4.3) holds;*
(ii) $\grave{e}_n(z_{m(n)+n}) \neq 0$ and $(p_n(z_{m(n)+n}) \neq 0$ or $q_n(z_{m(n)+n}) \neq 0)$;
(iii) $\grave{e}_n(z_{m(n)+n-1}) \neq 0$ and $(p_n(z_{m(n)+n-1}) \neq 0$ or $q_n(z_{m(n)+n-1}) \neq 0)$.

Proof for $n = 0$, $m > 0$. In this case all three statements hold if and only if $q_0 \neq 0$. First, $\grave{p}_0 q_0 - p_0 \grave{q}_0 = t_{m-1} q_0$. Second, $\grave{e}_0 = g$, and $g(z_{m-1}) = 0$ or $g(z_m) = 0$ imply by (1.4) that $q_0 = 0$. Conversely, $q_0 = 0$ implies by (1.4) that, for $i = 0$ and $i = 1$ either $g(z_{m-i}) = 0$ or $p(z_{m-i}) = 0$. □

Lemma 5. *If* (p_n, q_n) *is a weakly regular* $(m(n), n)$ *NPF (with* $n > 0$ *or* $m(n) > 0$*), then*

(i) $\partial q_n = n$ *and* $\partial \grave{p}_n = m(n) - 1$

or

(ii) $\partial p_n = m(n)$ *and* $\partial \grave{q}_n = n - 1$.

Proof for $n = 0$, $m > 0$. From the previous proof we know that $q_0 \neq 0$. But then, in view of (4.2), statement (i) is clearly true. □

Lemma 6. *Let* (p_n, q_n) *be an* $(m(n), n)$ *NPF (with* $n > 0$ *or* $m(n) > 0$*); then*

$$\det \begin{bmatrix} \grave{p}_n & p_n \\ \grave{q}_n & q_n \end{bmatrix} = \delta_n t_{m(n)+n-1}, \tag{4.9}$$

where δ_n *is the leading coefficient of* $\grave{p}_n q_n - \grave{q}_n p_n \in \mathcal{P}_{m(n)+n-1}$, *which does not vanish if and only if* (p_n, q_n) *is weakly regular.*

Proof for $n = 0$, $m > 0$. The determinant in (4.9) is equal to $t_{m-1} q_0$; hence, (4.9) holds with $\delta_0 = q_0$, which in this case is also the leading coefficient of $\grave{p}_0 q_0 - \grave{q}_0 p_0 \in \mathcal{P}_{m-1}$. □

Corollary 7. *If the* $(m(n), n)$ *NPF* (p_n, q_n) *(with* $n > 0$ *or* $m(n) > 0$*) is weakly regular, then any common zero of the polynomials* $\grave{p}_n$ *and* p_n *is a zero of* f *or an unattainable interpolation point of* p_n/q_n. *In particular, if* p_n/q_n *is a true* $(m(n), n)$ *interpolant and* f *does not vanish at any interpolation point, then* $\grave{p}_n$ *and* p_n *are relatively prime. These statement remain true when* $\grave{p}_n$, p_n, *and* f *are replaced by* $\grave{q}_n$, q_n, *and* g, *respectively.*

Proof. Assume $m, n > 0$ first. By (4.9), if (p_n, q_n) is weakly regular, any common zero of $\grave{p}_n$ and p_n must be a zero of $t_{m(n)+n-1}$. On the other hand, since $\partial w = 0$ in the factorization (1.5) of (p_n, q_n) (because of uniqueness), such a zero must be either a zero of $s_{m,n}$, *i.e.*, an unattainable point, or a zero of $\hat{p}_{m,n}$, in which case it must be a zero of f; see (1.4). The statement about exchanging $\grave{p}_n$, p_n, and f for $\grave{q}_n$, q_n, and g holds by symmetry. If $n = 0$, a zero of $\grave{p}_n = t_{m-1}$ is by (4.4) also a zero of $g p_0 - f q_0$. Hence, a common zero of $\grave{p}_0$ and p_0 must be a zero of f unless $q_0 = 0$, in which case a zero of p_0 is an unattainable point. □

Finally, we need to justify the later usage of the pair of residuals $(e_n, \grave{e}_n)$ as data of a Newton-Padé or multipoint Padé problem.

Lemma 8. *Assume that* (p_n, q_n) *is a weakly regular* $(m(n), n)$ *NPF (with* $n > 0$ *or* $m(n) > 0$*), and let* $Z(n) := \{z_j \in Z \,;\, j \geq m(n) + n - 1\}$. *Then* $\grave{e}_n$ *and* $\tau_n e_n$ *are relatively prime elements of* $\mathcal{N}_{Z(n)}$.

Proof. In [23] we established this lemma under the additional assumption of distinct interpolation points. Here we give a short proof that does not require this assumption. Let $n > 0$, and let (p_n, q_n) be a weakly regular NPF. Then, after normalizing an appropriate coefficient to 1, the other coefficients of these two polynomials and likewise those of $\grave{p}_n$ and $\grave{q}_n$ are determined by a nonsingular linear system that depends on f and g. Clearly, under small perturbations of the data, the coefficients and values of these polynomials and of t_{m+n-1} depend continuously on the data, and, hence, the same is true for $\grave{e}_n$ and $\hat{\tau}_n e_n$. (Note that under such a perturbation t_{m+n-1} remains a factor of $g\grave{q}_n + f\grave{p}_n$, and t_{m+n+1} remains a factor of $gq_n + fp_n$.) By such small perturbations we can make that none of the points $z_j \in Z(n)$ is a zero of t_{m+n-1}. (Actually, we could keep the zeros of t_{m+n-1} fixed; but this is not required.) Applying Cramer's rule to (4.6) and making use of (4.9) we get, after canceling t_{m+n-1},

$$g = \frac{1}{\delta_n} \det \begin{bmatrix} \grave{e}_n & \tau_n e_n \\ \grave{q}_n & q_n \end{bmatrix}, \quad f = \frac{1}{\delta_n} \det \begin{bmatrix} \grave{p}_n & p_n \\ \grave{e}_n & \tau_n e_n \end{bmatrix}. \tag{4.10}$$

These formulas remain true if some points z_j converge to zeros of t_{m+n-1}. Therefore, clearly, at any point where both $\grave{e}_n$ and e_n vanish, f and g must vanish also, in contrast to the assumption that f and g are relatively prime. Finally, if $n = 0$ and $m > 0$, then, for $i \in \mathbf{N}^+$, $0 = \grave{e}_0(z_{m+i}) = g(z_{m+i})$ implies that $gp_0 + fq_0$ does not vanish at z_{m+i} since f and g are relatively prime and q_0 is a nonzero constant. □

Note that the formulas (4.10) can be summarized as

$$[g \;\; f] = \frac{1}{\delta_n} [\grave{e}_n \;\; \tau_n e_n] \begin{bmatrix} q_n & -p_n \\ -\grave{q}_n & \grave{p}_n \end{bmatrix}. \tag{4.11}$$

Both for the general NPF recurrence and, of course, for the MPF recurrence to be discussed later, we need analogous definitions and statements for MPFs. For a $[\mu(n); n]$ MPF $(u_n, v_n) := (u_{\mu(n);n}, v_{\mu(n);n})$ of $(\hat{f}, \hat{g}; f, g)$, we let

$$(\grave{u}_n, \grave{v}_n) := (u_{\mu(n)-1;n-1}, v_{\mu(n)-1;n-1}) \tag{4.12}$$

denote its upper left neighbor. We call the MPF (u_n, v_n) *weakly regular* if

$$\frac{u_n}{v_n} \neq \frac{\grave{u}_n}{\grave{v}_n}, \quad i.e., \quad \grave{u}_n v_n - u_n \grave{v}_n \neq 0 \in \mathcal{L}. \tag{4.13}$$

The *residual* $(\hat{e}_n, e_n)$ of the MPF (u_n, v_n) consists of a formal Laurent series $\hat{e}_n \in \mathcal{L}_0^\star$ and a formal Newton series e_n for the points $z_{\mu(n)+n+i}$, $i = 1, 2, \ldots$, defined implicitly by

$$\hat{g}(z)u_n(z) + \hat{f}(z)v_n(z) = z^{\mu(n)} \hat{e}_n(z), \tag{4.14a}$$
$$g(z)u_n(z) + f(z)v_n(z) = t_{\mu(n)+n+1}(z)\, e_n(z). \tag{4.14b}$$

The definition of this residual and of the residual $(\grave{\hat{e}}_n, \grave{e}_n)$ of $(\grave{u}_n, \grave{v}_n)$ can be summarized in

$$\begin{bmatrix} \hat{g} & \hat{f} \\ g & f \end{bmatrix} \begin{bmatrix} \grave{u}_n & u_n \\ \grave{v}_n & v_n \end{bmatrix} = \begin{bmatrix} z^{\mu(n)-1} \grave{\hat{e}}_n & z^{\mu(n)} \hat{e}_n \\ t_{\mu(n)+n-1} \grave{e}_n & t_{\mu(n)+n+1}\, e_n \end{bmatrix}$$
$$= \begin{bmatrix} z^{\mu(n)-1} & 0 \\ 0 & t_{\mu(n)+n-1} \end{bmatrix} \begin{bmatrix} \grave{\hat{e}}_n & z\hat{e}_n \\ \grave{e}_n & \hat{\tau}_n e_n \end{bmatrix}, \tag{4.15}$$

where

$$\hat{\tau}_n(z) := (z - z_{\mu+n-1})(z - z_{\mu+n}) = t_2^{(\mu+n-2)}(z). \tag{4.16}$$

However, in the following we first consider $[\kappa+1; k]$ MPFs ($k \geq 1$); *i.e.*, we have to replace μ and n by $\kappa+1$ and k in all formulas. We let $(p_n, q_n) := (p_{m(n),n}, q_{m(n),n})$ be a fixed weakly regular (m, n) NPF. We assume that its upper left neighbor $(\grave{p}_n, \grave{q}_n)$ and the residuals e_n and $\grave{e}_n$ of this pair are also at our disposal. Then we can consider the following $[\kappa + 1; k]$ multipoint Padé problem for the polynomials $u_k^{(n)}$ and $v_k^{(n)}$, which will play the role of "recurrence coefficients":

$$\begin{aligned} z^{-m+1}[\grave{p}_n(z)u_k^{(n)}(z) + p_n(z)v_k^{(n)}(z)] &= O_-(z^{\kappa+1}), &(4.17a)\\ \grave{e}_n(z)u_k^{(n)}(z) + \tau_n(z)e_n(z)v_k^{(n)}(z) &= O(t_{\kappa+k+2}^{(m+n-2)}(z)). &(4.17b)\end{aligned}$$

The polynomial $t_{\kappa+k+2}^{(m+n-2)}$ is still defined by (4.5). In our notation from §2 we have for this problem to substitute $(\hat{f}, \hat{g}; f, g)$ by

$$(z^{-m+1}p_n(z), z^{-m+1}\grave{p}_n(z); \tau_n(z)e_n(z), \grave{e}_n(z)) \in \mathcal{L}_1^\star \times \mathcal{L}_0^\star \times \mathcal{N}_{Z(n)} \times \mathcal{N}_{Z(n)}. \tag{4.18}$$

(Recall that $Z(n)$ was defined in Lemma 8.) By Lemma 5 we know that $\partial(z^{-m+1}\grave{p}_n) = 0$ if (p_n, q_n) is weakly row-regular, and that $\partial(z^{-m+1}p_n) = 1$ if (p_n, q_n) is weakly column-regular. Hence, the assumption of §2 that $\hat{\gamma}_0 \neq 0$ or $\hat{\phi}_\iota \neq 0$ is fulfilled if we let $\iota = 1$. In addition to this $[\kappa + 1; k]$ MPF $(u_k^{(n)}, v_k^{(n)})$ we consider its upper left neighbor in the multipoint Padé table for the data (4.18), denoting it by $(\grave{u}_k^{(n)}, \grave{v}_k^{(n)})$.

We restrict κ to $-k \leq \kappa \leq k$, so that according to (2.3)

$$(\grave{u}_k, \grave{v}_k) \in \mathcal{P}_k \times \mathcal{P}_{k-1}, \quad (u_k^{(n)}, v_k^{(n)}) \in \mathcal{P}_{k+1} \times \mathcal{P}_k. \tag{4.19}$$

According to (4.15) the residual $(\hat{e}_k^{(n)}, e_k^{(n)})$ of $(u_k^{(n)}, v_k^{(n)})$ and the residual $(\grave{\hat{e}}_k^{(n)}, \grave{e}_k^{(n)})$ of $(\grave{u}_k^{(n)}, \grave{v}_k^{(n)})$ satisfy

$$\begin{bmatrix} z^{-m+1}\grave{p}_n & z^{-m+1}p_n \\ \grave{e}_n & \tau_n e_n \end{bmatrix} \begin{bmatrix} \grave{u}_k^{(n)} & u_k^{(n)} \\ \grave{v}_k^{(n)} & v_k^{(n)} \end{bmatrix} = \begin{bmatrix} z^\kappa & 0 \\ 0 & t_{\kappa+k}^{(m+n-2)} \end{bmatrix} \begin{bmatrix} \grave{\hat{e}}_k^{(n)} & z\,\hat{e}_k^{(n)} \\ \grave{e}_k^{(n)} & \tau_{n+k}\, e_k^{(n)} \end{bmatrix}, \tag{4.20}$$

where

$$\tau_{n+k}(z) := (z - z_{\kappa+k+m+n-1})(z - z_{\kappa+k+m+n}) = t_2^{(\kappa+k+m+n-1)}(z). \tag{4.21}$$

Note that τ_n is a factor of $u_k^{(n)}$ since $\kappa + k \geq 0$, and a factor of $\grave{u}_k^{(n)}$ if $\kappa + k \geq 2$. In particular, if $\kappa = k$ (≥ 1), then this multipoint Padé problem with $\iota = 1$ can be reduced to one with $\iota = -1$, which means reducing the degree of freedom by 2.

In the rest of this section, f and g will again refer to the data for the Newton-Padé problem for which (p_n, q_n) is an (m, n) NPF.

The general recurrence formula for NPFs from [23], which is stated next, assumes that (p_n, q_n), $(\grave{p}_n, \grave{q}_n)$, and $(e_n, \grave{e}_n)$ are known, and relies on the pair of MPFs just considered.

Theorem 9. *(General NPF recurrence) Let a relatively prime pair $(f,g) \in \mathcal{N}_Z \times \mathcal{N}_Z$ be given. Let $(m,n) \in \mathbf{N}^+ \times \mathbf{N}$, and let $[\kappa;k] \in \mathbf{Z} \times \mathbf{N}^+$ be such that $-k \le \kappa \le k$, $m+\kappa > 0$. Assume that (p_n,q_n) is a weakly regular (m,n) NPF of (f,g) with residual e_n, and let $(\grave{p}_n,\grave{q}_n)$ be an $(m-1,n-1)$ NPF of (f,g) with residual $\grave{e}_n$. Moreover, let $(u_k^{(n)},v_k^{(n)})$ be a $[\kappa+1;k]$ MPF with residual $(\hat{e}_k^{(n)},e_k^{(n)})$ of the data (4.18), and let $(\grave{u}_k^{(n)},\grave{v}_k^{(n)})$ be a $[\kappa;k-1]$ MPF with residual $(\grave{\hat{e}}_k^{(n)},\grave{e}_k^{(n)})$ for the same data. Then, the formulas*

$$\begin{bmatrix} \grave{p}_{n+k} & p_{n+k} \\ \grave{q}_{n+k} & q_{n+k} \end{bmatrix} := \begin{bmatrix} \grave{p}_n & p_n \\ \grave{q}_n & q_n \end{bmatrix} \begin{bmatrix} \grave{u}_k^{(n)} & u_k^{(n)} \\ \grave{v}_k^{(n)} & v_k^{(n)} \end{bmatrix} \tag{4.22}$$

and[1]

$$t_{m+n+k+\kappa-1}[\grave{e}_{n+k} \quad \tau_{n+k}e_{n+k}] := t_{m+n-1}[\grave{e}_n \quad \tau_n e_n] \begin{bmatrix} \grave{u}_k^{(n)} & u_k^{(n)} \\ \grave{v}_k^{(n)} & v_k^{(n)} \end{bmatrix} \tag{4.23}$$

yield an $(m+\kappa,n+k)$ NPF (p_{n+k},q_{n+k}) and an $(m+\kappa-1,n+k-1)$ NPF $(\grave{p}_{n+k},\grave{q}_{n+k})$ of (f,g), as well as the corresponding residuals e_{n+k} and $\grave{e}_{n+k}$, which are equal to $e_k^{(n)}$ and $\grave{e}_k^{(n)}$, respectively. The new NPF (p_{n+k},q_{n+k}) is weakly regular if and only if also $(u_k^{(n)},v_k^{(n)})$ is weakly regular.

Proof. The proof of this crucial result from [23] is so short that we can repeat it here. Since $(p_n,q_n) \in \mathcal{P}_m \times \mathcal{P}_n$ and $(\grave{p}_n,\grave{q}_n) \in \mathcal{P}_{m-1} \times \mathcal{P}_{n-1}$, (4.19) indicates that the pair (p_{n+k},q_{n+k}) defined by (4.22) lies in $\mathcal{P}_{m+k} \times \mathcal{P}_{n+k}$. However, by definition of $(u_k^{(n)},v_k^{(n)})$ as a $[\kappa+1;k]$ MPF of (4.18), we have actually $p_{n+k} \in \mathcal{P}_{m+\kappa}$, cf. (4.17a)–(4.17b). Analogously, it can be verified that $(\grave{p}_{n+k},\grave{q}_{n+k})$ defined by (4.22) lies in $\mathcal{P}_{m+\kappa-1} \times \mathcal{P}_{n+k-1}$. Moreover, by (4.6) and (4.20),

$$\begin{aligned} [g \quad f] \begin{bmatrix} \grave{p}_{n+k} & p_{n+k} \\ \grave{q}_{n+k} & q_{n+k} \end{bmatrix} &= [g \quad f] \begin{bmatrix} \grave{p}_n & p_n \\ \grave{q}_n & q_n \end{bmatrix} \begin{bmatrix} \grave{u}_k^{(n)} & u_k^{(n)} \\ \grave{v}_k^{(n)} & v_k^{(n)} \end{bmatrix} \\ &= t_{m+n-1}[\grave{e}_n \quad \tau_n e_n] \begin{bmatrix} \grave{u}_k^{(n)} & u_k^{(n)} \\ \grave{v}_k^{(n)} & v_k^{(n)} \end{bmatrix} \\ &= t_{m+n-1} \left[t_{k+\kappa}^{(m+n-2)} \grave{e}_k^{(n)} \quad t_{k+\kappa+2}^{(m+n-2)} e_k^{(n)} \right] \\ &= t_{m+n+k+\kappa-1} \left[\grave{e}_k^{(n)} \quad \tau_{n+k} e_k^{(n)} \right]. \end{aligned}$$

By (4.23), the expression on the second line is equal to $t_{m+n+k+\kappa-1}\,[\grave{e}_{n+k} \quad \tau_{n+k}e_{n+k}]$. Altogether, this shows that $(\grave{p}_{n+k},\grave{q}_{n+k})$ and (p_{n+k},q_{n+k}) are an $(m+\kappa-1,n+k-1)$ and an $(m+\kappa,n+k)$ NPF of (f,g), respectively, and that $\grave{e}_{n+k}$ and e_{n+k} defined by (4.23), as well as $\grave{e}_k^{(n)}$ and $e_k^{(n)}$ defined by (4.20), are equal to their residuals. Finally, if and only if both (p_n,q_n) and $(u_n^{(k)},v_n^{(k)})$ are weakly regular, the determinants[2] of

[1] On the left-hand side of Eq. (5.20) in [23], τ_n should be replaced by τ_{n+k} too.

[2] In the proof in [23] we also meant the determinants of the matrices, not the matrices themselves.

both matrices on the right-hand side of (4.22) do not vanish identically, hence the same is true for their product, which means that (p_{n+k}, q_{n+k}) is weakly regular. □

As pointed out in [23], Theorem 9 has an interpretation in terms of continued fractions. For example, for a diagonal sequence of weakly regular NPFs, p_n/q_n is a convergent (*i.e.*, a 'partial sum') of the *diagonal G-fraction* [25] of $-f/g$, and $e_n/\grave{e}_n$ is a representation of the corresponding tail of the G-fraction. Hence, to get p_{n+k}/q_{n+k} we just combine the nth convergent of the continued fraction with the kth convergent of its tail. In the Padé case, the diagonal G-fraction becomes a (diagonal) P-fraction [35]. This interpretation is not restricted to diagonal sequences, however.

5. General Recurrence Relations for Proper MPAs

Let us return to the multipoint Padé problem of §2. Again, consider a $[\mu(n); n]$ MPF $(u_n, v_n) := (u_{\mu(n);n}, v_{\mu(n);n})$ of $(\hat{f}, \hat{g}; f, g)$ and its upper left neighbor $(\grave{u}_n, \grave{v}_n)$. Recall from (4.13)–(4.15) the definitions of weak regularity and of the residuals of this pair of MPFs. First, we want to establish analogues of the Lemmas 4–6.

Lemma 4'. *Let (u_n, v_n) be a proper $[\mu(n); n]$ MPF (with $n > |\mu_0| \geq 0$), and let $(\grave{\hat{e}}_n, \grave{e}_n)$ be the residual of an $[\mu(n) - 1; n - 1]$ MPF $(\grave{u}_n, \grave{v}_n)$. Then the following statements are equivalent:*

(i) (u_n, v_n) is weakly regular, i.e., *(4.13) holds;*

(ii) $\grave{e}_n(z_{\mu(n)+n}) \neq 0$ and $(u_n(z_{\mu(n)+n}) \neq 0$ or $v_n(z_{\mu(n)+n}) \neq 0)$;

(iii) $\grave{e}_n(z_{\mu(n)+n-1}) \neq 0$ and $(u_n(z_{\mu(n)+n-1}) \neq 0$ or $v_n(z_{\mu(n)+n-1}) \neq 0)$.

Proof. If (u_n, v_n) is weakly regular, one can conclude from the definition (2.4a)–(2.4b) or from Theorem 1' that $(\grave{u}_n, \grave{v}_n)$ does not have $z_{\mu+n-1}$ or $z_{\mu+n}$ as an extra interpolation point. If, say, $z_{\mu+n}$ were one, then $(u_n, v_n) := ((z - z_{\mu+n-1})\grave{u}_n, (z - z_{\mu+n-1})\grave{v}_n)$ would be a $[\mu; n]$ MPF, possibly with $z_{\mu+n-1}$ as an unattainable point, and consequently, (4.13) would not hold. Similarly we can conclude that u_n and v_n cannot vanish simultaneously at $z_{\mu+n-1}$ or $z_{\mu+n}$. Because, if they did, say at $z_{\mu+n}$, then we could obtain a $[\mu - 1; n - 1]$ MPF $(\grave{u}_n, \grave{v}_n)$ by canceling the common factor $z - z_{\mu+n}$ in (u_n, v_n). The reverse directions of these two conclusions are verified easily too. For assume that (u_n, v_n) is not weakly regular and, say, $z_{\mu+n}$ is not an extra interpolation point of $(\grave{u}_n, \grave{v}_n)$. Then, $z_{\mu+n}$ is an unattainable point of u_n/v_n, and hence, by Theorem 1', $z - z_{\mu+n}$ is a factor of $s_{\mu;n}$, and therefore a common factor of u_n and v_n. □

Lemma 5'. *If (u_n, v_n) is a weakly regular proper $[\mu(n); n]$ MPF (with $n > |\mu_0| \geq 0$), then*

(i) $\partial v_n = n$ and $\partial \grave{u}_n = \mu(n) - 1$

or

(ii) $\partial u_n = \mu(n)$ and $\partial \grave{v}_n = n - 1$.

Proof. Assume (u_n, v_n) is weakly regular. Then not only the upper left neighbor of the $[\mu; n]$ entry $r_n := u_n/v_n$, but also the $[\mu - 1; n]$ entry or the $[\mu; n - 1]$ entry (or both) are different from r_n. If, say, the $[\mu - 1; n]$ entry differs, then clearly $\partial u_n = \mu$; otherwise (u_n, v_n) would qualify as an $[\mu - 1; n]$ MPF. Hence, (ii) holds unless $\partial \grave{v}_n < n-1$, in which case the pair $((z-z_{\mu+n-1})\grave{u}_n, (z-z_{\mu+n-1})\grave{v}_n)$ qualifies as

$[\mu; n-1]$ MPF, which means that also the left neighbor differs from r_n; consequently, $\partial v_n = n$. Therefore, (i) holds unless $\partial \grave{u}_n < \mu - 1$. In the latter case we could conclude that the same pair is a $[\mu-1; n]$ MPF, and that the $[\mu-1; n-1]$, the $[\mu; n-1]$, and the $[\mu-1; n]$ MPA are all the same; this would imply that they agree with the $[\mu; n]$ MPA, in contrast to the weak regularity of the latter. □

Lemma 6'. *Let (u_n, v_n) be a proper $[\mu(n); n]$ MPF (with $n > |\mu_0| \geq 0$); then*

$$\det \begin{bmatrix} \grave{u}_n & u_n \\ \grave{v}_n & v_n \end{bmatrix} = \hat{\delta}_n t_{\mu(n)+n-1}, \tag{5.1}$$

where $\hat{\delta}_n$ is the leading coefficient of $\grave{u}_n v_n - \grave{v}_n u_n \in \mathcal{P}_{\mu(n)+n-1}$, which does not vanish if and only if (u_n, v_n) is weakly regular.

Proof. The determinant is equal to $\grave{u}_n v_n - \grave{v}_n u_n \in \mathcal{P}_{\iota+2n-1}$, which, by the definition of weak regularity, is identically zero if and only if the MPF (u_n, v_n) is not weakly regular. Multiplying (4.14a)–(4.14b) by $\grave{v}_n$ and $\grave{u}_n$, and their analogues for $(\grave{u}_n, \grave{v}_n)$ by v_n and u_n, we obtain by pairwise subtraction

$$\hat{g}(\grave{u}_n v_n - \grave{v}_n u_n) = O_-(z^{\mu+n-1}), \qquad g(\grave{u}_n v_n - \grave{v}_n u_n) = O(t_{\mu+n-1}),$$
$$\hat{f}(\grave{u}_n v_n - \grave{v}_n u_n) = O_-(z^{\iota+\mu+n-1}), \quad f(\grave{u}_n v_n - \grave{v}_n u_n) = O(t_{\mu+n-1}).$$

Since f and g are relatively prime by assumption, and since $\hat{\phi}_\iota \neq 0$ or $\hat{\gamma}_0 \neq 0$ in (2.1), it follows that

$$\grave{u}_n v_n - \grave{v}_n u_n = O_-(z^{\mu+n-1}), \quad \grave{u}_n v_n - \grave{v}_n u_n = O(t_{\mu+n-1}).$$

In other words, $\grave{u}_n v_n - \grave{v}_n u_n \in \mathcal{P}_{\mu+n-1}$, and $t_{\mu+n-1}$ is a polynomial factor of it, the quotient being a scalar $\hat{\delta}_n$, which, since $t_{\mu+n-1}$ is monic, is equal to the leading coefficient of $\grave{u}_n v_n - \grave{v}_n u_n$ in $\mathcal{P}_{\mu+n-1}$. □

Corollary 7'. *If the $[\mu(n); n]$ MPF (u_n, v_n) (where $n > |\mu_0| \geq 0$) is weakly regular, then any common zero of the polynomials $\grave{u}_n$ and u_n is a zero of f or an unattainable interpolation point of u_n/v_n. In particular, if u_n/v_n is a true $[\mu(n); n]$ interpolant and f does not vanish at any interpolation point, then $\grave{u}_n$ and u_n are relatively prime. These statements remain true when $\grave{u}_n$, u_n, and f are replaced by $\grave{v}_n$, v_n, and g, respectively.*

Proof. By (5.1), if (u_n, v_n) is weakly regular, any common zero of $\grave{u}_n$ and u_n must be a zero of $t_{\mu(n)+n-1}$. On the other hand, since $\partial w = 0$ in the factorization (2.6) of (u_n, v_n) (because of uniqueness), such a zero must be either a zero of $s_{\mu;n}$, *i.e.*, an unattainable point, or a zero of $\hat{u}_{\mu;n}$, in which case it must be a zero of f; see (2.4b). The statement about exchanging $\grave{u}_n$, u_n, $\hat{f}$, and f for $\grave{v}_n$, v_n, $\hat{g}$, and g holds by symmetry. □

Finally, we need to justify the later usage of the pair of residuals $(e_n, \grave{e}_n)$ as data of another multipoint Padé problem. Recall the definition of $\hat{\tau}_n$ from (4.16).

Lemma 8'. *Assume that (u_n, v_n) is a weakly regular $[\mu(n); n]$ MPF (where $n > |\mu_0| \geq 0$), and let $\hat{Z}(n) := \{z_j \in Z\,;\, j \geq \mu(n) + n - 1\}$. Then $\grave{e}_n$ and $\hat{\tau}_n e_n$ are relatively prime elements of $\mathcal{N}_{\hat{Z}(n)}$, and* $\max\{\partial \hat{\grave{e}}_n, \partial \grave{e}_n\} = 0$.

Proof. By complete analogy to the proof of Lemma 8 we can conclude from (4.15) that the formulas

$$g = \frac{1}{\hat{\delta}_n} \det \begin{bmatrix} \grave{e}_n & \hat{\tau}_n e_n \\ \grave{v}_n & v_n \end{bmatrix}, \quad f = \frac{1}{\hat{\delta}_n} \det \begin{bmatrix} \grave{u}_n & u_n \\ \grave{e}_n & \hat{\tau}_n e_n \end{bmatrix}. \tag{5.2}$$

hold, even if some of the points z_j converge to zeros of $t_{\mu+n-1}$. Moreover, by applying Cramer's rule to the first row of (4.15) we get

$$\hat{g} = \frac{\ell_{\mu;n}}{\hat{\delta}_n} \det \begin{bmatrix} \grave{\hat{e}}_n & z\hat{e}_n \\ \grave{v}_n & v_n \end{bmatrix}, \quad \hat{f} = \frac{\ell_{\mu;n}}{\hat{\delta}_n} \det \begin{bmatrix} \grave{u}_n & u_n \\ \grave{\hat{e}}_n & z\hat{e}_n \end{bmatrix}, \tag{5.3}$$

where $\ell_{\mu;n}$ denotes the Laurent series at ∞ of the function $z^{\mu-1}/t_{\mu+n-1}(z)$. This Laurent series belongs to $\mathcal{L}^\star_{-n}$ and has exact degree $-n$, *i.e.*, the coefficient of z^{-n} does not vanish. In view of $u_n \in \mathcal{P}_{\iota+n}$, $v_n \in \mathcal{P}_n$, $\grave{u}_n \in \mathcal{P}_{\iota+n-1}$, $\grave{v}_n \in \mathcal{P}_{n-1}$, $z\hat{e}_n \in \mathcal{L}^\star_1$, and $\grave{\hat{e}}_n \in \mathcal{L}^\star_0$, we see that the first determinant is in $\mathcal{L}^\star_n$ and the second one in $\mathcal{L}^\star_{\iota+n}$, so that the formulas confirm that $\hat{g} \in \mathcal{L}^\star_0$ and $\hat{f} \in \mathcal{L}^\star_\iota$. Clearly, if the leading coefficients of $\grave{\hat{e}}_n$ and $\hat{e}_n$ both vanished, those of $\hat{g}$ and $\hat{f}$ would vanish also, contrary to the assumption for (2.1). □

The Eqs. (5.2) and (5.3) can be combined into

$$\begin{bmatrix} \hat{g} & \hat{f} \\ g & f \end{bmatrix} = \frac{1}{\hat{\delta}_n} \begin{bmatrix} \ell_{\mu;n} & 0 \\ 0 & 1 \end{bmatrix} \begin{bmatrix} \grave{\hat{e}}_n & z\hat{e}_n \\ \grave{e}_n & \hat{\tau}_n e_n \end{bmatrix} \begin{bmatrix} v_n & -u_n \\ -\grave{v}_n & \grave{u}_n \end{bmatrix}. \tag{5.4}$$

Now assume that (u_n, v_n) is weakly regular and that, moreover, $(\grave{u}_n, \grave{v}_n)$ and the residuals $(\grave{\hat{e}}_n, \hat{e}_n)$ and $(\grave{e}_n, e_n)$ are at our disposal, so that we can additionally consider $[\kappa+1; k]$ MPFs of the following multipoint Padé problem for the "recurrence coefficients" $u_k^{(n)}$ and $v_k^{(n)}$:

$$\begin{aligned} \grave{\hat{e}}_n(z) u_k^{(n)}(z) + z\hat{e}_n(z) v_k^{(n)}(z) &= O_-(z^{\kappa+1}), & \text{(5.5a)} \\ \grave{e}_n(z) u_k^{(n)}(z) + \hat{\tau}_n(z) e_n(z) v_k^{(n)}(z) &= O(t_{\kappa+k+2}^{(\mu+n-2)}(z)). & \text{(5.5b)} \end{aligned}$$

Hence, here the data are

$$(z\hat{e}_n(z), \grave{\hat{e}}_n(z); \hat{\tau}_n(z) e_n(z), \grave{e}_n(z)) \in \mathcal{L}^\star_1 \times \mathcal{L}^\star_0 \times \mathcal{N}_{\hat{Z}(n)} \times \mathcal{N}_{\hat{Z}(n)}. \tag{5.6}$$

($\hat{Z}(n)$ was defined in Lemma 8', and the polynomial $t_{\kappa+k+2}^{(\mu+n-2)}$ is still defined by (4.5).) From Lemma 5' we recall that $\partial(\grave{\hat{e}}_n) = 0$ or $\partial(z\hat{e}_n) = 1$. Hence, the assumption of §2 that $\hat{\gamma}_0 \neq 0$ or $\hat{\phi}_\iota \neq 0$ is again fulfilled for $\iota = 1$.

In addition to this $[\kappa+1; k]$ MPF $(u_k^{(n)}, v_k^{(n)})$ we consider its upper left neighbor, denoting it by $(\grave{u}_k^{(n)}, \grave{v}_k^{(n)})$. Restricting κ to $-k \leq \kappa \leq k$, we see that (4.19) still holds. The residuals $(\hat{e}_k^{(n)}, e_k^{(n)})$ of $(u_k^{(n)}, v_k^{(n)})$ and $(\grave{\hat{e}}_k^{(n)}, \grave{e}_k^{(n)})$ of $(\grave{u}_k^{(n)}, \grave{v}_k^{(n)})$ now satisfy

$$\begin{bmatrix} \grave{\hat{e}}_n & z\hat{e}_n \\ \grave{e}_n & \hat{\tau}_n e_n \end{bmatrix} \begin{bmatrix} \grave{u}_k^{(n)} & u_k^{(n)} \\ \grave{v}_k^{(n)} & v_k^{(n)} \end{bmatrix} = \begin{bmatrix} z^\kappa & 0 \\ 0 & t_{\kappa+k}^{(\mu+n-2)} \end{bmatrix} \begin{bmatrix} \grave{\hat{e}}_k^{(n)} & z\hat{e}_k^{(n)} \\ \grave{e}_k^{(n)} & \hat{\tau}_{n+k} e_k^{(n)} \end{bmatrix}, \tag{5.7}$$

where

$$\hat{\tau}_{n+k}(z) := (z - z_{\kappa+k+\mu+n-1})(z - z_{\kappa+k+\mu+n}) = t_2^{(\kappa+k+\mu+n-2)}(z). \tag{5.8}$$

Theorem 9'. *(General proper MPF recurrence) Let a relatively prime pair $(f,g) \in \mathcal{N}_Z \times \mathcal{N}_Z$ and a pair $(\hat{f},\hat{g}) \in \mathcal{L}_\iota^\star \times \mathcal{L}_0^\star$ with $\iota \geq -1$ and $\hat{\phi}_\iota \neq 0$ or $\hat{\gamma}_0 \neq 0$ be given. Let $[\mu;n]$, $[\kappa;k] \in \mathbb{Z} \times \mathbb{N}^+$ be such that $-n-1 \leq \mu \leq \iota+n$ and $-k \leq \kappa \leq k$. Assume that (u_n, v_n) is a weakly regular $[\mu;n]$ MPF of $(\hat{f},\hat{g};f,g)$ with residual $(\hat{e}_n, e_n)$, and let $(\grave{u}_n, \grave{v}_n)$ be a $[\mu-1;n-1]$ MPF of $(\hat{f},\hat{g};f,g)$ with residual $(\grave{\hat{e}}_n, \grave{e}_n)$.*

Moreover, let $(u_k^{(n)}, v_k^{(n)})$ be a $[\kappa+1;k]$ MPF with residual $(\hat{e}_k^{(n)}, e_k^{(n)})$ of the data (5.6), and let $(\grave{u}_k^{(n)}, \grave{v}_k^{(n)})$ be a $[\kappa;k-1]$ MPF with residual $(\grave{\hat{e}}_k^{(n)}, \grave{e}_k^{(n)})$ for the same data. Then, the formulas

$$\begin{bmatrix} \grave{u}_{n+k} & u_{n+k} \\ \grave{v}_{n+k} & v_{n+k} \end{bmatrix} := \begin{bmatrix} \grave{u}_n & u_n \\ \grave{v}_n & v_n \end{bmatrix} \begin{bmatrix} \grave{u}_k^{(n)} & u_k^{(n)} \\ \grave{v}_k^{(n)} & v_k^{(n)} \end{bmatrix} \tag{5.9}$$

and

$$\begin{bmatrix} z^\kappa & 0 \\ 0 & t_{k+\kappa}^{(\mu+n-2)} \end{bmatrix} \begin{bmatrix} \grave{\hat{e}}_{n+k} & z\hat{e}_{n+k} \\ \grave{e}_{n+k} & \hat{\tau}_{n+k} e_{n+k} \end{bmatrix} := \begin{bmatrix} \grave{\hat{e}}_n & z\hat{e}_n \\ \grave{e}_n & \hat{\tau}_n e_n \end{bmatrix} \begin{bmatrix} \grave{u}_k^{(n)} & u_k^{(n)} \\ \grave{v}_k^{(n)} & v_k^{(n)} \end{bmatrix} \tag{5.10}$$

yield a $[\mu+\kappa;n+k]$ MPF (u_{n+k}, v_{n+k}) and a $[\mu+\kappa-1;n+k-1]$ MPF $(\grave{u}_{n+k}, \grave{v}_{n+k})$ of $(\hat{f},\hat{g};f,g)$, as well as the corresponding residuals $(\hat{e}_{n+k}, e_{n+k})$ and $(\grave{\hat{e}}_{n+k}, \grave{e}_{n+k})$, which are equal to $(\hat{e}_k^{(n)}, e_k^{(n)})$ and $(\grave{\hat{e}}_k^{(n)}, \grave{e}_k^{(n)})$, respectively.

The new MPF (u_{n+k}, v_{n+k}) is weakly regular if and only if also $(u_k^{(n)}, v_k^{(n)})$ is weakly regular.

Proof. Since $(u_n, v_n) \in \mathcal{P}_{\iota+n} \times \mathcal{P}_n$ and $(\grave{u}_n, \grave{v}_n) \in \mathcal{P}_{\iota+n-1} \times \mathcal{P}_{n-1}$, (4.19) implies that the pairs (u_{n+k}, v_{n+k}) and $(\grave{u}_{n+k}, \grave{v}_{n+k})$ defined by (5.9) lie in $\mathcal{P}_{\iota+n+k} \times \mathcal{P}_{n+k}$ and $\mathcal{P}_{\iota+n+k-1} \times \mathcal{P}_{n+k-1}$, respectively. Moreover, by (5.9), (4.15), and (5.7),

$$\begin{aligned}
\begin{bmatrix} \hat{g} & \hat{f} \\ g & f \end{bmatrix} \begin{bmatrix} \grave{u}_{n+k} & u_{n+k} \\ \grave{v}_{n+k} & v_{n+k} \end{bmatrix} &= \begin{bmatrix} \hat{g} & \hat{f} \\ g & f \end{bmatrix} \begin{bmatrix} \grave{u}_n & u_n \\ \grave{v}_n & v_n \end{bmatrix} \begin{bmatrix} \grave{u}_k^{(n)} & u_k^{(n)} \\ \grave{v}_k^{(n)} & v_k^{(n)} \end{bmatrix} \\
&= \begin{bmatrix} z^{\mu-1} & 0 \\ 0 & t_{\mu+n-1} \end{bmatrix} \begin{bmatrix} \grave{\hat{e}}_n & z\hat{e}_n \\ \grave{e}_n & \hat{\tau}_n e_n \end{bmatrix} \begin{bmatrix} \grave{u}_k^{(n)} & u_k^{(n)} \\ \grave{v}_k^{(n)} & v_k^{(n)} \end{bmatrix} \\
&= \begin{bmatrix} z^{\kappa+\mu-1} & 0 \\ 0 & t_{\kappa+k+\mu+n-1} \end{bmatrix} \begin{bmatrix} \grave{\hat{e}}_k^{(n)} & z\,\hat{e}_k^{(n)} \\ \grave{e}_k^{(n)} & \hat{\tau}_{n+k}\, e_k^{(n)} \end{bmatrix},
\end{aligned}$$

while applying (5.7) instead of the last step yields

$$\begin{bmatrix} z^{\kappa+\mu-1} & 0 \\ 0 & t_{\kappa+k+\mu+n-1} \end{bmatrix} \begin{bmatrix} \grave{\hat{e}}_{n+k} & z\,\hat{e}_{n+k} \\ \grave{e}_{n+k} & \hat{\tau}_{n+k} e_{n+k} \end{bmatrix}.$$

This shows that $(\grave{u}_{n+k}, \grave{v}_{n+k})$ and (u_{n+k}, v_{n+k}) are a $[\mu+\kappa-1;n+k-1]$ and a $[\mu+\kappa;n+k]$ MPF of $(\hat{f},\hat{g};f,g)$, and that $(\grave{\hat{e}}_{n+k}, \grave{e}_{n+k})$ and $(\hat{e}_{n+k}, e_{n+k})$ defined by (5.10) as well as $(\grave{e}_k^{(n)}, \grave{e}_k^{(n)})$ and $(\hat{e}_k^{(n)}, e_k^{(n)})$ defined by (5.7) are equal to their residuals.

Again, if and only if also $(u_n^{(k)}, v_n^{(k)})$ is weakly regular, the determinants of both matrices on the right-hand side of (5.9) do not vanish identically, and thus the same is true for their product, which means that (u_{n+k}, v_{n+k}) is weakly regular. □

6. Comments and Conclusions

6.1. Product Representation of Basic Pairs

For the recursive construction of a regular NPF, say (p_{n_J}, q_{n_J}), we apply Theorem 9 repeatedly. Not only yields it such an NPF, but it also provides us with a representation of (p_{n_J}, q_{n_J}) and its upper left neighbor in terms of a product of 2×2 matrices whose elements are polynomials:

$$\begin{bmatrix} \grave{p}_{n_J} & p_{n_J} \\ \grave{q}_{n_J} & q_{n_J} \end{bmatrix} = \begin{bmatrix} \grave{p}_0 & p_0 \\ \grave{q}_0 & q_0 \end{bmatrix} \prod_{j=0}^{J-1} \begin{bmatrix} \grave{u}_{k_j}^{(n_j)} & u_{k_j}^{(n_j)} \\ \grave{v}_{k_j}^{(n_j)} & v_{k_j}^{(n_j)} \end{bmatrix}. \tag{6.1}$$

Here, $k_j := n_{j+1} - n_j$ and $n_0 := 0$. The index j in the product increases from left to right. In exact arithmetic we can choose k_j in each step as small as possible, namely just such that the determinant of the new factor is not identically zero. If $h = -f/g$ does not represent a rational function, we can let n_J go to infinity. The formal infinite product that is then obtained by proceeding, say, along a diagonal (*i.e.*, by choosing $\kappa = k$) is a matrix representation of the diagonal G-fraction.

In floating-point arithmetic, one has to avoid those factors in (6.1) that are "nearly singular". For special cases, the question what "nearly singular" means, has been addressed in [10], [37], [28], [30], [29], but it still requires further investigation.

6.2. A Connection Between NPFs and MPFs

By the close analogy between the theory of weakly regular NPFs and MPFs, and, in particular, between the general recurrences of Theorems 9 and 9' most of what can be said on the computation of NPFs holds with minor modifications for the computation of MPFs. In particular, the product representation (6.1) has an MPF analogue. This is no surprise since the 2×2 polynomial matrices that are used to update the pair of NPFs at the same time update a pair of MPFs, $(u_{N_J}^{(0)}, v_{N_J}^{(0)})$ and $(\grave{u}_{N_J}^{(0)}, \grave{v}_{N_J}^{(0)})$:

$$\begin{bmatrix} \grave{p}_{n_J} & p_{n_J} \\ \grave{q}_{n_J} & q_{n_J} \end{bmatrix} = \begin{bmatrix} \grave{p}_0 & p_0 \\ \grave{q}_0 & q_0 \end{bmatrix} \begin{bmatrix} \grave{u}_{N_J}^{(0)} & u_{N_J}^{(0)} \\ \grave{v}_{N_J}^{(0)} & v_{N_J}^{(0)}. \end{bmatrix}. \tag{6.2}$$

The data of the corresponding multipoint Padé problem are

$$(z^{-m+1}p_0(z), z^{-m+1}t_{m-1}(z); \tau_0(z)e_0(z), \grave{e}_0(z)) \in \mathcal{L}_1^\star \times \mathcal{L}_0^\star \times \mathcal{N}_{Z(0)} \times \mathcal{N}_{Z(0)}, \tag{6.3}$$

where $\grave{e}_0 = g$, $\tau_0 e_0 = (gp_0 + fq_0)/t_{m-1}$.

One might guess that most results for the MPF recurrence can be proved by determining first a Newton-Padé problem such that (6.2) holds. However, the data

(6.3) do not look general enough so that any data $(\hat{f}, \hat{g}; f, g)$ admitted for a multipoint Padé problem could be interpreted in this way. But we could establish the MPF results in this way for any such quadruple with $\iota = -1$, *finite* series $\hat{f}$ and $\hat{g}$ with at most m terms and $\gamma_0 := \hat{g}(\infty) = 1$, and a Newton series f satisfying $f(z_m) = f(z_{m+1}) = 0$. Setting

$$\begin{aligned} p_0(z) &:= z^{m-1}\hat{f}(z), & \grave{p}_0(z) &:= t_{m-1}(z) := z^{m-1}\hat{g}(z), \\ q_0(z) &:\equiv 1, & \grave{q}_0(z) &:\equiv 0, \\ e_0(z) &:= f(z)/\tau_0(z), & \grave{e}_0(z) &:= g(z). \end{aligned} \tag{6.4}$$

we could find from (4.11) data $(f^{\mathrm{NP}}, g^{\mathrm{NP}})$ [not to be mixed up with the given multipoint Padé data $(\hat{f}, \hat{g}; f, g)$] for a Newton-Padé problem related to the given multipoint Padé problem by (6.3).

6.3. Levinson-type and Schur-type Algorithms

As long as a small step size k is used in Theorem 9, the recurrence coefficients only depend on few initial coefficients of the Newton series for the residuals (or equivalent data) and the few highest coefficients of the numerators p_n and $\grave{p}_n$. There are two ways to deal with the data of the initial problem. In each step we either transform all the remaining data, *i.e.*, compute a full representation of the residuals (either as Newton series or as functions values and, possibly, derivatives), or we have to compute the data needed for the next step from the definition (4.6) of the residuals. Adopting the terminology used for fast Toeplitz solvers, which correspond to recurrences along two adjacent rows in the Padé table, we can say that in the first case we obtain a Schur-type algorithm, and in the second case a Levinson-type algorithm. The superfast algorithms discussed next combine the merits of both approaches.

Again, these remarks also apply to the multipoint Padé problem, except that the numerators $\grave{p}_n$ and p_n are replaced by the Laurent series residuals $\grave{e}_n$ and e_n.

6.4. Superfast Rational Interpolation

The straightforward way of evaluating the product (6.1) consists in multiplying it out from left to right. In other words, the partial product represented by the polynomials $\grave{p}_{n_j}$, $\grave{q}_{n_j}$, p_{n_j}, and q_{n_j} (whose degrees are growing) is multiplied by the next factor with the polynomials $\grave{u}_{k_j}^{(n_j)}$, $u_{k_j}^{(n_j)}$, $\grave{v}_{k_j}^{(n_j)}$, and $v_{k_j}^{(n_j)}$ (whose degrees are normally small). In the generic situation, where the Newton-Padé table is normal, *i.e.*, has no singular blocks, we do $N := N_J$ steps of total cost $O(\sum_{n=1}^{N} n) = O(N^2)$.

The fastest way of evaluating the product (6.1) is by a binary tree. Assuming a normal table and $N = 2^M$, we compute at the lowest level $\frac{1}{2}N$ matrix products at cost $O(1)$ each. At the highest level, there is just one matrix product involving polynomials of degree $\sim \frac{1}{2}N$. Using the fast Fourier transform (FFT) this last product costs $O(N \log N)$. At an intermediate level j, there are $2^{-j}N$ products costing $O(2^j \log 2^j)$ each; so the cost of this level is $O(Nj)$. Since j runs from 1 to $M = \log_2 N$, total cost is $O(N \log^2 N)$.

However, before we can evaluate product (6.1), we need to determine its factors. The individual factors are found by solving small, particularly structured linear systems. In the generic case they have order 2, and since there are N such factors, the total cost for solving these systems is $O(N)$. But before we can solve them, we need to prepare the corresponding data, and this is the most cost-critical part of the whole recursive process. The ***divide-and-conquer strategy*** allows us to reduce the given data according to a binary tree: whenever we have solved a problem of size $n = 2^j$, we can use the definitions (4.20) or (5.7), respectively, to find the corresponding new residuals of which we need to compute $O(n)$ terms in order to have the data for a new step of size $k = n$ ready. (The exact number of terms needed depends on the path taken in the Newton-Padé table.) Unless $k = 1$, this step will be divided further, but this is of no concern. The cost of computing residuals clearly depends on how the data and the residuals are represented. If this is by Newton series, then one can not expect, in general, to compute n coefficients of a residual in less than $O(n^2)$ operations. In contrast, in the case of Padé approximation or two-point Padé approximation, where data and residuals are represented by power and Laurent series, FFTs can be applied and reduce the cost to $O(n \log^2 n)$ operations. This gives rise to algorithms of total complexity $O(N \log^2 N)$. This principle was first explored by Brent, Gustavson, and Yun [6] for Padé recursions along a staircase, which lead to a superfast Hankel solver; see also [18]. Since 1980, it has appeared in many forms in superfast Hankel and Toeplitz solvers; for references see, *e.g.*, [8], [28], [30], [31]. It is a natural question to ask whether this operation count persists for certain rational interpolation problems involving more than two interpolation points. (Any two finite points could be mapped into 0 and ∞ by a fractional linear transformation that leaves $\mathcal{R}_{n,n}$ invariant.) Of particular interest are the diagonal recurrences ($m = n - 1$ or $m = n$, $\kappa = k$), which generalize the Hankel solvers, and the row recurrences ($m = \frac{1}{2}N - 1$, $\kappa = 0$), which generalize the Toeplitz solvers.

For interpolation in a few, say L, points $\hat{z}_\ell$ that all carry the same amount of data to be taken into account in a cyclic manner, a superfast rational interpolation algorithm indeed exists. Let in each point the data be given as a finite power series in $z - z_\ell$. In all our computation we want to store L different representations of the NPFs and MPFs involved, namely as power series in $z - z_\ell$. To switch between them would be more expensive than to build them up in parallel. Since L is independent of N, this does not increase the order of complexity. Assuming a normal Newton-Padé table, we can concentrate on step sizes where $\kappa + k$ is a multiple of L. Then the polynomials $t_{\kappa+k}^{(m+n-2)}$ and $t_{\kappa+k}^{(\mu+n-2)}$ in (4.20) and (5.7), by which we need to divide, are powers of $t_L(z) = (z - z_0) \cdots (z - z_{L-1})$. The FFT can be applied to build up these powers by recursive doubling and to perform the division.

Rational interpolation at the $2N$th roots of unity by $r \in \mathcal{R}_{N-1,N}$ is another interesting case that can be solved in $O(N \log^2 N)$ operations. At the lowest level of the row recurrence we interpolate at one new root (*i.e.*, $\kappa = 0$, $k = 1$). For the diagonal recurrence we add at the lowest level two new roots (*i.e.*, $\kappa = k = 1$) that lie on opposite sides of the unit circle; they are zeros of $z^2 = z_\ell$, where z_ℓ is any zero of $z_\ell^{N/2} = -1$. At the jth level of the row recurrence, and at the $(j-1)$th level of the diagonal recurrence, we interpolate additionally at the $n = 2^{j-1}$ zeros of $z^n = z_\ell$, where z_ℓ is any zero of $z_\ell^{N/n} = -1$. Since all interpolation points are distinct, the data will normally be given as function values. The computation of the values of the residuals at all nth roots of z_ℓ, which is done according to the definitions (4.20) and (5.7), requires the evaluation of the NPFs or MPFs at these points. Once more, with FFTs of length $2n$ this is accomplished in $O(n \log^2 n)$ operations. Altogether one obtains again the bound $(N/n)\, O(n \log n) = O(N \log n) = O(Nj)$ for level j, and an overall complexity of $O(N \log^2 N)$ for the algorithm.

An alternative is to apply the Kronecker algorithm [33], [45], [2] to this problem: first, one has to compute the interpolation polynomial of degree $2N - 1$ by an FFT of length $2N$; then one applies the $O(N \log^2 N)$ version of the Euclidean algorithm [6] to this polynomial and $z^{2N} - 1$.

The treatment of non-normal tables and the inclusion of look-ahead for avoiding unstable intermediate results makes these superfast algorithms more complicated, but the complexity remains $O(N \log^2 N)$ as long as the look-ahead step size remains bounded independent of N.

References

1. A. C. Antoulas, On recursiveness and related topics in linear systems, IEEE Trans. Automatic Control, AC-31 (1986), pp. 1121–1135.
2. ———, Rational interpolation and the Euclidean algorithm, Linear Algebra Appl., 108 (1988), pp. 157–171.
3. A. C. Antoulas and B. D. O. Anderson, On the scalar rational interpolation problem, IMA J. Math. Control Inform., 3 (1986), pp. 61–88.
4. B. Beckermann, A reliable method for computing M-Padé approximants on arbitrary staircases, J. Comput. Appl. Math., 40 (1992), pp. 19–42.
5. B. Beckermann and G. Labahn, A uniform approach to the fast computation of matrix-type Padé approximants, SIAM J. Matrix Anal. Appl., to appear.
6. R. P. Brent, F. G. Gustavson, and C. Y. Y. Yun, Fast solution of Toeplitz systems of equations and computation of Padé approximants, J. Algorithms, 1 (1980), pp. 259–295.
7. S. Cabay, M. H. Gutknecht, and R. Meleshko, Stable rational interpolation?, in Proceedings MTNS-'93, Regensburg, to appear.
8. S. Cabay and G. Labahn, A superfast algorithm for multi-dimensional Padé systems, Numerical Algorithms, 2 (1992), pp. 201–224.
9. S. Cabay, G. Labahn, and B. Beckermann, On the theory and computation of non-perfect Padé-Hermite approximants, J. Comput. Appl. Math., 39 (1992), pp. 295–313.
10. S. Cabay and R. Meleshko, A weakly stable algorithm for Padé approximants and the inversion of Hankel matrices, SIAM J. Matrix Anal. Appl., 14 (1993), pp. 735–765.
11. G. Claessens, On the Newton-Padé approximation problem, J. Approx. Theory, 22 (1978), pp. 150–160.
12. S. C. Cooper, A. Magnus, and J. H. McCabe, On the non-normal two-point Padé table, J. Comput. Appl. Math., 16 (1986), pp. 371–380.
13. M. Fiedler, Hankel and Loewner matrices, Linear Algebra Appl., 58 (1984), pp. 75–95.
14. R. W. Freund, M. H. Gutknecht, and N. M. Nachtigal, An implementation of the look-ahead Lanczos algorithm for non-Hermitian matrices, SIAM J. Sci. Comput., 14 (1993), pp. 137–158.
15. P. A. Fuhrmann, A matrix Euclidean algorithm and matrix continued fractions, System Control Letters, 3 (1983), pp. 263–271.
16. ———, On the partial realization problem and the recursive inversion of Hankel and Toeplitz matrices, Contemporary Math., 47 (1985), pp. 149–161.
17. M. A. Gallucci and W. B. Jones, Rational approximations corresponding to Newton series (Newton-Padé approximants), J. Approx. Theory, 17 (1976), pp. 366–392.
18. W. Gragg, F. Gustavson, D. Warner, and D. Yun, On fast computation of superdiagonal Padé fractions, Math. Programming Stud., 18 (1982), pp. 39–42.
19. W. B. Gragg, The Padé table and its relation to certain algorithms of numerical analysis, SIAM Rev., 14 (1972), pp. 1–62.
20. P. Graves-Morris, Efficient reliable rational interpolation, in *Padé Approximation and its Applications*, Amsterdam 1980, M. B. de Bruin and H. van Rossum, eds., vol. 888 of Lecture

Notes in Mathematics, Springer-Verlag, 1981, pp. 28–63.

21. P. R. GRAVES-MORRIS, Practical, reliable, rational interpolation, J. Inst. Maths. Applics., 25 (1980), pp. 267–286.
22. P. R. GRAVES-MORRIS AND T. R. HOPKINS, Reliable rational interpolation, Numer. Math., 36 (1981), pp. 111–128.
23. M. H. GUTKNECHT, Block structure and recursiveness in rational interpolation, in *Approximation Theory VII*, E. W. Cheney, C. K. Chui, and L. L. Schumaker, eds., Academic Press, Boston, 1992. To appear.
24. ———, A completed theory of the unsymmetric Lanczos process and related algorithms, Part II, SIAM J. Matrix Anal. Appl. To appear.
25. ———, Continued fractions associated with the Newton-Padé table, Numer. Math., 56 (1989), pp. 547–589.
26. ———, The rational interpolation problem revisited, Rocky Mountain J. Math., 21 (1991), pp. 263–280.
27. ———, A completed theory of the unsymmetric Lanczos process and related algorithms, Part I, SIAM J. Matrix Anal. Appl., 13 (1992), pp. 594–639.
28. ———, Stable row recurrences in the Padé table and generically superfast lookahead solvers for non-Hermitian Toeplitz systems, Linear Algebra Appl., 188/189 (1993), pp. 351–421.
29. M. H. GUTKNECHT AND M. HOCHBRUCK, Look-ahead Levinson- and Schur-type recurrences in the Padé table, in preparation.
30. ———, Look-ahead Levinson and Schur algorithms for non-Hermitian Toeplitz systems, IPS Research Report 93-11, IPS, ETH-Zürich, August 1993.
31. G. HEINIG AND K. ROST, *Algebraic Methods for Teoplitz-like Matrices and Operators*, Akademie-Verlag, Berlin, DDR, and Birkhäuser, Basel/Stuttgart, 1984.
32. U. HELMKE AND P. A. FUHRMANN, Bezoutians, Linear Algebra Appl., 122/123/124 (1989), pp. 1039–1097.
33. L. KRONECKER, Zur Theorie der Elimination einer Variabel aus zwei algebraischen Gleichungen, Monatsber. Königl. Preuss. Akad. Wiss., (1888), pp. 535–600.
34. H. MAEHLY AND C. WITZGALL, Tschebyscheff-Approximationen in kleinen Intervallen II, Stetigkeitssätze für gebrochen rationale Approximationen, Numer. Math., 2 (1960), pp. 293–307.
35. A. MAGNUS, Certain continued fractions associated with the Padé table, Math. Z., 78 (1962), pp. 361–374.
36. J. MEINGUET, On the solubility of the Cauchy Interpolation problem, in *Approximation Theory*, A. Talbot, ed., Academic Press, 1970, pp. 137–163.
37. R. J. MELESHKO, *A stable algorithm for the computation of Padé approximants*, PhD thesis, Department of Computing Science, University of Alberta (Canada), 1990.
38. G. OPITZ, Steigungsmatrizen, Z. Ang. Math. Mech., 44 (1964), pp. T52–T54.
39. B. N. PARLETT, D. R. TAYLOR, AND Z. A. LIU, A look-ahead Lanczos algorithm for unsymmetric matrices, Math. Comp., 44 (1985), pp. 105–124.
40. M. VAN BAREL AND A. BULTHEEL, The "look-ahead" philosophy applied to matrix rational interpolation problems, in Proceedings MTNS-'93, Regensburg, to appear.
41. ———, A matrix Euclidean algorithm for minimal partial realization, Linear Algebra Appl., 121 (1989), pp. 674–682.
42. ———, A new approach to the rational interpolation problem, J. Comput. Appl. Math., 32 (1990), pp. 281–289.
43. ———, The computation of non-perfect Padé-Hermite approximants, Numerical Algorithms, 1 (1991), pp. 285–304.
44. ———, A new formal approach to the rational interpolation problem, Numer. Math., 62 (1992), pp. 87–122.
45. D. D. WARNER, *Hermite Interpolation with Rational Functions*, PhD thesis, University of California at San Diego, 1974.
46. H. WERNER, A reliable method for rational interpolation, in *Padé Approximation and its Applications*, L. Wuytack, ed., vol. 765 of Lecture Notes in Mathematics, Springer-Verlag, 1979, pp. 257–277.
47. ———, Algorithm 51: A reliable and numerically stable program for rational interpolation in Lagrange data, Computing, 31 (1983), pp. 269–286.

MATRIX RATIONAL INTERPOLATION WITH POLES AS INTERPOLATION POINTS

M. VAN BAREL
Department of Computing Science, K.U.Leuven, Celestijnenlaan 200A, B-3001 Heverlee, Belgium

B. BECKERMANN
Institut für Angewandte Mathematik, Universität Hannover, Welfengarten 1, D-30167 Hannover, Germany

A. BULTHEEL
Department of Computing Science, K.U.Leuven, Celestijnenlaan 200A, B-3001 Heverlee, Belgium

and

G. LABAHN
Department of Computing Science, University of Waterloo, Waterloo, Ontario, Canada

Abstract. In this paper, we show the equivalence between matrix rational interpolation problems with poles as interpolation points and no-pole problems. This equivalence provides an effective method for computing matrix rational interpolants having poles as interpolation points. However, this equivalence is only valid in those cases where enough pole information is known. It is an open problem on how one can transform the pole problem to a no-pole problem in other cases.

Key words: Matrix rational interpolation, pole information

1. Introduction

In this paper, we consider a matrix rational interpolation problem with data given at one or more interpolation points. We also allow these interpolation points to be poles of our matrix rational interpolant. In the case of the scalar pole-problem (also allowing confluent interpolation points and ∞ as an interpolation point), Gutknecht [5] gives an algorithm computing the solution as a continued fraction along a diagonal or a staircase in the Newton-Padé table. Additional information about scalar rational interpolants (for example concerning their existence and uniqueness) can be found in Gutknecht [6] and Stahl [8]. In [13], Van Barel and Bultheel generalize their algorithm of [9] also allowing confluent interpolation points, poles and ∞ as an interpolation point.

The algorithm of [13] is based on the fact that all solutions of an equivalent linearized interpolation problem can be written in terms of some basis vectors of an $\mathbb{F}[z]$-submodule of the module of all polynomial vectors $\mathbb{F}^n[z]$. To add a new interpolation condition, we only have to update this set of basis vectors. This can be done in a fast and reliable way leading to $O(k^2)$ FLOPS for k interpolation conditions. For the pole-free problem, this was generalized to the vector case [1, 4,

A. Cuyt (ed.), Nonlinear Numerical Methods and Rational Approximation II, 137–148.

10, 11] and also to the matrix case [2, 3, 12]. Once we have the basis vectors, we can parametrize all solutions of the original rational interpolation problem having minimal complexity [12]. In this paper, we show how to transform the pole-problem to a pole-free problem. The latter can be solved using one of the methods referred to before.

In the next section, we define the rational interpolation problem that we want to solve. This section also discusses the $\vec{s}$-McMillan degree, used as a measure of the complexity of a rational interpolant. Sections 3 and 4 introduces the concept of (α, σ)-characteristics of a formal Laurent series around a point α, and shows how they can be computed. In Section 5, we show that our original rational interpolation problem with pole information is equivalent to another rational interpolation problem *without* poles as interpolation points. However, as we show in Section 6, the equivalence of the pole and the pole-free problem is only true when enough interpolation data are known at each pole. It remains an open problem how to solve the rational interpolation problem in an efficient way when this condition is not met.

2. Rational interpolation with poles

For the rational interpolation problem considered here, we have interpolation information at each point (possibly a pole) $\alpha \in I$, with I a non-empty, finite subset of $\mathbb{F}$, $\mathbb{F}$ a field. We will use the following notations: $\mathbb{F}[z]$ the set of polynomials, $\mathbb{F}[[z]]_\alpha$ the set of formal power series around $\alpha \in \mathbb{F}$, $\mathbb{F}(z)$ the field of rational functions and $\mathbb{F}((z))_\alpha$ the field of formal Laurent series

$$\mathbb{F}((z))_\alpha = \{\frac{1}{(z-\alpha)^k} \cdot f : f \in \mathbb{F}[[z]]_\alpha, k \in \mathbf{Z} \text{ (the set of integer numbers)}\}.$$

We also have the corresponding matrix versions, for example $\mathbb{F}^{p\times q}((z))_\alpha$ where p, q are positive integers. Note that $\mathbb{F}[z] \subset \mathbb{F}(z) \subset \mathbb{F}((z))_\alpha$.

In the sequel, the relations $\leq, \geq, \min$ between integer vectors have to be understood componentwise. The vector $\vec{e} := (1, 1, \ldots, 1, 1)$ and e_j is the j-th column of the identity matrix.

DEFINITION 1 *The* α-order, $\alpha \in \mathbb{F}$, *of* $\mathbf{T} \in \mathbb{F}^{p\times q}((z))_\alpha$ *is defined by*

$$\mathrm{ord}_\alpha \mathbf{T} := \sup\{\sigma \in \mathbf{Z} : \mathbf{T} = (z-\alpha)^\sigma \cdot \mathbf{R} \textit{ with } \mathbf{R} \in \mathbb{F}^{p\times q}[[z]]_\alpha\},$$

i.e., $\mathbf{R}$ *is a formal matrix power series around* α. *Note that* $\mathrm{ord}_\alpha \mathbf{0} = +\infty$. *For* $\mathbf{T} \in \mathbb{F}^{p\times q}((z))_\alpha$ *with rows* $\mathbf{T}_1, \ldots, \mathbf{T}_p \in \mathbb{F}^{1\times q}((z))_\alpha$ *the* α-row order $\mathrm{rord}_\alpha \mathbf{T}$ *is a vector with components* $\mathrm{ord}_\alpha \mathbf{T}_i$. *Similarly we define the* α-column order $\mathrm{cord}_\alpha \mathbf{T}$ *with respect to the columns of* $\mathbf{T}$. □

For a vector $\vec{\sigma} = (\sigma_1, \ldots, \sigma_m)$ with components from $\mathbf{Z} \cup \{+\infty\}$ we denote $(z-\alpha)^{\vec{\sigma}} := \mathrm{diag}\big((z-\alpha)^{\sigma_1}, \ldots, (z-\alpha)^{\sigma_m}\big)$, where $(z-\alpha)^{+\infty} := 0$. In this case we can write for any $\mathbf{T}$ and α

$$\mathbf{T}(z) = (z-\alpha)^{\mathrm{rord}_\alpha \mathbf{T}} \cdot \mathbf{R}_1(z) \quad \text{and} \quad \mathbf{T}(z) = \mathbf{R}_2(z) \cdot (z-\alpha)^{\mathrm{cord}_\alpha \mathbf{T}}, \tag{1}$$

where the rows of $\mathbf{R}_1$ and the columns of $\mathbf{R}_2$ corresponding to an α-order $\neq +\infty$ are unique power series around α with a non-zero value at α. The decomposition of (1) is then made unique by putting the remaining rows of $\mathbf{R}_1$ and the remaining columns of $\mathbf{R}_2$ equal to zero. We refer to (1) as the *α-row* and *α-column representation* of $\mathbf{T}$.

DEFINITION 2 *The projection operator $\Pi_{\alpha,\sigma}$ acting on $\mathbf{T} \in \mathbb{F}^{p\times q}((z))_\alpha$ is defined by*

$$\Pi_{\alpha,\sigma}\mathbf{T} := \sum_{k=\tau}^{\sigma-1} T_k(z-\alpha)^k, \quad \textit{where} \quad \mathbf{T}(z) = \sum_{k=\tau}^{\infty} T_k(z-\alpha)^k.$$

$\Pi_{\alpha,\sigma}\mathbf{T}$ will be called the (α,σ) main part of $\mathbf{T}$. □

The rational interpolation problem to be considered here is the following: Given $\mathbf{T}_\alpha \in \mathbb{F}^{p\times q}((z))_\alpha$, $\forall\alpha \in I$, find a rational function $Z \in \mathbb{F}^{p\times q}(z)$ such that the $(\alpha, \sigma(\alpha))$ main parts of Z and $\mathbf{T}_\alpha$ are the same $\forall\alpha \in I$. However, this problem has an infinite number of solutions and as such we only look for those rational functions which are in some sense solutions of minimal complexity. Complexity of a matrix rational function can be defined in several ways. In our case we use the notion of a $\vec{s}$-McMillan degree [12].

DEFINITION 3 *Let $\vec{s} = (s_1, s_2, \ldots, s_{p+q})$ with each (shift parameter) $s_i \in \mathbb{Z}$. The* $\vec{s}$-McMillan degree *of a pair of matrix polynomials (N, D), with $N \in \mathbb{F}^{p\times q}[z]$ and $D \in \mathbb{F}^{q\times q}[z]$, is defined as*

$$\vec{s}\text{-MM-deg } (N,D) = \max\left\{\text{deg det } P : P \textit{ is a } q\times q \textit{ submatrix of } z^{\vec{s}} \cdot \begin{bmatrix} N \\ D \end{bmatrix}\right\}.$$

For any rational function $Z \in \mathbb{F}^{p\times q}(z)$ let $Z = N\cdot D^{-1}$ be a right coprime polynomial matrix fraction description (RCPMFD). Then the $\vec{s}$-McMillan degree of Z is defined as

$$\vec{s}\text{-MM-deg } Z = \vec{s}\text{-MM-deg } (N,D).$$

□

Note that the RCPMFD of a rational function is unique up to a right unimodular factor U of both numerator and denominator [7, p. 441, Th. 6.5-4]); hence the definition of a $\vec{s}$-McMillan degree is well defined. When $\vec{s} = \vec{0}$ we get the classical McMillan degree. The McMillan degree is a well-known concept in linear system theory. It is a measure of the complexity of a linear system. (see e.g. [7, p. 439]).

The rational problem which we discuss is now given as follows.

DEFINITION 4 *Given $\mathbf{T}_\alpha \in \mathbb{F}^{p\times q}((z))_\alpha$, $\forall\alpha \in I$, a vector $\vec{s} \in \mathbb{Z}^{(p+q)\times 1}$ and the order indices $\sigma(\alpha)$, find a $Z \in \mathbb{F}^{p\times q}(z)$ with minimal $\vec{s}$-McMillan degree such that*

$$\text{ord}_\alpha(\mathbf{T}_\alpha - Z) \geq \sigma(\alpha) \quad \forall\alpha \in I. \tag{2}$$

□

We will show that the original rational interpolation problem given some formal Laurent series is equivalent to a rational interpolation problem with the same interpolation points $\alpha \in I$ but given formal power series (no poles anymore).

3. (α,σ)-characteristics

Since the rational interpolation problem studied in this paper only requires the (α,σ) main part of given formal Laurent series, consideration will have to be taken when we define characteristics limited to this main part.

DEFINITION 5 *A formal (matrix) Laurent series* $\mathbf{T} \in \mathbb{F}^{p\times q}((z))_\alpha$ *is called* α-column reduced *iff in the* α*-column representation* $\mathbf{T}(z) = \mathbf{R}(z)\cdot(z-\alpha)^{\mathrm{cord}_\alpha \mathbf{T}}$ *the nonzero columns of* $\mathbf{R}(\alpha)$ *are linearly independent.* $\mathbf{T} \in \mathbb{F}^{p\times q}((z))_\alpha$ *is called* (α,σ)-column reduced *iff* $\Pi_{\alpha,\sigma}\mathbf{T}(z)$ *is* α*-column reduced.* □

Clearly if $\mathbf{T}$ is (α,σ)-column reduced then so is its Laurent polynomial counterpart $\Pi_{\alpha,\sigma}\mathbf{T}$.

DEFINITION 6 *Given a formal (matrix) Laurent series* $\mathbf{T} \in \mathbb{F}^{p\times q}((z))_\alpha$, *a matrix polynomial* $D_\mathbf{T} \in \mathbb{F}^{q\times q}[z]$ *is called an* (α,σ)-column reducing denominator *of* $\mathbf{T}$ *iff*

$$\begin{array}{ll}(a) & \det D_\mathbf{T}(\alpha) \neq 0 \quad \textit{and hence } D_\mathbf{T}^{-1} \in \mathbb{F}^{q\times q}[[z]]_\alpha,\\ (b) & \mathbf{T}\cdot D_\mathbf{T} \textit{ is } (\alpha,\sigma)\textit{-column reduced.}\end{array} \tag{3}$$

□

Note that an (α,σ)-column reduced denominator does not depend on $\mathbf{T}$ but only on $\Pi_{\alpha,\sigma}\mathbf{T}$. This can be verified using the following lemma.

LEMMA 1 *For any matrix Laurent series* $\mathbf{T}_1$ *around* α *and any matrix power series* $\mathbf{T}_2$ *around* α*, both of suitable size, we have*

$$\Pi_{\alpha,\sigma}(\mathbf{T}_1\cdot\mathbf{T}_2) = \Pi_{\alpha,\sigma}(\Pi_{\alpha,\sigma}(\mathbf{T}_1)\cdot\mathbf{T}_2). \tag{4}$$

Proof: Trivial. □

Moreover, as seen in the following Theorem, there exist further invariances, namely so-called (α,σ)-characteristics.

THEOREM 1 *Let* $\mathbf{T} \in \mathbb{F}^{p\times q}((z))_\alpha$*, with corresponding* (α,σ)*-column reduced denominator* $D_\mathbf{T} \in \mathbb{F}^{q\times q}[z]$ *and* α*-column representation*

$$\Pi_{\alpha,\sigma}(\mathbf{T}\cdot D_\mathbf{T}) = N_\mathbf{T}\cdot(z-\alpha)^{\vec{\delta}(\mathbf{T})}.$$

Then the entries of $\vec{\delta}(\mathbf{T})$ *are unique up to permutation.*

Proof: Suppose D_1, D_2 are two (α,σ)-reducing denominators of $\mathbf{T}$, having the α-column representations

$$\Pi_{\alpha,\sigma}(\mathbf{T}\cdot D_1) = N_1\cdot(z-\alpha)^{\vec{\delta}_1}, \quad \text{and} \quad \Pi_{\alpha,\sigma}(\mathbf{T}\cdot D_2) = N_2\cdot(z-\alpha)^{\vec{\delta}_2}.$$

Without loss of generality we may assume that the columns of D_1, D_2 are ordered in such a manner that the components of both $\vec{\delta}_1$ and $\vec{\delta}_2$ are ascending. Suppose that there exists a k with $\delta_{1,j} = \delta_{2,j}$, $j = q, q-1, \ldots, k$ but $\delta_{1,k-1} > \delta_{2,k-1}$. Since

$D_1(\alpha)$ is regular, we can form $R := D_1^{-1} \cdot D_2$, a (matrix) power series about α. Set $\Lambda_j(z) := (z-\alpha)^{\vec{\delta}_j}$, $j = 1, 2$. Then using Lemma 1 we have

$$\begin{aligned} N_2 \cdot \Lambda_2 &= \Pi_{\alpha,\sigma}(\mathbf{T} \cdot D_2) = \Pi_{\alpha,\sigma}(\mathbf{T} \cdot D_1 \cdot D_1^{-1} \cdot D_2) \\ &= \Pi_{\alpha,\sigma}(\Pi_{\alpha,\sigma}(\mathbf{T} \cdot D_1) \cdot D_1^{-1} \cdot D_2) = \Pi_{\alpha,\sigma}(N_1 \cdot \Lambda_1 \cdot R). \end{aligned}$$

Since $N_j \cdot \Lambda_j = \Pi_{\alpha,\sigma}(N_j \cdot \Lambda_j)$, $j = 1, 2$ are both (α, σ)-column reduced, Definition 5 implies that $R(\alpha)$ has the following form

$$R(\alpha) = \begin{bmatrix} R_{11} & 0 & 0 \\ R_{21} & 0 & 0 \\ R_{31} & R_{3,2} & R_{3,3} \end{bmatrix}$$

where the rows and columns are partitioned as: $k-2, 1, q-k+1$. But R is regular at α because $D_1(\alpha)$ and $D_2(\alpha)$ are both regular, a contradiction. □

DEFINITION 7 *The entries of $\vec{\delta}(\mathbf{T})$ defined in the previous theorem are called the* (α, σ)-characteristics[1] *of* $\mathbf{T}$. *The number of finite (α, σ)-characteristics is called the* (α, σ)-index[2] *and denoted as $\kappa_{\alpha,\sigma}(\mathbf{T})$. In addition, the sum of the negative (α, σ)-characteristics of $\mathbf{T}$ is called the* (α, σ)-pole multiplicity *and will be abbreviated as* $\Delta(\alpha, \sigma, \mathbf{T})$. □

The dependency of (α, σ)-characteristics of $\mathbf{T}$ on σ can be described explicitly.

THEOREM 2 *Let $\mathbf{T} \in \mathbb{F}^{p\times q}((z))_\alpha$, $\sigma_1 < \sigma_2$ and $\vec{\delta}_1$ the (α, σ_1)-characteristics and $\vec{\delta}_2$ the (α, σ_2)-characteristics of $\mathbf{T}$, respectively. Then the entries of $\vec{\delta}_1$ and $\vec{\delta}_2$ smaller than σ_1 are unique up to permutation.*

Proof: Let $D_{\mathbf{T}}$ be an (α, σ_2)-reducing denominator of $\mathbf{T}$, then (after suitable permutation of columns of $D_{\mathbf{T}}$) we may write

$$\Pi_{\alpha,\sigma_2}(\mathbf{T} \cdot D_{\mathbf{T}}) = [N_{\mathbf{T}}^- \;\; N_{\mathbf{T}}^+] \cdot \begin{bmatrix} (z-\alpha)^{\vec{\delta}^-} & 0 \\ 0 & (z-\alpha)^{\vec{\delta}^+} \end{bmatrix}$$

where all the entries of $\vec{\delta}_2^-$ are smaller than σ_1 and $\vec{\delta}_2^+$ contains the remaining entries. Applying the projection operator Π_{α,σ_1} gives

$$\Pi_{\alpha,\sigma_1}(\mathbf{T} \cdot D_{\mathbf{T}}) = [\Pi_{\alpha,\sigma_1}\left(N_{\mathbf{T}}^- \cdot (z-\alpha)^{\vec{\delta}^-}\right)(z-\alpha)^{-\vec{\delta}^-} \quad 0] \cdot \begin{bmatrix} (z-\alpha)^{\vec{\delta}^-} & 0 \\ 0 & 0 \end{bmatrix}$$

which is obviously α-column reduced.[3]

□

As a consequence of Theorem 2, we may specify some cases where the (α, σ)-pole multiplicity does not depend on σ.

---

[1] Note the connection with the structural indices from linear system theory: see Kailath [7, p. 447]

[2] This number is bounded by $\min\{p, q\}$, the maximal number of columns of $N_{\mathbf{T}}(\alpha)$ which can be linearly independent.

[3] In particular, we may conclude that an (α, σ_2)-reducing denominator also is an (α, σ_1)-reducing denominator for $\sigma_1 < \sigma_2$.

COROLLARY 1 *Let* $\mathbf{T} \in \mathbb{F}^{p\times q}((z))_\alpha$ *with* $\kappa_{\alpha,\sigma} = \min\{p,q\}$ *or* $\sigma \geq 0$. *Then for any* $\mathbf{T}'$ *with* $\mathrm{ord}_\alpha(\mathbf{T}-\mathbf{T}') \geq \sigma$, *the* $(\alpha, 0)$*-characteristics of* $\mathbf{T}$ *and* $\mathbf{T}'$ *coincide. In particular, the pole multiplicities* $\Delta(\alpha,\sigma,\mathbf{T}')$, $\Delta(\alpha,\sigma,\mathbf{T})$, $\Delta(\alpha,0,\mathbf{T}')$ *and* $\Delta(\alpha,0,\mathbf{T})$ *all are equal.* □

Multiplication with $(z-\alpha)^\tau$ changes the α-characteristics in the following way:

COROLLARY 2 *Let* $\mathbf{T}$ *have the* (α,σ)*-characteristics* $\vec{\delta}$ *and let* $D_\mathbf{T}$ *be an* (α,σ)*-reducing denominator of* $\mathbf{T}$. *Then for any integer* τ, *the function* $\mathbf{T}_\tau := (z-\alpha)^\tau \cdot \mathbf{T}$ *has the* $(\alpha,\sigma+\tau)$*-characteristics* $\vec{\delta}+\tau\cdot\vec{e}$ *and* $D_\mathbf{T}$ *is an* $(\alpha,\sigma+\tau)$*-reducing denominator of* $\mathbf{T}_\tau$. □

4. Construction of (α,σ)-characteristics

How can we construct an (α,σ)-reduced denominator and, hence, the (α,σ)-characteristics of a given formal Laurent series $\mathbf{T}$? The following constructive method generates a finite sequence $(N_k, D_k, V_k)_k$ with

$$\begin{aligned} \text{Initialization:}\quad & D_0 := I_{q\times q} \\ \text{Recurrence:}\quad & N_k \in \mathbb{F}^{p\times q}((z))_\alpha, \quad N_k := \Pi_{\alpha,\vec{\sigma}}(\mathbf{T}\cdot D_k) \\ \text{Recurrence:}\quad & D_{k+1} := D_k \cdot V_k \in \mathbb{F}^{q\times q}[z], \quad D_{k+1}(\alpha) \ \text{ being regular,} \end{aligned}$$

and N_k being a Laurent polynomial around α. In the second recurence the polynomial factor V_k is determined as follows:

If N_k is not α-column reduced, we know from Definition 5 that $\exists a \in \mathbb{F}^{q\times 1}$ with

$$\mathbf{R}_k(\alpha)\cdot a = 0 \qquad \text{and} \qquad N_k(z) = \mathbf{R}_k(z)\cdot(z-\alpha)^{\mathrm{cord}_\alpha N_k}$$

where a has zero entries when the corresponding columns of $\mathbf{R}(\alpha)$ are zero. Let $\mathrm{cord}_{\max} N_k$ be equal to the maximum of the (finite) entries of $\mathrm{cord}_\alpha N_k$ corresponding to nonzero entries of N_k and let j be the index of this maximum in $\mathrm{cord}_\alpha N_k$. Construct the polynomial vector $v \in \mathbb{F}^{q\times 1}[z]$ as

$$v := (z-\alpha)^{-\mathrm{cord}_\alpha N_k + \mathrm{cord}_{\max} N_k \cdot \vec{e}}\, a$$

with the entries of v corresponding to zero entries of a also equal to zero. In this case we take $V_k = [e_1, e_2, \cdots, e_{j-1}, v, e_{j+1}, \cdots, e_q]$. Note that $V_k(\alpha)$ is regular.

If N_k is α-column reduced, D_k has the required properties of an (α,σ)-column reduced denominator of $\mathbf{T}$. That this process ends after a finite number of steps is clear since at each step we increase the α-order of one of the non-zero columns of N_k while at the same time the α-orders of the nonzero columns of $N_k = \Pi_{\alpha,\sigma}(N_k)$ are bounded by $\sigma - 1$.

We mention three important aspects of the above construction. First notice that if $\mathbf{T}$ is a power series around α, then all N_k are matrix polynomials. Secondly, we see that each V_k and therefore also each D_k may be assumed to be unimodular. The final point is stated in the next theorem. Its proof makes repeated use of Lemma 1.

THEOREM 3 *Let the matrix rational function Z have the RCPMFD $Z = N \cdot D^{-1}$ with matrix polynomials $N \in \mathbb{F}^{p\times q}[z]$, $D \in \mathbb{F}^{q\times q}[z]$, then there holds:*

(a) If $Z(\alpha)$ is finite, then the (α, σ)-characteristics of Z and the (α, σ)-characteristics of N coincide for all σ. In addition none of them are negative.

(b) With $\gamma > \mathrm{ord}_\alpha \det D$ we have $\kappa_{\alpha,\gamma}(D) = q$, and with D having the (α, γ)-characteristics $\delta_1, \ldots, \delta_{q'}, 0, \ldots, 0$, q' minimal, Z has the $(\alpha, 0)$-characteristics $-\delta_1, \ldots, -\delta_{q'}, +\infty, \ldots, +\infty$.

Proof: Note that in case (a) it is also true that $D(\alpha)$ is regular. Hence we may initialize the above procedure with $D_0 = D$, i.e., there exists a unimodular matrix $V \in \mathbb{F}^{q\times q}[z]$ such that $D \cdot V$ is an (α, σ)-reducing denominator and therefore $\Pi_{\alpha,\sigma}(N \cdot V)$ is (α, σ)-reduced, with its α-column representation yielding the (α, σ)-characteristics of Z.

In order to prove (b), we first notice that $\det D \neq 0$ and so the quantity γ is well-defined. Suppose that the unimodular matrix V is an (α, γ)-reducing denominator of D, i.e., there exist two polynomial matrices D_0, D_0' with

$$D \cdot V = D_0 \cdot (z - \alpha)^{\vec{\delta}} + (z - \alpha)^{\gamma} \cdot D_0',$$

$\vec{\delta}$ containing only nonnegative integers or $+\infty$, and the nonzero columns of D_0 are linearly independent at α. Taking determinants on both sides we see that none of the columns of D_0 may be different from zero , since otherwise the α-order of the right hand side would be greater or equal to γ which is greater than $\mathrm{ord}_\alpha \det (D \cdot V) = \mathrm{ord}_\alpha \det D < \gamma$. Consequently, $\kappa_{\alpha,\gamma}(D) = q$, and we may combine the matrices D_0, D_0' such that without loss of generality $D_0' = 0$. By assumption, the pair (N, D) and therefore the pair $(N \cdot V, D \cdot V)$ is right coprime. The first q' columns of $D \cdot V = D_0 \cdot (z - \alpha)^{\vec{\delta}}$ are zero at α, and hence the first q' columns of $N \cdot V$ denoted by N^+ must be linearly independent at α. We may conclude that D_0 is an $(\alpha, 0)$-reducing denominator of Z since with $\vec{\delta}' := (\delta_1, \ldots, \delta_{q'})$

$$\Pi_{\alpha,0}(Z \cdot D_0) = \Pi_{\alpha,0}(N \cdot V \cdot (z - \alpha)^{-\vec{\delta}}) = \begin{bmatrix} \Pi_{\alpha,0}(N^+ \cdot (z - \alpha)^{-\vec{\delta}'}) & 0 \end{bmatrix}$$

being obviously α-reduced. □

EXAMPLE 1 Given $\alpha = 0$, $\sigma = 2$, we use the method described above to determine the (α, σ)-characteristics of

$$\mathbf{T} = N_0 = \begin{bmatrix} 1 & -1 \\ 0 & 0 \end{bmatrix} z^{-1} + \begin{bmatrix} 1 & 2 \\ 0 & 0 \end{bmatrix} z^0 + \begin{bmatrix} 0 & 0 \\ 0 & 1 \end{bmatrix} z^1 + \cdots. \tag{5}$$

Because N_0 is not α-column reduced, we multiply to the right by

$$V_0 = \begin{bmatrix} 1 & 1 \\ 0 & 1 \end{bmatrix}. \tag{6}$$

We get after projection

$$N_1 = \begin{bmatrix} 1 & 0 \\ 0 & 0 \end{bmatrix} z^{-1} + \begin{bmatrix} 1 & 3 \\ 0 & 0 \end{bmatrix} z^0 + \begin{bmatrix} 0 & 0 \\ 0 & 1 \end{bmatrix} z^1. \tag{7}$$

Because N_1 still is not α-column reduced, we multiply to the right by

$$V_1 = \begin{bmatrix} 1 & -3z \\ 0 & 1 \end{bmatrix}. \tag{8}$$

We get after projection

$$N_2 = \begin{bmatrix} 1 & 0 \\ 0 & 0 \end{bmatrix} z^{-1} + \begin{bmatrix} 1 & 0 \\ 0 & 0 \end{bmatrix} z^{0} + \begin{bmatrix} 0 & -3 \\ 0 & 1 \end{bmatrix} z^{1}. \tag{9}$$

This last result $N_2 = \Pi_{\alpha,\sigma}(\mathbf{T} \cdot D_2)$ is α-column reduced. Therefore, the (α, σ)-characteristics are in this case: -1 and 1, the (α, σ)-pole multiplicity $\Delta(\alpha, \sigma, \mathbf{T}) = -1$ and the (α, σ)-index $\kappa_{\alpha,\sigma} = 2$. □

5. Equivalence

Before we can show equivalence between the original rational interpolation problem with pole information and another rational interpolation problem without poles as interpolation points, we need the following theorem.

THEOREM 4 *For each matrix rational function $Z \in \mathbb{F}^{p\times q}(z)$ with $(\alpha, 0)$-characteristics $\vec{\delta}$ and for each $\tau \geq 0$ with $\vec{\delta} + \tau \cdot \vec{e} \geq 0$,*

$$\vec{s}\text{-MM-deg } Z = \vec{r}\text{-MM-deg } \Big((z-\alpha)^{\tau} \cdot Z\Big) - \Delta(\alpha, 0, Z) \tag{10}$$

with $\vec{r} := \vec{s} - \tau \cdot (1, \ldots, 1, 0, \ldots, 0)$ (q zeros).

Proof: Because $\vec{\delta}_\tau := \vec{\delta} + \tau \cdot \vec{e} \geq 0$, by Corollary 2 the matrix rational function $Z_\tau := (z-\alpha)^\tau \cdot Z$ has the (α, τ)-characteristics $\vec{\delta}_\tau$, and $Z_\tau(\alpha)$ is finite. By the proof of Theorem 3(a) we get a RCPMFD $N_\tau \cdot D_\tau^{-1}$ for Z_τ with D_τ an (α, τ)-reducing denominator of Z_τ, i.e.,

$$Z_\tau \cdot D_\tau = N_\tau = N_\tau' \cdot (z-\alpha)^{\vec{\delta}_\tau} + (z-\alpha)^\tau \cdot N_\tau''$$

with N_τ'' being polynomial. Without loss of generality we may suppose that the components of $\vec{\delta}_\tau$ are either smaller than τ or equal to $+\infty$, and in addition ordered such that with $\vec{\delta}_\tau^-$ containing the finite entries of $\vec{\delta}_\tau$ (which then are non–negative and smaller than τ) we obtain the consistent partitions

$$N_\tau' = \begin{bmatrix} N_\tau^- \cdot (z-\alpha)^{\vec{\delta}^-} & 0 \end{bmatrix}, \quad D_\tau = \begin{bmatrix} D_\tau^- & D_\tau^+ \end{bmatrix}.$$

Notice that the matrix $N_\tau^-(\alpha)$ by definition of (α, τ)-characteristics has maximal rank, and $D_\tau(\alpha)$ is regular, hence also the matrix $D_\tau^+(\alpha)$ has maximal rank.

A RCPMFD $N \cdot D^{-1}$ for Z can be based upon N_τ and D_τ as follows. Define

$$\Lambda^+ := \begin{bmatrix} (z-\alpha)^{-\vec{\delta}_\tau^-} & 0 \\ 0 & (z-\alpha)^{-\tau} \cdot I \end{bmatrix}, \quad \Lambda^- := \begin{bmatrix} (z-\alpha)^{\tau\vec{e}-\vec{\delta}_\tau^-} & 0 \\ 0 & I \end{bmatrix},$$

where $\Lambda^- := (z-\alpha)^\tau \cdot \Lambda^+$ is polynomial. Because $Z = (z-\alpha)^{-\tau} \cdot Z_\tau$, a polynomial matrix fraction description of Z is given by the numerator $N := N_\tau \cdot \Lambda^+$ and the denominator $D := D_\tau \cdot \Lambda^-$. In order to show that (N, D) also is right coprime and therefore a RCPMFD for Z, we prove the equivalent condition that $\left[N(z)^T \;\; D(z)^T\right]^T$ has maximal rank $\forall z \in \mathbb{F}$ (cf., [7, Lemma 6.3-6, p. 379]). In case $z \neq \alpha$, this condition follows immediately from the right coprimeness of (N_τ, D_τ). For $z = \alpha$ we obtain

$$\begin{bmatrix} N \\ D \end{bmatrix}(\alpha) = \begin{bmatrix} N_\tau \cdot \Lambda^+ \\ D_\tau \cdot \Lambda^- \end{bmatrix}(\alpha) = \begin{bmatrix} N_\tau^-(\alpha) & 0 \\ X & D_\tau^+(\alpha) \end{bmatrix}$$

having also maximal rank (independently of the suitable matrix X) as shown above.

Hence, for the McMillan degree there holds

$$\begin{aligned} \vec{s}\text{-MM-deg } Z &= \vec{s}\text{-MM-deg } (N_\tau \cdot \Lambda^+, D_\tau \cdot \Lambda^-) \\ &= \vec{r}\text{-MM-deg } (N_\tau \cdot \Lambda^+ \cdot (z-\alpha)^\tau, D_\tau \cdot \Lambda^-) \\ &= \vec{r}\text{-MM-deg } (N_\tau, D_\tau) + \text{deg det } \Lambda^- \\ &= \vec{r}\text{-MM-deg } Z_\tau + \text{deg det } \Lambda^-. \end{aligned}$$

Note that, by Corollary 2, D_τ also is an $(\alpha, 0)$-reducing denominator of Z, such that $\Delta(\alpha, 0, Z) = -\text{deg det } \Lambda^-$. □

Now we are able to show the main assertion.

THEOREM 5 *Given the set I of interpolation points α, the formal Laurent series $\mathbf{T}_\alpha$, the order indices $\sigma(\alpha)$, $\forall \alpha \in I$ and the shift parameters $\vec{s}$. Let $\mathbf{R}_\alpha, \rho(\alpha), \tau(\alpha)$, $\forall \alpha \in I$, τ, ω be defined as follows:*

$$\begin{array}{ll} (a) & \text{choose } \tau_\alpha \text{ such that } -\tau(\alpha) \leq \min\{0, \text{ord}_\alpha(\mathbf{T}_\alpha)\}, \\ (b) & \tau := \sum_{\alpha \in I} \tau(\alpha), \\ (c) & \vec{r} := \vec{s} - \tau \cdot (1, \ldots, 1, 0, \ldots, 0) \quad (q \text{ zeros}), \\ (d) & \rho(\alpha) := \sigma(\alpha) + \tau(\alpha) \\ (e) & \mathbf{R}_\alpha := \omega \cdot \mathbf{T}_\alpha \in \mathbb{F}^{p \times q}[[z]]_\alpha \quad \text{with } \omega := \prod_{\alpha \in I} (z-\alpha)^{\tau(\alpha)}, \\ (f) & \Delta := \sum_{\alpha \in I} \Delta(\alpha, \sigma(\alpha), \mathbf{T}_\alpha). \end{array} \tag{11}$$

Moreover, let

$$\forall \alpha \in I: \quad \begin{array}{ll} (a) & \kappa_{\alpha,\sigma(\alpha)}(\mathbf{T}_\alpha) = \min\{p, q\} \\ & \text{or} \\ (b) & \sigma(\alpha) \geq 0. \end{array} \tag{12}$$

(a) For each Z satisfying (2) we have

$$\vec{s}\text{-MM-deg } Z = \vec{r}\text{-MM-deg } (\omega \cdot Z) - \Delta. \tag{13}$$

(b) If Z_τ is a solution to the rational interpolation problem with respect to $(\mathbf{R}_\alpha, \rho(\alpha)$, $\forall \alpha \in I$, $\vec{r})$ then $Z := \frac{1}{\omega} \cdot Z_\tau$ is a solution to the rational interpolation problem with respect to $(\mathbf{T}_\alpha, \sigma(\alpha), \forall \alpha \in I, \vec{s})$.

Proof: By Corollary 1 the $(\alpha, 0)$-characteristics of $\mathbf{T}_\alpha$ and Z coincide and $\Delta(\alpha$, $\sigma(\alpha)$, $\mathbf{T}_\alpha) = \Delta(\alpha, 0, Z)$. Hence, part (a) follows by successively applying Theorem 4. Assertion (b) is an immediate consequence. □

EXAMPLE 2 *Take interpolation in* 0, 1 *and* −1 *with*

$$\sigma(0)=2,\quad \sigma(1)=-1,\quad \sigma(-1)=0,\quad \mathbf{T}_\alpha=\mathbf{T}=\begin{bmatrix} \frac{-1}{(z+1)\cdot(z-1)^2} & z-2 \\ z & \frac{1}{(z-1)^3}\end{bmatrix} \tag{14}$$

We can apply Theorem 5 because $\sigma(0)\geq 0$, *the negative* $(\alpha,\sigma(\alpha))=(1,-1)$-*characteristics of* $\mathbf{T}$ *are* −3 *and* −2, *i.e. they are all known, and* $\sigma(-1)\geq 0$. *Let us choose the shift parameters* $\vec{s}=(1,1,0,0)$. *We can take* $\tau(0)=0$, $\tau(1)=3$ *and* $\tau(-1)=1$. *Hence,* $\rho(0)=2$, $\rho(1)=2$, $\rho(-1)=1$, $\omega(z)=(z-1)^3(z+1)$, $\vec{r}=(-3,-3,0,0)$ *and* $\mathbf{R}_\alpha=\mathbf{R}=\mathbf{T}\cdot\omega$, $\forall\alpha\in I=\{0,1,-1\}$. *To solve the pole-free problem, we use the module-theoretic framework of [12]. A basis for the submodule corresponding to* $(\mathbf{R},\rho(\alpha),\forall\alpha\in I,\vec{r})$ *is given by the columns of the polynomial matrix* $\mathbf{P}$:

$$\begin{bmatrix} -(3z-2)(z+1)(z-1)^2 & -(z-1) & -(z+1)(z-1) & (z+1)(z-1)^2(z-2) \\ z+1 & -(z+1)(z-1)^2z & -(z+1)(z-1)^2z & -(z+1)^2 \\ 0 & -1 & -(z+1) & 0 \\ -1 & 0 & 0 & z+1 \end{bmatrix}$$

with corresponding $\vec{r}$-*degrees* (1, 1, 1, 1). *All solutions* Z_τ *of the "no-pole" problem having minimal* $\vec{r}$-*MM-degree* 2 *can be written as* $Z_\tau=N_\tau\cdot D_\tau^{-1}$ *with*

$$\begin{bmatrix} N_\tau \\ D_\tau \end{bmatrix}=\mathbf{P}\cdot\begin{bmatrix} 1 & 0 \\ 0 & 1 \\ a & c \\ b & d \end{bmatrix} \qquad \text{with } a,b,c,d\in\mathbb{F}$$

and $D_\tau(0)$ *regular, i.e.* $b\neq 1$ *and* $c\neq -1$, $D_\tau(1)$ *regular, i.e.* $b\neq 2$ *and* $c\neq -2$, $D_\tau(-1)$ *regular, giving no further conditions on* a,b,c *or* d.

Hence, all solutions Z *of the "pole" problem can be written as* $Z=Z_\tau/\omega$ *having minimal* $\vec{s}$-*MM-degree*

$$\vec{s}\text{-MM-deg } Z=\vec{r}\text{-MM-deg } Z_\tau-\Delta=2-(-3-2-1)=8.$$

□

Note that the equivalence is also valid when we consider vector order indices $\vec{\sigma}(\alpha)$ instead of scalar ones.

COROLLARY 3 *Consider the rational interpolation problem of Definition 4 but with*

$$\text{cord}_\alpha(\mathbf{T}_\alpha-Z)\geq\vec{\sigma}(\alpha)\quad\forall\alpha\in I$$

instead of (2). Take the notation of Theorem 5 except for $\rho(\alpha)$ *for which a vector variant is defined as* $\vec{\rho}(\alpha):=\vec{\sigma}(\alpha)+\tau\vec{e}$. *Instead of (12), we assume that*

$$\forall\alpha\in I:\quad \begin{array}{l} (a)\ \kappa_{\alpha,\sigma_{\min}(\alpha)}(\mathbf{T}_\alpha)=\min\{p,q\} \\ or \\ (b)\ \vec{\sigma}(\alpha)\geq 0 \end{array} \tag{15}$$

with $\sigma_{\min}(\alpha) = \min_{i=1}^{p}\{\sigma_i(\alpha)\}$. Δ *is defined as*

$$\Delta = \sum_{\alpha \in I} \Delta(\alpha, \sigma_{\min}(\alpha), \mathbf{T}_\alpha).$$

In this case, the results of Theorem 5 are also valid.

Proof: The condition (15) guarantees that the $(\alpha, 0)$-indices of Z are fixed. □

6. Open problem

If condition (12) in Theorem 5 (or (15) of Corollary 3) is not satisfied, then it is an open problem how the original rational interpolation problem can be transformed into one without pole information. As an example we have:

EXAMPLE 3 Take interpolation in 0 and 1 with

$$\sigma(0) = 2, \quad \sigma(1) = -3, \quad \mathbf{T}_\alpha = \mathbf{T} = \begin{bmatrix} \frac{1}{1-z} & 0 \\ 0 & \frac{1}{(1-z)^5} \end{bmatrix} \tag{16}$$

Then we have

$$\omega(z) = (1-z)^5, \quad \tau = 5, \quad \rho(0) = 2, \quad \rho(1) = 2. \tag{17}$$

Note that $\mathbf{T}$ has the negative $(1, +\infty)$-characteristics -5 and -1, but $(1, \sigma(1))$ characteristics -5 and $+\infty$. For the submodule induced by the interpolation conditions due to $\rho(\alpha)$ and $\mathbf{R}_\alpha = \omega \cdot \mathbf{T}_\alpha$ we have as two bases the columns of the matrices

$$\mathbf{P}_1 = \begin{bmatrix} z^2(1-z)^2 & 0 & (1-z)^4 & 0 \\ 0 & z^2(1-z)^2 & 0 & 1+5z \\ 0 & 0 & 1 & 0 \\ 0 & 0 & 0 & 1 \end{bmatrix},$$

$$\mathbf{P}_2 = \begin{bmatrix} z^2(1-z)^2 & 0 & (1-z)^2(1-2z) & 0 \\ 0 & z^2(1-z)^2 & 0 & 1+5z \\ 0 & 0 & 1 & 0 \\ 0 & 0 & 0 & 1 \end{bmatrix}.$$

(a) Take $\vec{s} = (1,1,0,0)$ and therefore $\vec{r} = (-4,-4,0,0)$. Then both bases $\mathbf{P}_1$ and $\mathbf{P}_2$ are $\vec{r}$-column-reduced. We obtain the two "no-pole" solutions with minimal $\vec{r}$-McMillan degree

$$Z_1' = \mathbf{R} = \begin{bmatrix} (1-z)^4 & 0 \\ 0 & 1+5z \end{bmatrix}, \quad Z_2' = \begin{bmatrix} (1-z)^2(1-2z) & 0 \\ 0 & 1+5z \end{bmatrix} \tag{18}$$

which yields for our "pole-problem" the two solutions

$$Z_1 = \frac{1}{\omega} \cdot Z_1' = \mathbf{T} \text{ with } \vec{s}\text{-MM-deg } Z_1 = \vec{r}\text{-MM-deg } Z_1' + 6 = 0 + 6 = 6$$

$$Z_2 = \frac{1}{\omega} \cdot Z_2' \text{ with } \vec{s}\text{-MM-deg } Z_2 = \vec{r}\text{-MM-deg } Z_2' + 8 = 0 + 8 = 8$$

where we have applied Theorem 4 for computing the McMillan-degree. We see that two solutions of the "no-pole-problem" with minimal $\vec{r}$-MM-degree lead to solutions of the "pole-problem" with different $\vec{s}$-MM-degrees, Z_1 having minimal $\vec{s}$-MM-degree.

(b) Take $\vec{s} = (2, 2, 0, 0)$ and therefore $\vec{r} = (-3, -3, 0, 0)$. Then only the basis $\mathbf{P}_2$ is $\vec{r}$-column-reduced. We obtain the two "no-pole" solutions Z_1' and Z_2' of (18), but

$$Z_1 = \frac{1}{\omega} \cdot Z_1' = \mathbf{T} \text{ with } \vec{s}\text{-MM-deg } Z_1 = \vec{r}\text{-MM-deg } Z_1' + 6 = 1 + 6 = 7$$

$$Z_2 = \frac{1}{\omega} \cdot Z_2' \text{ with } \vec{s}\text{-MM-deg } Z_2 = \vec{r}\text{-MM-deg } Z_2' + 8 = 0 + 8 = 8.$$

Here the solution Z_1 of the "pole-problem" with minimal $\vec{s}$-MM-degree does not result from a solution of the "no-pole-problem" with minimal $\vec{r}$-MM-degree.

□

7. Conclusion

In this paper, we have shown that the matrix rational interpolation problem with poles as interpolation points can be transformed into a "pole-free" problem if enough pole-information is known. As a result, all solutions of the original problem having minimal complexity can be written down.

However, when not enough pole-information is available, it remains an open problem how we can represent all matrix rational functions satisfying the interpolation conditions.

References

1. B. Beckermann. A reliable method for computing M-Padé approximants on arbitrary staircases. *J. Comput. Appl. Math.*, 40:19–42, 1990.
2. B. Beckermann and G. Labahn. A uniform approach for the fast computation of Matrix-type Padé approximants. *SIAM J. Matrix Anal. Appl.* To appear.
3. B. Beckermann and G. Labahn. A uniform approach for Hermite-Padé and simultaneous Padé approximants and their matrix-type generalizations. *Numerical Algorithms*, 3:45–54, 1992.
4. S. Cabay, G. Labahn, and B. Beckermann. On the theory and computation of non-perfect Padé-Hermite approximants. *J. Comput. Appl. Math.*, 39:295–313, 1992.
5. M.H. Gutknecht. Continued fractions associated with the Newton-Padé table. *Numer. Math.*, 56:547–589, 1989.
6. M.H. Gutknecht. The rational interpolation problem revisited. *Rocky Mountain J. Math.*, 21(1):263–280, 1991.
7. T. Kailath. *Linear Systems*. Prentice-Hall, 1980.
8. H. Stahl. Existence and uniqueness of rational interpolants with free and prescribed poles. In E.B. Saff, editor, *Approximation Theory, Tampa, 1985-1986*, pages 180–208. Springer, 1987.
9. M. Van Barel and A. Bultheel. A new approach to the rational interpolation problem. *J. Comput. Appl. Math.*, 32(1-2):281–289, 1990.
10. M. Van Barel and A. Bultheel. A new approach to the rational interpolation problem: the vector case. *J. Comput. Appl. Math.*, 33(3):331–346, 1990.
11. M. Van Barel and A. Bultheel. The computation of non-perfect Padé-Hermite approximants. *Numer. Alg.*, 1:285–304, 1991.
12. M. Van Barel and A. Bultheel. A general module theoretic framework for vector M-Padé and matrix rational interpolation. *Numer. Alg.*, 3:451–461, 1992.
13. M. Van Barel and A. Bultheel. A new formal approach to the rational interpolation problem. *Numer. Math.*, 62:87–122, 1992.

PADÉ-TYPE APPROXIMANTS AND MULTIVARIATE POLYNOMIAL INTERPOLATION

PAUL SABLONNIÈRE
Laboratoire LANS
INSA, 20, avenue des Buttes de Coësmes
35043 RENNES Cédex (France)

Abstract. Padé-type approximants of a formal power series can be automatically derived from polynomial interpolants of some generating function of this series. Several examples are given, with special emphasis on Hakopian's multivariate polynomial interpolants of types I and II.

1. Introduction

This paper is a short survey on the connexion between polynomial interpolation and Padé-type approximation in several variables. It also discusses some new aspects of this topic in the case of Hakopian interpolation of type II. In section 2 we recall the definition of Padé-type approximants (PTA's) connected with univariate and tensor-product interpolation as described in chapters 1 and 4 of Brezinski's book [2]. In section 3, we recall the definition of PTA's associated with Hakopian interpolation of type I (see e.g. [4] or chapter 12 of [1]), i.e. mean value interpolation, as described by the author in [6] and [7].

Finally, in section 4, we recall the definition of multivariate interpolation of type II as described by Hakopian in [5] (see also chapter 13 of [1]). We define the corresponding PTA's for which we prove a result on the approximation order. We also give some examples showing the different types of rational approximants which can be obtained. The PTA's of section 3 seem to be simpler, but those of section 4 give the interesting possibility of introducing multiple pole lines. Only future applications will probably tell us which PTA's are the most useful.

2. The univariate and tensor product cases

2.1. The univariate case

Each formal power series (FPS) $f(t)$ can be written formally as:

$$f(t) = c\left(\frac{1}{1-xt}\right) = c\left(\sum_{i\geq 0} x^i t^i\right) = \sum_{i\geq 0} c_i t^i$$

where the functional c, acting on FPS in x, is defined by:

$$c(x^i) = c_i \qquad \text{for all } \ i \geq 0.$$

A. Cuyt (ed.), Nonlinear Numerical Methods and Rational Approximation II, 149–156.

The generating function $x \to g(x,t) = (1-xt)^{-1}$ is then interpolated by a polynomial of degree at most r on the set of points $X = \{x_0 < x_1 < \cdots < x_r\}$

$$P(x,t) = \sum_{i=0}^{r} \frac{\ell_i(x)}{1 - x_i t}$$

where $\{\ell_i(x), 0 \leq i \leq r\}$ is the classical Lagrange basis associated with X. (In case of multiple points, one can use Hermite interpolation polynomials).
Setting $A_i = c(\ell_i)$ for $0 \leq i \leq r$, one defines the PTA as the rational function in t:

$$c(P(\cdot,t)) = \sum_{i=0}^{r} \frac{A_i}{1 - x_i t} = (r/r+1)f(t)$$

whose numerator is of degree r and denominator of degree $r+1$. The main result is the following (see Brezinski [2], chapter 1, section 1.1).

$$f(t) - (r/r+1)f(t) = 0(t^{r+1}).$$

The complete Padé-type table and its properties are also described in [2]. The extension to Hermite interpolation is straightforward and gives PTA's with multiple poles. A convenient choice of poles leads to classical Padé approximants via a compact expression of the error term between f and its PTA's. In some applications, a good choice of poles leads to interesting results in the approximation of certain functions (see e.g. [2], sections 1.4 and 1.5).

2.2. The tensor-product case

One defines formally the double FPS $f(t,s)$ as:

$$f(t,s) = c\left(\frac{1}{(1-xt)(1-ys)}\right) = \sum_{i,j\geq 0} c_{ij} t^i s^i$$

where the functional c, acting on FPS in (x,y), is defined by:

$$c(x^i y^i) = c_{ij} \qquad \text{for all } \ i,j \geq 0.$$

The generating function $(x,y) \to g(x,y,t,s) = [(1-xt)(1-ys)]^{-1}$ is interpolated on the grid $X \times Y = \{(x_p, y_q), 0 \leq p \leq m, 0 \leq q \leq n\}$ by the tensor product polynomial

$$P(x,y,t,s) = \sum_{p=0}^{m}\sum_{q=0}^{n} \frac{\ell_p(x)\ell_q(y)}{(1-x_p t)(1-y_q s)}$$

where $\{\ell_p(x), 0 \leq p \leq m\}$ (resp.$\{\ell_q(y), 0 \leq q \leq n\}$) is the Lagrange basis associated with the set of points $X = \{x_0 < x_1 < \cdots < x_m\}$ ($Y = \{y_0 < y_1 < \cdots < y_n\}$).

Setting $A_{pq} = c(\ell_p(x)\ell_q(y))$, one defines the PTA $(m,n/m+1,n+1)f(t,s)$ as the rational function

$$c(P(\cdot,\cdot,t,s)) = \sum_{p=0}^{m}\sum_{q=0}^{n} \frac{A_{pq}}{(1-x_p t)(1-y_q s)}$$

and one deduces immediately the approximation property:

$$f(t,s) - (m, n/m+1, n+1)f(t,s) = 0(t^i s^{n+1}) + 0(t^{m+1} s^i), \quad \text{for} \quad i \geq 0$$

where the last term denotes an expansion of type:

$$\sum_{i\geq 0} \sum_{j\geq n+1} d_{ij} t^i s^t + \sum_{i\geq m+1} \sum_{j\geq 0} d_{ij} t^i s^j .$$

The definition and main properties of the complete Padé-type table are detailed in Brezinski [2], chapter 4.

The "poles" of these PTA's are straight lines which are parallel to the coordinate axes. Bivariate Hermite interpolation gives rise to multiple pole lines. A convenient choice of these pole lines eventually leads to some types of bivariate Padé approximants (see [2], chapter 4, section 4.2.3).

3. Multivariate Hakopian interpolation of type I

We first recall the main results on Hakopian interpolation of type I (see e.g. [4] and chapter 12 of [1]) and then the definition of associated PTA's given in [6] and [7] (see also [3], section 2.7). For sake of simplicity, the results are given in the bivariate case, but of course they hold in $\mathbb{R}^k$, for all $k \geq 2$.

3.1. Hakopian polynomial interpolation of type I

For $i \in I$ and, x_{i_1} and x_{i_2} in X define the line integral:

$$f\{x_i\} = f\{x_{i_1}, x_{i_2}\} = \int_{x_{i_1}}^{x_{i_2}} f((1-t)x_{i_1} + t x_{i_2}) dt.$$

Let $X = \{x_0, x_1, \cdots, x_r\}$ be a set of $r+1$ points of $\mathbb{R}^2$ **in general position**, i.e. satisfying, for each triple $\alpha \neq \beta \neq \gamma$: $x_\gamma \notin [x_\alpha, x_\beta]$, the segment joining x_α to x_β. For $I = \{i = (i_1, i_2) \in \mathbb{N}^2, 0 \leq i_1 < i_2 \leq r\}$, $\sharp I = \text{card } I = \frac{1}{2}r(r+1)$ is the number of segments joining two arbitrary points of X, and it is also equal to the dimension of the space $\mathbb{P}_{r-1}$ of bivariate polynomials of total degree at most $r-1$.

Hakopian proved the following result:

Theorem 1 . *Given $\sharp I$ real numbers $\{\gamma_i, i \in I\}$, there exists a unique polynomial $P \in \mathbb{P}_{r-1}$ satisfying the "interpolation conditions":*

$$P\{x_i\} = P\{x_{i_1}, x_{i_2}\} = \gamma_i \quad \text{for all} \quad i \in I.$$

In particular, one can construct a basis $\mathcal{L}_x = \{\ell_i, i \in I\}$ of $\mathbb{P}_{r-1}$ satisfying

$$\ell_i\{x_j\} = \delta_{ij} \qquad \text{for all} \quad (i,j) \in I \times I$$

and the expression of P is given by

$$P(x) = \sum_{i\in I} \gamma_i \ell_i(x) \qquad x \in \mathbb{R}^2.$$

3.2. Associated PTA's

Let $g(x,t) = (1 - x \cdot t)^{-2}$, where $x \cdot t$ is the euclidean scalar product of x and t in $I\!R^2$. For all $i \in I$, we take:

$$\gamma_i = g(\{x_i\}, t) \qquad \text{for all} \quad t \in I\!R^2.$$

Then it is easy to prove (see lemma 3.3 of [6]) that

$$\gamma_i = \frac{1}{(1 - x_{i_1} \cdot t)(1 - x_{i_2} \cdot t)},$$

and the interpolating polynomial of the generating function $x \to g(x,t) = (1 - x \cdot t)^{-2}$ can be written:

$$P(x,t) = \sum_{i \in I} \frac{\ell_i(x)}{(1 - x_{i_1} \cdot t)(1 - x_{i_2} \cdot t)}.$$

Now, any FPS $f(t)$ can be defined formally as:

$$\begin{aligned} f(t) = c(g(\cdot,t)) &= c\left(\sum_{n\geq 0}(n+1)(x\cdot t)^n\right) \\ &= \sum_{n\geq 0}\sum_{|i|=n}\binom{n}{i}(n+1)c(x^i)t^i \\ &= \sum_{n\geq 0}\sum_{|i|=n}\binom{n}{i}c_i t^i \end{aligned}$$

where $c_i = (n+1)\binom{n}{i}c(x^i)$ for $|i| = n, n \geq 0$. (we use the notations $|i| = i_1 + i_2$ and $x^i = x_1^{i_1} x_2^{i_2}$ for $i = (i_1, i_2) \in I\!N^2$ and $x = (x_1, x_2) \in I\!R^2$). This defines the functional c associated with f and acting on power series in x. Since $x \to g(x,t)$ is approximated by $P(x,t)$ for any fixed $t \in I\!R^2$, it seems natural to define (as in section 2) a PTA of f as follows:

$$c(P(\cdot,t)) = \sum_{i \in I} \frac{A_i}{(1 - x_{i_1} \cdot t)(1 - x_{i_2} \cdot t)} = (r - 1/r + 1)f(t).$$

We get a rational function in t whose numerator and denominator are respectively in $I\!P_{r-1}$ and $I\!P_{r+1}$, whence the notation above. Since the denominator is $Q_{r+1}(t) = \prod_{s=0}^{r}(1 - x_s \cdot t)$, we see that the "poles" are straight lines forming a "dual" configuration of the initial set of "interpolation" points X which define the polynomial interpolant.

The complete Padé-type table and some properties of these PTA's are given in [6]. Non rational PTA's associated with other multivariate series of functions are given in [7] together with some computational considerations and examples. For the approximation order, we get:

$$f(t) - (r - 1/r + 1)f(t) = O(t^r)$$

where the last expression denotes a FPS of type $\sum_{n\geq r}\sum_{|i|=n} d_i t^i$. (See [6], theorem 5.1 or [7], theorem 2).

4. Multivariate Hakopian interpolation of type II

As in the preceding section, we first recall the main results on bivariate Hakopian interpolation of type II and then we use them to construct another kind of bivariate PTA's. The results of this section being apparently new, we give more details than above.

4.1. Hakopian interpolation of type II

Let L be a set of $r+1$ straight lines $\{L_0, L_1, \cdots, L_r\}$ whose equations are given by:

$$1 - \lambda_s \cdot x = 0 \qquad 0 \leq s \leq r.$$

Let $I = \{i = (i_1, i_2), 0 \leq i_1 < i_2 \leq r\}$ as in section 3. We assume that L is **admissible** i.e. that for all $i \in I, L_{i_1} \cap L_{i_2} = x_i$ is a point of $\mathbb{R}^2$.

The multiplicity $m(x_i)$ is defined as the number of lines containing x_i. L is said to be **in general position** iff $m(x_i) = 2$ for all $i \in I$, i.e. iff each point of $X = \{x_i, i \in I\}$ is the intersection of exactly two lines of L: in this case $\sharp I = \sharp X = \frac{1}{2}r(r+1)$. If there is a least one point of multiplicity 3, we denote by $X' = \{x_j, j \in J \subset I\}$ the set of all distinct points from X. Hakopian proved the following

Theorem 2 . *For each arbitrary set of real numbers $\{\gamma_j^\alpha, j \in J, \alpha = (\alpha_1, \alpha_2), |\alpha| = \alpha_1 + \alpha_2 \leq m(x_j) - 2\}$, there exists a unique polynomial $P \in \mathbb{P}_{r-1}$ satisfying the interpolation conditions:*

$$D^\alpha P(x_j) = \gamma_j^\alpha$$

for all $j \in J$, and $|\alpha| \equiv m(x_j) - 2$.

Example 1. (pure Lagrange case)
L is in general position: let us define the Lagrange basis of $\mathbb{P}_{r-1}$ associated with X by:

$$\ell_i(x) = \prod_{\substack{s=0 \\ x_i \notin L_s}}^{r} \left(\frac{1 - \lambda_s \cdot x}{1 - \lambda_s \cdot x_i} \right) \qquad (i \in I)$$

It is easy to check that $\ell_i(x_j) = \delta_{ij}$ for all $(i,j) \in I \times I$, therefore the polynomial

$$P(x) = \sum_{i \in I} \gamma_i \ell_i(x)$$

satisfies $P(x_i) = \gamma_i$ for all $i \in I$.

Example 2. (mixed Lagrange-Hermite interpolation)
We take $r = 3$ and assume that the given set $L = \{L_0, L_1, L_2, L_3\}$ satisfies $L_1 \cap L_2 \cap L_3 = x_0$ (i.e. $m(x_0) = 3$) while L_0 intersects L_1, L_2 and L_3 at 3 distinct points x_1, x_2, x_3 respectively. Thus, we have:

$$X' = \{x_0, x_1, x_2, x_3\}, J = \{0, 1, 2, 3\}.$$

We define the following polynomials in $I\!P_2$:

$$\begin{aligned} \ell_1(x) &= \frac{(1-\lambda_2\cdot x)(1-\lambda_3\cdot x)}{(1-\lambda_2\cdot x_1)(1-\lambda_3\cdot x_1)} \\ \ell_2(x) &= \frac{(1-\lambda_1\cdot x)(1-\lambda_3\cdot x)}{(1-\lambda_1\cdot x_2)(1-\lambda_3\cdot x_2)} \\ \ell_3(x) &= \frac{(1-\lambda_1\cdot x)(1-\lambda_2\cdot x)}{(1-\lambda_1\cdot x_3)(1-\lambda_2\cdot x_3)} \end{aligned}$$

Now, let $\ell(x) = \dfrac{1-\lambda_0\cdot x}{1-\lambda_0\cdot x_0}$ and define

$$\begin{aligned} \ell_0(x) &= (1-(x-x_0)\nabla\ell(x_0))\ell(x) \\ \ell_0^{10}(x) &= ((x-x_0)\cdot e_1)\ell(x) \\ \ell_0^{01}(x) &= ((x-x_0)\cdot e_2)\ell(x) \end{aligned}$$

where $\nabla\ell(x) = \text{grad }\ell(x), e_1 = (1,0), e_2 = (0,1)$. It is easy to check that $\ell_i(x_j) = \delta_{ij}, \nabla\ell_i(x_j) = 0$ for $(i,j)\in J\times J$. Moreover, $\ell_0^{10}(x_0) = \ell_0^{01}(x_0) = 0, \nabla\ell_0^{10}(x_0) = e_1, \nabla\ell_0^{01}(x_0) = e_2$. Therefore the polynomial

$$P(x) = \sum_{j\in J}\gamma_j\ell_j(x) + \gamma_0^{10}\ell_0^{10}(x) + \gamma_0^{01}\ell_0^{01}(x)$$

obviously satisfies the Hermite conditions

$$P(x_j) = \gamma_j \text{ for all } j\in J, D^{10}P(x_0) = \gamma_0^{10}, D^{01}P(x_0) = \gamma_0^{01}.$$

Example 3. (pure Hermite case)

For $r = 3$, we assume that the four lines of L intersect at one point x of multiplicity 4. The associated basis of $I\!P_2$ is the Taylor basis

$$\ell^\alpha(x) = (x-x_0)^\alpha/\alpha!$$

for all $\alpha = (\alpha_1,\alpha_2), |\alpha|\le 2, \alpha! = \alpha_1!\alpha_2!$, and the interpolation polynomial

$$P(x) = \sum_{|\alpha|\le 2}\gamma^\alpha\ell^\alpha(x)$$

satisfies $D^\alpha P(x_0) = \gamma^\alpha$ for all α such that $|\alpha|\le 2$.

4.2. PTA's associated with Hakopian interpolation of type II

In this case, the generating function is $g(x,t) = (1-x\cdot t)^{-1}$ and $P(x,t)$ is the polynomial interpolant of type II of $x\to g(x,t)$ on the set X of points associated with the given set of lines L. More precisely, let

$$f(t) = c(g(\cdot,t)) = c\left(\sum_{n\ge 0}\sum_{|i|=n}\binom{n}{i}x^it^i\right) = \sum_{n\ge 0}\sum_{|i|=n}c_it^i,$$

where the functional c, acting on FPS in x, is defined by $c\left(\binom{n}{i}x^i\right) = c_i$ for $|i| = n, n \geq 0$. Therefore, the PTA of $f(t)$ associated with X is $c(P(\cdot,t)) = P_X(t)/Q_X(t)$. Here, the situation is more complicated than in section 3, (except for the Lagrange case): the degrees of P_X and Q_X depend on the multiplicities of points in X, as we shall see in examples below. But in all cases, we have the following result:

Theorem 3 . *For all $r \geq 1$, the approximation order is*

$$f(t) - c(P(\cdot,t)) = O(t^r)$$

where $O(t^r)$ has the same meaning as in section 3.

Proof. We give the proof in the pure Lagrange case (L is in general position), but it also holds for Hermite interpolation. For $|j| \leq r-1$, each monomial $e_j(x) = x^j$ has the unique representation:

$$e_j(x) = \sum_{i \in I} e_j(x_i)\ell_i(x)$$

where $\mathcal{L}_X = \{\ell_i, i \in I\}$ is the Lagrange basis of $I\!P_{r-1}$ associated with the set X of intersection points of L. The interpolant of $g(\cdot,t)$ is

$$P(x,t) = \sum_{i \in I} g(x_i,t)\ell_i(x) = \sum_{n \geq 0} \sum_{|j|=n} \binom{n}{j} \left\{ \sum_{i \in I} e_j(x_i)\ell_i(x) \right\} t^j.$$

Therefore, we obtain formally, setting $A_i = c(\ell_i)$ for $i \in I$

$$c(P(\cdot,t)) = \sum_{n \geq 0} \sum_{|j|=n} \binom{n}{j} \left\{ \sum_{i \in I} A_i e_j(x_i) \right\} t^j.$$

Now, we can prove that $\binom{n}{j}\sum_{i \in I} A_i e_j(x_i) = c_j$ for $|j| = n, n \leq r-1$.

By definition, we have $c_j = \binom{n}{j}c(e_j)$ for all j such that $|j| = n, n \geq 0$.

But for $0 \leq n \leq r-1$, we deduce $c(e_j) = \sum_{i \in I} A_i e_j(x_i)$, from the representation of e_j in $\mathcal{L}_X$, whence the result, q.e.d.

4.3. Some examples

(see the corresponding examples in section 4.1)

Example 1. L is in general position, therefore

$$P(x,t) = \sum_{i \in I} \frac{\ell_i(x)}{1 - x_i \cdot t}, \sharp I = R = \frac{1}{2}r(r+1)$$

$$c(P(\cdot,t)) = \sum_{i \in I} \frac{A_i}{1 - x_i \cdot t} = \frac{P_{R-1}(t)}{Q_R(t)}$$

where $Q_R(t) = \prod_{i \in I}(1 - x_i \cdot t) \in I\!P_R$.

In this case, the degrees of P_{R-1} and Q_R are very large,

Example 2. Here we have

$$P(x,t) = \sum_{i\in J} \frac{\ell_i(x)}{1 - x_i \cdot t} + \frac{t_1 \ell_0^{10}(x) + t_2 \ell_0^{01}(x)}{(1 - x_0 \cdot t)^2}$$

and the corresponding PTA is

$$c(P(\cdot,t)) = \sum_{i\in I} \frac{A_i}{1 - x_i \cdot t} + \frac{a_1 t_1 + a_2 t_2}{(1 - x_0 \cdot t)^2} = \frac{P_4(t)}{Q_5(t)},$$

where

$A_i = c(\ell_i), i \in J, a_1 = c(\ell_0^{10}), a_2 = c(\ell_0^{01}), Q_5(t) = (1-x_0 \cdot t)^2(1-x_1 \cdot t)(1-x_2 \cdot t)(1-x_3 \cdot t).$

In this case, we have a double pole line and 3 simple pole lines.

Example 3.

Since $D^\alpha \frac{1}{(1 - x \cdot t)^2} = \frac{|\alpha|! t^\alpha}{(1 - x \cdot t)^{|\alpha|+1}}$ for all α such that $|\alpha| \leq 2$, we get

$$P(x,t) = \sum_{|\alpha|\leq 2} \frac{|\alpha|!(x - x_0)^\alpha t^\alpha}{\alpha!(1 - x_0 \cdot t)^{|\alpha|+1}}$$

and

$$c(P(\cdot,t)) = \sum_{|\alpha|\leq 2} \frac{A_\alpha t^\alpha}{(1 - x_0 \cdot t)^{|\alpha|+1}} = \frac{P_2(t)}{Q_3(t)},$$

where $Q_3(t) = (1 - x_0 \cdot t)^3$. In this case, we get the minimum degree of P and Q for the order $f(t) - P_2(t)/Q_3(t) = 0(t^3)$.

Example 4. For $r = 4$ (cubic polynomial interpolant), there exist 5 possible configurations for L, which give 5 types of PTA's, from (9/10) in the Lagrange case to (3/4) in the pure Hermite case (with a pole line of order four). The intermediate cases (8/9), (7/8) and (6/7) have respectively one double pole line, two double pole lines and one triple pole line.

References

1. B.D. Bojanov, H.A. Hakopian, A.A. Sahakian. Spline functions and multivariate interpolations. Kluwer, Dordrecht, 1993.
2. C. Brezinski. Padé approximants and general orthogonal polynomials. ISNM 50, Birkhäuser Verlag, Basel, 1980.
3. C. Brezinski, J. Van Iseghem. Padé approximations. To appear in Handbook of Numerical Analysis, Vol III, P.G. Ciarlet and J.L. Lions eds, North Holland.
4. H.A. Hakopian. Multivariate divided differences and multivariate interpolation of Lagrange and Hermite type. J. Approx. Theory **34** (1982), 286-305.
5. H.A. Hakopian. Multivariate interpolation II of Lagrange and Hermite type. Studia Mathematica **80** (1984), 77-88.
6. P. Sablonnière. A new family of Padé-type approximants in $\mathbb{R}^k$. J. Comp. Applied Math. **9** (1983), 347-359.
7. P. Sablonnière. Padé-type approximants for multivariate series of functions. in Padé approximation and its applications. H. Werner and H.J. Bürger eds, LNM 1071, Springer-Verlag, Berlin, 1984, 238-251.

RATIONAL

APPROXIMATION

BEHAVIOR OF ALTERNATION POINTS IN BEST RATIONAL APPROXIMATION

D. BRAESS
Institut für Mathematik, Ruhr-Universität Bochum, 44780 Bochum, GERMANY

D.S. LUBINSKY
Department of Mathematics, Witwatersrand University, P.O. WITS 2050, Johannesburg, REPUBLIC OF SOUTH AFRICA

and

E.B. SAFF*
Institute for Constructive Mathematics, Department of Mathematics, University of South Florida, 4202 East Fowler Avenue, PHY 114, Tampa, Florida 33620-5700, USA

Abstract. The behavior of the equioscillation points (alternants) for the error in best uniform approximation on $[-1,1]$ by rational functions of degree n is investigated. In general, the points of the alternants need not be dense in $[-1,1]$, even when approximation by rational functions of degree (m,n) is considered and asymptotically $m/n \geq 1$. We show, however, that if more than $O(\log n)$ poles of the approximants stay at a positive distance from $[-1,1]$, then asymptotic denseness holds, at least for a subsequence. Furthermore, we obtain stronger distribution results when λn $(0 < \lambda \leq 1)$ poles stay away from $[-1,1]$. In the special case when a Markoff function is approximated, the distribution of the equioscillation points is related to the asymptotics for the degree of approximation.

1. Introduction

Let $f \in C[-1,1]$ be real-valued and $\mathcal{R}_{m,n}$ denote the collection of real rational functions of the form p/q, where $\deg p \leq m$, $\deg q \leq n$, $q \not\equiv 0$. It is known that for each pair of nonnegative integers (m,n) there exists a unique $r^*_{m,n} \in \mathcal{R}_{m,n}$ that is of best approximation to f on $[-1,1]$ in the sense that

$$\|f - r^*_{m,n}\|_{[-1,1]} < \|f - r\|_{[-1,1]} \quad \forall r \in \mathcal{R}_{m,n}, \quad r \neq r^*_{m,n},$$

where $\|\cdot\|_{[a,b]}$ denotes the sup norm on $[a,b]$. Writing

$$r^*_{m,n} = p^*_{m,n}/q^*_{m,n}, \tag{1.1}$$

where $p^*_{m,n}$ and $q^*_{m,n}$ have no common factors, we set

$$d = d(m,n) := \min(m - \deg p^*_{m,n}, n - \deg q^*_{m,n}). \tag{1.2}$$

Then there are $m+n+2-d$ points x_k:

$$-1 \leq x_1 < x_2 < \cdots < x_{m+n+2-d} \leq 1,$$

* The research of this author was supported, in part, by NSF grant DMS 920-3659.

A. Cuyt (ed.), Nonlinear Numerical Methods and Rational Approximation II, 159–172.

that satisfy

$$\lambda(-1)^k(f - r^*_{m,n})(x_k) = \|f - r^*_{m,n}\|_{[-1,1]}, \; k = 1, \ldots, m+n+2-d, \tag{1.3}$$

where $\lambda = 1$ or $\lambda = -1$, see e.g. [B, p.113]. Such a set of points x_k, called an *alternation set*, is in general not unique. For each pair (m, n) we let

$$A_{m,n} = A_{m,n}(f, [-1,1]) := \{x_k^{(m,n)}\}_{k=1}^{m+n+2-d} \tag{1.4}$$

denote an arbitrary, but fixed alternation set for best approximation from $\mathcal{R}_{m,n}$. Our results concern the denseness and asymptotic distribution of the sets $A_{m,n}$ as $m + n \to \infty$.

For the case of polynomial approximation, a celebrated result of Kadec [K] shows that for a given $f \in C[-1,1]$ there exists a sequence Λ of positive integers such that the sets $\{A_{m,0}\}_{m\in\Lambda}$ are asymptotically dense in $[-1,1]$. In fact, Kadec proved that such sets have limiting distribution equal to arcsine on $[-1,1]$; that is, the sets $\{A_{m,0}\}_{m\in\Lambda}$ behave asymptotically like the extreme points of the classical Chebyshev polynomials $T_n(x) = \cos(n \arccos x)$.

For rational approximation, Borwein, Grothmann, Kroó, and Saff [BGKS] have shown that denseness in $[-1,1]$ of a subsequence of alternation sets $A_{m,n}$ holds whenever $n = n(m)$ and $m/n \to \tau$ with $\tau > 1$. To state this result more precisely, we will refer to the *denseness of a set A in a set B* in the Hausdorff sense:

$$\rho(A, B) := \sup_{y\in B} \inf_{x\in A} |x - y|.$$

Specifically, we will consider

$$\rho_{m,n} := \rho(A_{m,n}, [-1,1]). \tag{1.5}$$

Then the result of [BGKS] is the following:

THEOREM 1.1. *Let $n = n(m)$, $m = 0, 1, 2, \ldots$, satisfy*

$$n(m) \le n(m+1) \le n(m) + 1, \quad n(m) \le m.$$

If $f \in C[-1,1]$ is not a rational function and

$$\frac{m - n(m)}{\log m} \to \infty \quad as \quad m \to \infty, \tag{1.6}$$

then

$$\liminf_{m\to\infty} \rho_{m,n(m)} = 0.$$

That is, for some sequence Λ of positive integers, the sets $A_{m,n(m)}, m \in \Lambda$, are asymptotically dense in $[-1,1]$.

Of course this theorem gives no information about denseness in the important case when $n = m$ or, more generally, when $n = m + \nu$ for some fixed integer ν. In fact, it is shown in [BGKS], [KP] that for each $\epsilon > 0$ there exists an $f \in C[-1,1]$ with the property that all the extreme points of $f - r^*_{n-1,n}$ lie in $[-1, -1+\epsilon]$ for every $n = 1, 2, \ldots$. Therefore, denseness of the alternation sets may not hold when $n(m) = m + 1$.

The purpose of the present paper is to give sufficient conditions for the denseness in $[-1, 1]$ of alternation sets for the case when $n = m + \nu$, with ν fixed. Roughly speaking we shall show that if more than $O(\log m)$ of the poles of $r^*_{m,m+\nu}$ stay at a positive distance from $[-1, 1]$, then the sets $A_{m,m+\nu}$ are asymptotically dense in $[-1, 1]$ for a subsequence of integers m. Furthermore, if all but $o(m)$ poles of $r^*_{m,m+\nu}$ approach infinity, then the limiting distribution of these alternation sets is the arcsine distribution. In the special case when f is a Markoff function, rather precise estimates for the number of alternation points in a subinterval of $[-1, 1]$ are obtained.

Our main results are stated in Section 2 with the proofs given in Sections 3 and 4.

2. Main Results

We assume throughout that f is real-valued on $[-1, 1]$. We denote by $\mathbf{C}$ the (finite) complex plane. We remark that a rational function p/q has a pole at infinity if $\deg p > \deg q$ and in such a case the multiplicity of this pole is $\deg p - \deg q$. These poles shall always be included when the poles outside a neighborhood of $[-1, 1]$ are considered. Furthermore, we shall count poles always with their multiplicities.

For convenience we state the results only for the case $n = m$, although the same statements hold for $n = m + \nu$, with ν fixed.

THEOREM 2.1. *Let $f \in C[-1, 1]$, f not a rational function. Let $\gamma_m(\epsilon)$ denote the number of those poles of its best approximant $r^*_{m,m}$ that lie outside an ϵ-neighborhood of $[-1, 1]$. If there exists an $\epsilon > 0$ such that*

$$\frac{\gamma_m(\epsilon)}{\log m} \to \infty \quad as \quad m \to \infty, \tag{2.1}$$

then asymptotic denseness of the alternation points holds; that is,

$$\liminf_{m \to \infty} \rho_{m,m} = 0, \tag{2.2}$$

where $\rho_{m,m}$ is defined in (1.5).

COROLLARY 2.2. *If $\liminf_{m\to\infty} \|f - r^*_{m,m}\|^{1/m}_{[-1,1]} < 1$ (in particular, if f is analytic on $[-1, 1]$), then Theorem 2.1 holds with assumption (2.1) replaced by*

$$\gamma_m(\epsilon) \to \infty \quad as \quad m \to \infty. \tag{2.3}$$

COROLLARY 2.3. *If f is analytic on $[-1, 1]$ and in $\mathbf{C}\backslash E$, where E has logarithmic capacity zero, and if it is not a rational function, then*

$$\liminf_{m \to \infty} \rho_{m,m} = 0.$$

REMARKS. (i) The assumptions of Corollary 2.3 imply that $r^*_{m,m} \to f$ in capacity in each bounded subset of $\mathbf{C}$, and this in turn can be used to show that (2.3) is satisfied. Unfortunately, one cannot say much more in general. A special case of the construction in [L1] shows that given any increasing sequence $\{\xi_m\}_{m=1}^{\infty}$, with $\xi_m \leq m$, $\lim_{m\to\infty} \xi_m = \infty$, there exists an entire function f and a subsequence $\mathcal{S}$ of integers, such that for $m \to \infty$, $m \in \mathcal{S}$, at least $m - \xi_m$ poles of $r^*_{m,m}$ approach $[-1, 1]$. In particular, we could choose $\xi_m = \log m$ or $\log\log m$, etc. This example illustrates the restrictiveness of (2.3) or (2.1), which involve all positive integers m.

(ii) A similar situation is encountered whenever $\{r^*_{m,m}\}_{m=1}^{\infty}$ converges in capacity in a neighborhood of $[-1, 1]$.

For the case when a positive proportion of poles of the $r^*_{m,m}$ stay away from $[-1, 1]$, there is a positive density of alternation points on each subinterval of $[-1, 1]$. To state this result we let

$$N_{m,n}(\alpha, \beta) := \text{number of points of } A_{m,n} \text{ lying in } [\alpha, \beta].$$

We will also write N_m instead of $N_{m,m}$.

THEOREM 2.4. *If, in Theorem 2.1, we have for some* $\epsilon > 0$,

$$\liminf_{m\to\infty} \frac{\gamma_m(\epsilon)}{m} > 0,$$

then there exists a constant $\Delta > 0$ *and a subsequence of integers* Λ *such that for all subintervals* $[\alpha, \beta] \subseteq [-1, 1]$

$$\liminf_{m\to\infty} \frac{N_m(\alpha, \beta)}{N_m(-1, 1)} \geq \Delta \left(\frac{\arccos \alpha - \arccos \beta}{\pi} \right), \quad m \in \Lambda. \tag{2.4}$$

The next result shows that if all but $o(m)$ poles of $r^*_{m,m}(z)$ tend to infinity, then a subsequence of alternation sets has the same distribution as that for the polynomial case.

THEOREM 2.5. *If, in Theorem 2.1, we have for each* $\tau > 0$,

$$\lim_{m\to\infty} \frac{\gamma_m(\tau)}{m} = 1, \tag{2.5}$$

then there exists a subsequence of integers Λ *such that for all* $[\alpha, \beta] \subseteq [-1, 1]$

$$\lim_{m\to\infty} \frac{N_m(\alpha, \beta)}{N_m(-1, 1)} = \frac{\arccos \alpha - \arccos \beta}{\pi}, \quad m \in \Lambda. \tag{2.6}$$

We remark that for the exponential function $f(x) = e^x$ it is shown in [BSW] that all the poles of r^*_m tend to infinity as $m \to \infty$. Hence (2.6) holds for this function (in fact with $\Lambda = \mathbf{N}$).

More precise results can be obtained for Markoff (Stieltjes) functions. Specifically, let T be a subinterval of $(-\infty, -1)$ and

$$f(z) = \int_T \frac{d\mu(t)}{t - z}, \tag{2.7}$$

where $d\mu$ is a nonnegative measure on T. Then f is analytic in $\mathbf{C}\backslash T$ and the best approximation $r^*_{n-1,n}$ is known to have n simple poles in T. The alternation set is unique and has $2n$ points, see e.g. [G1].

Consequently, if the cut T has a positive distance from the interval $[-1,1]$, then Theorem 2.4 applies. On the other hand, the distribution differs from (2.6) if T and $[-1,1]$ have a boundary point in common. Specifically, there is a strong connection between the degree of rational approximation on subintervals and the number of alternation points in it (i.e. in the subinterval).

For convenience, the degree of approximation on subintervals is denoted by

$$\operatorname{dist}(f,\mathcal{R}_{m,n})_{[\alpha,\beta]} := \inf_{r\in\mathcal{R}_{m,n}} \|f-r\|_{[\alpha,\beta]}.$$

THEOREM 2.6. *Assume that f is a Markoff function on $[-1,+1]$ and let $-1\le\alpha<\beta\le 1$. If $2m+2$ points of the alternation set $A_{n-1,n}$ are contained in $[\alpha,\beta]$ for some $m<n$ (i.e., if $N_{n-1,n}(\alpha,\beta)\ge 2m+2$), then*

$$\operatorname{dist}(f,\mathcal{R}_{m-1,m})_{[\alpha,\beta]} > \operatorname{dist}(f,\mathcal{R}_{n-1,n})_{[-1,+1]}. \tag{2.8}$$

When we refer here to approximation on (proper) subintervals, then we have a positive distance from the cut and the error of approximation decreases geometrically as $n\to\infty$.

COROLLARY 2.7. *Let f be a Markoff function on $[-1,+1]$ and let*

$$\epsilon_n := \|f-r^*_{n-1,n}\|_{[-1,+1]}, \quad n\in\mathbf{N}.$$

Moreover, assume that $-1\le\alpha<\beta\le 1$ and

$$\operatorname{dist}(f,R_{n-1,n})_{[\alpha,\beta]} \le 4\omega^{-2n}$$

for some $\omega=\omega(\alpha,\beta)>1$. Then we have

$$N_{n-1,n}(\alpha,\beta) \le 2+\frac{\log(4/\epsilon_n)}{\log\omega(\alpha,\beta)}. \tag{2.9}$$

We will make use of this corollary in the study of rational approximation of $|x|$ on $[-1,1]$. This approximation problem is easily seen to be equivalent to the approximation of $\sqrt{x}$ on $[0,1]$. Since $\sqrt{x}/x$ is a Stieltjes function, we consider the problem for $\mathcal{R}_{n,n-1}$. From [V], [B, p.153] or [S] we know that

$$\operatorname{dist}(\sqrt{x},\mathcal{R}_{n,n-1})_{[0,1]} \ge \text{const}\cdot e^{-\pi\sqrt{n+1}}.$$

If $\alpha>0$, then $\operatorname{dist}(\sqrt{x},\mathcal{R}_{n,n-1})_{[\alpha,1]}$ decreases exponentially since $\sqrt{x}$ is analytic in some neighborhood of the interval. Now it follows from Corollary 2.7 that there are at most

$$c(\alpha)[1+\sqrt{n}]$$

points of $A_{n,n-1}$ in $[\alpha,1]$. Here, $c(\alpha)$ is a constant which depends only on α.

For more precise results concerning the limiting distribution of extreme points for the class of functions $f(x)=x^\alpha$, $0<\alpha<1$, on [0, 1], see [SS].

3. Proofs

We will use arguments similar to those in [BGKS] together with the following simple lemmas.

LEMMA 3.1. *Let* $Q_\nu(\zeta) := \prod_{k=1}^{\nu}(1-\alpha_k\zeta)$, *where* $|\alpha_k| \le \delta < 1$ *for* $k = 1,\dots,\nu$. *Then for each* $1 \le m \le \nu$, *there exists a polynomial* $S_{m-1} \in \mathcal{P}_{m-1}$ *such that for all* $R > 1$,

$$\left\| \frac{S_{m-1}(\zeta)}{Q_\nu(\zeta)} - 1 \right\|_{|\zeta|=1} \le \frac{R}{R-1} \frac{(1+\delta R)^\nu}{(\delta+R)^m(1-\delta)^{\nu-m}}. \tag{3.1}$$

Here $\mathcal{P}_k$ denotes the collection of algebraic polynomials of degree $\le k$, and $\|\cdot\|_{|\zeta|=1}$ is the sup norm on the unit circle.

Proof. Choose $S_{m-1} \in \mathcal{P}_{m-1}$ to be the polynomial that interpolates Q_ν (in the Hermite sense) in the points $\overline{\alpha}_k$, $k = 1,\dots,m$. Then, by the Hermite formula (cf. [W]) we have, for each $R > 1$,

$$Q_\nu(\zeta) - S_{m-1}(\zeta) = \frac{1}{2\pi i}\int_{|t|=R} \frac{Q_\nu(t)\prod_{k=1}^{m}(\zeta-\overline{\alpha}_k)}{(t-\zeta)\prod_{k=1}^{m}(t-\overline{\alpha}_k)}\, dt, \quad |\zeta| < R.$$

Thus, for $|\zeta| = 1$,

$$1 - \frac{S_{m-1}(\zeta)}{Q_\nu(\zeta)} = \frac{1}{2\pi i}\int_{|t|=R} \prod_{k=1}^{m}\left(\frac{\zeta-\overline{\alpha}_k}{1-\alpha_k\zeta}\right) \prod_{k=1}^{m}\left(\frac{1-\alpha_k t}{t-\overline{\alpha}_k}\right) \prod_{k=m+1}^{\nu}\left(\frac{1-\alpha_k t}{1-\alpha_k\zeta}\right) \frac{dt}{t-\zeta}. \tag{3.2}$$

Since Blaschke products have modulus 1 on the unit sphere and

$$\left|\frac{1-\alpha_k t}{t-\overline{\alpha}_k}\right| \le \frac{1+\delta R}{\delta+R}, \quad \text{for } |t| = R,\ |\alpha_k| \le \delta,$$

a straightforward estimation of the integrand in (3.2) yields (3.1). ∎

COROLLARY 3.2. *Let* q_ν *be a real polynomial of degree* ν *whose zeros all lie outside the ellipse with focal points* -1 *and* $+1$

$$\mathcal{E}_\tau : \frac{x^2}{[\frac{1}{2}(\tau+\frac{1}{\tau})]^2} + \frac{y^2}{[\frac{1}{2}(\tau-\frac{1}{\tau})]^2} = 1, \quad \tau > 1. \tag{3.3}$$

Then for each $m \ge 1$, *there exists a real polynomial* $s_{m-1} \in \mathcal{P}_{m-1}$ *such that, for all* $R > 1$,

$$\left\| \frac{s_{m-1}}{q_\nu} - 1 \right\|_{[-1,1]} \le 2M + M^2, \tag{3.4}$$

where

$$M := \frac{R}{R-1} \frac{(1+\frac{R}{\tau})^\nu}{(R+\frac{1}{\tau})^m(1-\frac{1}{\tau})^{\nu-m}} \quad \text{if } m \le \nu, \tag{3.5}$$

and $M = 0$ *whenever* $m > \nu$.

Proof. Obviously, we may restrict ourselves to the case $m \le \nu$. Let $z = \psi(\zeta) = \frac{1}{2}(\zeta + \zeta^{-1})$ denote the Joukowski transformation that maps the exterior of the unit circle onto $\mathbf{C}\setminus[-1,1]$. Then $\psi(\zeta)$ maps the circle $|\zeta| = \tau$ onto the ellipse $\mathcal{E}_\tau$ of (3.3). Thus we can write, for some constant c,

$$q_\nu(z) = cQ_\nu(\zeta)Q_\nu(\zeta^{-1}),$$

where $Q_\nu \in \mathcal{P}_\nu$, and all zeros of Q_ν lie in $\{|\zeta| > \tau\}$. By Lemma 3.1, there exists $S_{m-1} \in \mathcal{P}_{m-1}$ that satisfies (3.1) with $\delta = 1/\tau$. Then

$$s_{m-1}(z) := cS_{m-1}(\zeta)S_{m-1}(\zeta^{-1})$$

is a real polynomial in $\mathcal{P}_{m-1}$ for which (3.4) holds. ∎

Finally we need the following lemma concerning peaking polynomials, which is a slight modification of Lemma 2.1 in [BGKS].

LEMMA 3.3. *Given $-1 \le \alpha < \beta \le 1$ and $j \in \mathbf{N}$, there exists a polynomial $\pi_j \in \mathcal{P}_j$ with*

$$||\pi_j||_{[-1,\alpha]\cup[\beta,1]} \le \frac{1}{3}, \tag{3.6}$$

and

$$\max_{x\in[\alpha,\beta]} \pi_j(x) > c_1 e^{c_2 j(\beta-\alpha)}, \tag{3.7}$$

where c_1, c_2 are positive constants independent of α, β, and j. More precisely, we can take $\pi_j(x) := q_j(x - (\alpha+\beta)/2)$, where

$$q_j(y) = \frac{1}{3}T_{[j/2]}\left(1 + \frac{\xi^2}{2} - \frac{y^2}{2}\right), \quad \xi := \frac{\beta-\alpha}{2},$$

and $T_{[j/2]}$ is the Chebyshev polynomial of degree $[j/2]$.

Proof of Theorem 2.1. We drop the double subscripts and write $\rho_m = \rho_{m,m}$,

$$r_m^* := r_{m,m}^* = \frac{p_m^*}{q_m^*}, \quad p_m^*, q_m^* \in \mathcal{P}_m,$$

where p_m^*, q_m^* have no common factors. Set

$$E_m := ||f - r_m^*||_{[-1,1]}. \tag{3.8}$$

Since f is not a rational function, $\{E_m\}_0^\infty$ is a nonincreasing sequence of positive numbers that converges to zero. From elementary properties of series it follows that

$$\sum_{m=0}^{\infty} \frac{E_m - E_{m+1}}{E_m + E_{m+1}}$$

is divergent. Thus there exists a subsequence $\Lambda \subset \mathbf{N}$ such that $E_m \ne E_{m+1}$, and

$$\frac{E_m + E_{m+1}}{E_m - E_{m+1}} \le m^2, \quad m \in \Lambda. \tag{3.9}$$

Following the argument in [K] and [BGKS], we set

$$R_m := \frac{r_m^* - r_{m+1}^*}{E_m - E_{m+1}}, \quad m \in \Lambda. \tag{3.10}$$

By the triangle inequality and (3.9), we obtain

$$\|R_m\|_{[-1,1]} \le \frac{E_m + E_{m+1}}{E_m - E_{m+1}} \le m^2, \quad m \in \Lambda. \tag{3.11}$$

Also, for the points $-1 \le x_1^{(m)} < x_2^{(m)} < \cdots < x_{2m+2-d(m)}^{(m)} \le 1$ of each alternation set A_m of $f - r_m^*$ we have

$$\lambda(-1)^{k-1} R_m(x_k^{(m)}) \ge \frac{E_m - E_{m+1}}{E_m - E_{m+1}} = 1, \quad k = 1, \ldots, 2m+2-d(m), \tag{3.12}$$

where $\lambda = +1$ or $\lambda = -1$ and

$$d(m) := \min(m - \deg p_m^*,\ m - \deg q_m^*). \tag{3.13}$$

The rational function R_m can be written in the form

$$R_m = \frac{P_m}{q_m^* q_{m+1}^*},$$

where $\deg P_m = 2m + 1 - d(m)$. Thus, if $\hat{P} \in \mathcal{P}_{2m+1-d(m)}$, $\hat{P} \not\equiv P_m$, then $R_m + \hat{P}/q_m^* q_{m+1}^*$ has at most $2m + 1 - d(m)$ zeros.

Now choose $\tau > 1$ so that the ellipse $\mathcal{E}_\tau$ of (3.3) lies in the ϵ-neighborhood of $[-1, 1]$. Write

$$q_m^* = q_{m,1}^* q_{m,2}^*, \tag{3.14}$$

where $q_{m,1}^*, q_{m,2}^*$ are polynomials with real coefficients and $q_{m,1}^*$ has all its zeros outside the ϵ-neighborhood of $[-1, 1]$ while $q_{m,2}^*$ has all zeros within ϵ of $[-1, 1]$. Let σ_m denote the total number of poles of r_m^* in $\overline{\mathbf{C}} = \mathbf{C} \cup \{\infty\}$. Then $\sigma_m = m - d(m)$ and

$$\deg q_{m,2}^* = \sigma_m - \gamma_m = m - d(m) - \gamma_m, \qquad \gamma_m = \gamma_m(\epsilon). \tag{3.15}$$

Next fix $R > 1$ and choose δ, $0 < \delta < 1$, so that

$$\mu := \frac{1 + \frac{R}{\tau}}{\left(R + \frac{1}{\tau}\right)^\delta \left(1 - \frac{1}{\tau}\right)^{1-\delta}} < 1. \tag{3.16}$$

Let $\nu = \nu(m) := \gamma_m + \gamma_{m+1}$ and $k = k(m) := [\delta\nu] + 1$. Therefore, $k \ge \delta(\gamma_m + \gamma_{m+1}) \ge \delta(\deg q_{m,1}^* + \deg q_{m+1,1}^*)$ and by Corollary 3.2 there exists $s_{k-1} \in \mathcal{P}_{k-1}$ such that

$$\left\| \frac{s_{k-1}}{q_{m,1}^* q_{m+1,1}^*} - 1 \right\|_{[-1,1]} \le 2M_\nu + M_\nu^2, \tag{3.17}$$

where

$$M_\nu := \frac{R}{R-1} \mu^\nu. \tag{3.18}$$

Since $\mu < 1$, the set $\Lambda' := \{m \in \Lambda : \; 2M_\nu + M_\nu^2 < 1/2\}$ contains all but a finite number of elements of Λ. Let

$$\hat{P}_{m,1} := s_{k-1} q^*_{m,2} q^*_{m+1,2}, \quad m \in \Lambda'. \tag{3.19}$$

Then $\hat{P}_{m,1}/q^*_m q^*_{m+1} = s_{k-1}/q^*_{m,1}q^*_{m+1,1}$ and from (3.17) we have for $m \in \Lambda'$

$$\frac{1}{2} < \frac{\hat{P}_{m,1}}{q^*_m q^*_{m+1}}(x) < \frac{3}{2}, \quad x \in [-1,1]. \tag{3.20}$$

The degree of the factors of $\hat{P}_{m,1}$ are known from (3.15) and from the construction of s_{k-1}. Since $d(m+1) \geq 0$, we have

$$\begin{aligned} \deg \hat{P}_{m,1} &\leq [\delta\gamma_m] + m - d(m) - \gamma_m + [\delta\gamma_{m+1}] + m + 1 - \gamma_{m+1} \\ &= [\delta\gamma_m] - \gamma_m + [\delta\gamma_{m+1}] - \gamma_{m+1} + 2m + 1 - d(m). \end{aligned} \tag{3.21}$$

Next, for $m \in \Lambda'$, let $x^*_m \in [-1,1]$ satisfy

$$\min_j \left| x^*_m - x_j^{(m)} \right| = \rho_m. \tag{3.22}$$

If $x^*_m \in [-1, x_1^{(m)}]$, let $\pi_{j(m)}$ be the polynomial of Lemma 3.3 with $\alpha = -1$, $\beta = x_1^{(m)}$, and $j(m) = \gamma_m - [\delta\gamma_m] + \gamma_{m+1} - [\delta\gamma_{m+1}]$. The degree of the polynomial $\hat{P}_{m,1}$ has been estimated in (3.21), and the product

$$\hat{P}_m := \pi_{j(m)} \hat{P}_{m,1} \tag{3.23}$$

has degree at most $2m+1-d(m)$. Therefore,

$$R_m \pm \frac{\hat{P}_m}{q^*_m q^*_{m+1}}$$

has at most $2m+1-d(m)$ (finite) zeros. From Lemma 3.3 and (3.20) it follows that

$$\left\| \frac{\hat{P}_m}{q^*_m q^*_{m+1}} \right\|_{[-1,\alpha]\cup[\beta,1]} \leq \frac{1}{2} \tag{3.24}$$

and

$$\left\| \frac{\hat{P}_m}{q^*_m q^*_{m+1}} \right\|_{[\alpha,\beta]} \geq \frac{c_1}{2} e^{c_2 j(m)(\beta-\alpha)}. \tag{3.25}$$

Now from (3.12) and (3.24) we see that $R_m \pm \hat{P}_m/q^*_m q^*_{m+1}$ has $2m+1-d(m)$ zeros in $[x_1^{(m)}, 1]$ and so must be zero-free in $[-1, x_1^{(m)}] = [\alpha, \beta]$. Consequently, upper and lower bounds for R_m on $[\alpha,\beta]$ are provided in (3.12) and (3.25), resp.

$$\frac{c_1}{2} e^{c_2 j(m)(\beta-\alpha)} \leq \|R_m\|_{[\alpha,\beta]} \leq m^2.$$

This together with (3.22) implies

$$\rho_m \le \frac{c_3}{j(m)} \log m \le \frac{c_3 \log m}{\gamma_m - [\delta\gamma_m]} \le \frac{c_3}{1-\delta}\frac{\log m}{\gamma_m}. \tag{3.26}$$

Here c_3 is a constant independent of α, β and m. Similarly, if $x_m^* \in \left[x_{2m+2-d(m)}^{(m)}, 1\right]$ we get (3.26).

For the case when $x_m^* \in (x_1^{(m)}, x_{2m+2-d(m)}^{(m)})$ we choose α, β as two consecutive zeros of R_m with $\beta - \alpha \ge \rho_m$. After replacing $R_m$by $-R_m$, if necessary, we may assume that R_m is positive in (α, β). Since the rational function

$$R_m - \frac{1}{2}\left(\frac{\hat{P}_m}{q_m^* q_{m+1}^*} - 1\right)$$

has at most $2m + 1 - d(m)$ zeros, it has no zeros in (α, β). Again we obtain (3.26).

So inequality (3.26) is valid for all $m \in \Lambda'$. On taking the limit, it follows from assumption (2.1) that (2.2) holds. ■

Proof of Corollary 2.2. With E_m as defined in (3.8), the assumption $\liminf_{m\to\infty} E_m^{1/m} < 1$ implies that $\liminf_{m\to\infty} E_{m+1}/E_m < 1$ and so there exists a subsequence $\Lambda \subset \mathbf{N}$ and a constant K such that

$$\frac{E_m + E_{m+1}}{E_m - E_{m+1}} \le \frac{2}{1-(E_{m+1}/E_m)} \le K, \quad m \in \Lambda.$$

Then from the preceding proof we deduce that the inequality (3.26) holds with $\log m$ replaced by a suitable constant. Hence, $\gamma_m \to \infty$ implies that $\liminf_{m\to\infty} \rho_m = 0$. ■

Proof of Corollary 2.3. First note that $f - r_m^*$ has at least $2m+1-d(m) \ge m+1$ zeros in $(-1, 1)$. This is sufficient to ensure that $r_m^* \to f$ in capacity in bounded subsets of $\mathbf{C}$. The proof of this fact is essentially the same as that of the Nuttall-Pommerenke theorem [P]. A formal statement and proof can be found in [L2].

We shall prove that (2.3) is true for some $\varepsilon > 0$. Since f is analytic on $[-1, 1]$ and not rational, one of the following situations holds:

(I) f has infinitely many singularities (points of non-analyticity) outside a neighborhood of $[-1, 1]$.

(II) f has finitely many singularities in $\bar{\mathbf{C}} \setminus [-1, 1]$, and at least one of them is an essential singularity. This includes the possibility that ∞ is an essential singularity.

Now, convergence in capacity means that $\forall\,\varepsilon > 0$, $\forall\, r \ge 1$, $\forall\, m$ large enough

$$|f - r_m^*|(z) \le \varepsilon^m, \quad |z| \le r,\ z \notin K_m\,,$$

where $\operatorname{cap} K_m \le \varepsilon$. It implies that each singularity of f attracts a pole of r_m as $m \to \infty$ since due to a result of Gončar [G2] the contrary would lead to a contradiction.

Consequently in Case I, each of the infinitely many singularities of f attracts poles of r_m^* and for small enough ε, $\gamma_m(\varepsilon)$ approaches ∞ as $m \to \infty$.

In Case II, an unbounded number of poles are attracted by an essential singularity of f as $m \to \infty$. (Assuming the contrary shows that f would have only poles of finite order; cf. [G1, G2]). When f has an essential singularity at ∞, this also remains also true. To see this, choose a Möbius transformation, say $w(z) := z/(z+1)$. Then $r_m^*(w(z))$ converges in capacity to $f(w(z))$ as $m \to \infty$, and $f(w(z))$ has an essential singularity at $z = -1$. Here -1 attracts infinitely many poles of $r_m^*(w(z))$ by Gončar's lemma. Consequently $w(-1) = \infty$ attracts infinitely many poles of r_m^*. ■

Proof of Theorem 2.4. Proceeding as in the proof of Theorem 2.1, we define p_m^*, q_m^*, R_m, Λ', s_{k-1}, and $\hat{P}_{m,1}$ as in the proof for that theorem. Now, set

$$t(m) := \gamma_m - [\delta\gamma_m] + \gamma_{m+1} - [\delta\gamma_{m+1}]. \tag{3.27}$$

By hypothesis,

$$\sigma := \liminf_{m\to\infty} \frac{\gamma_m}{m} > 0, \tag{3.28}$$

so that

$$\liminf_{n\to\infty} \frac{t(m)}{m} \geq 2\sigma(1-\delta) > 0. \tag{3.29}$$

Given $[\alpha, \beta] \subset [-1, 1]$, set $j(m) := (\log m)^2$ and

$$\tilde{P}_{m,2}(x) := \pi_{j(m)}(x) T_{t(m)-j(m)}(x), \tag{3.30}$$

where $\pi_{j(m)}$ is defined by the formula in Lemma 3.3 and $T_{t(m)-j(m)}$ is the Chebyshev polynomial of degree $t(m) - j(m)$. Let

$$L(m) := \text{card}\,\{x \in [\alpha, \beta] : |T_{t(m)-j(m)}(x)| = 1,\ \ \pi_{j(m)}(x) > 2m^2\}. \tag{3.31}$$

Since $j(m)/\log m \to \infty$, it is easily verified from the definition of $\pi_{j(m)}$ and the properties of Chebyshev polynomials that

$$\lim_{m\to\infty} \frac{L(m)}{t(m) - j(m)} = \frac{\arccos\alpha - \arccos\beta}{\pi}. \tag{3.32}$$

Next we set $\tilde{P}_m := \hat{P}_{m,1}\tilde{P}_{m,2}$ and observe that from (3.21), (3.27) and (3.30) we have

$$\deg \tilde{P}_m \leq 2m + 1 - d(m).$$

Hence $R_m - \tilde{P}_m/q_m^* q_{m+1}^*$ has at most $2m+1-d(m)$ zeros. From (3.20), (3.30), and Corollary 3.2 we know that

$$\left| \frac{\tilde{P}_m(x)}{q_m^*(x) q_{m+1}^*(x)} \right| \leq \frac{3}{2}\frac{1}{3} < 1 \quad \text{for } x \in [-1, \alpha] \cup [\beta, 1],$$

and so (cf. (3.13)) $R_m - \tilde{P}_m/q_m^* q_{m+1}^*$ has at least

$$N_m(-1, \alpha) + N_m(\beta, 1) - 2 = N_m(-1, 1) - N_m(\alpha, \beta) - 2$$

zeros in $(-1,\alpha)\cup(\beta,1)$. Also, from (3.11), (3.31) and (3.20), we see that $R_m - \tilde{P}_m/q_m^* q_{m+1}^*$ has at least $L(m)-1$ zeros in $[\alpha,\beta]$. Hence,

$$N_m(-1,1) - N_m(\alpha,\beta) - 2 + L(m) - 1 \le 2m+1-d(m) = N_m(-1,1)-1,$$

and so

$$\frac{L(m)-2}{2m+2} \le \frac{L(m)-2}{N_m(-1,1)} \le \frac{N_m(\alpha,\beta)}{N_m(-1,1)}.$$

Thus from (3.29) and (3.32) we obtain

$$\sigma(1-\delta)\left(\frac{\arccos\alpha - \arccos\beta}{\pi}\right) \le \liminf_{\substack{m\to\infty \\ m\in\Lambda}} \frac{N_m(\alpha,\beta)}{N_m(-1,1)}, \tag{3.33}$$

which proves that (2.4) holds with $\Delta = \sigma(1-\delta)$. ■

Proof of Theorem 2.5. With the hypothesis of (2.5) we can take in the preceding proof $\sigma = 1$ and δ arbitrarily close to zero since we obtain $\mu < 1$ in (3.16) by choosing τ sufficiently large. Hence from (3.33) we get

$$\frac{\arccos\alpha - \arccos\beta}{\pi} \le \liminf_{\substack{m\to\infty \\ m\in\Lambda'}} \frac{N_m(\alpha,\beta)}{N_m(-1,1)},$$

for some subsequence $\Lambda' \subset \Lambda$. Applying this inequality to the intervals $[-1,\alpha]$ and $[\beta,1]$ we deduce that

$$\limsup_{\substack{m\to\infty \\ m\in\Lambda'}} \frac{N_m(\alpha,\beta)}{N_m(-1,1)} \le \frac{\arccos\alpha - \arccos\beta}{\pi},$$

and (2.6) follows. ■

4. Markoff Functions

Assume that f is a Markoff function on $[-1,1]$. The interpolation of f by rational functions is well understood (see e.g. [G1] or [B, p. 124]). If $r_n \in \mathcal{R}_{n-1,n}$, then $f-r_n$ has at most $2n$ zeros. Moreover if $f-r_n$ has exactly $2n$ zeros, say $z_1 < z_2 < \ldots < z_{2n}$, then r_n has the form

$$r_n(x) = \sum_{k=1}^{n} \frac{a_k}{1-t_k x}, \tag{4.1}$$

with $a_k > 0$. This means that r_n is again a Markoff function with the measure concentrated on n points. Furthermore, the difference $f - r_n$ changes its sign at the zeros, i.e.

$$(-1)^i (f-r_n)(x) > 0 \quad \text{for} \quad z_i < x < z_{i+1},\ i = 0,1,\ldots,2n. \tag{4.2}$$

Here, $z_0 = -1$ and $z_{2n+1} = 1$.

Next, let r_n be as above and $m < n$. Assume that $r_m \in \mathcal{R}_{m-1,m}$ interpolates r_n at a subset of $2m$ consecutive points, say at $z_{2k+1}, z_{2k+2}, \ldots, z_{2k+2m}$. Then

$$(-1)^i (r_n - r_m)(x) > 0 \quad \text{for} \quad z_i < x < z_{i+1},\ 2k \le i \le 2k+2m. \tag{4.3}$$

Indeed, from (4.1) we know that r_n may be regarded as a Markoff function. If we replace n by m and f by r_n in (4.2), we obtain (4.3). Now from (4.2) and (4.3) we know that $f - r_n$ and $r_n - r_m$ have the same sign in $(z_{2k}, z_{2k+2m+1})$ and the following is immediate.

THEOREM 4.1 (Comparison Theorem). *Let r_m and r_n be as above. Then*

$$|f - r_m|(x) \geq |f - r_n|(x) \quad \text{for } z_{2k} < x < z_{2k+2m+1}, \tag{4.4}$$

with strict inequality outside the points of interpolation.

Proof of Theorem 2.6. Let $r^*_{n-1,n}$ be the best approximation to f from $\mathcal{R}_{n-1,n}$. Assume that $2m+2$ points of $A_{n-1,n}$ belong to the subinterval $[\alpha, \beta]$. After dropping the first or the last one we have

$$\alpha \leq x_{2k} < x_{2k+2m+1} \leq \beta$$

for some $k \geq 0$. Let $r_{m-1,m} \in \mathcal{R}_{m-1,m}$ be the rational function which interpolates f at the $2m$ zeros of $f - r_{n-1,n}$ in $[x_{2k}, x_{2k+2m+1}]$. The comparison theorem above yields

$$(-1)^i |f - r_{m-1,m}|(x_i) \geq \|f - r^*_{n-1,n}\| \text{ for } i = 2k, 2k+1, \ldots, 2k+2m+1. \tag{4.5}$$

By de La Vallée-Poussin's theorem the right-hand side of (4.5) is a lower bound of $\operatorname{dist}(f, \mathcal{R}_{m-1,m})_{[\alpha,\beta]}$ and the proof of (2.8) is complete. ∎

The proof of Corollary 2.7 is left to the reader.

From the Stieltjes functions we can see that the assumptions of Theorem 2.5 cannot be substantially relaxed. We consider the approximation of $\sqrt{x}$ from $\mathcal{R}_{n,n-1}$ on some interval $[a, b]$ with $0 < a < b$. Then it is known [A] that half of the alternation points (except possibly one) are located in the subinterval $[a, \sqrt{ab}]$. Consequently, in this case we have a deviation from the arcsine distribution. For this function the zeros and the alternation points are explicitly known (see [A]). There is an analytic map, which maps the plane with two cuts $\mathbf{C}\backslash(-\infty, 0)\backslash[a, b]$ onto an annular domain. Specifically, the mapping can be explicitly given in terms of elliptic functions. This map sends the point of the alternant to equally spaced points on a circle.

In the general case (2.7) we have only a weaker result. It is known that

$$\operatorname{dist}(f, \mathcal{R}_{n-1,n}) \leq \text{const} \cdot \omega^{-2m}$$

where $1/\log\omega$ is the capacity of the condensor whose one plate is the cut $T =: [c_1, c_2]$ and the other plate is the interval of approximation $[-1, 1]$. Moreover, if the measure satisfies a Szegő condition

$$\int_T \frac{\log|\mu'|\,dt}{\sqrt{(t-c_1)(c_2-t)}} > -\infty,$$

then one has

$$\lim_{n\to\infty} \operatorname{dist}(f, \mathcal{R}_{n-1,n})^{1/n} = \frac{1}{\omega^2}.$$

Now let ω' be defined in the same way when the interval $[1,+1]$ is replaced by $[\alpha,\beta]$. Obviously $\omega' < \omega$. From Theorem 2.6 we conclude that

$$\lim_{n\to\infty} \frac{N_{n-1,n}(\alpha,\beta)}{N_{n-1,n}(-1,1)} \le \frac{\omega'}{\omega}.$$

References

1. N.I. Akhieser (1947): Lectures on the Theory of Approximation. Frederick Ungar, New York, 1956.
2. D. Braess (1986): Nonlinear Approximation Theory. Springer-Verlag, Heidelberg – New York.
3. L. Baratchart, E.B. Saff, and F. Wielonsky (1993): *Rational interpolation of the exponential function* (manuscript).
4. P.B. Borwein, A. Kroó, R. Grothmann, and E.B. Saff (1989): *The density of alternation points in rational approximation.* Proc. Amer. Math. Soc. **105**, 881–888.
5. A.A. Gončar (1978): *On the speed of rational approximation of some analytic functions.* Math USSR Sbornik **34**, 131–146.
6. A.A. Gončar (1975): *On the convergence of generalized Padé approximants of meromorphic functions.* Mat. Sbornik **98**(140), 503–514 = Math. USSR. Sbornik **27**, 503–514.
7. M.I. Kadec (1960): *On the distribution of points of maximal deviation in the approximation of continuous functions by polynomials.* Uspekhi Mat. Nauk **15**, 199–202.
8. A. Kroó and F. Peherstorfer (1993): *On asymptotic distribution of oscillation points in rational approximation.* Analysis Mathematica (to appear).
9. D.S. Lubinsky (1983): *Divergence of complex rational approximations.* Pacific J. Math. **108**, 338–360.
10. D.S. Lubinsky (1984): *On convergence of rational and best rational approximations.* J. Math. Anal. Applns. **98**, 419–434.
11. Ch. Pommerenke (1978): *Padé approximants and convergence in capacity.* J. Math. Anal. Applns. **34**, 131–145.
12. H. Stahl (1992): *Best uniform rational of $|x|$ on $[-1,1]$.* Mat. Sbornik **183**, 85–112.
13. E.B. Saff and H. Stahl (1993): *Asymptotic distribution of poles and zeros of best rational approximants to x^α on $[0,1]$.* Proceedings of 'Semester on Function Theory', International Banach Center, Warsaw, 1992.
14. N.S. Vjačeslavov (1975): *On the uniform approximation of $|x|$ by rational functions.* DAN SSSR **220**, 512–515 = Soviet Math. Dokl. **16**, 100–104.
15. J.L. Walsh (1969), Interpolation and Approximation by Rational Functions in the Complex Domain (5th ed.), Colloquium Publication, Vol. XX, Amer. Math. Soc., Providence, R.I.

ASYMPTOTICS AND SUPER ASYMPTOTICS FOR BEST RATIONAL APPROXIMATION ERROR NORMS TO THE EXPONENTIAL FUNCTION (THE '1/9' PROBLEM) BY THE CARATHÉODORY-FEJÉR METHOD

ALPHONSE P. MAGNUS
Institut Mathématique, Université Catholique de Louvain
Chemin du Cyclotron 2
B-1348 Louvain-la-Neuve
Belgium
E-mail: magnus@anma.ucl.ac.be

Abstract. Let E_n be the error norm of the best L_∞ rational approximation of degree n to the exponential function $\exp(-t)$ on $[0,\infty)$. Grounds are given for setting the conjectured limit $E_n/q^n \to 2q^{1/2}$ when $n \to \infty$, where q is the known constant '1/9'= 1/9.2890254919208189187554494359517450610316948677..., based on the singular values and functions of the relevant Hankel operator (Carathéodory-Fejér's method).
Moreover, hints are given according to which a valuable asymptotic expansion of E_n should also contain n^{th} powers of new constants q_1 = '1/56', q_2 = '1/240', etc.

Key words: Rational approximation, Carathéodory Fejér's method, '1/9' problem.

1. Introduction

Best rational approximation to $\exp(-t)$ on $0 \le t < \infty$ has been much worked since [5]: as recalled in [3] and [26] chap. 2, the best L_∞ error norms

$$E_n = \min_{P_n, Q_n} \max_{0 \le t < \infty} \left| e^{-t} - \frac{P_n(t)}{Q_n(t)} \right|,$$

where the minimum is taken on the real polynomials P_n and Q_n of degree $\le n$, have been found to decrease geometrically. The rate of decrease, once thought to be 1/9, later called '1/9', has been determined in [11]. This will be recalled briefly in section 2.
On the other hand, quasioptimal rational approximations based on the Carathéodory-Fejér's (CF) method have been shown in [24] to be close to the optimal ones (see section 3). Table I shows values of E_n (from [3]) and eigenvalues λ_n of the appropriate Hankel matrix (from [15] and [24]): the relative closeness of the $\sigma_n = |\lambda_n|$'s to the E_n's is amazing, a tentative explanation will be given in section 5. Following [15], more asymptotic spectral properties of the relevant Hankel matrix are studied here. More precisely, a description of the asymptotic behaviour of the eigenfunctions of a spectrally equivalent integral Hankel operator will be attempted here (in section 4).
The following conjecture will be presented:

Conjecture 1 . *E_n/q^n and $|\lambda_n|/q^n \to 2q^{1/2}$ when $n \to \infty$, where q = '1/9', the constant described in [11] (Theorem 2) and [15].*

A. Cuyt (ed.), Nonlinear Numerical Methods and Rational Approximation II, 173–185.

TABLE I
Error norms and eigenvalues.

n	E_n	λ_n
0	5.000000000000000E-01	5.601715174207940E-01
1	6.683104216185045E-02	-6.680573308019967E-02
2	7.358670169580528E-03	7.355581867871742E-03
3	7.993806363356878E-04	-7.994517064498902E-04
4	8.652240695288851E-05	8.652095258749368E-05
5	9.345713153026646E-06	-9.345740936352446E-06
6	1.008454374899671E-06	1.008453857122026E-06
7	1.087497491375248E-07	-1.087497586430036E-07
8	1.172265211633491E-08	1.172265194363526E-08
9	1.263292483322314E-09	-1.263292486435714E-09
10	1.361120523345448E-10	1.361120522787731E-10
11	1.466311194937487E-11	-1.466311195036850E-11
12	1.579456837051239E-12	1.579456837033622E-12
13	1.701187076340353E-13	-1.701187076343463E-13
14	1.832174378254041E-14	1.832174378253495E-14
15	1.973138996612803E-15	-1.973138996612899E-15
16	2.124853710495224E-16	2.124853710495207E-16
17	2.288148563247892E-17	-2.288148563247895E-17
18	2.463915737765169E-18	2.463915737765169E-18
19	2.653114658063313E-19	-2.653114658063313E-19
20	2.856777383549094E-20	2.856777383549094E-20

2. Exponential behaviour of the error norm.

Let us first consider that the approximation P_n/Q_n interpolates $\exp(-t)$ at the positive numbers $t_1, \ldots, t_{n+1}$. From the Hermite-Walsh remainder formula,

$$e^{-t} - \frac{P_n(t)}{Q_n(t)} = \frac{Z_{n+1}(t)}{Q_n(t)} \frac{1}{2\pi i} \int_C \frac{Q_n(\tau)}{Z_{n+1}(\tau)} e^{-\tau} \frac{d\tau}{\tau - t},$$

where $Z_{n+1}(t) = (t-t_1)\ldots(t-t_{n+1})$, and where the parabola-like contour C encloses t and $[0,\infty)$. One can then choose Q_n, Z_{n+1} and C (which may depend on n) such that the absolute value of the rational function Z_{n+1}/Q_n is much smaller on $[0,\infty)$ than on C (Zolotarev problem). A geometric decrease of the error norms can then already be exhibited ([7] chap.6), but happens to be larger than the observed value. This is explained by the numerous phase changes of Q_n/Z_{n+1} on C, making the integral much smaller that the integral of the absolute value. The proper way to go further is to take into account n more interpolation points $t_{n+2}, \ldots, t_{2n+1}$, this amounts to the property that Q_n is orthogonal to polynomials of degree $< n$ with respect to the complex weight $\exp(-\tau)/Z_{2n+1}(\tau)$ on C, where $Z_{2n+1}(t) = (t-t_1)\ldots(t-t_{2n+1})$. We then have a more convenient remainder formula

$$e^{-t} - \frac{P_n(t)}{Q_n(t)} = \frac{Z_{2n+1}(t)}{Q_n^2(t)} \frac{1}{2\pi i} \int_C \frac{Q_n^2(\tau)}{Z_{2n+1}(\tau)} e^{-\tau} \frac{d\tau}{\tau - t}. \tag{1}$$

Using and extending results of Stahl ([21]; see [22]-[23] for other striking examples of Stahl's *savoir-faire*), Gončar and Rahmanov [11] showed that the interpolation points and the zeros of Q_n tend to be distributed for large n in such a way that $[Q_n^2(\tau)/Z_{2n+1}(\tau)]^{1/n}$ behaves essentially as $\exp(-2\mathcal{V}(\tau/n))$, where the real part of $\mathcal{V}$ is constant on $[0,\infty)$, the real part of $2\mathcal{V}(z)+z$ is another constant on a fixed cut F (which has to be found), and where the phase of $\exp(2\mathcal{V}(z)+z)$ takes opposite values on the two sides of F (from the Stahl-Gončar-Rahmanov symmetry property about the normal derivatives of the real part of $2\mathcal{V}+z$ and the Cauchy-Riemann conditions translating this as a property of the tangent derivatives of the imaginary part of $2\mathcal{V}+z$, i.e., the phase of the exponential). From the values on $[0,\infty)$ and F, the limit '1/9' of the nth root of the norm of (1) could be established (Theorem 2 of [11]).

Here is another way to recover an approximate description of the error function, using a change of variables which will be useful later in connection with the CF method: $Z_{2n+1}(t)/Q_n^2(t)$ is very close to equioscillating on $[0,\infty)$, let $\tilde{Q}_n$ be the polynomial whose zeros are the square roots with positive real parts of $-n^{-1}$ times the zeros of Q_n, i.e., if $p_1,\dots,p_n$ are the zeros of $\tilde{Q}_n$, $-np_1^2,\dots,-np_n^2$ are the zeros of Q_n, and $\tilde{Q}_n(u)\tilde{Q}_n(-u) = Q_n(-nu^2)$. Then, $Z_{2n+1}(-nu^2)/Q_n^2(-nu^2) \sim [\tilde{Q}_n(u)/\tilde{Q}_n(-u)]^2 + [\tilde{Q}_n(-u)/\tilde{Q}_n(u)]^2$ is a good guess relating the distribution of the interpolation points on $u \in$ the imaginary axis ($t = -nu^2 \geq 0$) to the poles of the approximation. According to [11], the p_k's tend to be distributed on a fixed locus Γ: $\tilde{Q}_n(-u)/\tilde{Q}_n(u) \sim \exp(n\Phi(u))$, where

$$\Phi(u) = \int_\Gamma \log\frac{-u-p}{u-p}\,d\mu(p), \tag{2}$$

where μ is a (still unknown) positive measure of unit total weight on Γ (still unknown too). Remark that Φ is an odd function defined outside Γ and $-\Gamma$, that $\Phi(u)$ is a pure imaginary number when u is imaginary, because the distribution μ is symmetric about the real axis: $d\mu(p) = d\mu(\overline{p})$, that its expansion about 0 starts as $\Phi(u) = 2\int_\Gamma p^{-1}d\mu(p)\,u + \cdots$, and that $\Phi(\infty) = \pi i$ (as $\int_\Gamma d\mu(p) = 1$). Then, (1) becomes, with $t = -nu^2$ and $\tau = -n\xi^2$:

$$\begin{aligned} e^{nu^2} - \frac{P_n(-nu^2)}{Q_n(-nu^2)} &= \frac{Z_{2n+1}(-nu^2)}{Q_n^2(-nu^2)}\frac{1}{2\pi i}\int_\Gamma \frac{Q_n^2(-n\xi^2)}{Z_{2n+1}(-n\xi^2)}e^{n\xi^2}\frac{2\xi}{\xi^2-u^2}\,d\xi \\ &\sim \left(e^{2n\Phi(u)} + e^{-2n\Phi(u)}\right)\frac{1}{2\pi i}\int_\Gamma \frac{e^{n\xi^2}}{e^{2n\Phi(\xi)}+e^{-2n\Phi(\xi)}}\frac{2\xi}{\xi^2-u^2}\,d\xi. \end{aligned} \tag{3}$$

The connection with the function $\mathcal{V}$ of [11] is $\Phi(u) = -\mathcal{V}(-u^2)+$ constant.

The following expression of Φ as an elliptic integral will be needed:

$$\Phi(u) = \pi i + \int_{+i\infty}^{u}\left(\frac{u^2}{\xi^2}-1\right)\left[\left(1-\frac{\xi^2}{\xi_1^2}\right)\left(1-\frac{\xi^2}{\xi_2^2}\right)\right]^{-1/2}d\xi, \tag{4}$$

where ξ_1 and ξ_2 are two complex conjugate numbers which will soon be determined, and where $\left[\left(1-\frac{\xi^2}{\xi_1^2}\right)\left(1-\frac{\xi^2}{\xi_2^2}\right)\right]^{-1/2} \sim |\xi_1|^2/\xi^2$ when ξ is large. In order to see that (4) has the desired properties, we first establish the link with the Jacobi notation for elliptic integrals through the change of variables ([17]) $\eta = 2i/[\xi/|\xi_1| - |\xi_1|/\xi]$, $v = 2i/[u/|\xi_1| - |\xi_1|/u]$, then, with $\xi_1 = |\xi_1|\exp(i\theta_1)$, $\xi_2 = |\xi_1|\exp(-i\theta_1)$, $k = \sin\theta_1$, $\xi = i|\xi_1|(1+\sqrt{1-\eta^2})/\eta$ if $|\xi| \geq |\xi_1|$,

TABLE II
'1/9' related data.

$k = \sin\theta_1$	=	0.90890855754854147823611890874479350490101396934041
$k' = \cos\theta_1$	=	0.41699548440604205639041957807087776692610248051382
$1/q = \exp(\pi K'/K)$	=	9.28902549192081891875544943595174506103169486775012
K	=	2.32104973253042114734283739983633918849213061106173
$E = K/2$	=	1.16052486626521057367141869991816959424606530553086
K'	=	1.64669144431946837372958069030713103423036178930922
E'	=	1.50010688965199892576311071782207995131063998866470
$\xi_1, \xi_2 = \pi(k' \pm ik)/K$	=	$0.564412701731271 \pm 1.230228033100522\, i$

$\xi = i|\xi_1|(1-\sqrt{1-\eta^2})/\eta$ if $|\xi| \le |\xi_1|$,
$\left[\left(1-\frac{\xi^2}{\xi_1^2}\right)\left(1-\frac{\xi^2}{\xi_2^2}\right)\right]^{-1/2} d\xi = \pm\frac{i}{2}|\xi_1|(1-k^2\eta^2)^{-1/2}(1-\eta^2)^{-1/2}d\eta$ (with the + sign if $|\xi| \ge |\xi_1|$, with the minus sign otherwise). Then, $\Phi(0) = \pi i - \int_{+i\infty}^{0}\left[\left(1-\frac{\xi^2}{\xi_1^2}\right)\left(1-\frac{\xi^2}{\xi_2^2}\right)\right]^{-1/2} d\xi = \pi i - i|\xi_1|K$, where $K = \int_0^1 (1-\eta^2)^{-1/2}(1-k^2\eta^2)^{-1/2}d\eta$ is the complete elliptic integral of first kind of modulus k. The condition $\Phi(0) = 0$ yields the first condition $|\xi_1|K = \pi$. Next, we want $u - \Phi'(u)$ to take opposite values on the sides of Γ, in particular $\Phi'(u)/u = 1$ at ξ_1:
$u^{-1}\Phi'(u) = -i|\xi_1|^{-1}\int_0^v [1+(1-\eta^2)^{1/2}]^{-2}\eta^2(1-\eta^2)^{-1/2}(1-k^2\eta^2)^{-1/2}d\eta =$

$$-\frac{i}{|\xi_1|}\left[2(1-k^2v^2)^{1/2}\frac{1-(1-v^2)^{1/2}}{v} + K(v;k) - 2E(v;k)\right],$$

where $K(v;k) = \int_0^v (1-\eta^2)^{-1/2}(1-k^2\eta^2)^{-1/2}d\eta$ and $E(v;k) = \int_0^v (1-\eta^2)^{-1/2}(1-k^2\eta^2)^{1/2}d\eta$ are the incomplete elliptic integrals of first and second kind (the results of [2] section 3.1 are used, but the notation is slightly different here [Akhiezer uses K as the argument of E]). At $u = \xi_1$, $\eta = 1/k$, where one has $K(k^{-1};k) = K+iK'$ and $E(k^{-1};k) = E+i(EK'-\pi/2)/K$ (see bottom of [2] section 3.1), so that $\Phi'(\xi_1)/\xi_1 = 1-(i/\pi)(K-2E)(K+iK')$ whence the condition $K = 2E$ discussed in various forms (going back to Halphen 1886!) in [11] [15] [26]. The condition $K = 2E$ could be obtained more easily ([17]) by expressing $\Phi'(\xi_1)/\xi_1 - \Phi'(\xi_2)/\xi_2 = 0$, but the present derivation showed that $\Phi'(\xi_1)/\xi_1 = 1$ as well. It is then clear that $\Phi'(u) - u$ takes opposite values on the sides of Γ, as $\Phi'(u) - u = \pm 2u\int_{\xi_1}^{u}\xi^{-2}\left[\left(1-\frac{\xi^2}{\xi_1^2}\right)\left(1-\frac{\xi^2}{\xi_2^2}\right)\right]^{-1/2} d\xi$.
If we call Φ_1 the analytic continuation of $\Phi = \Phi_0$ across Γ, we see that $\Phi'_1(u) = 2u - \Phi'(u)$, therefore $\Phi_1(u) = u^2 - \Phi(u) +$ constant. This constant is found from $\Phi_1(\xi_1) = \Phi(\xi_1)$, hence, by $\Phi(u) = \pi i + u\Phi'(u)/2 - i|\xi_1|K(v;k)/2$: $\Phi(\xi_1) = \pi i + \xi_1^2/2 - i|\xi_1|(K+iK')/2 = \pi i/2 + \xi_1^2/2 + \pi K'/(2K)$, so

$$\Phi_1(u) = u^2 - \Phi(u) + \pi i - \log q, \tag{5}$$

where $q = \exp(-\pi K'/K)$ is the famous '1/9' constant. Similarly, continuation of Φ across $-\Gamma$ yields

$$\Phi_2(u) = -u^2 - \Phi(u) + \log q + \pi i. \tag{6}$$

The real part of $\xi^2 - 2\Phi(\xi)$ is the constant $\log q$ on the arc Γ, as it should. The numerical values are recalled in Table II. The point ξ_1 ($\sqrt{-a}$ in [11]) is marked on figure 2; the arc

Γ is almost the rectilinear segment joining ξ_1 and $\xi_2 = \overline{\xi_1}$ (see also fig.1 ($m = 1$) of [15] where ξ_1 and ξ_2 are called X_2 and X_1). Interesting expansions of Φ are,

$$near\ 0 : \Phi(u) = 2u - (\xi_1^{-2} + \xi_2^{-2})u^3/3 - (\xi_1^{-4}/4 + \xi_1^{-2}\xi_2^{-2}/6 + \xi_2^{-4}/4)u^5/5 + \cdots \quad (7)$$

$$near\ +i\infty : \Phi(u) = \pi i + 2|\xi_1|^2/(3u) + |\xi_1|^2(\xi_1^2 + \xi_2^2)/(15u^3) + \cdots \quad (8)$$

3. Use of the Carathéodory-Fejér's method.

Carathéodory and Fejér (and also Schur and Takagi) studied several problems of estimation of Fourier and Laurent series and gave very constructive answers. Initially, the problem was: given a bounded complex sequence $g_0, g_1, \ldots$, estimate $\nu = \sup|\sum_0^\infty g_k\zeta^k|$ on $|\zeta| < 1$. Looking at the norms of the partial sums $\|\sum_0^N g_k\zeta^k\|_\infty$ would be most clumsy and would suggest wrong answers in some cases. The method of Carathéodory and Fejér is the following: for each $N > 1$, construct the expansion $\sum_0^{N-1} g_k\zeta^k + \sum_N^\infty g_k^{(N)}\zeta^k$ of *lowest possible* L_∞ norm ν_N. It is then shown that the ν_N's form an increasing sequence with limit ν, moreover, that ν_N is the largest singular value of a Hankel matrix constructed with $g_0, \ldots, g_{N-1}$ Other singular values become useful if n poles in $|\zeta| < 1$ are allowed, by adding an expression of the form $\sum_N^\infty g_k^{(N)}\zeta^k / \sum_0^n e_k\zeta^{n-k}$.

The appropriate setting for approximating a given series $f_+(z) = \sum_1^\infty a_k z^k$ with real coefficients is to look for a function of the form

$$r_n(z) = \sum_{-\infty}^n d_k z^k / \sum_0^n e_k z^k = \sum_{-\infty}^0 d_{k+n} z^k / \sum_0^n e_{n-k} z^{-k},$$

where the numerator series converges in $|z| > 1$, with n poles in $|z| > 1$, and such that the norm of $\sum_1^N a_k z^k - r_n(z)$ is the lowest possible on $|z| = 1$ ([24], section 1; the connection with the original CF problem is made by $z = 1/\zeta$, $g_k = a_{N-k}$). If the given function is continuous on the unit circle, we may consider infinite Hankel matrices ($N \longrightarrow \infty$) as compact operators ([1], [4], [16]). We then arrive at an error of the form

$$f_+(z) - r_n(z) = \lambda_n z \frac{u_1 + u_2 z + \ldots}{u_1 + u_2 z^{-1} + \ldots} \quad (9)$$

with real $\lambda_n, u_1, u_2, \ldots$, and where $u_1 + u_2 z + \cdots$ has exactly n zeros in $|z| < 1$, so that the error curve is a perfect circle about the origin of radius $\sigma_n = |\lambda_n|$ and winding number $2n + 1$. Factorizing $u_1 + u_2 z^{-1} + \cdots$ as $\sum_0^n e_{n-k} z^{-k}$ times a second factor, which is another expansion in z^{-1}, we see that gathering the positive powers in the product of (9) by $u_1 + u_2 z^{-1} + \cdots$, we have $a_k u_1 + a_{k+1} u_2 + \cdots = \lambda_n u_k$, $k = 1, 2, \ldots$, i.e., the eigenproblem $H\mathbf{u} = \lambda_n \mathbf{u}$, where H is the infinite Hankel matrix $[a_{i+j+1}]_{i,j=0}^\infty$, and where λ_n is the $(n+1)^{th}$ eigenvalue of H in decreasing order of the absolute values: $|\lambda_0| \geq |\lambda_1| \geq \ldots$ This brief description supposes that $|\lambda_n|$ is not repeated: $|\lambda_{n-1}| > |\lambda|_n > |\lambda_{n+1}|$, see [1], [9], [12], [16] for a more complete discussion including complex functions and degenerate cases.

The expression $(a_0 + r_n(z) + r_n(1/z))/2$ yields then an approximation with n poles outside $[-1, 1]$ in the $x = (z + 1/z)/2$ plane to $F(x) = (a_0 + f_+(z) + f_+(1/z))/2 = a_0/2 + \sum_1^\infty a_k T_k(x)$, with an error function which equi-oscillates exactly at $2n + 2$ points of $[-1, 1]$. Finally, a Chebyshev economization is performed on $(a_0 + r_n(z) + r_n(1/z))/2$ in order to get a rational function of x [24].

Rational approximations to $\exp(-t)$ on $[0, \infty)$ are first translated as rational functions of $x = (c - t)/(c + t)$, $c > 0$, so that we have to consider the Chebyshev expansion $F(x) =$

$\exp(c(x-1)/(x+1))$, or the Laurent coefficients of $F((z+1/z)/2) = \exp(c(z-1)^2/(z+1)^2)$. Any positive value of c yields a spectrally equivalent Hankel matrix H, but the most explicit setting appears when one considers $c \to \infty$:

Theorem 1 . *The eigenvalues of the Hankel matrix H are the same as those of the integral Hankel operator $\mathcal{H}$ given on $L_2(0,\infty)$ by*

$$(\mathcal{H}f)(x) = \frac{2}{\sqrt{\pi}} \int_0^\infty e^{-(x+y)^2} f(y)\, dy \qquad (10)$$

Indeed, the meaningful part of the integral $a_k = (\pi i)^{-1} \int_{|z|=1} \exp(c(z-1)^2/(z+1)^2) z^{-k-1} dz$ is a neighbourhood of $z = 1$, as $c(z-1)^2/(z+1)^2$ has a strongly negative real part when $z \neq 1$ on the unit circle. Let $z = 1 + 2iuc^{-1/2}$: $a_k \sim 2\pi^{-1}c^{-1/2} \int_{-\infty}^{\infty} \exp(-u^2 - 2ikc^{-1/2}u)du = 2(\pi c)^{-1/2}\exp(-k^2/c)$ when $c \to \infty$. The elements

$$(H\mathbf{u})_k = \sum_{m=0}^{\infty} a_{k+m}u_m \sim 2(\pi c)^{-1/2} \sum_{m=0}^{\infty} \exp(-(k+m)^2/c)u_m$$

behave essentially as values of a function of the variable $kc^{-1/2}$: let $x = kc^{-1/2}$, $y = mc^{-1/2}$, $f(y) = u_m c^{1/4}$, (so that $\sum_0^\infty u_m^2 = c^{-1/2}\sum_0^\infty [f(mc^{-1/2})]^2 \sim \int_0^\infty (f(y))^2 dy$ remains constant), $g(x) = (H\mathbf{u})_k c^{1/4} \sim 2(\pi c)^{-1/2}\sum_{m=0}^{\infty} \exp(-(k+m)^2/c)c^{1/4}u_m \sim 2\pi^{-1/2}\int_0^\infty \exp(-(x+y)^2)f(y)\, dy$. In particular, if $H\mathbf{u} = \lambda \mathbf{u}$, then $\mathcal{H}f = \lambda f$, with the same λ.

The eigenfunctions $f_0, \dots, f_9$ corresponding to $\lambda_0, \dots, \lambda_9$ of $\mathcal{H}$ are plotted in figure 1. The study of the asymptotic behaviour of f_n for large n should shed some light on the value of λ_n. If a uniform asymptotic formula $\tilde{f}_n$ could be found so that the L_2 norm $\|\mathcal{H}\tilde{f}_n - \tilde{\lambda}_n\tilde{f}_n\| < \varepsilon_n$ could hold, with $\|\tilde{f}_n\| = 1$ and $\varepsilon_n/|\tilde{\lambda}_n| \to 0$ when $n \to \infty$, then $\tilde{\lambda}_n$ would be a valuable asymptotic estimate of λ_n. Unfortunately, the subject is not yet so advanced.

4. Asymptotic behaviour of the eigenfunctions.

The typical oscillations of the eigenfunctions shown by figure 1 suggest that $f_n(x)$ contains the exponential of n times a function taking imaginary values (from 0 to πi) on a part of the real axis. Actually, as the transition point separating oscillatory behaviour from monotonously decreasing behaviour seems to increase like $n^{1/2}$, we expect an asymptotic formula involving the exponential of n times a fixed function of $x/n^{1/2}$ ... For a more accurate guess, we know that if the CF method is successful, the poles of the CF approximant are very close to the poles of the best rational approximation [24], so $u_1 + u_2 z^{-1} + \cdots$ and $\tilde{Q}_n(u)$ should have common factors, with $n^{1/2}u = c^{1/2}(z-1)/(z+1)$. When $c \to \infty$: $\sum_1^\infty u_k z^k \sim \sum_1^\infty f_n(kc^{-1/2})[1+2(n/c)^{1/2}u]^k \sim c^{1/2}\int_0^\infty f_n(\xi)\exp(-2n^{1/2}u\xi)d\xi$,i.e., the Laplace transform of f_n should contain at least the factor $\exp(-n\Phi(u))$. This asks for the inverse Laplace transform of $\exp(-n\Phi)$: $f_n(\xi)$ is expected to involve $\int_{Br} \exp[-n\Phi(u/(2n^{1/2}))+\xi u]du$ on some Bromwich contour. A saddlepoint analysis predicts this integral to behave like $\exp[-n\Phi(s/(2n^{1/2})) + \xi s]$, where s is a root of $\xi = (n^{1/2}/2)\Phi'(s/(2n^{1/2}))$. So, $s/n^{1/2}$ is a root x of the equation $2y = \Phi'(x/2)$ when $y = \xi/n^{1/2}$ is given. Let Ψ' be the inverse function of Φ', then $x = 2\Psi'(2y)$ and

$$\Phi(x/2) = \pi i + xy - \Psi(2y) \;\; \textit{if} \;\; 2y = \Phi'(x/2), \; \textit{i.e.}\; x = 2\Psi'(2y) \qquad (11)$$

indeed, $\Phi(x/2) - \pi i = \int_\infty^{x/2} \Phi'(\zeta)d\zeta = \int_0^{2y} \eta d\Psi'(\eta) = 2y\Psi'(2y) - \Psi(2y)$, so that $-\Phi(s/(2n^{1/2})) + \xi s = \Psi(2\xi/n^{1/2})$.

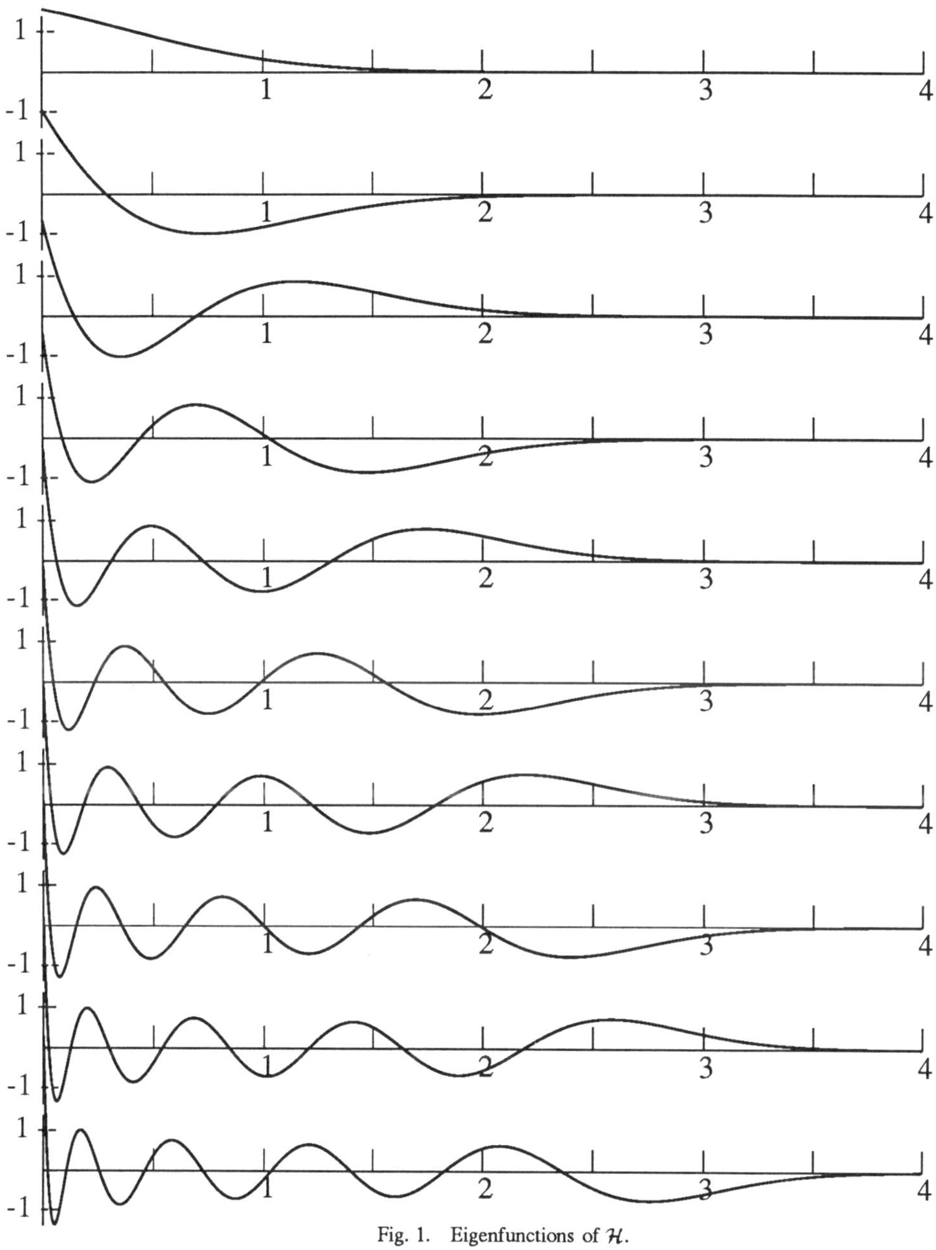

Fig. 1. Eigenfunctions of $\mathcal{H}$.

Figure 2 shows the images of horizontal and vertical lines of the upper right quarter plane Re$u \geq 0$ and Im$u \geq 0$ by Φ' and its analytic continuation $\Phi'_1(u) = 2u - \Phi'(u)$ across the cut $\Gamma = [\xi_1, \overline{\xi}_1]$. The net of solid lines (images of verticals Reu=0,0.25,0.50,etc. by

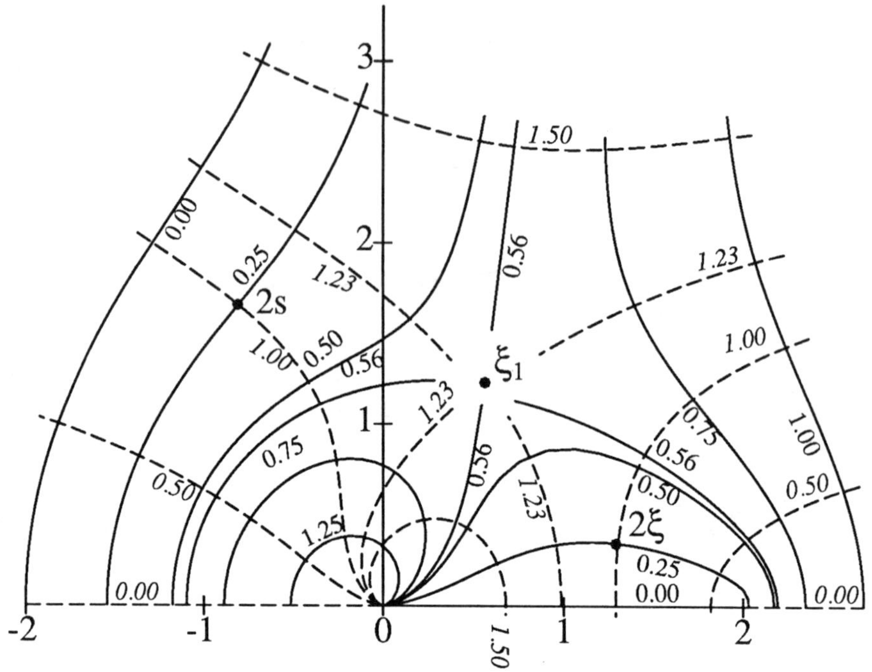

Fig. 2. Image of the upper right quarter plane by Φ'

Φ') and dashed lines (images of horizontals Imu=0,0.25,0.50, etc. by Φ') cover a region extending up to -2 to the left ($-2 = \Phi'_1(0)$). This means that Ψ' maps this region to the upper right quarter plane. For instance, the point $1.311 + 0.344i$ marked 2ξ in figure 2 is at the intersection of the solid line of label 0.25 and the dashed line of label 1.00: this means that $1.311 + 0.344i = \Phi'(0.25 + i)$ and also $0.25 + i = \Psi'(1.311 + 0.344i)$.

From the expansions (7) and (8):

$$near\ 0 : \Psi(v) = 2\left(-\frac{2|\xi_1|^2}{3}\right)^{1/2} v^{1/2} + \cdots, \tag{12}$$

$$near\ 2 : \Psi(v) = \pi i + \frac{2}{3}\left(-\frac{3}{2(\xi_1^{-2} + \xi_2^{-2})}\right)^{1/2} (v-2)^{3/2} + \cdots, \tag{13}$$

So, 0 and 2 are singular points of Ψ which is pure imaginary in $[0,2]$. This seems to explain the observed behaviour of the eigenfunctions f_n:

Conjecture 2 . *The eigenvalues λ_0, $\lambda_1, \ldots$ of the Hankel integral operator (10), with $|\lambda_0| \geq |\lambda_1| \geq \ldots$, satisfy*

$$\lambda_n \sim 2(-1)^n q^{n+1/2}\ when\ n \to \infty. \tag{14}$$

The corresponding eigenfunctions $f_0, f_1, \dots$ of (10) satisfy

$$f_n(x) \sim A(x/\nu^{1/2})\exp[\nu\Psi(2x/\nu^{1/2})] + B(x/\nu^{1/2})\exp[-\nu\Psi(2x/\nu^{1/2})]\,,\; n \to \infty. \tag{15}$$

with $\nu = n + 1/2$, and where A and B are (still unknown) fixed functions, and where Ψ is such that Ψ' is the inverse function of Φ' (with Φ given by (2) and (4)).

To explore the validity of this conjecture, let us introduce the proposed formula of $f_n(\nu^{1/2}y)$ in (10):

$$(\mathcal{H}f_n)(\nu^{1/2}\xi) \sim (2\nu/\pi^{1/2}) \int_0^\infty \exp(-\nu(\xi+\eta)^2)[A\exp(\nu\Psi(2\eta)) + B\exp(-\nu\Psi(2\eta))]d\eta.$$

We estimate these integrals by saddlepoint analysis:

$$(\mathcal{H}f_n)(\nu^{1/2}\xi) \sim 2A(s_1)(1-2\Psi''(2s_1))^{-1/2}\exp[-\nu((\xi+s_1)^2 - \Psi(2s_1)]+$$
$$+2B(s_2)(1+2\Psi''(2s_2))^{-1/2}\exp[-\nu((\xi+s_2)^2+\Psi(2s_2)],$$

where s_1 and s_2 are roots of $\xi + s = \pm\Psi'(2s)$, found through Φ' and its continuations across Γ and $-\Gamma$: let 2ξ be in the range of Φ', say $2\xi = \Phi'(w)$. Then, as $\Phi'(w) + \Phi'_1(w) = 2w$ and $\Phi'(w) + \Phi'_2(w) = -2w$, $s_1 = \Phi'_1(w)/2$ and $s_2 = \Phi'_2(w)/2$ are valid solutions. For instance, figure 2 shows a point $2\xi = 1.311 + 0.344i$ corresponding to $w = 0.25 + i$. The point marked $2s$ is $\Phi'_1(w) = 2s_1 = -0.811 + 1.656i$ ($2s_2$ is not on figure 2).

Now, using (5), (6) and (11): $-(\xi+s_1)^2+\Psi(2s_1) = -w^2-\Phi_1(w)+2s_1w+\pi i = -w^2+\Phi(w)-w^2+\log q-\pi i+2s_1w+\pi i = -\Psi(2\xi)+\log q+\pi i$, $-(\xi+s_1)^2-\Psi(2s_2) = -w^2+\Phi_2(w)-2s_1w-\pi i = -w^2-\Phi(w)-w^2+\log q+\pi i-2s_2w-\pi i = \Psi(2\xi)+\log q-\pi i$, so, we see that the exponentials of $\pm\nu\Psi(2\xi)$ are recovered. Eigenfunctions should therefore satisfy

$$2A(s_1)[1-2\Psi''(2s_1)]^{-1/2}\exp(\nu\pi i)q^\nu = \lambda_n B(\xi)\,, \tag{16}$$

$$2B(s_2)[1+2\Psi''(2s_2)]^{-1/2}\exp(-\nu\pi i)q^\nu = \lambda_n A(\xi)\,, \tag{17}$$

where $s_1, s_2 = \Phi_{1,2}(w)/2 = -\xi \pm \Psi'(2\xi)$ if $2\xi = \Phi'(w)$.

If this is enough to guess that λ_n will behave like $(-1)^n q^n$, these equations do not give clear indications on what the functions A and B should be. Moreover, these functions are probably discontinuous (on Stokes lines), as a result of representing the entire function f_n by an asymptotic formula involving functions with branch-points. So, as $\Psi(\xi) \to +\infty$ when $\xi \to +\infty$, and as $f_n \in L_2$, one must have $A(\xi) \equiv 0$ in a region containing $[2, \infty)$, but $A(\xi) \not\equiv 0$ in $[0, 2]$ where the two imaginary exponentials are needed in order to explain the oscillations of f_n. This state of things is common in differential equations discussions (JWKB-Liouville-Green-Steklov theory [6]), where one has connection formulas. Assuming similar tools to be valuable here, let us look for a model entire function sharing the properties of $f_n(\nu^{1/2}\xi)$ near $\xi = 1$: as the main behaviour is $\exp(\nu\Psi(2\xi)) \sim$ const. $\exp((\xi-1)^{3/2})$, let us choose the Airy function Ai which shares this behaviour, and is often found in asymptotic estimates ([6] [18] [19]): $f_n(\nu^{1/2}\xi) \sim$ const. $(-1)^n Ai((3\nu(\Psi(2\xi)-\pi i)/2)^{3/2}) \sim$ const. $(-1)^n(\Psi(2\xi)-\pi i)^{-1/6}\sin(\nu(-\Psi(2\xi)/i+\pi)+\pi/4)$ near $\xi = 1$ when n is large. This formula (with $\nu = n + 1/2$) agrees quite well with the numerical results, moreover, estimates of integrals in terms of Airy functions are known to be valid when two saddle points coalesce ([18], [19]), which is precisely true when $\xi = 1$ ($s_1 = s_2 = -1$). Near $\xi = 0$, $\Psi(2\xi) \sim$ const. $(-\xi)^{1/2}$ and a close model appears now to be the Bessel function const. $J_0(\nu\Psi(2\xi)/i) \sim$ const. $(\Psi(2\xi))^{-1/2}\cos(\nu\Psi(2\xi)/i - \pi/4)$. The satisfactory matching of the oscillating terms $(-1)^n\sin(\nu\Psi/i+\pi/4) = \cos(\nu\Psi/i-\pi/4)$ holds on the whole interval $\xi \in [0, 1]$ if $\nu = n+1/2$. This suggests that the ratio $A(\xi)/B(\xi)$ is the constant $\exp(-i\pi/2) = -i$ in a region containing

TABLE III
Differences between error norms and singular values.

n	$E_n - \sigma_n$	ratios	acceleration
0	-6.017152E-02	-16.619159	16.619159
1	2.530908E-05	-2377.467434	198.775020
2	3.088302E-06	8.195145	68.668900
3	-7.107011E-08	-43.454295	61.246073
4	1.454365E-09	-48.866753	58.541695
5	-2.778333E-11	-52.346699	57.460486
6	5.177776E-13	-53.658797	56.901326
7	-9.505479E-15	-54.471496	56.591766
8	1.726997E-16	-55.040521	56.417269
9	-3.113400E-18	-55.46979	56.32180
10	5.57717 E-20	-55.8240	56.2763
11	-9.9363E-22	-56.129	56.264
12	1.7617E-23	-56.402	56.275
13	-3.110E-25	-56.65	56.30
14	5.46E-27	-57.	56.3
15	-9.6E-29	-57.	56.
16	1.7E-30	-56.	56.
17	-3.E-32		

$[0,1]$. If this is true up to $\xi = \xi_1/2$, where $s_1 = \xi = \xi_1/2$, (and where $\Psi''(\xi_1) = 0$) (16) gives indeed $\lambda_n \sim -2i\exp((n+1/2)\pi i)q^{n+1/2}$, i.e., (14).

A more complete asymptotic expansion has been worked in [3], the most concise expression seems to be $E_n \sim 2q^{n+1/2}\exp[-1/(12(n+1/2)) + O(n^{-5})]$.

5. Super and hyper asymptotics: '1/56', etc.

The use of CF as quasi optimal rational approximation is justified by various results (inequalities in [9] [14]) showing how the E_n's can indeed be close to the σ_n's. But Table III shows that the matching is quite dramatic. Trefethen and Gutknecht have studied classes of functions with Chebyshev coefficients decreasing like powers of ε, to show that if $E_n \sim \sigma_n$ behaves like ε^{2n+1}, $|E_n - \sigma_n|$ could decrease as fast as ε^{5n+3}! ([24], sec.2). Table III shows indeed that the $|E_n - \sigma_n| = |E_n - |\lambda_n||$'s decrease like powers of a number close to 1/56, definitely smaller that '1/9'. How can we explain this phenomenon?

If we consider that the oscillation of the best error function is completely explained by the $\exp(2n\Phi) + \exp(-2n\Phi)$ factor of (3), the amplitude is therefore given by the integral on Γ of $\exp(n\xi^2)[\exp(2n\Phi(\xi)) + \exp(-2n\Phi(\xi)]^{-1} = \sum_0^\infty (-1)^k \exp[n(\xi^2 - 2(1+2k)\Phi(\xi))]$, as $\mathrm{Re}\Phi > 0$ on Γ. Of course, '1/9' is the constant modulus of $\exp(\xi^2 - 2\Phi(\xi))$ on Γ. Curiously enough, the integration of the other terms on Γ yield other decreasing exponentials of which the first appears indeed to be close to 1/56... (see Table IV: the exponential $\exp[n(\xi^2 - 2(1+2k)\Phi(\xi))]$ is evaluated at the saddlepoint s_k: $s_k - (1+2k)\Phi'(s_k) = 0$ for $k = 1, 2, \ldots$) It is not clear how to detect these new exponentially decreasing contributions from the numerical

TABLE IV
Other rates of decrease: '1/56' etc.

k	s_k	$\Phi'(s_k)/s_k$	$1/q_k$
1	0.72878+ 1.48300i	0.33333	56.690353
2	0.86860+ 1.70178i	0.20000	240.251663
3	0.97785+ 1.87517i	0.14286	846.908936

sequence $\{E_n\}$ alone, but the comparison of the E_n's and the σ_n's appears to be a lucky circumstance allowing to observe this new phenomenon. Can we find a formula for σ_n suggesting how it can be so close to E_n? The best starting point is probably this one: among other characterizations, σ_n is the smallest possible norm of a Hankel matrix of symbol $B_n\varphi$, where $\varphi(z)$ is here $\exp(c(z-1)^2/(z+1)^2)$ and $B_n(z)$ is a Blaschke product $\prod_1^n(z-a_k)/(1-za_k)$ ([1], [16]; the a_k's are supposed to be symmetrically placed with respect to the real axis($|a_k| < 1$)). This norm involves only the Laurent coefficients with negative index of $B_n\varphi$: it may be estimated as $(2\pi i)^{-1}\sum_{p=1}^{\infty}\int_\gamma B_n(\zeta)\varphi(\zeta)\zeta^{p-1}z^{-p}d\zeta = (2\pi i)^{-1}\int_\gamma B_n(\zeta)\varphi(\zeta)(z-\zeta)^{-1}d\zeta$ on $|z| = 1$. By the change of variable $n^{1/2}\xi = c^{1/2}(\zeta-1)/(\zeta+1)$, we have an integral involving $\prod_1^n(\xi-b_k)/(\xi+b_k)\exp(n\xi^2)$. This begins to look to $\exp(n\xi^2)\tilde{Q}_n(\xi)/\tilde{Q}_n(-\xi)$ in the notations of section 2. It should be possible to use the orthogonality arguments of [21] in order to exhibit the square of $\tilde{Q}_n(\xi)/\tilde{Q}_n(-\xi)$ instead. The study of the CF method of approximation of analytic functions would then be on the same level than the existing theory of rational approximation.

Finally, work is presently done on asymptotic expansions involving several exponentially decreasing term: see [18], [20] for super- and hyper-asymptotics. It would be interesting to see if these theories can cover the present phenomenon, in particular if *convergent* asymptotic expansions can be produced:

Conjecture 3 . *The best rational approximation error norms E_n and the appropriate CF singular values σ_n have hyper-asymptotic (perhaps convergent) expansions*

$$E_n \sim \sum_{k=0}^{\infty} q_k^n \exp[S_k(n+1/2)], \tag{18}$$

and

$$\sigma_n \sim \sum_{k=0}^{\infty} q_k^n \exp[U_k(n+1/2)], \tag{19}$$

when $n \to \infty$, with $S_k(x) \sim \sum_{m=0}^{\infty} s_{k,m}x^{-2m-1}$, $U_k(x) \sim \sum_{m=0}^{\infty} u_{k,m}x^{-2m-1}$, $U_0(x) \equiv S_0(x)$, and where $q_k = \exp[s_k^2 - 2(1+2k)\Phi(s_k)]$, s_k being the root of $s_k - (1+2k)\Phi'(s_k) = 0$, $k = 0, 1, \ldots$

6. Acknowledgements

Many thanks to Ph. Fortemps who provided references.

The eigenfunctions of Figure 3 have been computed with the LAPACK library (program DSPEVX) installed on the Convex C3820 of the University.

Van al die niks te zeggen hebben
zijn die die zwijgen 't aangenaamst. [1]
(From the programme 'Antwerp,
Cultural Capital of Europe 1993').

Special thanks are also due A. Cuyt and L. Wuytack who organized the first Antwerp Conference on Nonlinear Numerical Methods and Rational Approximation in 1987, and all the people who then gave me warm and kind advices on the use of the CF method in the study of the '1/9' problem. At the very least, I should have supplied the missing proofs in [15] for these Proceedings of the Second Conference. I had more than 5 years for achieving that. Of course, I did nothing of the sort, and years passed by without disturbing the peace of my mind. Then, notice came that the abstracts for the Second Conference should be ready for April 1st (no joke) 1993. As all I had to do was to establish an asymptotic formula exhibiting an exponential behaviour with respect to n, I proceeded to fill the following template:

Theorem. *The required eigenfunctions behave like AX^n when n is large, where the functions A and X are A =* so and so, *and* X = so and so.

But I did not know neither A nor X on April 1st. Fortunately, the contributors were allowed the new deadline May 1st, it seems that it was known that was in deep trouble. After painful weeks, I got the X function, it is the eerie function $\exp\Psi$ discussed in Section 4. Considering the technical difficulties, I decided to present this triumphal finding in a more cautious way:

Proposition. *The required eigenfunctions behave like AX^n when n is large, where the functions A and X are A =* so and so (still to be found) , *and* $X = \exp\Psi$.

(A theorem is a statement for which I think I have a proof, a proposition is a statement for which I have no proof, but pretend to have one).

On September 1st (the conference was to start on September 5th), I still did not know A, so I decided to be honest:

Conjecture. *The required eigenfunctions behave like* etc. (basically, Conjecture 2 in Section 4).

I don't know if I shall still be *persona grata* on the Third Conference, but I can promise at most one **Theorem**, perhaps several **Propositions**, quite a number of **Conjectures** and a lot of **Problems**.

References

1. V.M. ADAMYAN, D.Z. AROV, M.G. KREIN, Analytic properties of Schmidt pairs for a Hankel operator and the generalized Schur-Takagi problem, *Mat. Sb.* **86** (128) (1971) 34-75 = *Math. USSR Sb.* **15** (1971) 31-73.
2. N.I. AKHIEZER, *Elements of the Theory of Elliptic Functions*, 2nd ed., "Nauka", Moscow, 1970 (in Russian) = Translations of Math. Monographs **79**, Amer. Math. Soc. , Providence, 1990.
3. A.J. CARPENTER, A. RUTTAN, and R.S. VARGA, Extended numerical computations on the "1/9" conjecture in rational approximation theory, pp. 383-411 *in Rational Approximation and Interpolation*, (P.R.GRAVES-MORRIS, E.B.SAFF, and R.S.VARGA, editors), *Lecture Notes Math.* **1105**, Springer-Verlag, 1984.

[1] Of all those who have nothing to say, the most agreeable are those who are silent.

4. C.K. CHUI, X. LI, J.D. WARD, On the convergence rate of s–numbers of compact Hankel operators, *Circuits, Systems, and Signal Processing* **11** (1992) 353-362.
5. W.J. CODY, G. MEINARDUS, and R.S. VARGA, Chebyshev rational approximation to e^{-x} on $[0,+\infty)$ and application to heat-conduction problems, *J. Approx. Theory* **2** (1969), 50-65.
6. A. ERDÉLYI, *Asymptotic Expansions*, Dover, New York, 1956.
7. T. GANELIUS, Degree of rational approximation, pp. 9-78 *in* T. GANELIUS *et al. Lectures on Approximation and Value Distribution, Sém. Math. Sup. - Sém. Sci. OTAN*, Université de Montréal, Québec, 1982.
8. C. GLADER, G. HÖGNÄS, P.M. MÄKILÄ and H.T. TOIVONEN, Approximation of delay systems–a case study, *Int. J. Control*, **53** (1991) 369-390.
9. K. GLOVER, All optimal Hankel-norm approximations of linear multivariable systems and their L^∞-error bounds, *Int. J. Control*, **39** (1984) 1115-1193.
10. K. GLOVER, Model reduction: a tutorial on Hankel-norm and lower bounds on L^2 errors, pp. 288-293 *in* 10th *World Congress on Automatic Control Preprints*, IFAC, Munich 1987.
11. A.A. GONCHAR, E.A. RAKHMANOV, Equilibrium distribution and the degree of rational approximation of analytic functions, *Mat. Sb.* **134** (176) (1987) 306-352 = *Math. USSR Sbornik* **62** (1989) 305-348.
12. E. HAYASHI, L.N. TREFETHEN, M. GUTKNECHT, The CF Table, *Constr. Approx.* **6** (1990) 195-223.
13. J.W. HELTON, H.J. WOEDERMAN, Symmetric Hankel operators: minimal norm extensions and eigenstructures, *Lin. Alg. and its Appl.* **185** (1993) 1-19.
14. J. KARLSSON, Rational approximation in *BMOA* and H^∞, Preprint 1987-14 ISSN 0347-2809 Dept. of Math. Chalmers Univ. of Technology and Univ. of Göteborg, 1987.
15. A.P. MAGNUS, On the use of the Carathéodory-Fejér for investigating '1/9' and similar constants, pp. 105-132 *in Nonlinear Numerical Methods and Rational Approximation*, (A. CUYT, editor), D.Reidel, Dordrecht, 1988.
16. J. MEINGUET, A simplified presentation of the Adamjan-Arov-Krein approximation theory, pp. 217-248 *in Computational Aspects of Complex Analysis*, (H. WERNER, L. WUYTACK, E. NG and H.J. BÜNGER, editors), Reidel, Dordrecht 1983.
17. J. MEINGUET, Private communication 30-31 Dec. 1985.
18. A. OLDE DAALHUIS, *Uniform, Hyper-, and q-Asymptotics*, Ph.D. Univ. Amsterdam, CWI, Amsterdam, 1993.
19. A. OLDE DAALHUIS, N. TEMME, Uniform Airy type expansions of integrals, *SIAM J. Math. Anal.* March 1994.
20. A. OLDE DAALHUIS, Hyperasymptotic expansions of confluent hypergeometric functions, *IMA J. Appl. Math.* **49** (1992) 203-216.
21. H. STAHL, Orthogonal polynomials with complex-valued weight function, I, *Constr. Approx.* **2** (1986) 225-240; II, *ibidem* **2** (1986) 241-251.
22. H. STAHL, Uniform rational approximation of $|x|$, p. 110-130 *in Methods of Approximation Theory in Complex Analysis and Mathematical Physics, Euler Institute, 1991.* (A.A. GONCHAR and E.B. SAFF, editors), "Nauka", Moscow, 1992 and Springer (*Lecture Notes Math.* **1550**), Berlin, 1993.
23. H. STAHL, Best uniform rational approximation of x^α on $[0,1]$, *Bull. AMS*, **28** (1993) 116-122.
24. L.N. TREFETHEN, M. GUTKNECHT, The Carathéodory-Fejér method for real rational approximation, *SIAM J. Numer. Anal.* **20** (1983) 420-436.
25. L.N. TREFETHEN, Private communication of MIT CF memo 15 (25 Sept. 1984).
26. R.S. VARGA, *Scientific Computation on Mathematical Problems and Conjectures*, CBMS-NSF Reg. Conf. Series in Appl. Math. **60**, SIAM, Philadelphia, 1990.

FREQUENCY WEIGHTED H_2 MODEL REDUCTION

G. SCHELFHOUT,* P. VAN OVERSCHEE and B. DE MOOR[†]
ESAT–SISTA
Katholieke Universiteit Leuven
Kard. Mercierlaan 94
B-3001 Leuven
BELGIUM
Tel.: 00-32-16-220931, FAX: 00-32-16-221855
e-mail : geert.schelfhout@esat.kuleuven.ac.be

Abstract. We consider the problem of approximating a discrete time single-input single-output system by one of order less than r, minimizing the frequency weighted H_2 norm of the modeling error.
Using Lagrange multipliers, we derive a set of necessary conditions, which are solved iteratively in a way inspired by the inverse power iteration algorithm for singular value decomposition.

Keywords: H_2 norm, frequency weighting, inverse iteration

1. Introduction

H_2-optimal model reduction was one of the first model reduction techniques considered and has continued to raise interest in the last thirty years. Most approaches boil down to the calculation of a gradient and subsequent optimization using a descent algorithm, e.g. (Aplevich, 1973a; Galiana, 1973; Spanos *et al.*, 1992; Wilson *et al.*, 1979; Baratchart *et al.*, 1988a). Computing the gradient is generally costly however, therefore non-gradient algorithms (Lepschy *et al.*, 1991a) are potentially interesting. The method presented here is a novel non-gradient algorithm.

Frequency weighting is a very important asset of a model reduction method for control design purposes, and in general it is hard to name applications where the quality of the model has to be equally good for all frequencies. Ample motivation for frequency weighted model reduction in a control context can be found e.g. in (Anderson *et al.*, 1989; Doyle, 1984). Nevertheless, frequency weighted extensions for model reduction algorithms are rarely given in the literature (Noldus *et al.*, 1976; Halevi, 1990; Enns, 1984; Latham *et al.*, 1985). We have included frequency weighting with a stable and minimum phase weighting function.

* GS and PVO are research assistants with the NFWO (Belgian National Fund for Scientific Research)
† BDM is a research associate with the NFWO

A. Cuyt (ed.), Nonlinear Numerical Methods and Rational Approximation II, 187–199.

2. Definitions and notation

T is the unit circle in the complex plane.

Let $x(z) \triangleq \sum_{-\infty}^{\infty} x_i z^{-i}$

$[x(z)]_i \triangleq x_i$: i-th Markov parameter of $x(z)$
$[x(z)]_+ \triangleq \sum_{i=1}^{\infty} x_i z^{-i}$: strictly proper part of $x(z)$
$[x(z)]_{0\ldots r}$: first $r+1$ Markov parameters of $x(z)$
$\|x(z)\|_2 \triangleq \sqrt{\sum_{i=-\infty}^{\infty} h_i^2}$
$L_2 \triangleq \{x(z) |\ \|x(z)\|_2 < \infty\}$
$L_2^+ \triangleq \{x(z) \in L_2 |\ x(z) = [x(z)]_+\}$ (the set of strictly proper functions $\in L_2$)
$L_2^{0+} \triangleq \{x(z) \in L_2 |\ z^{-1}x(z) \in L_2^+\}$ (i.e. the Hardy space H_2)
RL_2 : $\{x(z) \in L_2 \mid x(z)$ converges to a rational function on $T\}$
RL_2^+ : $RL_2 \cap L_2^+$

It is important to note that if a series $x(z) \in L_2^+$, it converges for all z satisfying $|z| > 1$; if $x(z) \in RL_2^+$, for $|z| > \alpha$ (for some $\alpha < 1$); and if $x(z) \in RL_2$, it converges wherever $\alpha < |z| < \beta$, (where α, β are real numbers satisfying $\alpha < 1 < \beta$).

$x \star y$: convolution of series x and y ($(x \star y)_i \triangleq \sum_{j=-\infty}^{\infty} x_j y_{i-j}$)
If two series $x(z)$ and $y(z)$ converge in open regions S_x resp. S_y such that $S_x \cap S_y$ is nonempty, then the convolution $x(z)y(z)$ has well-defined coefficients and converges in $S_x \cap S_y$.

3. Necessary conditions

Let $w(z) \in RL_2^+$ be a stable and minimum phase (i.e. $w^{-1}(z) \in RL_2^+$) frequency weighting function[1] of order w. Let $h(z) \triangleq \sum_{i=0}^{\infty} h_i z^{-i} \in RL_2^+$ be the impulse response of a strictly proper stable system of order n. Suppose that $w(z)$ has no multiple poles and zeroes, that none of these coincide with a pole of $h(z)$, and that $h(z)$ has no multiple poles either[2].
We want to find the system with impulse response $\hat{h}(z) = \sum_{i=1}^{\infty} \hat{h}_i z^{-i} \in RL_2^+$ such that $\frac{1}{2}\sum_{i=1}^{\infty}([w \star h]_i - [w \star \hat{h}]_i)^2$ is minimized[3], and the $\infty \times (r+1)$ Hankel matrix $\hat{H}$ constructed from the Markov parameters $\hat{h}_i$, is not full rank, i.e. $\hat{h}(z)$ should have order less or equal to r. The rank-deficiency constraint on $\hat{H}$ is equivalent to the existence of a nonzero null vector, i.e. : $\exists y : \|y\|_2 = 1$ and $\hat{H}y = 0$.

---

[1] The assumption that $w(z)$ be stable and minimum phase is not restrictive, since only $w(z^{-1})w(z)$ is significant.

[2] An extension to the case where $w(z)$ has multiple poles or zeros is possible (using the same techniques), but omitted for the sake of brevity. The same remark is in order for $h(z)$.

[3] Note that one does not obtain the same minimum if one minimizes over all $\hat{h} = \sum_{i=0}^{\infty} \hat{h}_i z^{-i}$, as one would in the unweighted case.

Using Lagrange multipliers l_k and λ, one may formulate the problem as

$$\text{ext}_{\hat{h},y,l,\lambda}\frac{1}{2}\sum_{i=1}^{\infty}([w\star h]_i - [w\star\hat{h}]_i)^2 + \sum_{k=1}^{\infty} l_k \sum_{j=0}^{r}\hat{h}_{k+j}y_j + \lambda(y^T y - 1)$$

Taking the derivatives of the Lagrangian w.r.t. $\hat{h}, l, y$ and λ yields the corresponding four conditions, written in terms of formal series (see appendix A) :

$$\left[w(z^{-1})w(z)(h(z)-\hat{h}(z))\right]_+ = l(z)y(z),$$

$$\left[\hat{h}(z)y(z^{-1})\right]_+ = 0,\ \left[\hat{h}(z)l(z^{-1})\right]_{0\ldots r} = 0 \text{ and } \|y\| = 1$$

$\hat{h}(z) \in RL_2^+$, therefore also $l(z) \in RL_2^+$, so that by Lemma 1 the first condition may be rewritten as :

$$h(z) - \hat{h}(z) = w^{-1}(z)\left[w^{-1}(z^{-1})l(z)y(z)\right]_+ \tag{1}$$

By convolution with $y(z^{-1})$ and $l(z^{-1})$ respectively, and taking the causal part, $\hat{h}(z)$ (and $\hat{H}$) is eliminated from the above necessary conditions:

$$\begin{aligned}\left[h(z)y(z^{-1})\right]_+ &= \left[w^{-1}(z)\left[w^{-1}(z^{-1})l(z)y(z)\right]_+ y(z^{-1})\right]_+ \\ \left[h(z)l(z^{-1})\right]_{0\ldots r} &= \left[w^{-1}(z)\left[w^{-1}(z^{-1})l(z)y(z)\right]_+ l(z^{-1})\right]_{0\ldots r}\end{aligned}$$

If these equations can be solved for $y(z)$ and $l(z)$, $\hat{h}(z)$ can be recomputed from $y(z)$ and $l(z)$ using eq. (1).

4. Iteration scheme

Note that in the unweighted case the equations above reduce to:

$$\begin{aligned}\left[h(z)y(z^{-1})\right]_+ &= \left[y(z)y(z^{-1})l(z)\right]_+ \\ \left[h(z)l(z^{-1})\right]_{0\ldots r} &= \left[l(z)l(z^{-1})y(z)\right]_{0\ldots r}\end{aligned}$$

In (De Moor, 1993b), $y(z)$ and $l(z)$ are represented as vectors of Markov parameters y_i and l_i, and these two equations are written as $Hy = T_y T_y^T l \triangleq D_y l$ and $H^T l = T_l T_l^T y \triangleq D_l y$, where T_y and T_l are Toeplitz matrices depending affinely on y_i and l_i respectively, and having only zeroes below the diagonal, such that D_y and D_l are invertible. If one rescales $l = x\sigma$ such that x is a unit vector as well, the equations read : $Hy = T_y T_y^T x\sigma$ and $H^T x = T_x T_x^T y\sigma$. The error $\|h - \hat{h}\|_2$ can be expressed as $\|x * y\|_2\sigma$. The similarity with an SVD problem and the dependence of the error criterion on the 'singular value' σ then prompts the following inverse power iteration scheme :

$$\tilde{A}_{k-1}\begin{bmatrix} y_k \\ l_k\end{bmatrix} \triangleq \begin{bmatrix} H & -D_{y_{k-1}} \\ 0 & H^T\end{bmatrix}\begin{bmatrix} y_k \\ l_k\end{bmatrix} = \begin{bmatrix} 0 \\ D_{l_{k-1}}y_{k-1}\end{bmatrix}$$

Note that if l_k is represented as a vector, it must necessarily be truncated to length p (as well as H and D_y). $\tilde{A}_{k-1}$ is invertible whenever $D_{y_{k-1}}$ is, hence there is always

just one solution for y_k and l_k.
This way one solves the problem for a finite time horizon H_2 norm $\|h(z)\|^2 \triangleq \sum_i^p h_i^2$. When using this scheme to find an approximate solution for the infinite time horizon scheme, one has a tradeoff between memory and time requirements on the one hand (small p is desirable), and accuracy requirements on the other (large p desirable). Letting $p \to \infty$, these equations read (in the previous notation):

$$\begin{aligned}
\left[h(z)y^{[k]}(z^{-1})\right]_+ &= \left[l^{[k]}(z)y^{[k-1]}(z)y^{[k-1]}(z^{-1})\right]_+ \\
\left[h(z)l^{[k]}(z^{-1})\right]_{0\ldots r} &= \left[l^{[k-1]}(z)y^{[k-1]}(z)l^{[k-1]}(z^{-1})\right]_{0\ldots r}
\end{aligned}$$

By analogy, we propose the following iteration for the frequency weighted case :

$$\left[h(z)y^{[k]}(z^{-1})\right]_+ = \left[w^{-1}(z)\left[w^{-1}(z^{-1})l^{[k]}(z)y^{[k-1]}(z)\right]_+ y^{[k-1]}(z^{-1})\right]_+ \tag{2}$$

$$\left[h(z)l^{[k]}(z^{-1})\right]_{0\ldots r} = \left[w^{-1}(z)\left[w^{-1}(z^{-1})l^{[k-1]}(z)y^{[k-1]}(z)\right]_+ l^{[k-1]}(z^{-1})\right]_{0\ldots r} \tag{3}$$

It is possible to find a solution $y^{[k]}(z)$ and $l^{[k]}(z) \in RL_2^+$ for these equations, and to express the denominator of $l^{[k]}(z)$ in terms of previously computed quantities, such that all the unknowns left to be computed each iteration are 'numerator quantities' (numerator coefficients or residuals in the poles) appearing linearly in the equations above (this statement is elaborated in more detail in the sequel).
The iteration could now be implemented with the polynomial techniques used for the unweighted case in (De Moor *et al.*, 1993a) (i.e. all series $h(z)$, $y^{[k]}(z)$, $l^{[k]}(z)$, $w(z)$, $[w^{-1}(z^{-1})l^{[k]}(z)y^{[k-1]}(z)]_+$ are represented as ratios of polynomials in z) resulting in a $(3n+2w+2r)$-by-$(3n+2w+2r)$ linear system to be solved each iteration (see section 4.2). To circumvent the problem of having to solve a linear system this size, a partial fraction representation of $l(z)$ is used. This has the advantage that equating causal parts of two expressions is equivalent with equating the residuals in the stable poles. As these residuals are exactly the unknowns, far fewer unknowns are needed, and only an $(n+w+r)$-by-$(n+w+r)$ linear system[4] has to be solved each time (cf. infra).

4.1. More definitions

The following definitions are needed in the sequel: $h(z) \triangleq \frac{b(z)}{a(z)} \triangleq \sum_{i=1}^n \frac{H_i}{z-\alpha_i}$,
$\hat{h}(z) \triangleq \frac{\hat{b}(z)}{\hat{a}(z)} \triangleq \sum_{i=1}^r \frac{\hat{H}_i}{z-\hat{\alpha}_i}$, $\qquad w(z) \triangleq \frac{n_w(z)}{d_w(z)} \triangleq \sum_{i=1}^w \frac{W_i}{z-\alpha_{wi}}$
$w^{-1}(z) = (w(z))^{-1}$
$a_i^k \triangleq \alpha_i \quad (i = 1, \ldots, n), \qquad a_i^k \triangleq \alpha_{wi} \quad (i = n+1, \ldots, n+r),$
$a_i^k \triangleq \hat{\alpha}_i^k \quad (i = n+r+1, \ldots, n+r+w)$
$\beta_{wi} \triangleq$ roots of $n_w(z)$ (= zeroes of $w(z)$)

[4] With complex coefficients this time, and a solution with complex conjugate elements.

From $\left[\hat{h}(z)y(z^{-1})\right]_+ = 0$, we find that $\hat{a}(z) = \eta y(z^{-1})$ (where η is a constant), therefore define:
$\hat{a}^{[k]}(z) \triangleq y^{[k]}(z^{-1}) \Rightarrow y^{[k]}(z) = \hat{a}^{[k]}(z^{-1}) \qquad \deg \hat{a}^{[k]}(z) = r$
$\hat{a}_r^{[k]}(z) \triangleq z^r \hat{a}^{[k]}(z^{-1})$
Finally define $\xi^{[k]}(z) \triangleq [w^{-1}(z^{-1})l^{[k]}(z)y^{[k-1]}(z)]_+$

4.2. Some useful properties

Assume $l^{[k]}(z) \in RL_2$, then $\xi^{[k]}(z) \in RL_2$ as well. Since $\left[w^{-1}(z)\xi^{[k]}(z)y^{[k-1]}(z^{-1})\right]_+ = \left[h(z)y^{[k]}(z^{-1})\right]_+ \in RL_2^+$ (2), $\xi^{[k]}(z) \in RL_2^+$ as well, since if it had any unstable pole, $w^{-1}(z)\xi^{[k]}(z)$ and $\left[w^{-1}(z)\xi^{[k]}(z)y^{[k-1]}(z^{-1})\right]_+$ would have the same pole. Rewriting (2) as $\frac{d_w(z)}{n_w(z)}\xi^{[k]}(z)\hat{a}^{[k-1]}(z) = \left[\frac{b(z)}{a(z)}y^{[k]}(z^{-1})\right]_+ + p(z)$, where $p(z)$ is a polynomial of degree $r-1$, and multiplying both sides with $w(z)\frac{1}{\hat{a}^{[k-1]}(z)}$, it is found that $\xi^{[k]}(z)$ must converge to a rational function of the form[5] :

$$\xi^{[k]}(z) = \frac{n_w(z)\phi^{[k]}(z)}{a(z)\hat{a}^{[k-1]}(z)d_w(z)}, \qquad \deg \phi^{[k]}(z) \leq n + r - 1$$

at least in an open region containing T, and since $\xi^{[k]}(z) \in RL_2^+$, in an open region containing $\{|z| \geq 1\}$.
Now since $\xi^{[k]}(z) = [w^{-1}(z^{-1})y_r^{[k-1]}(z^{-1})z^{-r}l^{[k]}(z)]_+$, Lemma 1 yields:
$z^{-r}l^{[k]}(z) = \left[w(z^{-1})\frac{1}{y_r^{[k-1]}(z^{-1})}\xi^{[k]}(z)\right]_+$, so in an open region containing $\{|z| \geq 1\}$:

$$l^{[k]}(z) = \frac{z^r f^{[k]}(z)}{a(z)\hat{a}^{[k-1]}(z)d_w(z)}, \qquad \deg f^{[k]}(z) \leq n + w - 1$$

These are the representations used in the polynomial version of (De Moor *et al.*, 1993a). $[z^{-1}h(z)l^{[k]}(z^{-1})]_+$ and $[z^{-1}w^{-1}(z)[w^{-1}(z^{-1})l^{[k-1]}(z)y^{[k-1]}(z)]_+ l^{[k-1]}(z^{-1})]_+$ need only be equal in the first $r+1$ Markov parameters, so the difference can be expressed as $\frac{s^{[k]}(z)}{a(z)\hat{a}^{[k-2]}(z)n_w(z)}$, where $s^{[k]}(z)$ is an unknown polynomial with $\deg s^{[k]}(z) = n + w - 2$.
So in this version, the unknowns are the coefficients of $\hat{a}^{[k]}(z), f^{[k]}(z), \phi^{[k]}(z)$ and $s^{[k]}(z)$, giving $3n + 2w + 2r$ unknowns.

4.3. Partial fraction representation

As mentioned earlier, one may express $\xi^{[k]}(z)$ and $l^{[k]}(z)$ as sums of pole-residue fractions to avoid the rather large number of unknowns and other numerical problems connected with these polynomial representations.

---

[5] During the first few iterations of the algorithm, $\hat{a}^{[k-1]}(z)$ may have unstable roots, in this case only the 'stable root' factors of $\hat{a}^{[k-1]}(z)$ should be included in the denominator of $\xi^{[k]}(z)$. In such a case, $l^{[k]}(z)$ has the 'stable root' factors of $\hat{a}^{[k-1]}(z)$ and $\hat{a}_r^{[k-1]}(z)$ in its denominator.

- First define the polynomials $p_i(z) \triangleq \frac{a(z)}{z-\alpha_i}$, then write

$$\hat{a}^{[k]}(z) = \sum_i^n A_i^k p_i(z), \quad \Rightarrow \quad \begin{bmatrix} 0_{n-r-1} \\ \hat{a}_k \end{bmatrix} = \Gamma A_k$$

Here the columns of Γ store the coefficients of $p_i(z)$
Note that $\deg p_i(z) = n-1$, so $\deg \hat{a}^{[k]}(z) = r$ gives $n-r-1$ conditions on A_k :

$$[\, I_{n-r-1} \;\; 0 \,]\, \Gamma A_k \triangleq \Gamma_r A_k = 0 \tag{4}$$

$$[\, 0 \;\; I_{r+1} \,]\, \Gamma A_k \triangleq \Gamma_a A_k = \hat{a}_k \tag{5}$$

- We write $l^{[k]}(z)$ as:

$$l^{[k]}(z) \triangleq \sum_{i=1}^{n} \frac{C_{\alpha i}^k}{z-\alpha_i} + \sum_{i=1}^{r} \frac{C_{\hat{\alpha} i}^k}{z-\hat{\alpha}_i^{[k-1]}} + \sum_{i=1}^{w} \frac{C_{wi}^k}{z-\alpha_{wi}} \triangleq \sum_i \frac{C_i^k}{z-a_i^{k-1}}$$

- By splitting $w^{-1}(z^{-1})l^{[k]}(z)y^{[k-1]}(z) \in RL_2$ into partial fractions and retaining only the stable terms, one finds that $\xi^{[k]}(z)$ converges to (see Lemma 2) :

$$\xi^{[k]}(z) = \left[w^{-1}(z^{-1})l^{[k]}(z)y^{[k-1]}(z)\right]_+ = \sum_{i=1}^{n} \frac{w^{-1}(\alpha_i^{-1})\hat{a}^{[k-1]}(\alpha_i^{-1})C_{\alpha i}^k}{z-\alpha_i}$$

$$+ \sum_{i=1}^{r} \frac{w^{-1}(\hat{\alpha}_i^{-1})\hat{a}^{[k-1]}(\hat{\alpha}_i^{-1})C_{\hat{\alpha} i}^k}{z-\hat{\alpha}_i} + \sum_{i=1}^{w} \frac{w^{-1}(\alpha_{wi}^{-1})\hat{a}^{[k-1]}(\alpha_{wi}^{-1})C_{wi}^k}{z-\alpha_{wi}} \tag{6}$$

- The dependence of A_i^k on $C_{\alpha i}^k$ is found using Lemma 2 :

$$\left[h(z)y^{[k]}(z^{-1})\right]_+ = \left[\frac{b(z)}{a(z)}\hat{a}^{[k]}(z)\right]_+ = \left[b(z)\sum_i \frac{A_i^k}{z-\alpha_i}\right]_+$$

$$= \left[w^{-1}(z)\hat{a}^{[k-1]}(z)\left[w^{-1}(z^{-1})l^{[k]}(z)\hat{a}^{[k-1]}(z^{-1})\right]_+\right]_+$$

i.e. defining $D^{k-1} \triangleq \operatorname{diag}\left(\frac{w^{-1}(\alpha_i)w^{-1}(\alpha_i^{-1})\hat{a}^{[k-1]}(\alpha_i)\hat{a}^{[k-1]}(\alpha_i^{-1})}{b(\alpha_i)}\right)$, one finds :
$A_k = D^{k-1}C_\alpha^k$, which yields (combined with (4)(5)) :

$$\Gamma_a D^{k-1} C_\alpha^k = \hat{a}_k \tag{7}$$

$$\Gamma_r D^{k-1} C_\alpha^k = 0 \tag{8}$$

Equation (8) gives $n-r-1$ conditions on C^k.

- The fact that $l^{[k]}(z)$ has an r-multiple zero in $z=0$ yields a further r conditions on C^k (with an obvious definition for P_α^{k-1}, $P_{\hat{\alpha}}^{k-1}$, P_w^{k-1}) :

$$\frac{d^{s-1}}{dz^{s-1}} l^{[k]}(z)\Big|_{z=0} = 0, \qquad s = 1 \ldots r \qquad \Rightarrow \sum_i \frac{C_{\alpha i}^k}{\alpha_i^s} + \sum_i \frac{C_{\hat{\alpha} i}^k}{\hat{\alpha}_i^s} + \sum_i \frac{C_{wi}^k}{\alpha_{wi}^s} = 0$$

$$\text{i.e.} \qquad P_\alpha^{k-1}C_\alpha^k + P_{\hat{\alpha}}^{k-1}C_{\hat{\alpha}}^k + P_w^{k-1}C_w^k = 0 \tag{9}$$

- Likewise, the knowledge of w zeroes of $\xi^{[k]}(z)$ gives w conditions on C^k, i.e. plugging eq. (6) into $\xi^{[k]}(\beta_{wi}) = 0$ yields : $\sum_{j=1}^{n} \frac{w^{-1}(\alpha_j^{-1})\hat{a}^{[k-1]}(\alpha_j^{-1})C_{\alpha j}^k}{\beta_{wi}-\alpha_j} + \sum_{j=1}^{r} \frac{w^{-1}(\hat{\alpha}_j^{-1})\hat{a}^{[k-1]}(\hat{\alpha}_j^{-1})C_{\hat{\alpha} j}^k}{\beta_{wi}-\hat{\alpha}_j} + \sum_{j=1}^{w} \frac{w^{-1}(\alpha_{wj}^{-1})\hat{a}^{[k-1]}(\alpha_{wj}^{-1})C_{wj}^k}{\beta_{wi}-\alpha_{wj}} = 0$

$$\text{i.e.} \qquad G_\alpha^{k-1}C_\alpha^k + G_{\hat{\alpha}}^{k-1}C_{\hat{\alpha}}^k + G_w^{k-1}C_w^k = 0 \tag{10}$$

- Finally, the first $r+1$ Markov parameters of $\left[h(z)l^{[k]}(z^{-1})\right]_+$ and $\left[w^{-1}(z)\left[w^{-1}(z^{-1})l^{[k-1]}(z)y^{[k-1]}(z)\right]_+ l^{[k-1]}(z^{-1})\right]_+$ are equal :

$$\left[h(z)l^{[k]}(z^{-1})\right]_{0\ldots r} = \left[\sum_i \frac{H_i}{z-\alpha_i} l^{[k]}(z^{-1})\right]_{0\ldots r} = \sum_i l^{[k]}(\alpha_i^{-1})H_i \left[\frac{1}{z-\alpha_i}\right]_{0\ldots r}$$

The second expression $\left[w^{-1}(z)\left[w^{-1}(z^{-1})l^{[k-1]}(z)y^{[k-1]}(z)\right]_+ l^{[k-1]}(z^{-1})\right]_+$ does not contain any unknowns. First its partial fraction decomposition $\sum_i \frac{R_{\beta i}}{z-\beta_{wi}} + \sum_i \frac{R_{\alpha i}}{z-\alpha_i} + \sum_i \frac{R_{\hat{\alpha} i}}{z-\hat{\alpha}_i^{k-2}}$ is computed by application of Lemma 2 in the poles β_{wi}, α_i and $\hat{\alpha}_i$, which yields :

$$\begin{aligned}
R_{\beta i} &\triangleq l^{[k-1]}(\beta_{wi}^{-1}) \sum_j \frac{w^{-1}((a_j^{k-1})^{-1})\hat{a}^{[k-1]}((a_j^{k-1})^{-1})C_j^{k-1}}{\beta_{wi}^{-1} - a_j^{k-1}} W_i \\
R_{\alpha i} &\triangleq w^{-1}(\alpha_i)w^{-1}(\alpha_i^{-1})l^{[k-1]}(\alpha_i^{-1})\hat{a}^{[k-1]}(\alpha_i^{-1})C_{\alpha i}^{k-1} \\
R_{\hat{\alpha} i} &\triangleq w^{-1}(\hat{\alpha}_i^{k-2})w^{-1}((\hat{\alpha}_i^{k-2})^{-1})l^{[k-1]}((\hat{\alpha}_i^{k-2})^{-1})\hat{a}^{[k-1]}((\hat{\alpha}_i^{k-2})^{-1})C_{\hat{\alpha} i}^{k-1}
\end{aligned}$$

So we find for $s = 0, \ldots, r$ (with an obvious definition for J_*) :

$$\sum_i \frac{H_i}{\alpha_i}\alpha_i^s \sum_j \frac{C_j^k}{\alpha_i^{-1} - a_j^k} = \sum_i R_{\beta i}\beta_{wi}^{s-1} + \sum_i R_{\alpha i}\alpha_i^{s-1} + \sum_i R_{\hat{\alpha} i}(\hat{\alpha}_i^{k-2})^{s-1} \triangleq V_s^{k-1}$$

$$\text{i.e.} \qquad J_\alpha^{k-1}C_\alpha^k + J_{\hat{\alpha}}^{k-1}C_{\hat{\alpha}}^k + J_w^{k-1}C_w^k = V^{k-1} \tag{11}$$

These are $r+1$ conditions still required to be able to solve for C^k .

- Summing it all up, we find equations (8)(9)(10)(11) yield $n+r+w$ linear conditions on $n+r+w$ unknowns C_i^k, i.e. the following $(n+r+w)$-by-$(n+r+w)$ linear system with complex coefficients has to be solved each time :

$$\begin{bmatrix} \Gamma_r D^{k-1} & 0 & 0 \\ P_\alpha^{k-1} & P_{\hat{\alpha}}^{k-1} & P_w^{k-1} \\ G_\alpha^{k-1} & G_{\hat{\alpha}}^{k-1} & G_w^{k-1} \\ J_\alpha^{k-1} & J_{\hat{\alpha}}^{k-1} & J_w^{k-1} \end{bmatrix} \begin{bmatrix} C_\alpha^k \\ C_{\hat{\alpha}}^k \\ C_w^k \end{bmatrix} = \begin{bmatrix} 0 \\ 0 \\ 0 \\ V^{k-1} \end{bmatrix} \tag{12}$$

5. Computation of $\hat{h}(z)$

The computation of $\hat{h}(z)$ is extremely simple:

$$\hat{h}(z) = \sum_{i=1}^{n} \frac{\hat{H}_i}{z - \hat{\alpha}_i} = h(z) - w^{-1}(z)\left[w^{-1}(z^{-1})l(z)y(z)\right]_+$$

$$\Rightarrow \hat{H}_i = -w^{-1}(\hat{\alpha}_i)w^{-1}(\hat{\alpha}_i^{-1})\hat{a}(\hat{\alpha}_i)C_{\hat{\alpha}i} \qquad (13)$$

6. Initialization

For the initial $\hat{h}(z)$ one can take the model produced by any model reduction method, e.g. frequency weighted balanced truncation (FWBT) à la Enns (Enns, 1984). This immediately yields the initial $y(z) = \hat{a}(z^{-1})$.

As in the optimum one has : $\left[w(z^{-1})w(z)(h(z) - \hat{h}(z))\right]_+ = l(z)y(z)$, one could hope that this relation will hold approximately for a good approximation of the optimal model. So the initial $l(z)$ is computed as $l_0(z) \triangleq \left[y^{-1}(z)w(z^{-1})w(z)\left(h(z) - \hat{h}(z)\right)\right]_+$.

7. Algorithm

Initialization:

Compute a good approximation $\hat{h}_0(z) = \frac{\hat{b}_0(z)}{\hat{a}_0(z)}$

Take $y(z) = \hat{a}_0(z^{-1})$ and $l(z) = \left[y^{-1}(z)w(z^{-1})w(z)(h(z) - \hat{h}(z))\right]_+$

Do {

1. Solve the linear system (12) $\Rightarrow l^{[k]}(z)$
2. Find $\hat{a}_k$ from (7) $\Rightarrow y^{[k]}(z)$
3. $l^{[k]} \leftarrow \frac{l^{[k]}}{\|\hat{a}_k\|}$
4. $\hat{a}_k \leftarrow \frac{\hat{a}_k}{\|\hat{a}_k\|}$
5. $k \leftarrow k + 1$

} **while** $\|\hat{a}_k - \hat{a}_{k-1}\| \geq \epsilon$ [6]

Compute $\hat{h}(z)$ using (13)

8. Example

Consider the following hydraulic actuator control system :

[6] other convergence tests may be used, e.g. one may stop when the decrease of the object function, or when the quantity $\hat{H}^T l$ becomes smaller than a certain tolerance.

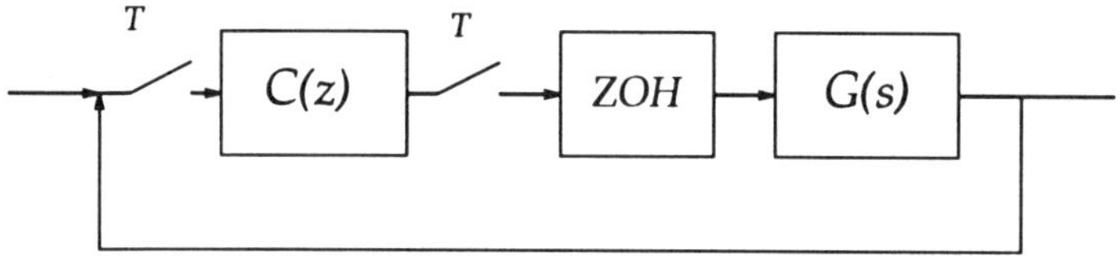

$$\text{Actuator transfer function } G(s) = \frac{1000}{s^3 + 40s^2 + 800s + 1000}$$

With a sampling period of $T = 0.01$s, one finds that the discrete-time system describing the zero order hold and the plant $G(s)$ is

$$G(z) = 10^{-4}\frac{1.51z^2 + 5.44z + 1.23}{z^3 - 2.6z^2 + 2.28z - 0.67}$$

An H_∞ design with suitable weighting functions on the sensitivity function $(I + GC)^{-1}$ and the complementary sensitivity function $GC(I + GC)^{-1}$ yields the following controller :

$$C(z) = \frac{160z^6 - 373.6z^5 + 108.7z^4 + 334.7z^3 - 279.1z^2 + 29.8z + 19.5}{z^6 - 1.522z^5 + 0.166z^4 + 0.29z^3 + 0.045z^2 + 0.018z + 0.0037}$$

Choosing $w(z) = 1$, the reduced-order controller of smallest order that is still stabilizing has $r = 3$:

$$\text{The } H_2 \text{ optimally reduced controller : } \hat{C}_{H_2} = \frac{160z^3 - 95.6z^2 - 141.6z + 107.9}{z^3 + 0.21z^2 + 0.014z + 0.066}$$

$$\text{The balanced truncation (BT) reduced controller } \hat{C}_{BT} = \frac{160z^3 - 186.1z^2 - 139.3z + 171.8}{z^3 - 0.34z^2 - 0.38z - 0.23}$$

is not stabilizing.
Convergence is reached in 11 iterations and after 317660 flops, compared to 73100 flops for BT.

$$\text{Let } H_{H_2} \triangleq (I + GC_{H_2})^{-1}GC_{H_2}, \text{ then } \frac{\|H_{H_2} - H\|_\infty}{\|H\|_\infty} = 0.25$$

The Bode plots, error magnitude versus frequency and convergence of $\|\hat{a}_k - \hat{a}\|$ are shown in figure 1.
If we choose $w(z)$ to be[7] $(I+GC)^{-2}G$ (reduced to order 5 by BT, to avoid common poles with G), the smallest order yielding a stabilizing controller is $r = 2$, both for the optimal controller $\hat{C}_{wH_2}$ and the controller $\hat{C}_{wBT}$ reduced by frequency weighted balanced truncation (FWBT) (Enns, 1984).

$$\hat{C}_{wH_2} = \frac{160z^2 - 272.6z + 123.3}{z^2 - 0.544z + 0.01} \Rightarrow \frac{\|H_{wH_2} - H\|_\infty}{\|H\|_\infty} = 0.12$$

[7] This weighting function aims at minimizing the variation in the closed loop transfer function $(I + GC)^{-1}GC$

$$\hat{C}_{wBT} = \frac{160z^2 - 294z + 140.1}{z^2 - 0.853z + 0.0074} \Rightarrow \frac{\|H_{wBT} - H\|_\infty}{\|H\|_\infty} = 0.36$$

This time convergence is reached in 26 iterations and after 1.5 10^6 flops, compared to 325000 flops needed for FWBT.
The closed loop magnitude Bode plots and errors are shown in figure 2.
In general, (frequency weighted) H_2-reduced systems yield better closed loop behavior than FWBT, the advantage however decreases as r approaches n.

9. Conclusions

We have presented an algorithm for frequency weighted H_2 model reduction with a computational cost which is comparable with that of frequency weighted balanced truncation (the most widely used frequency weighted model reduction method), but which often yields better reduced order controllers in the sense that stabilization is more likely to be maintained and the loss of performance is smaller.

References

ANDERSON, B.D.O., and J.B. MOORE, 1989, Optimal Control - Linear Quadratic Methods. *Prentice-Hall,* Chapter 10

APLEVICH, J.D., 1973a, Approximation of discrete linear systems. *Int. J. Control,* **17**, 565-575

APLEVICH, J.D., 1973b, Gradient methods for optimal linear systems reduction. *Int. J. Control,* **18**, 767-772

BARATCHART, L., and M. OLIVI, 1987, Recent and new results in rational L_2 approximation. *Modeling, Robustness and Sensitivity Reduction in Control Systems, R.F. Curtain Ed.*, 119-126

BARATCHART, L., and M. OLIVI, 1988a, New tools in rational L_2 approximation. *Proc. IFAC Symp. Identification and System Parameter Estimation, Beijing,* **2**, 1014-1019

BARATCHART, L., M. OLIVI, and F. WIELONSKI, 1990, Asymptotic properties in rational L_2 approximation. *Analysis and Optimization of Systems, A. Bensoussan and J.L. Lions, ed., Springer Verlag,* **144**, 477-486

BRYSON, A.E., and A. CARRIER, 1990, Second-order algorithm for optimal model order reduction. *J. Guidance Control Dynam.*, **13**, 887-892

DE MOOR, B., P. VAN OVERSCHEE, and G. SCHELFHOUT, 1993a, Rational H_2 model reduction for SISO systems. *Proc. IFAC 1993, Sydney, Vol. II-227*

DE MOOR, B., 1993b, Total least squares for affinely structured matrices and the noisy realization problem. *To appear in IEEE Trans. Sign. Proc.*

DOYLE, J.C., and G. STEIN, 1981, Multivariable feedback design: Concepts for a classical/modern synthesis. *IEEE Tr AC,* **26**, 4-16

ENNS, D.F., 1984, Model reduction with balanced realization: An error bound and frequency weighted generalization. *Proc. IEEE CDC, Las Vegas,* 127-132

GALIANA, F.D., 1973, On the Approximation of Multiple-Input Multiple-Output Constant Linear Systems. *Int. J. Control,* **17**, 1313-1324

HALEVI, Y., 1990, Frequency weighted model reduction via optimal projection. *Proc. 29th CDC, Honolulu, Hawaii,* 2906-2911

HYLAND, D.C., and D.S. BERNSTEIN, 1985, The Optimal Projection Equations for Model Reduction and the Relationships Among the Methods of Wilson, Skelton, and Moore *IEEE Tr AC,* **30**, 1201-1211

KABAMBA, P.T., 1985, Balanced Gains and Their Significance for L_2 Model Reduction. *IEEE Tr AC,* **30**, 690-693

LATHAM, G.A., and B.D.O. ANDERSON, 1985, Frequency-weighted optimal Hankel-norm approximation of stable transfer functions. *Syst. Control Lett.*, **5**, 229-236

LEPSCHY, A., S. LORO, G.A. MIAN, and U. VIARO, 1991a, A numerical procedure for rational L_2 approximation. *MTNS 91, Kobe, Japan*, 271-272
LEPSCHY A., G.A. MIAN, G. PINATO, and U. VIARO, 1991b, Rational L_2 approximation: A non-gradient algorithm. *Proc. 30th CDC, Brighton, UK*, 2321-2323
MEIER, L., and D.G. LUENBERGER, 1967, Approximation of linear constant systems. *IEEE Tr AC*, **12**, 585-587
MISHRA, R.N., and D.A. WILSON, 1980, A new algorithm for optimal reduction of multivariable systems. *Int. J. Control*, **31**, 443-466
NOLDUS, E., and M. DECOSTER, 1976, Reduction of linear systems by weighted frequency response matching. *Regelungstechnik*, 122-128
SPANOS, J.T., M.H. MILMAN, and D.L. MINGORI, 1992, A New Algorithm for L_2 Optimal Model Reduction. *Automatica*, **28**, 897-909
WILSON, D.A., 1970, Optimum solution of model-reduction problem. *Proc. IEE-D*, **117**, 1161-1165
WILSON, D.A., 1974, Model Reduction of Multivariable Systems. *Int. J. Control*, **20**, 57-64
WILSON, D.A., and R.N. MISHRA, 1979, Optimal Reduction of Multivariable Systems. *Intl. J. Control*, **29**, 267-278

Appendix

A. Taking the derivative of the Lagrangian

Taking the derivative w.r.t. l, y and λ is straightforward.
Taking the derivative w.r.t. $\hat{h}_i$:
$\frac{\partial}{\partial \hat{h}_i}\left(l^T \hat{H} y\right) = [l(z)y(z)]_i$

$$\begin{aligned}
&\frac{\partial}{\partial \hat{h}_i}\left(\frac{1}{2}\sum_{i=1}^{\infty}\left([w\star h]_i - \left[w\star \hat{h}\right]_i\right)^2\right)\\
&= \frac{\partial}{\partial \hat{h}_i}\left(\frac{1}{4j\pi}\oint_T z^{-1}w(z^{-1})w(z)\left(h(z^{-1})-\hat{h}(z^{-1})\right)\left(h(z)-\hat{h}(z)\right)dz\right)\\
&= -\frac{1}{4j\pi}\oint_T w(z^{-1})w(z)\left(z^{i-1}\left(h(z)-\hat{h}(z)\right)+z^{-i-1}\left(h(z^{-1})-\hat{h}(z^{-1})\right)\right)dz\\
&= -\frac{1}{2j\pi}\oint_T z^{i-1}w(z^{-1})w(z)\left(h(z)-\hat{h}(z)\right)dz\\
&= -\left[w(z^{-1})w(z)\left(h(z)-\hat{h}(z)\right)\right]_i \quad \text{(using Laurent's theorem)}
\end{aligned}$$

Therefore we get $-\left[w(z^{-1})w(z)\left(h(z)-\hat{h}(z)\right)\right]_+ + l(z)y(z) = 0$

From this we can derive a generalized orthogonality condition between the approximant and the error, as in (De Moor *et al.*, 1993a) :

$$\begin{aligned}
&\frac{1}{2j\pi}\oint_T z^{-1}\hat{h}(z^{-1})w(z^{-1})w(z)\left(h(z)-\hat{h}(z)\right)dz\\
&= \frac{1}{2j\pi}\sum_{i=1}^{\infty}\hat{h}_i\oint_T z^{i-1}w(z^{-1})w(z)\left(h(z)-\hat{h}(z)\right)dz\\
&= \sum_{i=1}^{\infty}\hat{h}_i\,[l(z)y(z)]_i = \left[\hat{h}(z)l(z^{-1})y(z^{-1})\right]_0 = \left[\left[\hat{h}(z)y(z^{-1})\right]_+ l(z^{-1})\right]_0 = 0
\end{aligned}$$

$$\Rightarrow \sum_{i=1}^{\infty} \left(w \star \hat{h}\right)_i \left(w \star (h - \hat{h})\right)_i = 0$$

B. Lemma's

Lemma 1

$$x(z) \in L_2^+,\ w(z),\ w^{-1}(z) \in RL_2^{0+} \ \Rightarrow \ \left[w^{-1}(z^{-1})\left[w(z^{-1})x(z)\right]_+\right]_+ = x(z),\ \|z\| > 1$$

Proof :
Since $x(z) \in L_2^+$, it converges in $\{z|\ 1 < \|z\|\}$; since $w(z), w^{-1}(z) \in RL_2^{0+}$, $\exists \beta > 1$ such that $w(z^{-1})$, $w^{-1}(z^{-1})$ converge in $\{z|\ \|z\| < \beta\}$.
$w(z^{-1})$ is bounded on T, hence $w(z^{-1})x(z) \in L_2$ (conv. in $\{z|\ 1 < \|z\| < \beta\}$) and $w(z^{-1})x(z) = q(z) + l(z^{-1}) \quad (1 < \|z\| < \beta)$, where $q(z) \in L_2^+$, and $l(z) \in L_2^{0+}$.
Therefore $w^{-1}(z^{-1})q(z) = x(z) - w^{-1}(z^{-1})l(z^{-1}) \qquad (1 < \|z\| < \beta)$,
and $\left[w^{-1}(z^{-1})q(z)\right]_+ = \left[w^{-1}(z^{-1})\left[w(z^{-1})x(z)\right]_+\right]_+ = x(z)$, whenever $\|z\| > 1$. □

Lemma 2 *Let* $|\alpha| < 1$, $x(z), y(z) \in L_2^+$, *where* $y(z) = \sum_{i=1}^{l} \frac{Y_i}{(z-\alpha)^i}$ *in* $\{|z| > |\alpha|\}$

Then $\left[x(z^{-1})y(z)\right]_+ = \sum_{i=1}^{l} \frac{C_i}{(z-\alpha)^i}$ *for* $|z| > |\alpha|$, $C_i \triangleq \sum_{r=0}^{l-i} \frac{Y_{i+r}}{r!}\left[\frac{d^r}{dz^r}(x(z^{-1}))\right]_{z=\alpha}$

Proof :
Expanding $x(z^{-1})$ (conv. $|z| < 1$) as a Taylor series of $(z - \alpha)$, we find:

$$\begin{aligned}
x(z^{-1}) &= \sum_{i=0}^{\infty} \frac{1}{i!}\frac{d^i}{dz^i}\left(x(z^{-1})\right)_{z=\alpha}(z-\alpha)^i \qquad (|z-\alpha| < 1 - |\alpha|) \\
\frac{x(z^{-1})}{(z-\alpha)^s} &= \sum_{i=0}^{\infty} \frac{1}{i!}\frac{d^i}{dz^i}\left(x(z^{-1})\right)_{z=\alpha}(z-\alpha)^{i-s} \qquad (0 < |z-\alpha| < 1 - |\alpha|) \\
\left[\frac{x(z^{-1})}{(z-\alpha)^s}\right]_+ &= \sum_{i=1}^{s} \frac{1}{(s-i)!}\frac{d^{s-i}}{dz^{s-i}}\left(x(z^{-1})\right)_{z=\alpha}\frac{1}{(z-\alpha)^i} \qquad (|z| > |\alpha|) \\
\Rightarrow \left[x(z^{-1})y(z)\right]_+ &= \sum_{s=1}^{l} Y_s \sum_{i=1}^{s} \frac{1}{(s-i)!}\frac{d^{s-i}}{dz^{s-i}}\left(x(z^{-1})\right)_{z=\alpha}\frac{1}{(z-\alpha)^i} \\
&= \sum_{i=1}^{l}\sum_{r=0}^{l-i} \frac{Y_{r+i}}{r!}\frac{d^r}{dz^r}\left(x(z^{-1})\right)_{z=\alpha}\frac{1}{(z-\alpha)^i}
\end{aligned}$$
□

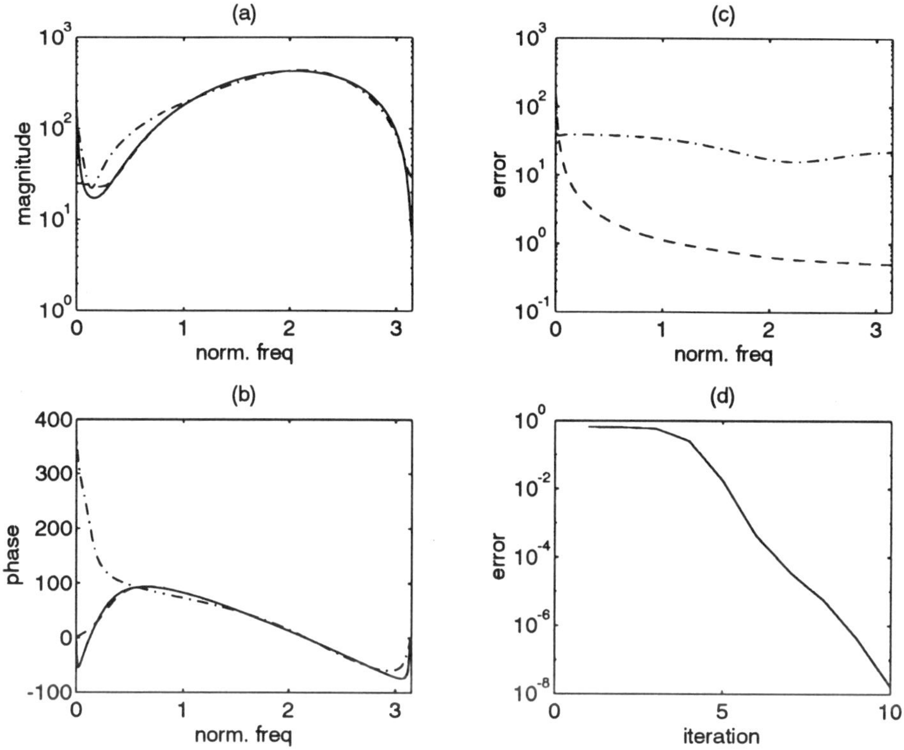

Fig. 1. Dashed line : H_2 reduced model, dash-dotted line : BT model . (a) magnitude Bode plots, (b) phase Bode plots, (c) error magnitude, (d) convergence of $||\hat{a}_k - \hat{a}||$

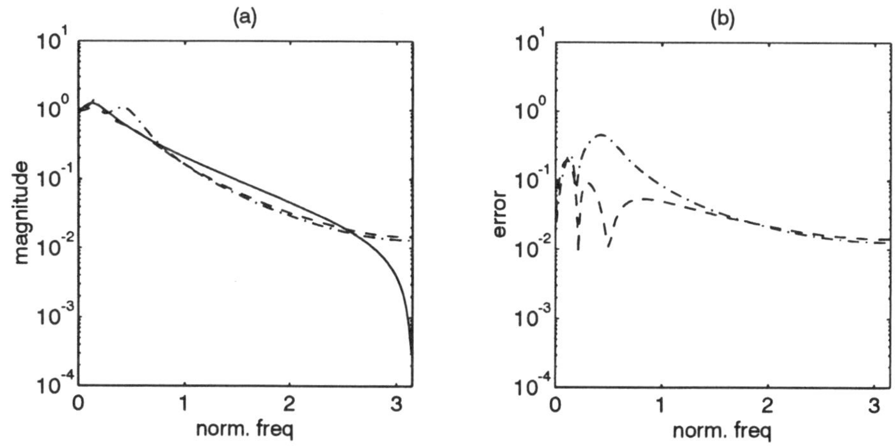

Fig. 2. Closed loop magnitude Bode plots and errors (dashed line : FW H_2 reduced model, dash-dotted line : FWBT model).

AN APPROXIMATION METHOD BASED ON THE SECOND KIND CHEBYSHEV POLYNOMIALS

M. BOZZINI and M. ROSSINI
Dipartimento di Matematica, Università di Milano
Via C. Saldini 50, 20133 Milano, Italy

Abstract. In this work we present a method based on the Chebyshev polynomials for approximating unknown functions with constraints. The properties of the approximating function are studied and numerical result are achieved.

Key words: Orthogonal polynomials, smoothing, approximation, shape-preserving.

1. Introduction

In these last years many works have been devoted to the problem concerning function approximations with constraints. Given a set of noisy data, the problem is that of constructing an approximation of the unknown function with specific characteristics as, for example, monotonicity and convexity.

In literature we find some authors that have dealt with this problem. In particular we remember [5] in which the teoretical results of interpolation procedures preserving some shape, defined in the form of a convex constraint, are extended to the case of smoothing spline but no algorithm is given for the construction of the approximating function; in [6] the authors present an algorithm which determines a spline function whose behaviour depends on a parameter that has to be choosen by the user in order to obtain the desired solution.

Finally we mention two recent notes [7], [8] presented in Algoritm for Approximation III (Oxford 1992) and in Curves and Surfaces (Chamonix 1993) respectively; [7] attemptes to solve the problem using the rational tensioned splines, while, in [8] a method based on the ν-splines is presented .

In this work we present a computational and efficient method which, exploiting the well-known properties of the Chebyshev polynomials of the first and second kind, determines an approximation of the unknown function as juxtaposition of polynomials $\sum_{i=0}^{g} d_i T_i(t)$ of fixed degree g: these polynomials are obtained by integration from the polynomials $\sum_{i=0}^{g-1} b_i U_i(t)$ which approximate the first derivative of $f(x)$ $x \in [a, b]$. The coefficients b_i are computed using the well-known Clenshaw algorithm [3].

This method permits to construct an algorithm which is entirely automatic. Moreover theorems on convergence and on convergence rate are given; more precisely we have proved that the approximating function $\Phi(x)$ of $f(x)$, arising from the above mentioned juxtaposition, is C^1 on $[a, b]$ and it is monotone with probability tending to one.

A. Cuyt (ed.), Nonlinear Numerical Methods and Rational Approximation II, 201–216.

2. Description of the Method

2.1. First Derivative Values Estimate

Let be assigned N discrete observations of the function $f(x) \in C^1[a,b]$

$$y_i = f_i + e_i, \quad i = 1, \ldots, N, \tag{1}$$

where $f_i = f(x_i)$ and e_i random variables indipendent identically distributed with $E(e_i) = 0, \quad E(e_i^2) = \sigma^2 < \infty$.

Using these quantities, we determine a sequence $Y_i, \quad i = 1, \ldots, N$ which estimates the first derivative values $f_i' = f'(x_i), \quad i = 1, \ldots, N$. For each point $x_i, \quad i = 1, \ldots, N$, we consider a suitable centered neighborhood I_i of x_i whose measure is μ_i and compute the quantities

$$Y_i = \frac{\sum_{j=1}^{n_i} (y_j - \bar{y}) d_{ij}}{\sum_{j=1}^{n_i} d_{ij}^2}, \qquad i = 1, \ldots, N \tag{2}$$

where n_i is the number of observations which are in I_i, d_{ij} is equal to $x_j - x_i$ and $\bar{y} = \frac{1}{n_i} \sum_{j=1}^{n_i} y_j$.

It is easy to prove that (2) approximates f_i', $i = 1, \ldots, N$, in fact the following relation holds

$$Y_i = f_i' + O(\mu_i) + O_p(\frac{1}{\sqrt{n_i \mu_i^3}}). \tag{3}$$

2.2. Subdivision of the data set

Now we consider the new data set $\{x_i, Y_i\}, \quad i = 1, \ldots, N$ and divide the interval $[a,b]$ into subintervals $J_1, J_2, \ldots, J_k$ in order to scan the function in parts where the first derivative is almost a constant.

The subdivision is obtained by a criterion that verifies the null hypothesis

$$H_r: \quad E(Y_i) = \eta^r \qquad \forall i \in J_r$$

in opposition to the alternative

$$H_a^r: \quad E(Y_i) \neq \eta^r \qquad \forall i \in J_r.$$

More precisely, we consider the estimator $\Delta_0 = \frac{1}{N} \sum_{i=1}^N Y_i$, whose expected value $E(\Delta_0)$ is equal to η^0, while $E(Y_i) = \eta_i$. If $\eta_i = \eta^0 + o(1), \quad \forall i = 1, \ldots, N$, the expected value of the statistics

$$V_0 = \frac{1}{N-1} \sum_{i=1}^N (Y_i - \Delta_0)^2$$

is

$$E(V_0) = \Psi \sigma^2 + o(1),$$

where

$$\Psi \simeq \frac{1}{N} \sum_{i=1}^N n_i \mu_i,$$

therefore the ratio

$$R_0 = \frac{V_0}{\Psi\sigma^2}$$

will assume values next to one.

If R_0 assumes large values then the null-hypothesis $E(Y_i) = \eta^0, \quad i = 1, \ldots, N$ is rejected. So we consider the values $Y_1, \ldots, Y_j$ such that only one of these two cases is verified

$$Y_s \leq \Delta_0, \quad \forall s = 1, \ldots, j, \quad Y_{j+1} > \Delta_0$$

or

$$Y_s \geq \Delta_0, \quad \forall s = 1, \ldots, j, \quad Y_{j+1} < \Delta_0.$$

Then

$$\begin{aligned} \Delta_1 &= \frac{1}{j}\sum_{s=1}^{j} Y_s \\ V_1 &= \frac{1}{j-1}\sum_{s=1}^{j}(Y_s - \Delta_1)^2 \end{aligned}$$

and we repeat the test in the hypothesis $E(Y_i) = \eta^1, \quad i = 1, \ldots, j$. The process ends when we find a set $Y_i, \quad i = 1, \ldots, j_r$ such that the inequality $|R - 1| < tol$ holds, where tol is a tolerance fixed in advance.

Observe that in general the noise variance σ^2 is unknown. We can get an estimation of σ^2 by the statistics

$$\hat{\sigma}_r^2 = \frac{1}{n_1 - 2}\sum_{i=1}^{n_1}(y_i - \hat{A}x_i - \hat{B})^2$$

of the regression line $r(x) = Ax + B$ in the neighbourhood I_1 of x_1 where we assume that the first derivative variation is sufficiently small. In fact

$$\begin{aligned} \sigma^2 &= E(\frac{1}{n_1}\sum_{i=1}^{n_1}(y_i - f_i)^2) \\ &= E(\frac{1}{n_1}\sum_{i=1}^{n_1}(y_i - r_i + r_i - f_i)^2) \\ &= \sigma_r^2 - E(\frac{1}{n_1}\sum_{i=1}^{n_1}(r_i - f_i)^2) \end{aligned}$$

and putting

$$\rho = \frac{cov(f, x)}{\sigma_x \sigma_f}$$

with

$$\sigma_x = E(\frac{1}{n_1}\sum_{i=1}^{n_1}(x_i - \bar{x})^2)$$

$$\sigma_f = E(\frac{1}{n_1}\sum_{i=1}^{n_1}(f_i - \bar{f})^2)$$
$$cov(f,x) = E(\frac{1}{n_1}\sum_{i=1}^{n_1}(x_i - \bar{x})(f_i - \bar{f}))$$

we get

$$\sigma^2 = \sigma_r^2 + (1-\rho^2)\sigma_f^2.$$

Therefore, from the previous considerations, $\rho^2 \simeq 1$ and so we have $\sigma^2 \simeq \sigma_r^2$.

2.3. Approximation of $f(x)$

Consider the generic interval J_r, and the set $\{x_i, Y_i\}_{i=1}^{m_r}$ with $x_i \in J_r$.

We transform the data from the interval $J_r : [t_r, t_{r+1}]$ to the interval $[-1,1]$ by

$$X = \frac{2t - t_r - t_{r+1}}{t_{r+1} - t_r}, \quad X \in [-1,1], \quad t \in J_r.$$

In this way we have to consider the values $Y_i^T, \quad i = 1,\dots,m_r$ obtained from the transformation of first derivative estimates $Y_i, \quad i = 1,\dots,m_r$ by

$$Y_i^T = \frac{\mu(J_r)}{2} Y_i$$

where $\mu(J_r)$ is the measure of J_r.

Using the least square technique we approximate $f'(X), \quad X \in [-1,1]$, by the polynomial of fixed degree $g-1$

$$G^r(X) = \sum_{j=0}^{g-1} c_j P_j(X) \tag{4}$$

where

$$c_j = \frac{\sum_{i=1}^{m_r} Y_i^T P_j(X_i)}{\sum_{i=1}^{m_r} P_j^2(X_i)} \tag{5}$$

and the polynomials $P_0(X),\dots,P_{g-1}(X)$ are orthogonal on the (transformed) data abscissae $\{X_i\}$. It is well-known that the orthogonal polynomials on $\{X_i\}$ are generated by the recurrence relation proposed by [1], [2]

$$P_{j+1}(X) = 2(X - \alpha_{j+1})P_j(X) - \beta_j P_{j-1}(X) \tag{6}$$
$$j = 0,\dots,g-2$$

where $P_0(x) = \frac{1}{2}$, $\beta_0 = 0$ and

$$\alpha_{j+1} = \frac{\sum_{i=1}^{m_r} X_i P_j^2(X_i)}{\sum_{i=1}^{m_r} P_j^2(X_i)}, \quad j = 0,\dots,g-2$$
$$\beta_j = \frac{\sum_{i=1}^{m_r} P_j^2(X_i)}{\sum_{i=1}^{m_r} P_{j-1}^2(X_i)}, \quad j = 1,\dots,g-2$$

On the other hand, also the equation (4) can be generated by the recurrence relation

$$G_s(x) = G_{s-1}(X) + c_s P_s(X) \quad s = 0, \ldots, g-1. \tag{7}$$

Moreover we can represent the polynomials $P_j(X)$ and $G_s(X)$ by their Chebyshev expansions

$$P_j(X) = \sum_{i=0}^{j} a_i^j T_i(X), \quad j = 1, \ldots, g-1 \tag{8}$$

$$G_s(X) = \sum_{k=0}^{s} b_k^s T_i(X), \quad s = 1, \ldots, g-1 \tag{9}$$

Substitution of (8) into (6) yields the following recurrence relation for the coefficients $\{a_i^j\}$:

$$a_i^{j+1} = a_{|i-1|}^j + a_{i+1}^j - 2\alpha_{j+1} a_i^j - \beta_j a_i^{j-1} \tag{10}$$

where $j = 0, \ldots, g-2$, $i = 0, \ldots, j$, $a_i^j = 0$ if $i > j$ and $a_j^j = 1$.

Now substituting, (8) and (9) into (7) we get the following relation for the coefficients b_k^s:

$$b_k^s = \sum_{i=k}^{s} c_i a_k^i, \quad s = 0, \ldots, g-1. \tag{11}$$

In this way we write (4) as linear combination of the first kind Chebyshev polynomials

$$G^r(X) = \sum_{k=0}^{g-1} b_k^{g-1} T_i(X) \tag{12}$$

Recalling that

$$U_{2k}(X) = 1 + 2\sum_{i=1}^{k} T_{2i}(X)$$

$$U_{2k-1}(X) = 2\sum_{i=1}^{k} T_{2i-1}(X)$$

$$T'_{k+1}(X) = (k+1)U_k(X)$$

where the polynomials $U_k(X)$ are the second kind Chebyshev polynomials, we can rewrite (12) as

$$G^r(X) = \sum_{i=0}^{g-1} d_i U_i(X) \tag{13}$$

with

$$d_i = \frac{b_i^{g-1} - b_{i+2}^{g-1}}{2} \tag{14}$$

where $b_i^{g-1} = 0$ if $i > g-1$.

Therefore we obtain the approximation of the unknown function $f(X)$ on $[-1,1]$ corresponding to the interval $J_r \subset [a,b]$, integrating (13)

$$\Phi^r(X) = \int_{-1}^{X} G^r(\tau)d\tau + \phi_r \tag{15}$$

then

$$\Phi^r(X) = \sum_{j=0}^{g-1} \frac{d_j}{j+1}(T_{j+1}(X) - T_{j+1}(-1)) + \phi_r \tag{16}$$

2.4. Continuity conditions

We put the integration constants ϕ_r, which appear in the approximations $\Phi^r(X)$ corresponding to J_r, $r = 1,\ldots,k$, equal to

$$\phi_r = \Phi^{r-1}(1), \quad r = 2,\ldots,k \tag{17}$$

and

$$\phi_1 = \frac{\sum_{i=1}^{n_1} y_i}{n_1}.$$

Therefore the approximating function is $C^0[a,b]$.

Since we want an approximating function at least $C^1[a,b]$, it is necessary that the approximation of the first derivative is $C^0[a,b]$. Precisely, the following conditions must be satisfied

$$\begin{aligned} \frac{2}{\mu(J_r)}(G^r(1) + \alpha_r(1)) &= \frac{2}{\mu(J_{r+1})}(G^{r+1}(-1) + \alpha_{r+1}(-1)) \\ r &= 1,\ldots,k \end{aligned} \tag{18}$$

where $\alpha_r(X)$ depends on the constraints assigned on the approximating function. If monotonicity is requested we will consider

$$\alpha_r(X) = \lambda_r,$$

while if convexity is requested, it is more suitable to choose a linear function as

$$\alpha_r(X) = \lambda_r X.$$

When the expression of $\alpha_r(X)$ is assigned, (18) permits to determine the unknown parameters λ_r $r = 1,\ldots,k$.

So the approximations of $f'(X)$ and $f(X)$, $X \in [-1,1]$ are

$$G^r(X) = \sum_{i=0}^{g-1} d_i U_i(X) + \alpha_r(X) \tag{19}$$

$$\Phi^r(X) = \int_{-1}^{X} (G^r(\tau) + \alpha_r(\tau))d\tau + \phi_r \tag{20}$$

2.5. The Algorithm

In this subsection we summarize the main steps of the algorithm for the construction of the approximating function, obtained from the method presented in the previous sections.

Input (x_i, y_i), $i = 1, \dots, N$, g =degree of the approximating function, n_i, $i = 1, \dots, N$.

Step 1. Computation of the first derivative values Y_i, $i = 1, \dots, N$ by relation (2).

Step 2. Computation of $\hat{\sigma}_r^2$ using the data y_i, $i = 1, \dots, n_1$.

Step 3. Subdivision of the data set into k intervals $J_1, \dots, J_k$ in which the first derivative is almost a constant by the test $\hat{R} = \frac{V_i}{\Phi \hat{\sigma}_r^2} \leq 1$.

Step 4. Construction of the approximating function in each interval J_r, $r = 1, \dots, k$:

for $r = 1, \dots, k$
compute the coefficients d_i, $i = 0, \dots, g-1$
next r
for $r = 1, \dots, k$ compute $\alpha_r(t)$ and ϕ_r
next r

3. Asymptotic Properties

In this section we prove the following important result:

Theorem 1. *$\Phi^r(t)$ is monotone with probability tending to one, $\forall r = 1, \dots, k$.*

To this purpose we premise some theorems on the L_1, L_2, L_∞ convergence of the approximating function.

We suppose $f \in C^1[a,b]$ and write the quantities Y_i, defined in section 2.1, as

$$Y_i = f_i' + \theta_i \mu_i + \delta_i, \quad \theta_i \in R, \quad i = 1, \dots, N, \tag{21}$$

where it is easy to prove that the random variables δ_i have moments

$$E(\delta_i) = 0, \quad E(\delta_i \delta_j) \simeq 0 \quad i \neq j, \quad E(\delta_i^2) \simeq \frac{\sigma^2}{n_i \mu_i^3}. \tag{22}$$

Therefore, by the relation (21) the function

$$G^r(t) = \sum_{i=0}^{g-1} d_i U_i(t) + \alpha_r(t), \quad t \in J_r, \quad r = 1, \dots, k \tag{23}$$

is decomposed as

$$\begin{aligned} G^r(t) &= \sum_{i=0}^{g-1} \hat{d}_i U_i(t) + \hat{\alpha}_r(t) + \sum_{i=0}^{g-1} \tilde{d}_i U_i(t) + \tilde{\alpha}_r(t) + O(\mu) \\ &= \hat{g}^r(x) + \hat{\alpha}^r(t) + \tilde{g}^r(t) + \tilde{\alpha}_r(t) + O(\mu) \end{aligned} \tag{24}$$

where $\mu = \max_{1 \leq i \leq m_r} \mu_i$, and hence

$$\begin{aligned} \Phi^r(t) &= \int_{t_r}^t (\hat{g}^r(\tau) + \hat{\alpha}_r(t)) d\tau + \hat{\phi}_r + \int_{t_r}^t (\tilde{g}^r(\tau) + \tilde{\alpha}_r(t)) d\tau + \tilde{\phi}_r \\ &= \hat{\Phi}^r(t) + \tilde{\Phi}^r(t) + O(\mu^2(J_r)). \end{aligned} \tag{25}$$

Theorem 2. *$\forall r = 1, \ldots, k$ we have that*

$$||f'(t) - \hat{g}^r(t) - \hat{\lambda}_r||_2^2 = O(\mu(J_r)), \tag{26}$$

and

$$||f'(t) - \hat{g}^r(t) - \hat{\lambda}_r||_1 = O(\mu(J_r)). \tag{27}$$

Proof [1]

For what is mentioned in 2.4 the function $\hat{g}^r(t) + \hat{\lambda}_r$ is also equal to

$$\sum_{j=0}^{g-1} c_j P_j(t) + \hat{\lambda}_r \tag{28}$$

where

$$c_j = \frac{\sum_{i=1}^{m_r} f_i' P_j(x_i)}{\sum_{i=1}^{m_r} P_j^2(x_i)} \tag{29}$$

$$\hat{\lambda}_r = \frac{\mu(J_r)}{2}\left(\sum_{j=2}^{r}(\hat{g}^{j-1}(t_j) - \hat{g}^j(t_j))\right) \tag{30}$$

For what we have said in the previous sections, the behaviour of the first derivative in each interval $J_r \quad r = 1, \ldots, k$, is almost equal to a constant η^r therefore, when $m_r \to \infty$, we have

$$\begin{aligned} \hat{\lambda}_r &\simeq \mu(J_r)\left(\frac{\eta^1 - \eta^r}{4}\right) \\ &\leq \mu(J_r)\left(\frac{\eta^{max} - \eta^{min}}{4}\right) \end{aligned} \tag{31}$$

where $\eta^{max} = max_{r=1,\ldots k}\eta^r$ and $\eta^{min} = min_{r=1,\ldots k}\eta^r$. Let us consider

$$||f'(t) - \hat{g}^r(t) - \hat{\lambda}_r||_2^2 \leq ||f'(t) - \hat{g}^r(t)||_2^2 + ||\hat{\lambda}_r||_2^2 \tag{32}$$

$$||f'(t) - \hat{g}^r(t)||_2^2 = \int_{J_r} (f'(t))^2 dt - \sum_{i=0}^{g} c_i^2, \tag{33}$$

when $m_r \to \infty$ we have

$$c_i \simeq \gamma\mu(J_r)\frac{(\int_{J_r} f'(t) l_i(t) dt)}{(\int_{J_r} l_i^2(t) dt)^2} = \gamma\mu(J_r)\bar{c}_i, \quad \gamma \in R^+, \tag{34}$$

where $l_i(x)$ are the legendre polynomials of degree i. Hence

$$||f'(t) - \hat{g}^r(t)||_2^2 = O(\mu(J_r)), \tag{35}$$

and since

$$||\hat{\lambda}_r||_2^2 = O(\mu(J_r)^3), \tag{36}$$

[1] All the demonstrations of this section are carried on considering $\alpha_r(t) = \lambda_r$. The results are true also in the case $\alpha_r(t) = \lambda_r t$.

we have that

$$||f'(t) - \hat{g}^r(t) - \hat{\lambda}_r||_2^2 = O(\mu(J_r)).$$

Then

$$||f'(t) - \hat{g}^r(t) - \hat{\lambda}_r||_1 = O(\mu(J_r)).$$

Theorem 3. $\forall r = 1, \ldots, k$

$$||f(t) - \hat{\Phi}^r(t)||_\infty = O(\mu(J_r)) \tag{37}$$

and

$$||f(t) - \hat{\Phi}^r(t)||_2 = O(\mu^{\frac{3}{2}}(J_r)). \tag{38}$$

Proof

$$\begin{aligned} ||f(t) - \hat{\Phi}^r||_\infty &= Sup_{t \in J_r} |f(t) - \hat{\Phi}^r(t)| \\ &\le Sup_{t \in J_r} \{ \int_{t_r}^{t} |f'(\tau) - \hat{g}^r(\tau) - \hat{\lambda}_r| d\tau + |f(t_r) - \hat{\phi}_r| \} \\ &\le ||f'(x) - \hat{g}^r(x) - \hat{\lambda}_r||_1 + |f(t_r) - \phi_r|. \end{aligned} \tag{39}$$

Now we consider

$$|f(t_r) - \hat{\phi}_r|$$

we prove that

$$|f(t_r) - \hat{\phi}_r| = O(\mu_1).$$

In fact in each interval J_{r-1} $f(t)$ is approximated by

$$f(t_{r-1}) + \sum_{i=0}^{g} c_i \int_{t_r - 1}^{t} P_i(\tau) d\tau \tag{40}$$

which converges uniformly to $f(t)$ when g, m_r go to ∞. Then

$$f(t_r) = f(t_{r-1}) + \sum_{i=0}^{\infty} \bar{c}_i \int_{t_r - 1}^{t_r} l_i(\tau) d\tau. \tag{41}$$

Moreover from (17) we have also

$$\hat{\phi}_r \simeq \hat{\phi}_{r-1} + \sum_{i=0}^{g} \bar{c}_i \int_{t_r - 1}^{t_r} l_i(\tau) d\tau, \quad r = 2, \ldots, k. \tag{42}$$

Then we obtain

$$|f(t_r) - \hat{\phi}_r| = |f(t_{r-1}) - \hat{\phi}_{r-1}| \quad \forall r \tag{43}$$

and so

$$|f(t_r) - \hat{\phi}_r| = |f(t_1) - \hat{\phi}_1|. \tag{44}$$

As consequence of the continuity of $f(t)$ and of the mean value theorem, we have

$$|f(t_1) - \hat{\phi}_1| = O(\mu_1). \tag{45}$$

From (27), (39) and (45), when $\mu(J_r) \to 0$, $r = 1, \ldots, k$ and $\nu_i \to 0$, $i = 1, \ldots, N$, we obtain

$$||f(t) - \hat{\Phi}_g^r||_\infty = O(\mu(J_r))$$

and, for the well-known norms relations, we have

$$||f(t) - \hat{\Phi}_g^r||_2 = O(\mu^{\frac{3}{2}}(J_r)).$$

Theorem 4. *The random variable $\tilde{g}^r(t)$ is such that*

$$E(\tilde{g}^r(t)) = 0 \tag{46}$$

and

$$E(\tilde{g}^r(t))^2 = O(\frac{1}{n_l \mu_l^3 m_r \mu(J_r)}), \tag{47}$$

where $n_l \mu_l^3 = \min_{i=1,\ldots,m_r} n_i \mu_i^3$.

Proof

(46) is trivial. Consider

$$E((\tilde{g}^r(t))^2) = E((\sum_{i=0}^{g-1} \tilde{c}_i P_i(t))^2) \tag{48}$$

$$= \sum_{i=0}^{g-1} E(\tilde{c}_i^2) P_i(t)^2 + \sum_{i \neq j} E(\tilde{c}_i \tilde{c}_j) P_i(t) P_j(t) \tag{49}$$

with

$$\tilde{c}_i = \frac{\sum_{j=1}^{m_r} \delta_j P_i(x_j)}{\sum_{j=1}^{m_r} P_i^2(x_j)}$$

When $m_r \to \infty$ and consequently $n_i \to \infty$, $i = 1, \ldots, m_r$, recalling that

$$\frac{1}{n_i} \sum_{j=1}^{n_i} d_{ij} \to \int_{\frac{-\mu_i}{2}}^{\frac{\mu_i}{2}} x dx = 0,$$

we have

$$E(\tilde{c}_i \tilde{c}_j) \simeq 0,$$

while

$$E(\tilde{c}_i^2) = O(\frac{1}{n_l \mu_l^3 m_r \mu(J_r)}) \tag{50}$$

where where $n_l \mu_l^3 = \min_{i=1,\ldots,m_r} n_i \mu_i^3$. Hence we conclude that

$$E(\tilde{g}^r(t))^2 = O(\frac{1}{n_l \mu_l^3 m_r \mu(J_r)}), \tag{51}$$

Corollary 1. *The random variable $\tilde{\Phi}^r(t)$ is such that*

$$E(\tilde{\Phi}^r(t)) = 0, \tag{52}$$

$$E(\tilde{\Phi}^r(t))^2 = O(\frac{1}{n_l \mu_l^3 m_r \mu(J_r)}), \tag{53}$$

where $n_l \mu_l^3 = \min_{i=1,\ldots,m_r} n_i \mu_i^3$.

Proof

The proof is trivial.

Theorem 5. *The random variable*

$$\tilde{\lambda}_r = \frac{\mu(J_r)}{2}(\sum_{j=2}^{r} \tilde{g}^{j-1}(t_j) - \tilde{g}^j(t_j)), \tag{54}$$

has expected value

$$E(\tilde{\lambda}_r) = 0, \tag{55}$$

and variance

$$E(\tilde{\lambda}_r^2) \simeq \frac{\mu(J_r)^2}{2}\frac{1}{n_l\mu_l^3}\sum_{j=2}^{r}\frac{1}{m_j\mu(J_j)}, \tag{56}$$

where $n_l\mu_l^3 = \min_{i=1,\ldots,m_r} n_i\mu_i^3$. **Proof**
(55) is trivial. Then consider the variance of (54)

$$E(\tilde{\lambda}_r)^2 = \frac{\mu^2(J_r)}{4}\{\sum_{j=2}^{r}(\tilde{g}^{j-1}(t_j) - \tilde{g}^j(t_j))^2 + \sum_{j\neq i}^{r}(\tilde{g}^{i-1}(t_i) - \tilde{g}^i(t_i))(\tilde{g}^{j-1}(t_j) - \tilde{g}^j(t_j))\}. \tag{57}$$

From (47) we obtain

$$E(\tilde{\lambda}_r^2) \simeq \frac{\mu(J_r)^2}{2}\frac{1}{n_l\mu_l^3}\sum_{j=2}^{r}\frac{1}{m_j\mu(J_j)}. \tag{58}$$

where $n_l\mu_l^3 = \min_{i=1,\ldots,m_r} n_i\mu_i^3$.

Now we can prove the theorem mentioned at the beginning of this section.

Proof of Theorem 1

From (24) and considering (27), (47), (56) we get that

$$E||f'(t) - G^r(t)||_2^2 = O(\mu(J_r)) + O(\frac{1}{n_l\mu_l^3 m_r\mu(J_r)}),$$

hence when

$$\mu(J_r) \rightarrow 0,$$

consistentely with

$$m_r\mu(J_r) \rightarrow \infty$$

and consequently

$$n_i\mu_i^3 \rightarrow \infty,$$

it is proved that the sequence of the first derivative approximating functions converges in quadratic mean to $f'(x)$, $x \in [a, b]$. Therefore $\Phi^r(t)$ is monotone $\forall r = 1, \ldots, k$ with probability tending to one

Corollary 2. *The convergence rate of $\Phi^r(t)$ to $f(t)$ is given by the following relation*

$$\Phi^r(t) = f(t) + O(\mu(J_r)) + O_p(\frac{1}{\sqrt{n_l\mu_l^3 m_r \mu(J_r)}}),$$

where $n_l\mu_l^3 = \min_{i=1,\ldots,m_r} n_i\mu_i^3$.

Proof

The thesis cames from (37), (53).

4. Consideration on the the choice of the degree g

The algorithm presented in section 2.3, provides the set of coefficients d_i. These coefficients gives indications on the best choice of the degree s. In fact, since the approximating functions of $f'(x)$ and $f(x)$ converge in quadratic mean, we would expect the d_i decrease with increasing i. In practice the rounding errors and the errors on the data, affect the behaviour of the d_i. Then for values of i exceeding a certain s, the coefficients will fluctuate around zero in an apparently random manner. The degree s may be then accepted as optimal degree and, the choice of a degree $l > s$, will not provide a better approximation. In fact the change in the d_i will be merely of the order of unwanted noise.

Moreover it is better to choose a degree not very large, otherwise the errors carried on in the computation can dominate the solution.

5. Numerical Results

In this section we present some numerical results drew from the literature. First of all we provide an example which shows the different results obtained when we approximate directly the data $y_i \quad i = 1, \ldots, 11$ by mean of the same least square technique based on the Clenshaw algorithm (fig 1) and when we approximate, by the method described in 2.3, the first derivative (fig 2). The degree g of the two approximations is equal to 4. From the two graphics we observe that in the first case the approximating function is not monotone while in the second case it is monotone.

In the following examples we have used the simple precision and we have choosen $n_i = 5, \quad i = 1, \ldots, N$. We have estimated σ^2 by $\hat{\sigma}_r^2$ using the first five points.

Example 1

We have considered $N = 21$ equispaced observations of $y = artg(5x)$ $x \in [-1, 1]$ coming from a normally distributed population with $\sigma = 0.05$ and ratio signal to noise $\frac{\sigma}{f} = \frac{1}{50}$. In fig 3 the approximating function of degree $g = 4$ is shown.

Example 2

The example in fig 4 refers to $f(x) = (1 - x^4)^{(\frac{1}{4})}$ which has a sharp variation in the first derivative. In this case we have considered $N = 50$ equispaced observations coming from a normally distributed population with $\sigma = 0.01$ and ratio signal to noise $\frac{\sigma}{f} = \frac{1}{100}$. The degree of the approximating function is 3.

Example 3

As last example, we have considered the classical titanium data drew from [4] pag 222. In fig 5 we can see the approximating function of degree 4 to the first 26 data while in fig 6 the complete approximating function of degree 4 is shown.

References

1. Forsythe C.W. *Generation and Use of Orthogonal Polynomials for Data Fitting with Digital Computer* J. Soc. Ind. Appl. Math., Vol. 5, pp 74 (1957).
2. Ascher M. and Forsythe G. E. *SWAC Experiments on the use of Orthogonal Polynomials for Data Fitting* J. ACM, vol 5 pp 9-21 (1958).
3. Clenshaw C. W. *Curve Fitting with Digital Computer* Computer J., Vol. 4, pp 260-264 (1961).
4. De Boor C. *A Practical Guide to Spline*, Springer Verlag, Berlin, New York, (1978) .
5. Utreras F. *Smoothing noisy data under monotonicity constrained Existence, Characterization and Convergence Rates* Numer.Math. Vol. 47 pp 611-625 (1985).
6. Elfving T. and Andersson L. E. *An Algorithm for Computing Constrained Smoothing Spline Functions* Num. Math. Vol. 52 pp 583-595 (1988).
7. Elliott G. H. *Least Squares Data Fitting using Shape preserving Piecewise Approximations* presented in Algorithm for Approximation III, July 27-31, 1992, Oxford.

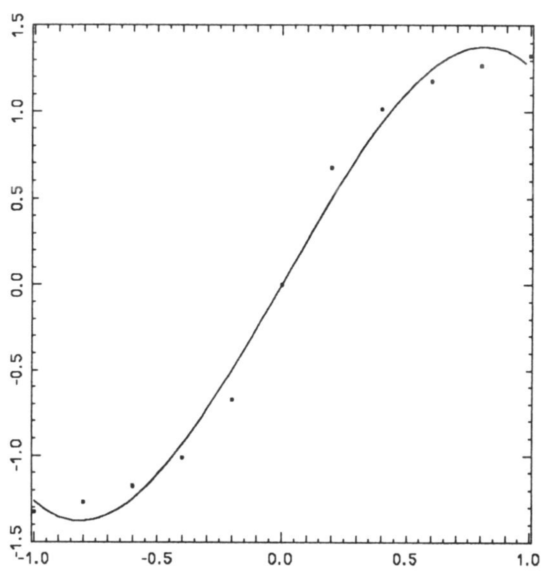

Fig. 1.

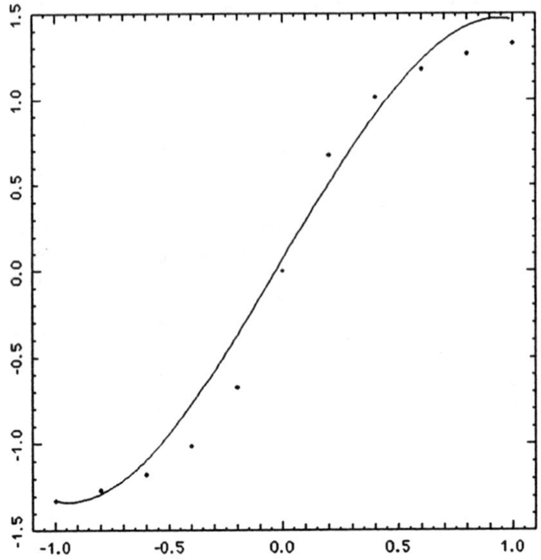

Fig. 2.

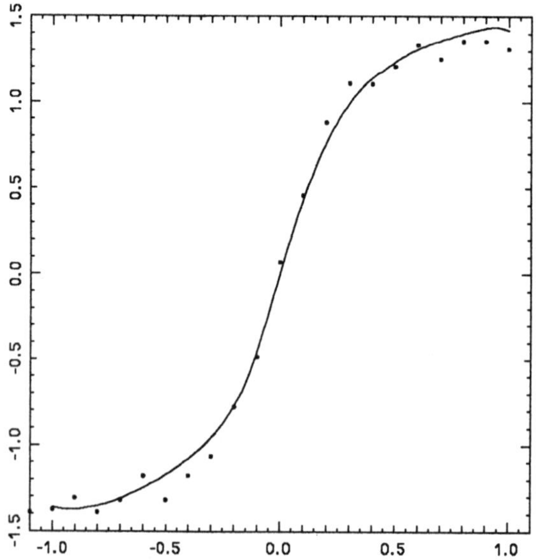

Fig. 3.

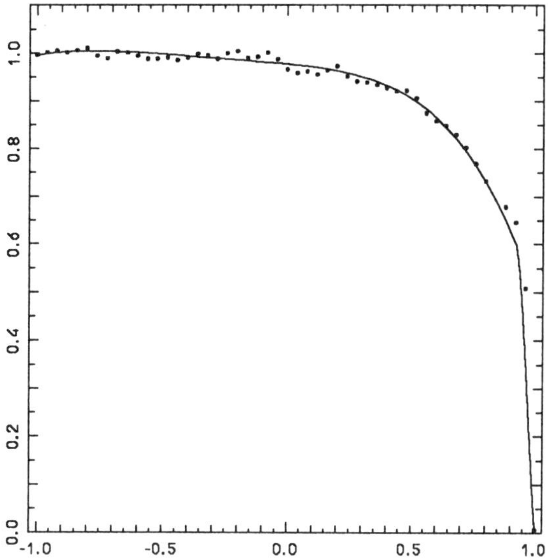
0.0
0.2
0.4
0.6
0.8
1.0
-1.0
-0.5
0.0
0.5
1.0

Fig. 4.

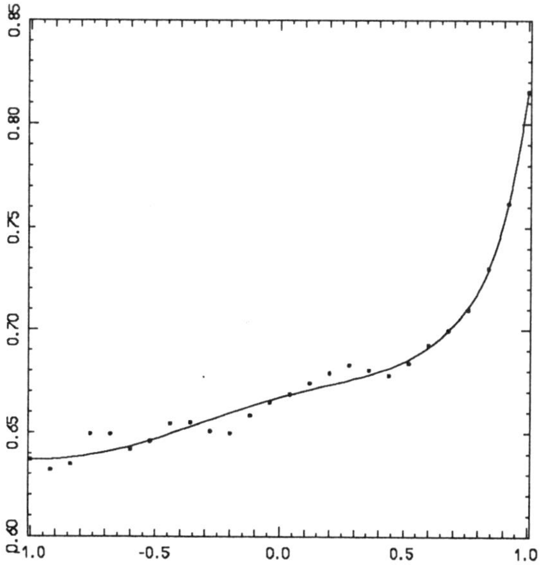
0.60
0.65
0.70
0.75
0.80
0.85
-1.0
-0.5
0.0
0.5
1.0

Fig. 5.

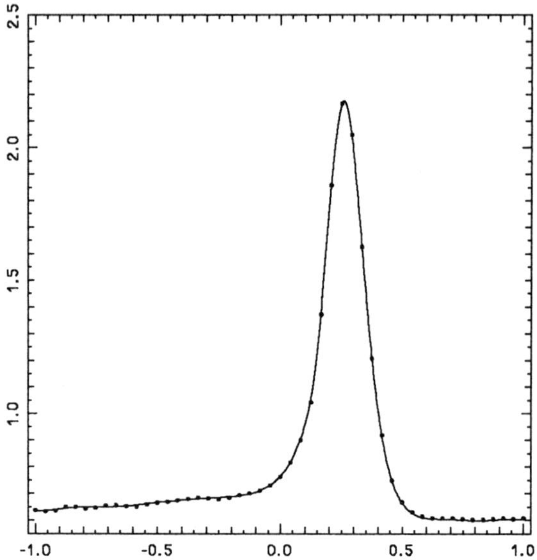

Fig. 6.

PADÉ

APPROXIMATION

ON THE EXTENSION OF EIERMANN'S THEOREM TO SEVERAL COMPLEX VARIABLES

CATARINA RUDÄLV
Department of Mathematics
University of Umeå
S-901 87 Umeå
Sweden

Abstract. Daras [2] has extended a result by Eiermann [3] to several complex variables and applied it to Padé-type approximants. Our main purpose is to give a short background and a new proof of Daras' theorem.

Key words: Summability methods, Padé-type approximants, Convergence results, Error formula

1. Introduction and notations

For a given summability method, the Okada theorem ([4],[6]) describes a domain, into which an arbitrary power series can be analytically continued, if such a domain is known for the geometric series. Eiermann has extended this to a general summability theorem [3] and applied it to Padé-type approximants. This generalized form is a criterion for the efficiency of Padé-type approximants. In [1] Cala-Rodriguez and Wallin gave a new simpler proof of Eiermann's theorem based on an integral representation of the approximation error which made it easier to understand Eiermann's theorem. They concentrated on the case given by the Padé-type approximants but remarked that the method works in a straightforward way for the general summability method by Eiermann. In [2] Daras shows that two extensions of Eiermann's criterion can be derived in the case of an open polydisc in $\mathbb{C}^n$. The techniques that he uses are similar to those of Eiermann. In this note we give a alternative method to prove Daras' theorem. We will concentrate on the application to Padé-type approximants which seems to be the most interesting case. The alternative proof is inspired by the method in [1] and is based on an integral representation of the error (Theorem 2). This representation, also, makes it easy to state other convergence results (Section 4).
First we will give some basic notations which will be used throughout this paper. $\mathbf{N}^n$ will denote the cartesian product of n copies of $\mathbf{N} = \{0, 1, ...\}$. D will denote an open, connected subset of $\subset \mathbb{C}^n$, called a *domain*. We shall denote by $\mathcal{H}(D)$ the *set of all holomorphic functions on D*. The *polydisc* $\mathbf{P}(a, r)$ of *multiradius* $r = (r_1, ..., r_n)$, $r_j > 0$, and *center* $a \in \mathbb{C}^n$ is the product of n open discs in $\mathbb{C}$:

$$\mathbf{P}(a,r) = \{z \in \mathbb{C}^n : |z_j - a_j| < r_j, j = 1, ..., n\}$$

A. Cuyt (ed.), Nonlinear Numerical Methods and Rational Approximation II, 219–228.

which sometimes will be denoted $\mathbf{P}$. Let $\mathbf{P}_i$ denote the projection of the polydisc to the z_i-plane, i.e.

$$\mathbf{P}_i = \{z_i \in \mathbb{C} : |z_i - a_i| < r_i\}, \quad i = 1, ..., n$$

We also use the abbreviation

$$(w-z) = (w_1 - z_1)\cdots(w_n - z_n), \quad w^1 = w_1 \cdot \ldots \cdot w_n$$
$$f(w) = f(w_1, ..., w_n), \quad dw = dw_1 ... dw_n$$

2. Daras' theorem for Padé-type approximants

For the sake of simplicity we will henceforth restrict ourselves to the case when $n = 2$. The extension to a more general case will follow in a straightforward way. Suppose $f \in \mathcal{H}(D)$, $D \subset \mathbb{C}^2$ is a domain, containing zero. Formally we have

$$f(z) = \sum_{\alpha \in \mathbf{N}^2} c_\alpha z^\alpha \tag{1}$$

Given a polynomial Q_M satisfying

$$Q_M(z_1, z_2) = \sum_{(i,j) \in M} b_{ij} z_1^i z_2^j, \; b_{00} = 1 \tag{2}$$

$$M = [0, m_1] \times [0, m_2] \cap \mathbf{N}^2, \quad m_1, m_2 \in \mathbf{N}$$

We now choose the unique polynomial, P_{M-1} such that P_{M-1} satisfies

$$\exists \; a_{ij} : \; P_{M-1}(z_1, z_2) = \sum_{(i,j) \in M-1} a_{ij} z_1^i z_2^j \tag{3}$$

$$\exists \; e_{ij} : \; (fQ_M - P_{M-1})(z_1, z_2) = \sum_{(i,j) \in \mathbf{N}^2 \setminus M-1} e_{ij} z_1^i z_2^j \tag{4}$$

or with the notation used in [2]

$$(fQ_M - P_{M-1})(z_1, z_2) = \mathcal{O}(z_1^i z_2^{m_2}) + \mathcal{O}(z_1^{m_1} z_2^j), \; i, j \geq 0, \quad z_1, z_2 \to 0$$

where we have used the abbreviation

$$M - 1 = [0, m_1 - 1] \times [0, m_2 - 1] \cap \mathbf{N}^2$$

By definition, P_{M-1}/Q_M is the *Padé-type approximant* (P.T.A) of f, of type $(M - 1/M)$, with prescribed poles at the points where $Q_M(z_1, z_2) = 0$ and Q_M are the *generating polynomials*. It is easy to see that P_{M-1}/Q_M satisfies

$$\frac{P_{M-1}}{Q_M}(z) = \sum_{k_1=0}^{m_1} \sigma^1_{m_1,k_1}(z) \cdot \sum_{i=0}^{k_1} \left(\sum_{k_2=0}^{m_2} \sigma^2_{m_2,k_2}(z) \cdot \sum_{j=0}^{k_2} c_{ij} z_1^i z_2^j \right)$$

where $\sigma^i_{m_i,k_i}$ are rational functions in z uniquely determined by b_{ij}. Thus, P.T.A can be seen as one of the two summability methods introduced in [2], by Daras. If we introduce the following functions

$$q_{m_i}(z_i) := \prod_{j=1}^{m_i}(1 - z_i\beta^{(j)}_{i,m_i}), \quad \beta^{(j)}_{i,m_i} \in C, \quad \tilde{q}_{m_i}(z_i) := z_i^{m_i} q_{m_i}(\frac{1}{z_i}), \quad i = 1, 2 \quad (5)$$

and assume, for the moment, that Q_M satisfies

$$Q_M(z_1, z_2) = q_{m_1}(z_1) \cdot q_{m_2}(z_2) \quad (6)$$

then, when D is a polydisc with center in zero and biradius $r = (r_1, r_2)$, Daras has the following application, of his theorem, to Padé-type approximants.

Theorem 1 ([2],Th 2.1) *Let $f \in \mathcal{H}(\mathbf{P}(0,r)$. If the polynomials Q_M satisfy (5), (6) and*

$$\lim_{\substack{m_1\to\infty \\ m_2\to\infty}} \frac{\tilde{q}_{m_1}(z_1^{-1})\tilde{q}_{m_2}(w_2) + \tilde{q}_{m_1}(w_1)\tilde{q}_{m_2}(z_2^{-1}) - \tilde{q}_{m_1}(w_1)\tilde{q}_{m_2}(w_2)}{\tilde{q}_{m_1}(z_1^{-1}) \cdot \tilde{q}_{m_2}(z_2^{-1})} = 0$$

uniformly for (w_1, w_2, z_1, z_2) on any compact subset of an open set $\Lambda \subseteq \mathbb{C}^4$, containing $\mathbb{C}^2 \times \{0, 0\}$, then

$$\lim_{\substack{m_1\to\infty \\ m_2\to\infty}} (P_{M-1}/Q_M)(z_1, z_2) = f(z_1, z_2)$$

uniformly on any compact subset of

$$\Bigg\{(z_1, z_2) \in \mathbb{C}^2 :$$

$$\lim_{\substack{m_1\to\infty \\ m_2\to\infty}} \frac{\tilde{q}_{m_1}(z_1^{-1})\tilde{q}_{m_2}(\xi_2^{-1}) + \tilde{q}_{m_1}(\xi_1^{-1})\tilde{q}_{m_2}(z_2^{-1}) - \tilde{q}_{m_1}(\xi_1^{-1})\tilde{q}_{m_2}(\xi_2^{-1})}{\tilde{q}_{m_1}(z_1^{-1}) \cdot \tilde{q}_{m_2}(z_2^{-1})} = 0,$$

$$\forall \xi_i \in \bar{\mathbb{C}} \setminus \mathbf{P}_i, \quad i = 1, 2\Bigg\}$$

Daras proves this theorem in the same way as Eiermann. Both of them introduces a linear functional Ω_f, acting on the space of polynomials, defined by $\Omega_f(x^\nu) = c_\nu$. The main part of the proof, in the case of a polydisc, is contained in a lemma ([2],Lemma 1.2) which gives us an local estimation of the error. We will give an alternative proof of this theorem, not using this functional Ω_f. This new proof makes it easier to understand why certain assumptions need to be made, concerning Q_M, and it also gives us an idea how to extend this to more general domains.

3. A new proof of Daras' theorem

3.1. The integral representation

Our proof of Theorem 1 will be a straightforward consequence of the following error formula. Let E_{M-1} be defined as

$$E_{M-1}(z_1,z_2)=\left(f-\frac{P_{M-1}}{Q_M}\right)(z_1,z_2)$$

If we introduce the following two functions

$$\tilde{f}(z_1,z_2):=\frac{1}{z_1}\frac{1}{z_2}f(\frac{1}{z_1},\frac{1}{z_2}) \tag{7}$$

$$\tilde{Q}_M(z_1,z_2):=z_1^{m_1}z_2^{m_2}Q_M(\frac{1}{z_1},\frac{1}{z_2}) \tag{8}$$

and use the notation

$$\Gamma_i^{-1}:=\{z_i^{-1}:\ z_i\in\Gamma_i\} \tag{9}$$

then, if $D=\mathbf{P}(0,r)$, $r=(r_1,r_2)$, we can prove

Theorem 2 *Let $f\in\mathcal{H}(\mathbf{P})$ and P_{M-1}/Q_M be the Padé-type approximant, of type (M-1/M), of f with generating polynomial Q_M given by (2). Let $z\in\mathbf{P}$ be such that $Q_M(z)\neq 0$, and Γ_i be a contour, in $\mathbf{P}_i$, around zero, containing z_i in its interior. Then*

$$E_{M-1}(z_1,z_2)=\frac{1}{(2\pi i)^2}\int\limits_{\Gamma_1^{-1}}\int\limits_{\Gamma_2^{-1}}\frac{\tilde{f}(w)}{(1-wz)}\left[\frac{\tilde{Q}_M(z_1^{-1},w_2)+\tilde{Q}_M(w_1,z_2^{-1})-\tilde{Q}_M(w_1,w_2)}{\tilde{Q}_M(z_1^{-1},z_2^{-1})}\right]dw \tag{10}$$

As an example, we can choose Γ_i to be a circle with center at zero and radius ρ_i where ρ_i is any number between $|z_i|$ and r_i.

Proof. First we notice that, using (7),

$$\frac{\left[\tilde{Q}_M(z_1^{-1},w_2)+\tilde{Q}_M(w_1,z_2^{-1})-\tilde{Q}_M(w_1,w_2)\right]}{\tilde{Q}_M(z_1^{-1},z_2^{-1})}=\frac{(z_2w_2)^{m_2}Q_M(z_1,\frac{1}{w_2})+(z_1w_1)^{m_1}Q_M(\frac{1}{w_1},z_2)+(z_1w_1)^{m_1}(z_1w_2)^{m_2}Q_M(\frac{1}{w_1},\frac{1}{w_2})}{Q_M(z_1,z_2)} \tag{11}$$

If $(z_1,z_2)=(0,0)$, then (11) equals zero and by checking with (2)-(3)and recalling the definition of E_{M-1}, we see that also the left-hand side of (10) equals zero, thus,

(10) is true in this case.
If $z_1 = 0,\ z_2 \neq 0$ then, by checking with (2)-(3) , we see that the left-hand side of (10), becomes

$$E_{M-1}(0, z_2) = \mathcal{O}(z_2^{m_2}),\ z_2 \to 0$$

i.e. it reduces to the one-variable case. On the other hand, the expression in the brackets on the right-hand side reduces to

$$\frac{(z_2 w_2)^{m_2} Q_M(0, \frac{1}{w_2})}{Q_M(0, z_2)}$$

and thus, the right-hand side will be equal to

$$\frac{1}{(2\pi i)^2} \int\!\!\int_{\Gamma_1^{-1}\Gamma_2^{-1}} \frac{\tilde{f}(w_1, w_2)}{(1 - w_2 z_2)} \frac{(z_2 w_2)^{m_2} Q_M(0, \frac{1}{w_2})}{Q_M(0, z_2)} dw_1 dw_2 \tag{12}$$

But since this an iterated integral, keeping z_2 and w_2 fixt, we can use Cauchy's formula in one variable and thus, using (7), we can reduce (12) to

$$\frac{1}{2\pi i} \int_{\Gamma_2^{-1}} \frac{-f(0, \frac{1}{w_2})}{(1 - w_2 z_2) w_2} \frac{(z_2 w_2)^{m_2} Q_M(0, \frac{1}{w_2})}{Q_M(0, z_2)} dw_2$$

If we introduce the functions

$$f_2(z_2) := f(0, z_2), \qquad \tilde{f}_2(z_2) := \frac{1}{z_2} f_2(\frac{1}{z_2})$$

$$q_{m_2}(z_2) := Q_M(0, z_2), \qquad \tilde{q}_{m_2}(z_2) := z_2^{m_2} q_{m_2}(\frac{1}{z_2})$$

then, finally, the right-hand side of (10) will be equal to

$$-\frac{1}{2\pi i} \int_{\Gamma_2^{-1}} \frac{\tilde{f}_2(w_2)}{(1 - w_2 z_2)} \frac{\tilde{q}_{m_2}(w_2)}{\tilde{q}_{m_2}(\frac{1}{z_2})} dw_2 = \frac{1}{2\pi i} \int_{-\Gamma_2^{-1}} \frac{\tilde{f}_2(w_2)}{(1 - w_2 z_2)} \frac{\tilde{q}_{m_2}(w_2)}{\tilde{q}_{m_2}(\frac{1}{z_2})} dw_2$$

It can be shown that $-\Gamma_2^{-1}$ satisfies the index condition in ([1], Th1), and thus, we have received the one-variable case, which has already been investigated by Cala-Rodriguez and Wallin. The case $z_1 \neq 0, z_2 = 0$ follows in the same way.
Thus, we may assume that $z_1, z_2 \neq 0$. Using the definition of E_{M-1}, (3) and Cauchy's formula, it can be shown (see for instance [5]) that E_{M-1} satisfies

$$E_{M-1}(z_1, z_2)$$
$$= \frac{1}{(2\pi i)^2} \int\!\!\int_{\Gamma_1\Gamma_2} \frac{(fQ_M)(w)}{(w - z)Q_M(z)} \left[\left(\frac{z_1}{w_1}\right)^{m_1} + \left(\frac{z_2}{w_2}\right)^{m_2} - \left(\frac{z_1}{w_1}\right)^{m_1} \left(\frac{z_2}{w_2}\right)^{m_2} \right] dw \tag{13}$$

The next step is to show that the right-hand side of (13), which also can be written

$$\frac{1}{(2\pi i)^2}\iint_{\Gamma_1\Gamma_2}\frac{f(w)}{(w-z)}\left[\frac{Q_M(w_1,w_2)\left(\frac{z_1}{w_1}\right)^{m_1}+Q_M(w_1,w_2)\left(\frac{z_2}{w_2}\right)^{m_2}-}{Q_M(z_1,z_2)}\right.$$
$$\left.\frac{Q_M(w_1,w_2)\left(\frac{z_1}{w_1}\right)^{m_1}\left(\frac{z_2}{w_2}\right)^{m_2}}{}\right]dw \quad (14)$$

is equivalent to

$$\frac{1}{(2\pi i)^2}\iint_{\Gamma_1\Gamma_2}\frac{f(w)}{(w-z)}\left[\frac{Q_M(w_1,z_2)\left(\frac{z_1}{w_1}\right)^{m_1}+Q_M(z_1,w_2)\left(\frac{z_2}{w_2}\right)^{m_2}-}{Q_M(z_1,z_2)}\right.$$
$$\left.\frac{Q_M(w_1,w_2)\left(\frac{z_1}{w_1}\right)^{m_1}\left(\frac{z_2}{w_2}\right)^{m_2}}{}\right]dw \quad (15)$$

We notice that the two formulas, (14) and (15), differ only in the first two terms, in the brackets, of the integrands. We have to show that the difference between the first terms (and also the second terms) is equal to zero. But this is a consequence of the following two lemmas.

Lemma 1

$$\frac{Q_M(w_1,w_2)-Q_M(w_1,z_2)}{(w_2-z_2)}$$

is a polynomial in w_2 *and* z_2 *for fixt* w_1.

Proof. By definition, Q_M satisfies (2). This implies

$$\begin{aligned} Q_M(w_1,w_2)-Q_M(w_1,z_2) &= \sum_{j=1}^{m_2}(w_2^j-z_2^j)\cdot\sum_{i=0}^{m_1}b_{ij}w_1^i \\ &= (w_2-z_2)\sum_{i=0}^{m_1}b_{i1}w_1^i+\ldots+(w_2^{m_2}-z_2^{m_2})\sum_{i=0}^{m_1}b_{im_2}w_1^i \end{aligned}$$

Thus, we get

$$\begin{aligned} \frac{Q_M(w_1,w_2)-Q_M(w_1,z_2)}{(w_2-z_2)} &= \sum_{j=1}^{m_2}\frac{(w_2^j-z_2^j)}{(w_2-z_2)}\sum_{i=0}^{m_1}b_{ij}w_1^i \\ &= \sum_{j=1}^{m_2}(w_2^{j-1}+w_2^{j-2}z_2+\ldots+z_2^{j-1})\sum_{i=0}^{m_1}b_{ij}w_1^i \end{aligned}$$

which, for each fixed w_1 is a polynomial in w_2 and z_2. □

With the help of this lemma we can prove the following.

Lemma 2

$$\iint\limits_{\Gamma_1\Gamma_2} \frac{f(w)}{(w-z)} \left[Q_M(w_1,w_2) - Q_M(w_1,z_2)\right] \left(\frac{z_1}{w_1}\right)^{m_1} dw = 0 \tag{16}$$

Proof. Lemma 1 gives us the property that $f(w_1,w_2)\left[\frac{Q_M(w_1,w_2)-Q_M(w_1,z_2)}{(w_2-z_2)}\right]$ is analytic inside Γ_2, for each fixed w_1. Since the integral in (16) is an iterated integral we get, by using Fubini's theorem and Cauchy's theorem

$$\int_{\Gamma_1} \left[\int_{\Gamma_2} f(w_1,w_2) \frac{[Q_M(w_1,w_2) - Q_M(w_1,z_2)]}{(w_2-z_2)} dw_2\right] \frac{\left(\frac{z_1}{w_1}\right)^{m_1}}{(w_1-z_1)} dw_1 = 0$$

which proves the lemma. □

Using the same procedure we can also prove

$$\iint\limits_{\Gamma_1\Gamma_2} \frac{f(w)}{(w-z)} \left[Q_M(w_1,w_2) - Q_M(z_1,w_2)\right] \left(\frac{z_2}{w_2}\right)^{m_2} dw = 0$$

which proves that (14) and (15) are equivalent and, thus, the first part of the proof, of the theorem, is completed. The second and last part of the proof is done by making a change of variables. The correct choice is to put $\tilde{w}_1 = \dfrac{1}{w_1}$ and $\tilde{w}_2 = \dfrac{1}{w_2}$. If we use the notations (7)-(9) then (15), which can be written as

$$\frac{1}{(2\pi i)^2} \iint\limits_{\Gamma_1\Gamma_2} \frac{f(w_1,w_2)}{(w-z)} \left[\frac{\frac{Q_M(w_1,z_2)}{w_1^{m_1} z_2^{m_2}} + \frac{Q_M(z_1,w_2)}{z_1^{m_1} w_2^{m_2}} - \frac{Q_M(w_1,w_2)}{w_1^{m_1} w_2^{m_2}}}{Q_M(z_1,z_2)\frac{1}{z_1^{m_1} z_2^{m_2}}}\right] dw_1 dw_2$$

is equal to

$$\frac{1}{(2\pi i)^2} \iint\limits_{\Gamma_1\Gamma_2} \frac{\tilde{f}(\frac{1}{w_1},\frac{1}{w_2})}{(1-\frac{z_1}{w_1})(1-\frac{z_2}{w_2})} \left[\frac{\tilde{Q}_M(\frac{1}{w_1},\frac{1}{z_2}) + \tilde{Q}_M(\frac{1}{z_1},\frac{1}{w_2}) - \tilde{Q}_M(\frac{1}{w_1},\frac{1}{w_2})}{\tilde{Q}_M(\frac{1}{z_1},\frac{1}{z_2})}\right] dw_1 dw_2$$

$$= \frac{1}{(2\pi i)^2} \iint\limits_{\Gamma_1^{-1}\Gamma_2^{-1}} \frac{\tilde{f}(\tilde{w}_1,\tilde{w}_2)}{(1-\tilde{w}_1 z_1)(1-\tilde{w}_2 z_2)}$$

$$\left[\frac{\tilde{Q}_M(\tilde{w}_1,\frac{1}{z_2}) + \tilde{Q}_M(\frac{1}{z_1},\tilde{w}_2) - \tilde{Q}_M(\tilde{w}_1,\tilde{w}_2)}{\tilde{Q}_M(\frac{1}{z_1},\frac{1}{z_2})}\right] d\tilde{w}_1 d\tilde{w}_2$$

$$= \frac{1}{(2\pi i)^2} \iint\limits_{\Gamma_1^{-1}\Gamma_2^{-1}} \frac{\tilde{f}(\tilde{w})}{(1-\tilde{w}z)} \left[\frac{\tilde{Q}_M(\tilde{w}_1,\frac{1}{z_2}) + \tilde{Q}_M(\frac{1}{z_1},\tilde{w}_2) - \tilde{Q}_M(\tilde{w}_1,\tilde{w}_2)}{\tilde{Q}_M(\frac{1}{z_1},\frac{1}{z_2})}\right] d\tilde{w}$$

After changing names of the variables we have proved the assertion. □

3.2. Daras' theorem on Padé-type approximants

In this section we are going to use Theorem 2 to prove Theorem 1. Since we do not make the assumption that Q_M satisfies (6) we will have a little bit more freedom in the choice of generating polynomials. "Our" version of Theorem 1 will be the following

Theorem 3 *Let $f \in \mathcal{H}(\mathbf{P}(0,r))$. Let b_{ij}, be given complex numbers such that Q_M given by (2) satisfies:*

$$\lim_{\substack{m_1\to\infty\\ m_2\to\infty}} \frac{\tilde{Q}_M(z_1^{-1},w_2)+\tilde{Q}_M(w_1,z_2^{-1})-\tilde{Q}_M(w_1,w_2)}{\tilde{Q}_M(z_1^{-1},z_2^{-1})} = 0 \tag{17}$$

uniformly for (w_1,w_2,z_1,z_2) on any compact subset of an open set $\Lambda \subseteq \mathbb{C}^4$, containing $\mathbb{C}^2 \times \{0,0\}$. Then

$$\lim_{\substack{m_1\to\infty\\ m_2\to\infty}} (P_{M-1}/Q_M)(z_1,z_2) = f(z_1,z_2)$$

uniformly on any compact subset of

$$\mathcal{F}(\Lambda,\mathbf{P}) := \left\{(z_1,z_2)\in\mathbb{C}^2: \quad (\xi_1^{-1},\xi_2^{-1},z_1,z_2)\in\Lambda, \quad \forall \xi_i \in \bar{\mathbb{C}}\setminus\mathbf{P}_i, \quad i=1,2\right\}$$

Proof. It can be shown (see [2],Prop 1.1 ii)) that $\mathcal{F}(\Lambda,\mathbf{P})$ is an open subset of $\mathbb{C}^2$. Then, we observe that $\mathcal{F}(\Lambda,\mathbf{P}) \subset \mathbf{P}$ since if there exists a point $z^0 = (z_1^0,z_2^0) \in \mathcal{F}(\Lambda,\mathbf{P})\setminus\mathbf{P}$ then $z_i^0 \in \mathbb{C}\setminus\mathbf{P}_i$, $i=1,2$ and thus, by the definition of $\mathcal{F}(\Lambda,\mathbf{P})$, $((z^0)^{-1},z^0)\in\Lambda$ which contradicts (17). Take a point $z^0 \in \mathcal{F}(\Lambda,\mathbf{P})$ and introduce the sets

$$K_i := (\bar{\mathbb{C}}\setminus\mathbf{P}_i)^{-1} = \bar{\mathbb{C}}\setminus\mathbf{P}_i^{-1}, \quad i=1,2$$
$$K := K_1\times K_2$$

K is a compact set in $\mathbb{C}^2$ containing $(0,0)$ and we also notice that $\xi_i \in \bar{\mathbb{C}}\setminus\mathbf{P}_i$ if and only if $\xi_i^{-1}\in K_i$, $i=1,2$. Thus, $(\xi_1^{-1},\xi_2^{-1},z_1^0,z_2^0)\in\Lambda$ if $\xi_i^{-1}\in K_i$, i.e.

$$K\times\{z_1^0,z_2^0\}\subset\Lambda$$

Since $K\times\{z_1^0,z_2^0\}$ is a compact in $\mathbb{C}^4$ and Λ is open in $\mathbb{C}^4$, there exists an open polydisc $\mathcal{O}=\mathcal{O}(0,\rho)\subset\mathbb{C}^2$ and a closed polydisc $\bar{\Delta}=\bar{\Delta}(z^0,\epsilon)\subset\mathbb{C}^2$ so that

$$\mathcal{O}\times\bar{\Delta}\subset\Lambda \tag{18}$$

We choose $\mathcal{O}$ and $\bar{\Delta}$ so small that $\bar{\Delta}\subset\mathbf{P}$ and $\bar{\Delta}^{-1}\subset\bar{\mathbb{C}}^2\setminus\mathcal{O}$; the last condition is possible because $\mathbf{P}_i^{-1}$ is open and $(z_i^0)^{-1}\in\mathbf{P}_i^{-1}$. Now we choose circles C_i such that $C_i\in\mathcal{O}_i\setminus K_i$, $i=1,2$ where $\mathcal{O}_i := \{z_i\in\mathbb{C}: \ |z_i|<\rho_i\}$, some ρ_i, i.e.

$$C_i := \{z_i\in\mathbb{C}: \ |z_i|=\tau_i\} \quad \text{where} \quad \tau_i : \frac{1}{r_i}<\tau_i<\rho_i$$

By (18), $C_1\times C_2\times\bar{\Delta}$ is a compact subset of Λ and we can use Theorem 2 with C_i^{-1} as Γ_i and $z\in\bar{\Delta}$ to conclude that if (17) is satisfied then E_{M-1} converges uniformly to zero on $\bar{\Delta}$. □

4. Other convergence results

From the integral representation of the error in Theorem 2 and the proof of Theorem 3 it is easy to prove alternative convergence results. Again, assume that $f \in \mathcal{H}(\mathbf{P})$, $\mathbf{P} = \mathbf{P}(0, r)$, Q_M and $\tilde{Q}_M$ are as before, and that P_{M-1}/Q_M is the P.T.A, of f, of type $(M - 1/M)$.

Proposition 1 *Assume that , for some bounded neighbourhoods $\mathcal{O}_i$ of K_i, where $K_i = \bar{\mathbb{C}} \setminus \mathbf{P}_i^{-1}$, $K = K_1 \times K_2$ and some compact product set $F = F_1 \times F_2$, $F \subset \mathbf{P}$,*

$$\lim_{\substack{m_1\to\infty\\ m_2\to\infty}} \left(\max_{\substack{w\in\bar{\mathcal{O}}\\ z\in F}} \left| \frac{\left[\tilde{Q}_M(z_1^{-1}, w_2) + \tilde{Q}_M(w_1, z_2^{-1}) - \tilde{Q}_M(w_1, w_2)\right]}{\tilde{Q}_M(z_1^{-1}, z_2^{-1})} \right| \right) = 0 \qquad (19)$$

Then

$$\lim_{\substack{m_1\to\infty\\ m_2\to\infty}} (P_{M-1}/Q_M)(z_1, z_2) = f(z_1, z_2)$$

uniformly on F.

Proof. The set F^{-1} is closed in $\bar{\mathbb{C}}^2$ and by (19), each F_i^{-1} has a positive distance to $\bar{\mathcal{O}}_i$. Consequently, it is possible to find $\Gamma_i \subset \mathcal{O}_i \setminus K_i$ such that we can use Theorem 2 for all $z = (z_1, z_2) \in F$, and the proposition is proved. □

Proposition 2 *Assume that $\mathcal{O}$ and F as in Proposition 1, E_{M-1} defined as in Section 2 and*

$$\limsup_{\substack{m_1\to\infty\\ m_2\to\infty}} \left(\max_{\substack{w\in\bar{\mathcal{O}}\\ z\in F}} \left| \frac{\left[\tilde{Q}_M(z_1^{-1}, w_2) + \tilde{Q}_M(w_1, z_2^{-1}) - \tilde{Q}_M(w_1, w_2)\right]}{\tilde{Q}_M(z_1^{-1}, z_2^{-1})} \right|^{1/m_1\cdot m_2} \right) \le r < 1$$

Then

$$\limsup_{\substack{m_1\to\infty\\ m_2\to\infty}} \left(\max_{z\in F} |E_{M-1}(z_1, z_2)|^{1/m_1\cdot m_2} \right) \le r \qquad (20)$$

Proof. The proof is done in the same way as in Proposition 1. □

Using Proposition 2 we can give the following example

Example If Q_M satisfies (6) where $\beta_{i,m_i}^{(j)} = \beta_i$, $i = 1, 2$ then one can show that

$$\limsup_{\substack{m_1\to\infty\\ m_2\to\infty}} \left(\max_{w\in\bar{\mathcal{O}}} \left| \frac{\left[\tilde{Q}_M(z_1^{-1}, w_2) + \tilde{Q}_M(w_1, z_2^{-1}) - \tilde{Q}_M(w_1, w_2)\right]}{\tilde{Q}_M(z_1^{-1}, z_2^{-1})} \right|^{1/m_1\cdot m_2} \right)$$

$$\le \max_{w\in\bar{\mathcal{O}}} \left(\max\left(\frac{|w_1 - \beta_1|}{|z_1^{-1} - \beta_1|}, \frac{|w_2 - \beta_2|}{|z_2^{-1} - \beta_2|}\right) \right)$$

if z_i^{-1} is different from β_i. When $\bar{O}_i$ shrinks to K_i, the right-hand side becomes

$$\max_{w\in K}\left(\max(\frac{|w_1-\beta_1|}{|z_1^{-1}-\beta_1|},\frac{|w_2-\beta_2|}{|z_2^{-1}-\beta_2|})\right) = \max_{\substack{(\xi_1,\xi_2)\\ \xi_i\in\bar{\mathbb{C}}\setminus\mathbf{P}_i}}\left(\max(\frac{|\xi_1^{-1}-\beta_1|}{|z_1^{-1}-\beta_1|},\frac{|\xi_2^{-1}-\beta_2|}{|z_2^{-1}-\beta_2|})\right)$$

which gives the right value of r in (20) with $F = \{(z_1, z_2)\}$. In particular, the Padé-type approximants converge to f at $z = (z_1, z_2)$ if $z^{-1} = (z_1^{-1}, z_2^{-1})$ lies outside the polydisc with center in $\beta = (\beta_1, \beta_2)$ and polyradius $\rho = (\rho_1, \rho_2)$ where

$$\rho_i = \max_{\xi_i\in\bar{\mathbb{C}}\setminus\mathbf{P}_i} |\xi_i^{-1} - \beta_i|$$

References

1. F.Cala-Rodriguez, H.Wallin *Padé-type approximants and a summablitity theorem by Eiermann* J.Comput.Appl.Math 39 (1992) 15-21
2. N.Daras *The convergence of Padé-type approximants to holomorphic functions of several variables* Appl.Num.Math 6 (1989/1990) 341-360
3. M.Eiermann *On the convergence of Padé-type approximants to analytic functions* J.Comput.Appl.Math. 10 (1984) 219-227
4. W.Gawronski, R.Trautner, *Verschärfung eines Satzes von Borel-Okada über Summierbarkeit von Potenzreihen*, Period.Math.Hungar.7 (1976) 201-211
5. R.Orive, P.Gonzalez-Vera *On convergence of certain multivariate rational approximants* J.Inst.Math.Comp.2 (1991) 31-40
6. Y.Okada, *Über die Annäherung analytischer Funktionen*, Math.Z. 23 (1925) 62-71

ON THE ZEROS OF PADÉ APPROXIMANTS AND PADÉ ERROR FUNCTIONS FOR SOME CLASSES OF FUNCTIONS

RALITZA K. KOVACHEVA*
Institute of Mathematics
Bulgarian Academy of Sciences

Abstract. In the present paper, we deal with functions $f(z) := \sum_{n=0}^{\infty} a_n z^n$ whose coefficients satisfy the smoothness condition $a_{j+1} \cdot a_{j-1}/a_j^2 \to \eta$ as $j \to \infty$, $\eta \neq \infty$. Theorems concerning the asymptotic behaviour as $n \to \infty$, m - fixed, of the normalized in an appropriate way Padé approximants $\pi_{n,m}$ and of the normalized Padé error functions $(f - \pi_{n,m})$ are provided. As a consequence, results concerning the limiting distribution of the zeros of both sequences are deduced.

Key words: Padé approximants, Padé error function

1. Introduction and main Results

Let

$$f(z) := \sum_{j=0}^{\infty} a_j z^j \tag{1}$$

be a function with $a_j \neq 0$ for all nonnegative integers j, $(j \in \mathbf{N})$ large enough. We set

$$\eta_j := a_{j+1} \cdot a_{j-1}/a_j^2, \qquad j = j_0, j_1, \ldots.$$

The basic assumption throughout the present work is that

$$\eta_j \to \eta, \text{ as } j \to \infty. \tag{2}$$

We note that (1) represents an entire function of order zero, when $\eta < 1$; when $\eta = 1$, the radius $\rho(f)$ of convergence is zero, finite or infinite. When $\eta > 1$, the series (1) has zero radius of convergence.

Further, we assume that the numbers η_j tend to η in a prescribed "smooth way", namely there exist complex numbers $\{c_i\}_{i=1}^{\infty}$ with $c_1 \neq 0$ such that for each positive integer $N, N > 1$, the representation

$$\eta_n = \eta \cdot (1 + c_1/n + \sum_{i=2}^{N} c_i/n^i + o(n^{-N})). \tag{3}$$

holds.

This kind of convergence has been introduced and studied by D. Lubinsky in [1], where important theorems resulting from (3) with respect to the asymptotic of

* This work is supported by Bulgarian Ministery for Science and Education Grant 69-91

A. Cuyt (ed.), Nonlinear Numerical Methods and Rational Approximation II, 229–236.

Toeplitz determinants and uniform convergence of the row in the table of classical Padé approximants are proved. Therefore, in what follows condition (3) will be called "Lubinsky's smoothness condition".

Introduce the function $H_\eta(z)$ with

$$H_\eta(z) := \sum_{j=0}^{\infty} \eta^{j(j+1)/2} z^j .$$

It is clear that for $\eta < 1, H_\eta(z)$ is an entire function. If $\eta = 1$, then $H_\eta(z)$ is holomorphic in the unit disk; in the case when $\eta > 1$, the radius of convergence is zero.

Note that $H_\eta(z) = h(\eta z)$, where $h(z)$ is the partial theta function. Its properties (natural boundary, domains omiting zeros ect) have been studied in [2].

Denote by $S_n(z) = S_n(f, z)$ the n-th partial sum of the function $f(z)$:

$$S_n(z) := \sum_{j=0}^{n} a_j z^j .$$

The first result in the present paper refers to the asymptotic behaviour as $n \to \infty$ of the normalized in an appropriate way sequences of the partial sums $\{S_n\}$ and of the differences $\{f - S_n\}$.

Set

$$Z_n(u) := S_n(ua_n/a_{n+1})/a_n(ua_n/a_{n+1})^n$$

and

$$W_n(u) := (f - S_n)(ua_n/a_{n+1})/a_{n+1}(ua_n/a_{n+1})^{n+1}.$$

Theorem 1: *Assume that (2) holds with (i): $\eta < 1$ and (ii): with $\eta = 1$ in a way that $\eta_n \leq 1$ for all positive integers n large enough.*

Then

i:)

$$lim Z_n(u) \to H_\eta(1/u), \tag{4i}$$

as $n \to \infty$ locally uniformly in $\bar{\mathbf{C}} - \{O\}$ and

$$W_n(u) \to H_\eta(u) \tag{5i}$$

locally uniformly in $\mathbf{C}$;

ii:)

$$Z_n(u) \to H_\eta(1/u), \tag{4ii}$$

locally uniformly in $\{u, u > 1\}$ and

$$W_n(u) \to H_\eta(u) \tag{5ii}$$

locally uniformly in $\{u, u < 1\}$.

As usual, "locally uniformly" means an uniform convergence on compact subsets in the metric of Chebyshev.

Theorem 1 was proved by E. B. Saff (see [3], Theorem 1.1).

Let now m be a fixed positive integer. In our further considerations, we shall assume that f is holomorphic at the zero (in a neighbourhood) and is not a rational function having less or equal than m finite poles (multiplicities included) in $\mathbf{C}$ (we write $f \notin \mathcal{R}_m$).

For each $n, n \in \mathbf{N}$, let $\pi_{n,m}(= \pi_{n,m}(f))$ be the Padé approximant to the function f of order (n, m). Recall that $\pi_{n,m} = p/q, \deg p \leq n, \deg q \leq m, q \not\equiv 0$, where the polynomials p and q are determined by the condition $(f \cdot q - p)(z) = O(z^{n+m+1})$. For each pair (n, m) the function $\pi_{n,m}$ always exists and is uniquely determined (see, for example, [4]). We set

$$\pi_{n,m} = P_{n,m}/Q_{n,m},$$

where $Q_{n,m}(0) = 1$ and both polynomials $P_{n,m}$ and $Q_{n,m}$ do not have a common divisor.

Let $D(n, m) = \det\{a_{n-j+k}\}_{j,k=1}^{m}$ be the Toeplitz determinant formed from the Maclaurin coefficients of the function f. From the nonrationality of f, it follows that the sequence Λ of those positive integers n for which $D(n, m) \cdot D(n + 1, m + 1) \neq 0$, is infinite (see, [5], [6]) and the equalit y $\pi_{n,m} \equiv \pi_{k(n),m}$, where $k(n) := \max\{k, k \leq n, k \in \Lambda\}$ is valid. Without loss of generality we shall assume that $\Lambda \equiv \mathbf{N}$. In this case there holds (see [4])

$$(f.Q_{n,m} - P_{n,m})(z) = z^{n+m+1} \cdot (-1)^m \cdot D(n + 1, m + 1)/D(n, m) + \ldots,$$

$$Q_{n,m}(z) = 1 + \ldots + z^m \cdot (-1)^m D(n + 1, m)/D(n, m)$$

and

$$P_{n,m}(z) = z^m \cdot D(n, m + 1)/D(n, m) + \ldots + d_{n,m}.$$

In what follows, we shall call the difference $f - \pi_{n,m}$ the Padé error function.

Denote by $E_{n,m}(u)$ the error function normalized as follows:

$$E_{n,m}(u) := \frac{(f - \pi_{n,m})(ua_n/a_{n+1})}{(ua_n/a_{n+1})^{n+m+1} \cdot (-1)^m \cdot D(n + 1, m + 1)/D(n, m)}$$

The next result supplies information about the asymptotic behaviour as $n \to \infty$ of the sequence $E_{n,m}(u)$ in the case when the numbers η_n tend to $\eta = 1$ and satisfy Lubinsky's smoothness condition (3).

Theorem 2: *Let $m \in \mathbf{N}$ be fixed and $f \notin \mathcal{R}_m$. Assume that $a_j \neq 0$ for j large enough; assume, further that η_n admits the expansion (3) with $\eta = 1$, $c_1 \neq O$ and $\eta_n \leq 1$ for all $n \in \mathbf{N}$ sufficiently large.*

Then

$$E_{n,m}(u) \to (1 - u)^{-2m-1} \text{ as } n \to \infty$$

locally uniformly in $\{u, u < 1\}$.

From Theorem 2, we have

Corollary 1: *With the assumptions of Theorem 2, for each fixed $m \in \mathbf{N}$ and any $\varepsilon, 0 < \varepsilon < 1$, the Padé error function $(f - \pi_{n,m})(z)$ has no zeros in $0 < z < a_n/a_{n+1} \cdot (1 - \varepsilon)$ for n sufficiently large.*

We note that in the case being considered the normalized Padé error functions $E_{n,m}(u)$ are, starting with a number n_0, holomorphic in the unit disk D_1 (see (6) and (7) below) and the unit circle is the natural boundary (m is fixed). Combining Corollary 1 and this observation, we conclude that all the nonzero zeros of the error functions (provided there are some) are situated in the annulus $\mathcal{E}_{n,m}(\varepsilon) := a_n/a_{n+1} \cdot (1-\varepsilon) < z < a_n/a_{n+1}$ for n sufficiently large.

In [3], it was shown that

Corollary A: *With the assumption of Theorem 2, for each fixed $m \in \mathbf{N}$ and any positive ε, the Padé approximant $\{\pi_{n,m}(z)\}$ has no zeros $z > a_n/a_{n+1} \cdot (1+\varepsilon)$ for n sufficiently large.*

The next results provide more precise information concerning the behaviour of the zeros of the sequence of the normalized Padé error functions $E_{n,m}(u)$ as $n \to \infty$ for the special when the first coefficient c_1 in (3) is a real negative number.

Theorem 3: *If $c_1 < 0$, then $u = 1$ is a limit point of zeros of $\{E_{n,m}(u)\}_{n=1}^{\infty}$.*

From here, we obviously have

Corollary 2: *In the conditions of Theorem 3, any point u with $u = 1$ is a limit point of zeros of $\{E_{n,m}(u)\}_{n=1}^{\infty}$.*

In [3], the validity of the same result with respect to the zeros of Padé approximants $\pi_{n,m}$, (m-fixed) was established (see Theorem 1.4).

For any $\varepsilon, 0 < \varepsilon < 1$, denote by $\mathcal{P}_{n,m}(\varepsilon)$ the annulus $a_n/a_{n+1} \cdot (1-\varepsilon) < z < a_n/a_{n+1} \cdot (1+\varepsilon)$.

Combining Theorem 3 and the result mentioned above, we come to

Corollary 3: *In the conditions of Theorem 3, for each fixed $m \in \mathbf{N}$, any $\varepsilon, 0 < \varepsilon < 1$, and n large enough*

a): the Padé approximant $\pi_{n,m}$ has at least one zero in the annulus $\mathcal{P}_{n,m}(\varepsilon)$ and

b): the Padé error function $(f - \pi_{n,m})$ has at least one zero in the annulus $\mathcal{E}_{n,m}(\varepsilon)$.

Assuming in addition f to be an entire function, we conclude that in the conditions of Theorem 3 there is a sequence $\Xi, \Xi \subset \mathbf{N}$, such that all the nonzero zeros of the error functions $(f - \pi_{n,m})(z)$ go to infinity as $n \to \infty, n \in \Xi$ at the same rate as the corresponding sequence $\{a_n/a_{n+1}\}, n \in \Xi$.

Set

$$R := \liminf_{n \to \infty} a_n/a_{n+1}.$$

Coming back to the last considerations, we notice that if $R = \infty$ and f is an entire function, then $\lim_{n\to\infty} a_n/a_{n+1} = \infty$ and $\Xi \equiv \mathbf{N}$.

Further, if $0 < R < \infty$, (recall that in accordance with the general assumption concerning f we have $\rho_0(f) > 0$) there is an infinite sequence $\Theta, \Theta \subset \mathbf{N}$ such that all the zeros of $\pi_{n,m}$ have their accumulation points on the closure of the disk $D_R := \{z, z < R\}$. On the other hand, it has been established in [1] that $\pi_{n,m} \to f$ as $n \to \infty$ locally uniformly inside D_R, so that any compact set in D_R contains not more than finitely many accumulation points of the zeros of $\pi_{n,m}$. Consequently, the circle $\{z, z = R\}$ attracts all the zeros (except a finite number) of $\pi_{n,m}$, as $n \in \Theta$. At the same time, $D_R - \{0\}$ does not contain, in view of Theorem 2, accumulation points of the zeros of $(f - \pi_{n,m})$ as $n \to \infty$ (recall that each Padé error function has a zero at $z = 0$ of order $m+n+1$). If in addition the assumptions of Theorem 3 are fulfilled, then $\{z, z = R\}$ contains all the accumulation points of the zeros of $(f - \pi_{n,m})(z)$ as $n \to \infty, n \in \Theta$. If $R = \limsup_{n\to\infty} a_n/a_{n+1} > 0$, then $R = \rho(f)$ and $\Theta \equiv \mathbf{N}$ (compaire Corollary 3).

Important functions to which Theorem 3 may be applied are the exponential function

$$f(z) = \exp z = \sum_{j=0}^{\infty} z^j / j|$$

and the Mittag-Leffler function of order $\lambda, \lambda > 0$,

$$f(z) = \sum_{j=0}^{\infty} z^j / \Gamma(1 + j/\lambda), \qquad \lambda > 0.$$

Following [2], we introduce the polynomials $B_m(u) := B_m(u, q), m \in \mathbf{N}$ fixed, as follows:

$B_o(u) := 1$ and for $m = 1, 2, \ldots$

$$B_m(u) := B_{m-1}(u) - u \cdot q^{m-1} \cdot B_{m-1}(u/q).$$

When q is not a root of unity, then

$$B_m(-u) = \sum_{j=0}^{m} \frac{u^j \prod_{k=1}^{j}(1 - q^{m+1-k})}{\prod_{k=1}^{j}(1 - q^k)};$$

furthermore, $B_m(u) = (1 - u)^m$, when $q = 1$.

These polynomials are of importance in the investigation of the distribution of the zeros of Padé error functions $f - \pi_{n,m}$ and Padé approximants $\pi_{n,m}$ in the case when the number η in (3) is not a root of unity.

For $0 < q < 1$, the polynomials B_m (suitably normalized) are orthogonal with respect to a nonnegative weight on the unit circle (see [7]), so that all their zeros lie in $\{z, z \leq 1\}$. For $q = 1$ results concerning the distribution of the zeros of the polynomials $B_m(u), m = 0, 1, \ldots$ can be found in [2].

For our goal, we introduce an appropriate normalization of the error functions. Set

$$e_{n,m} := \frac{(f - \pi_{n,m})(ua_n/a_{n+1})/a_{n+1}}{(ua_n/a_{n+1})^{n+1}}.$$

The following theorem is valid:

Theorem 4: *Assume that (3) holds for a number η with $\eta^{m_o+1} = 1$ for a number $m_0, m_0 \geq 1$ and $\eta^m \neq 1$ for $m = 1, \ldots, m_o,$. Then for any $m, m \leq m_0$ there holds*

$$e_{n,m}(u) \to H_\eta(u) + \sum_{k=0}^{m-1} \frac{\prod_{j=1}^{k}(1 - \eta^j)(-1)^{k+1} u^k}{B_k(u) B_{k+1}(u)} \quad as\ n \to \infty$$

locally uniformly in the domains described by Theorem 5i),5ii) excluding $\mathcal{B}$, where $\mathcal{B}$ is the set of the zeros of the polynomials $B_k = B_k(z, \eta), k = 1, \ldots, m$

The analogous result concerning the zeros of Padé approximant is (see [3]):

Corollary B: *In the conditions of Theorem 4 it is valid that*

$$\lim \frac{P_{n,m}(ua_n/a_{n+1})}{a_n(ua_n/a_{n+1})^n} = B_m(u) \cdot \{H_\eta(1/u) - \sum_{k=0}^{m-1} \frac{(-u)^m \prod_{j=1}^{k}(1 - \eta^j)}{B_k(u) B_{k+1}(u)}\}$$

for an appropriate domain described by Theorem 1,4i), 4ii).

Set $\delta(m,\eta) := \min\{z, B_m(z,\eta) = 0, k = 1,\ldots,m\}$.

In accordance with the results of [2] and in view of Theorem 4, we may write

Corollary 4: *With the assumptions of Theorem 4, for any* $\varepsilon, 0 < \varepsilon < 1$, *the Padé error function* $f - \pi_{n,m}$ *has for n sufficiently large not more than a finite number of zeros in* $0 < z < \delta(m,\eta)(1-\varepsilon)\cdot a_n/a_{n+1}$.

2. Proofs of the Results

Set

$$D_{n,m} := \frac{D(n,m)}{a_n^m}$$

The following lemma is of essential importance for all the considerations in the present paper.

Lemma 1 (see [1]): *Let f be a formal power series, wi th* $a_j \neq O$ *for j large enough. Assume that* η_j *has the asymptotic expansion (3) with* $c_1 \neq 0$. *Then for* $m = 1, 2, \ldots$ *we have*

$$D_{n,m} = (-c_1/n)^{m(m-1)/2} \cdot \prod_{j=1}^{m-1} j^{m-j} \cdot \{1 + \alpha(1,m)/n + o(1/n)\} \text{ as } n \to \infty.$$

If (3) holds for a number η *that is not a root of unity then*

$$\lim_{n\to\infty} D(n,m)/a_n^m = \prod_{j=1}^{m-1} (1-\eta^j)^{m-j}$$

and

$$\lim_{n\to\infty} Q_{n,m}(ua_n/a_{n+1}) = B_m(u).$$

The proofs of Theorem 2 and Theorem 3 will appear in [8], so that in the present paper we will confine ourselves on sketching the general ideas only.

For any n, the normalized Padé error function $E_{n,m}$ could be rewritten in the form

$$E_{n,m}(u) = \frac{1 + \sum_{j=1}^{\infty} d_{n,j} F_{n,m,j} u^j}{Q_{n,m}(ua_n/a_{n+1}),}. \tag{6}$$

In the last formula

$$d_{n,j} := \prod_{l=1}^{j} \eta_{n+l}^{j+1-l}$$

and

$$F_{n,m,j} =$$

$$(-1)^m \frac{D_{n,m}}{D_{n+1,m+1}} \cdot \sum_{k=0}^{m} \prod_{l=1}^{j+1} \eta_{n+l}^{m-k} \prod_{l=j-2}^{m-j-k} \eta_{n+l}^{m+j+1-k-l} \cdot \frac{a_n^k}{a_{n+1}^k} \cdot q_{m-k,n,m},$$

where the numbers $q_{m-k,n,m}$ are the coefficients of the polynomial $Q_{n,m}$, namely

$$Q_{n,m}(z) := 1 + \sum_{k=1}^{m} q_{m-k,n,m} \cdot z^k .$$

The following lemma describes the asymptotic behaviour of the products $\prod_{l=1}^{j+1} \eta_{n+l}$ as $n \to \infty$. Before presenting it, we introduce for a given function g and a number $p, p \in \mathbf{N}$ the operator

$$\nabla^p g(x) := \sum_{i=0}^{p} (-1)^i \binom{p}{i} g(x-i).$$

Lemma 2 (see [3]): *Assume that η_n is of the form (3) with $\eta = 1$ and $c_1 \neq 0$. Let N be a fixed positive integer.*

a) Then for each $j, j < n/(N+1)$ we have

$$\prod_{l=1}^{j} \eta_{n+l} = 1 + \sum_{s=1}^{N} j \cdot \mathrm{P}_{s-1}(j)/n^s + \mathcal{L}_{N+1}(j,n), \quad \textit{as } n \to \infty,$$

where $\mathrm{P}_s, s = 1, \ldots, N-1$ are polynomials which do not depend on n and j but only on the coefficients $c_s, s = 1, \ldots, N-1$; the degree of each P_s does not exceed s and $\mathrm{P}_1(j) = c_1$;

b) For any nonnegative integer p and for $j < (n-p)/(N+1+2p)$ it is valid that

$$n^{N+p+1} \nabla^p \mathcal{L}_{N+1}(j,n) < C_1(p, N+1) \cdot j^{N+1},$$

where $C_1(p, N+1)$ is a constant that does not depend on j and n.

In the case being considered, the connection between $F_{\ldots,m,\ldots}$ and $F_{\ldots,m-1,\ldots}$ is given by

$$F_{n,m,j} :=$$

$$\frac{-D(n,m)D(n+1,m)}{D(n+1,m+1)D(n,m-1)} \cdot \frac{a_n}{a_{n+1}} \cdot \{F_{n,m-1,j+1} \cdot \prod_{l=1}^{j+1} \eta_{n+l} - F_{n-1,m-1,j+1}\}.$$

Using this formula and Lemma 2, one can show by induction that for $j < n$ there holds asymptotically

$$F_{n,m,j} = \prod_{l=1}^{m} \frac{(j+l)}{m|} + \frac{\mathcal{P}_m(j)}{n} + L_m(j,n), \text{ as } n \to \infty, \tag{7}$$

where $\mathcal{P}_m(j)$ is a polynomial of degree not exceeding m and

$$n^2 \cdot L_m(j,n) < C_2(m) \cdot j^{m+1} \text{ as } n \to \infty.$$

with $C_2(m)$ a positive constant not depending on n.

The proof of Theorem 2 now follows easily.

We prove Theorem 3 arguing on a contrary. Suppose that $u = 1$ is not a limit point of zeros of the sequence $\{E_{n,m}\}$, as $n \to \infty$; then the functions $E_{n,m}^{1/n}$ ($E_{n,m}^{1/n}(0) = 1$) form a normal family in some neighbourhood $\mathcal{U}$ of $u = 1$ which tends uniformly to 1 on the clocure of $\mathcal{U}$.

Then we show that for δ small enough $\Re E_{n,m}(\exp\delta) > \exp(k\delta)$ with k a positive constant.

Proof of Theorem 4:

From the equality (see [4])

$$(\pi_{n,k} - \pi_{n,k+1})(z) = (-1)^{k+1} \frac{z^{n+k+1} D(n+1,k+1)}{Q_{n,k}Q_{n,k+1}(z)D(n,k)}$$

we get

$$(f - \pi_{n,m})(z) = (f - S_n)(z) + \sum_{k=0}^{m-1} (\pi_{n,k} - \pi_{n,k+1})(z)$$

$$= (f - S_n)(z) + \sum_{k=0}^{m-1} (-1)^{k+1} \cdot \frac{D(n+1,k+1)}{D(n,k)} \cdot \frac{z^{n+k+1}}{Q_{n,k}(z)Q_{n,k+1}(z)},$$

so that

$$e_{n,m}(u) = W_n(u) + \sum_{k=0}^{m-1} \frac{(-1)^{k+1} u^k D_{n+1,k+1}}{D_{n,k} \cdot (Q_{n,k}Q_{n,k+1})(ua_n/a_{n+1})}.$$

The statement of Theorem 4 follows now from Lemma 1.

Acknowledgements

The author of this paper is very oblidged to Prof. K. Ivanov for the useful discussions and to Prof. V. Z. Hristov for the help in prepairing and editing the manuscript.

References

1. Lubinsky, D.S. :1987, Uniform Convergence of Rows of Padé Table for Functions with Snooth Maclaurin Coefficients, *Constr. Approx* Vol no **3**, pp. 307–330.
2. Lubinsky, D.S. and Saff, E.B.: 1987, Convergence of Padé Approximants of Partial Theta Function and Rogers-Szego Polynomials, *Constr. Approx.* Vol no**3**, pp. 331–361.
3. Kovacheva, R.K. and Saff, E.B.: 1993, Zeros of Padé Approximants for Entire Functions with Smooth Maclaurin Coefficients, *to appear*
4. Perron, O.: 1957, Die Lehre von den Kettenbruchen, 3rd ed., *Chelsea,* New York.
5. Baker, G.A.: 1975, Essentials of Padé Approximants, *Academic Press,* New York.
6. Baker Jr., G.A. and Graves-Morris, P.: 1981, Padé Approximants, Part 1: Basic Theory, in *Encyclopedia of Mathematics and Its Applications*, Cambridge Univ. Press, Cambridge.
7. Askey, R. and Ismail, M.: 1984, Reccurence Relations, Continued Fractions and Orthogonal Polynomials, *Mem. Amer. Soc.*, Vol. no **49** (300).
8. Kovacheva, R.K.: 1993, Zeros of Padé Error Functions for a Special Class of Functions, *Institute of Mathematics, Bulgarian Academy of Sciences*, Preprint.

LINEARIZATION OF POLYNOMIAL MAPS AND SINGULARITY ANALYSIS WITH EXTENDED PRECISION PADÉ APPROXIMANTS

L. BILLI
SISSA
Via Beirut 4, Miramare 34013 Trieste
ITALY

and

G. TURCHETTI
Dipartimento di Fisica dell'Università
Via Irnerio 46, Bologna
INFN Sezione di Bologna
ITALY

Abstract.
Padé approximants (P.A.) have been recently used to explore the singularity structure of the normalizing transformations of hamiltonian maps. We report here on a systematic investigation based on P.A. to detect the singularities of the analytic diffeomorphisms which linearize a polynomial map. Extended precision algorithms were used to compute high order P.A. (200 decimal digits for [100/100] P.A.), in order to check the stability of the results and to avoid the presence of poles and zeroes on the convergence circle or within it due to the noise introduced by round off.

1. Introduction

Padé approximants have been used for a long time in several areas of theoretical and mathematical physics in order to accelerate the convergence or to sum divergent series namely to provide the analytic continuation of functions known only through their Taylor series expansion [1]. Occasionally the pattern of zeroes and poles of the P.A. has been explored in order to understand the global properties of the function. This is numerically a hard task since high order P.A. need to be computed to obtain recognizable patterns and to check their stability.
Presently a strong motivation comes from the area of dynamical systems where the borderline between regular (*integrable*) and chaotic behaviour is related to the global analyticity properties of the flow in the complex time plane for differential equations, of the normalizing transformations for maps.
There are conjectures relating the non integrability to the presence of natural boundaries and some preliminary numerical explorations based on the use of P.A. for hamiltonian maps seem to confirm these [2–6]. However in this search it is crucial to check the reliability of the numerical results by controlling the effect of round off, which is known to produce natural boundaries on the convergence disk [7–10]. To this end we have chosen to analyze by P.A. the singularity pattern of the linearizing transformations of polynomial maps at an attractive or indifferent fixed point in or-

A. Cuyt (ed.), Nonlinear Numerical Methods and Rational Approximation II, 237–240.

der to explore the global analyticity properties. Using adequate accuracy allowed by the package MPA [11] in order to avoid the presence of zeroes and poles due to noise, it is shown that singularity patterns consisting of families of branch points, with the corresponding cuts, or natural boundaries with fractal structure are reproduced [12–15]

2. Results on Polynomial Maps

We first introduce a few basic definitions and results of holomorphic dynamics in the case of polynomial maps of $\mathbf{C}$ [16]. Let $F : \mathbf{C} \to \mathbf{C}$ be such that

$$F(z) = \lambda z + \sum_{n \geq 2} f_n z^n \tag{1}$$

with $|\lambda| < 1$. An *orbit* is the set $\{z_0, F(z_0), F(F(z_0)), \ldots\}$ for $\forall z_0 \in \mathbf{C}$.
The origin is a *fixed point* and attracts nearby orbits; there is a neighbourhood D of the origin such that $F \circ F \circ \cdots \circ F(D) \to 0$ uniformly when the number n of iterates diverges. In this case there exists a biholomorphic change of coordinate (conjugation) $\zeta = \Psi(z) = z + O(z^2)$ such that the orbits in the new system are the iterations of the linear part of the map, namely $\{\zeta_0, \lambda\zeta_0, \ldots, \lambda^n\zeta_0, \ldots\}$.
The functions Ψ and its inverse Φ satisfy the following functional equations

$$\Psi \circ F(z) = \lambda\Psi(z) \qquad \Phi(\lambda\zeta) = F \circ \Phi(\zeta) \tag{2}$$

We consider the special case of a quadratic map $F(z) = \lambda z + z^2$ where the only attractive points are $z = 0$ and $z = \infty$; the boundary of their basin of attraction is a closed connected set $\mathcal{J}$ known as Julia set. We have computed the P.A. for the Taylor series of the functions Ψ and Φ. The function Ψ is analytic in a domain $\mathcal{F}$, whose boundary is $\mathcal{J}$, and cannot be continued beyond it. We report in fig. 1 the pattern of poles of the [100/100] approximant, computed with 200 digits accuracy, and the set $\mathcal{J}$; in spite of the fractal nature of $\mathcal{J}$, the agreement is good. The scenario changes for the inverse function Φ, which turns out to be analytic on the cut ζ plane. The singularities of Φ are branch points located at $\{\zeta_c, \zeta_c/\lambda^{-1}, \ldots, \zeta_c/\lambda^{-n}, \ldots\}$, where $\zeta_c = \Psi(z_c)$ and $F'(z_c) = 0$. The Padé approximants choose one possible way of cutting the plane. Gluing all the sheets we construct the global Riemann surface; this is one to one with $\mathcal{F}$, which can be tasselated so that each subset is one to one with a sheet [17].
All the poles of the P.A. are on the real half line and their residues all have the same sign suggesting that $\Phi(\zeta)$ is a Stieltjes function. Indeed this can be rigorously proved.
If λ is complex $\lambda = e^{2\pi i p/q}$ where p, q are integers with no common divisor, then the poles (and zeroes) of P.A. are located on q symmetric half lines emerging from a sequence of q points lying on a spiral $\zeta_k = \zeta_c\lambda^{-k}$ for $k = 0, \ldots, q$. This suggests the presence of q cuts, a way of constructing the Riemann function and tasselating $\mathcal{F}$.
If $|\lambda| < 1$ but its phase is $2\pi\omega$ where ω is irrational then the poles accumulate on half lines (corresponding to the denominators of the continued fraction truncations of ω) issued from the points of a spiral $\eta_k = \zeta_c\lambda^{-k}$.

Finally in the limit case of an indifferent fixed point $\lambda = e^{2\pi i\omega}$ all the zeroes and poles of Φ are on the circle of radius $|\zeta_c|$, which is a natural boundary of Φ. The poles and zeroes of Ψ are on a fractal curve, corresponding to the boundary of the Siegel disc.

A crucial point in the previous analysis is that the singularity pattern which turned out always to be consistent with known rigorous results, whenever available, would become unrecognizable if the accuracy is decreased. Indeed the effect of truncation if we use d decimal digits amounts to replace the exact function $\Psi(z)$ with

$$\Psi_\epsilon(z) = \sum_{k=0}^{\infty} \psi_n(1 + \epsilon r_n) z^n \tag{3}$$

where r_n are random numbers in $[-1, 1]$ and $\epsilon = 10^{-d}$ if d decimal digits are used. Since Ψ_ϵ has a natural boundary on the convergence disc of Ψ the $[N/N]$ P.A. will locate zeroes and poles on it as $N \to \infty$ and within it for finite N. As an example we show in fig. 2 the poles of a high order P.A. of Ψ and Φ computed with ordinary accuracy. The difference with fig. 1 for Ψ is evident; in the case of Φ the poles on the noise disc mask the simple pattern of a Stieltjes function.

To conclude, the use of P.A. to explore the presence of natural boundaries of conjugation functions of dynamical systems seems to be founded, provided a very careful analysis is made of the stability of the results using high accuracy.

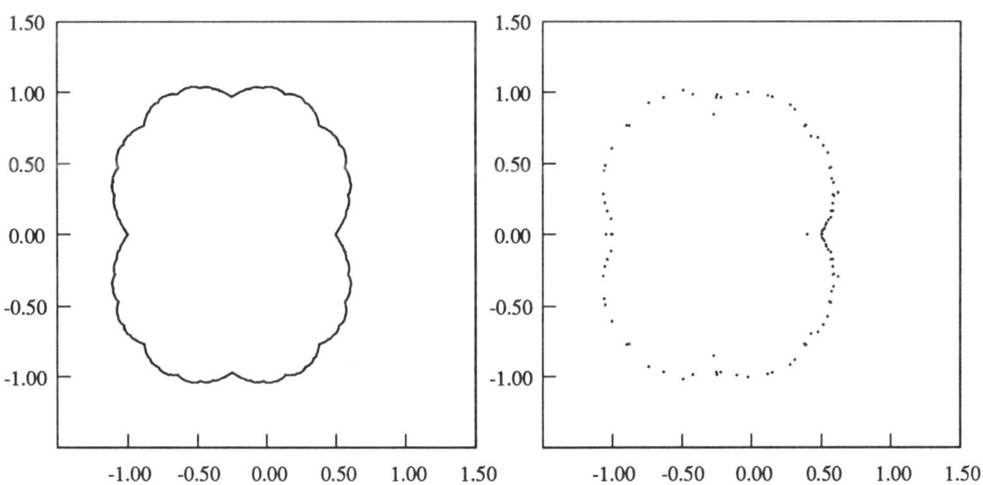

Fig. 1. Julia set of $F(z) = 1/2\, z + z^2$ (*on the left*) and respective poles of $[100/100]_\Psi$ approximant (*on the right*) computed with 200 digits.

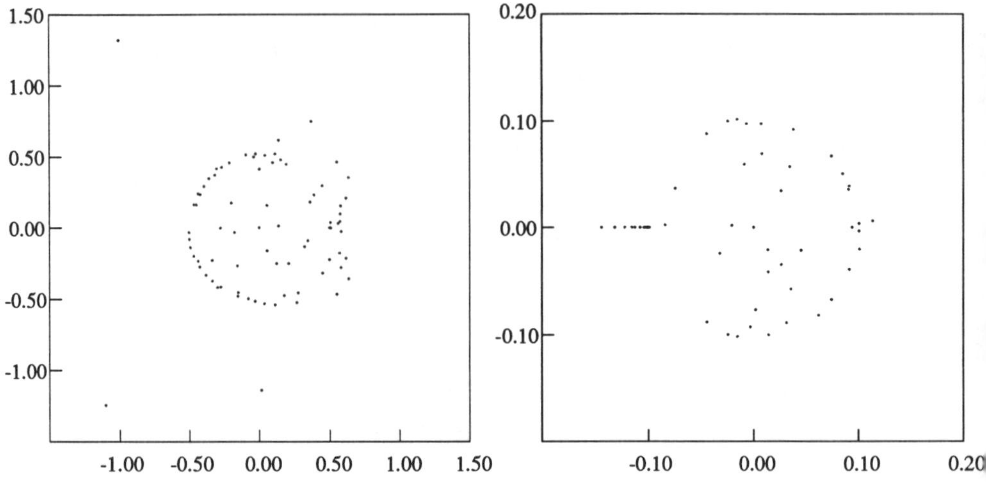

Fig. 2. Poles of $[100/100]_\Psi$ approximant (*on the left*) and poles of $[100/100]_\Phi$ approximant (*on the right*), both computed with 20 digits.

References

1. J. Baker, P. Graves Morris, "Padé Approximants" in *Encyclopedia of Mathematics*, vol 13, 14 Add. Wesley (1981)
2. G. Servizi, G. Turchetti "ormal forms singularities in area preserving maps and analytic continuation by Padé Approximants" Physics Letters **151**, 485 (1990)
3. A. Berretti, L. Chierchia "On the complex analytic structure of the golden invariant curve for the standard map" Nonlinearity **3**, 39 (1990)
4. C. Falcolini, R. de La Llave "Numerical calculation of domains of analyticity for perturbation theories in presence of small divisors" J. Stat. Phys. **67**, 645 (1992)
5. A. Berretti, S. Marmi "Standard map at complex rotation numbers: creation of natural boundaries" Phys. Rev. Lett **68**, 1443 (1992)
6. L. Billi, M. Malavasi, G. Turchetti "Natural boundaries of normalizing transformations" in *Integrability and chaotic behaviours*, ed. J.Seimenis, Plenum
7. J. Gilewicz, B. Truong-Van "Froissart doublets in the Pad/'e approximation and noise" in *Constructive theory of functions* Sophia 1988
8. A. T. Bharucha-Reid, M. Sambandham "Random Polynomials" Academic Press 1986
9. J. L. Gammel, J. Nuttal "Convergence of Padé Approximants to quasianalytic functions beyond natural boundaries" *Jour. Math. Analysis and Appl.* **43** (1973) 694.
10. J. L. Gammel "Continuation of functions beyond natural boundaries" *Rocky Mountain Journal of Mathematics* **4**, 203 (1974)
11. Brent "MPA: multiple precision floating point arithmetic" CERN Library A105
12. R. Xie "Padé approximants applied to the singularity structure for conjugation problems" Chaos : theory and practice Ed. T. Bountis, Springer Verlag (1992)
13. L. Billi "Analisi di singolarità in problemi di dinamica olomorfa" Phys. Dep. thesis, Bologna University (1992)
14. L. Billi, G. Turchetti, E. Todesco " Singularity analysis by Padé Approximants of some holomorphic maps" Bologna preprint (1993).
15. P. Blanchard "Complex analytic dynamics on the Riemann sphere" *Bull. Am. Math. Soc.* **11** (1984) 85-141.
16. L. Billi, G. Turchetti, R. Xie "Natural boundaries of hamiltonian maps and the genesis of the Siegel disk" Phys. Rev. Lett. in press.

ON ALGEBRAIC COMPUTATION OF NUMBER OF POLES OF MEROMORPHIC FUNCTIONS IN THE UNIT DISK

B.GLEYSE AND V.KALIAGUINE*
I.N.S.A. de Rouen
76 131 Mont-Saint-Aignan, Cedex
FRANCE

Abstract. We propose a software to calculate the number of poles of an meromorphic functions in the unit disk. The base of the algorithm is the Montessus de Ballore's theorem and a formal algebra algorithm to detect the number of zeros of rational polynomial in the unit disk.

Key words: Schur algorithm, poles of meromorphic functions, REDUCE system

1. Introduction

Let $f(z)$ be a function meromorphic in the unit disk Δ and analytic on the unit circle $|z| = 1$. We suppose that all the coefficients in the power series expansion

$$f(z) = c_0 + c_1 z + c_2 z^2 + c_3 z^3 + \dots \tag{1}$$

are complex rationals, e.g. $c_n = a_n + ib_n$ with $a_n, b_n \in Q$, $n = 0, 1, 2, \dots$. Our problem is to detect the exact number of poles (with multiplicity) of the function $f(z)$ in Δ. This problem is important for some signals and systems applications ([3]). Indeed, in many cases the transfert function in signal processing is meromorphic and the question of the localisation of their poles is of interest. Note that the coefficients in (1) are all rational and we are looking for an algebraic algorithm that can be implemented using some computer algebra software (Reduce, Maple, Mathematica or others). A similar problem is the one of detecting the number of zeros of rational polynomial $P \in Q[z]$ in a given domain in the complex plane. It is well known problem in the theory of stability of dynamical system ([7], [5]). Different algorithms were proposed and studied ([9], [2], [10], [6]). The situation for meromorphic functions is quite different. Because an philosophical reason it is clear that a finite algorithm to detect the number of poles of meromorphic function in the unit disk does not exist, so we are looking for an practical algorithm adaptable for applications. To construct it we use the Pade table of rational approximants to $f(z)$, the Montessus de Ballore's convergence theorem and a new algorithm to calculate number of zeros of rational polynomials in the unit disk. The software was developped in the REDUCE environment and tested for various meromorphic functions.

* On leave from Nizhniy Novgorod State University, Russia

A. Cuyt (ed.), Nonlinear Numerical Methods and Rational Approximation II, 241–246.

2. Algorithm

Let M be the number of poles of $f(z)$ in Δ. We fix $\{N_k\}$, a strongly increasing sequence of positive integers, $k = 1, 2, \ldots$. We denote by $[n/m]$ the Padé approximant P/Q to the function $f(z)$ at the point $z_0 = 0$ with restrictions $\deg P \leq n$, $\deg Q \leq m$ (see [1]). From modern version of the Montessus de Ballore's convergence theorem (see appendix 1) it is known that the M-th row in the Pade table of $f(z)$ converges uniformly (in the sense of the metric on the Riemann sphere) on the unit disk to $f(z)$ and the distances between the poles of function and the poles of $[n/M]$ approximants decay with a geometric rate. So for some $n > n_0$ all the poles of the approximants are in Δ. In general it is not true for the next row in the Pade table. This observation is the starting point of the algorithm. To apply it we start with the first row in the Pade table and calculate the denominators of

$$[1/1], [2/1], \ldots, [N_1/1]$$

Then we apply the algorithm of the next section to detect the number of zeros of denominators in the unit disk and we obtain the sequence

$$\alpha_1^{(1)}, \alpha_2^{(1)}, \ldots, \alpha_{N_1}^{(1)}$$

where $\alpha_k^{(1)} = 0, 1$. The next step is to calculate the denominators of the approximants

$$[N_1/2], [N_1 + 1/2], \ldots, [N_2/2]$$

and the associated sequence of integers

$$\alpha_{N_1}^{(2)}, \alpha_{N_1+1}^{(2)}, \ldots, \alpha_{N_2}^{(2)}$$

where $\alpha_k^{(2)} = 0, 1, 2$. On the same way we obtain the sequence

$$\alpha_{N_2}^{(3)}, \alpha_{N_2+1}^{(3)}, \ldots, \alpha_{N_3}^{(3)}$$

and so on. We stop the algorithm when we observe that for the m-th step the numbers $\alpha_k^{(m)}$ are stable and $\alpha_k^{(m)} = m$, for the next step the numbers $\alpha_k^{(m+1)}$ are stable too but $\alpha_k^{(m+1)} \leq m$ or the sequence $\alpha_k^{(m+1)}$ is stochasticaly instable. The conclusion is that in this case $M = m$. We need to calculate the denominators of the Padé approximants along a step-line (defined by the sequence $\{N_k\}$) in the general Padé table. There are many possibilities to do it. One is to use the usual algorithms for algebraic computation of Pade approximants (possible to be implemented in REDUCE, MAPLE or MACSYMA) or the direct determinant formulas. Once the denominator is calculated we use a version of the Schur algorithm to detect the number of zeros of rational polynomials in the unit disk (see the next section). Of course, the proposed algorithm is not ideal, it depends on the choice of the "observation" sequence $\{N_k\}$. Our stop rôle is justified (by the properties of Padé approximants) only in the case where all poles of $f(z)$ are simple and lie on the same distance from the point $z_0 = 0$, in the other cases the stop rôle has to be modified. The results of the application of the algorithm to various meromorphic functions are presented in the appendix 2. The time required depends on the choice of $\{N_k\}$ and on the time of computation of the number of zeros of polynomials.

3. Number of zeros of polynomials

In this section we present different methods to determine the exact number of zeros (counted with their multiplicities) of polynomials in $R[x]$ in the unit disk using Schur, Marden and Cohn transformations [4], [6], [5], [7]. Let $p(x) = \sum_{i=0}^{n} a_i x^i$ be a polynomial of degree n, $\deg(p) = n$. We consider the Schur sequence $\{p_k\}$ of polynomials denoted by $p_k = \sum_{i=0}^{n-k} a_i^{(k)} x^i$. It is defined by the following recurrence relation :

$$p_{k+1} = a_0^{(k)} p_k - a_{n-k}^{(k)} p_k^* \qquad k = 0, 1, ..., n-1.$$

$p_0 = p$, where $p_k^*(x) = x^{n-k} p_k\left(x^{-1}\right)$, the reciprocal polynomial of p_k. The coefficients $a_0^{(k)}$ $(= p_k(0))$ denoted by γ_k , $k = 0, 1, ..., n$ are called the Schur constants.

3.1. Regular case

We find a recurrence relation for the number of zeros of consecutive terms in the Schur sequence (with Rouché's theorem). Let us denote by $N_D(p)$ the number of zeros of p within the unit disk and $N_C(p)$ on the unit disk,

$$N_D(p_k) = \begin{cases} N_D(p_{k+1}) & \text{if } \gamma_{k+1} > 0 \\ n - k - N_D(p_{k+1}) & \text{if } \gamma_{k+1} < 0 \end{cases}$$

$k = 0, 1, \ldots, n-1$. By property of p on the unit circle, we have : $N_C(p_k) = N_C(p_{k+1})$. This argument can be repeated. If γ_1, γ_2,..., $\gamma_k \neq 0$ and let those indices k_j, $j = 1, 2, ..., m$, where $k_1 < k_2 < ... < k_m$ for which $\gamma_{k_j} < 0$ with the notation $S(p_k) = \sum_{j=1}^{m} (-1)^{j-1}(n + 1 - k_j)$, then $N_D(p) = S(p_k) + (-1)^m N_D(p_k)$ and $N_C(p) = N_C(p_k)$.We use the Schur-Cohn test $\gamma_{k+1} = 0$ to go from one term to the next one in the Schur sequence. When it is satisfied, we stop the construction of the sequence at the last calculated term, p_{k+1}.

3.2. Singular case

If γ_{k+1} $(= a_0^{(k+1)})$ is identically zero, we obtain : $a_{n-k}^{(k)} = u a_0^{(k)}$, where $u = \pm 1$. Then the coefficients of p_{k+1} can be written in the form

$$a_i^{(k+1)} = a_0^{(k)} (a_i^{(k)} - u a_{n-k-i}^{(k)}) \qquad i = 0, 1, ..., n - (k+1).$$

We suppose that γ_0, γ_1,..., $\gamma_k \neq 0$. thus $a_{n-k}^{(k)} \neq 0$. This implies that p_{k+1} is identically zero if and only if we have :

$$a_{n-k-i}^{(k)} = u a_i^{(k)} \qquad i = 1, ..., n - (k+1).$$

Cohn transformation $(p_{k+1} = 0)$.
The Cohn polynomial transformation is denoted $p^{(1)}$, defined by :

$$p^{(1)} = \left(\frac{dp_k}{dx}\right)^*$$

we have $N_D(p^{(1)}) = N_D(p_k)$ and $N_C(p_k) = \deg(p_k) - 2N_D(p^{(1)})$.
Marden transformation [7] (p_{k+1} not identically zero).

There exists $q \leq (n-k)/2$ such that :

$$a_{n-k-i}^{(k)} = ua_i^{(k)} \qquad i = 1, ..., n-(k+q-1) \quad \text{and} \quad a_{n-(k+q)}^{(k)} \neq ua_q^{(k)}$$

Let

$$a = \frac{(a_{n-(k+q)}^{(k)} - ua_q^{(k)})}{a_{n-k}^{(k)}}$$

The Marden polynomial transform denoted $p^{(2)}$, is defined by :

$$p^{(2)} = \begin{cases} ((x^q+2)p_k(x))_1 & \text{if } a > 0 \\ ((x^q-2)p_k(x))_1 & \text{if } a < 0 \end{cases}$$

where $()_1$ is the first iterated Schur sequence. The degrees of Marden transform and p_k are the same and the first Marden transform Schur constant is negative. We have $N_D(p^{(2)}) = N_D(p_k)$ and $N_C(p^{(2)}) = N_C(p_k)$. If p_k is not constant in x then these two polynomial transforms are not identically zero.

3.3. Description of the algorithm

Input : a polynomial p in $Q[x]$ of degree n in x ($n \geq 1$).
Output : its number of zeros in the unit disk (with multiplicity).
For $k = 1, 2, ..., n$ Computation of p_k and γ_k and Evaluation of $N_D(p)$ and $N_C(p)$ Schur-Cohn criterion and specification of polynomial transforms with respect to the numbers of zeros within the unit disk and on the unit circle.
The experiments were run on a Sun Sparc workstation using the REDUCE computer algebra system to calculate $N_{\overline{D}}(p)$ and to use it for the computation of the number of poles of meromorphic functions.

Appendix

A. The Montessus de Ballore's theorem

Let as below $f(z)$ be a meromorphic function in $\Delta = \{z : |z| < 1\}$ analytic on the unit circle $|z| = 1$. Then the application $f : \overline{\Delta} \to \overline{C}$ is a continous mapping from the closed unit disk to the closed complex plane with the Riemann spherical metric. Let M be the number of poles of $f(z)$ in Δ. Then the following statement is known as the Montessus de Ballore's convergence theorem ([1]):

Theorem 1 *The sequence $[n/M]$ converges uniformly on the closed unit disk to the function $f(z)$ (in the sense of spherical metric). Moreover each pole of $f(z)$ attracts the same number of poles of approximants (counting with multiplicity) and the distances between the poles of the $f(z)$ and of the $[n/M]$ decay with a geometric rate.*

B. Examples

The algorithm was tested for various meromorphic functions, given essentially as a sum $f(z) = \phi_1(z) + \phi_2(z) + \phi_3(z)$, where $\phi_1(z)$ is usually a special function, $\phi_2(z)$ is rational with poles in Δ and $\phi_3(z)$ is rational with poles outside the $\overline{\Delta}$. We choose $N_k = 4k$, $k = 1, 2, \ldots$. The results are:

1) $f(z) = \exp(z) + 1/(z^3 + 1/8) + z/(z^2 + 4)$ then

1 0 1 0

2 1 2 1

3 3 3 3

3 4 3 4 Stop

$M = 3$.

2) $f(z) = \cos(z) + (z + 1)/((2z + 1)(4z^2 - 4/5 - 3/5i))$ then

1 1 0 1

2 2 2 2

3 3 3 3

3 3 3 3 Stop

$M = 3$.

3) $f(z) = \log(1 + z/2) + z/(z^2 + 1/4) + (z + 1)/(z^2 + 2i)$ then

0 1 0 1

2 2 2 2

2 2 3 2 Stop

$M = 2$.

4) $f(z) = \sin(z) + (z + 1)/((z^2 + 1/4)(z^3 - 1/4))$, then

1 0 1 0

2 2 2 2

3 2 3 2

4 3 4 3

5 5 5 5

6 5 6 5 Stop

$M = 5$.

5) $f(z) = \exp(z) + z/((z^2 + 1/4)(z - 1/2)^2) + (z + 1)/(z^2 + 2i)$, then

1 1 1 1

1 2 1 2

3 3 3 3

4 4 4 4

4 5 4 4 Stop

$M = 4$.

References

1. G.Baker and P.Graves-Morris "Padé approximants, Part 1.Basic theory", *Addison-Wesley Publ. Comp.*, 1981.
2. P.Chenin and B.Gleyse "Nombre de racines d'un polynôme de R[x] dans le disque unité" *Rapp. Rech.*, n.458, TIM3/IMAG, Grenoble, 1984.
3. P.Delsart and Y.Genin "On the role of orthogonal polynomials on the unit circle in digital signal processing applications", *in (P.Nevai editor) Orthogonal Polynomials: Theory and Practice*, p.115-134, Kluwer Academic Publishers, 1990.

4. J. Dieudonné, "Calcul Infinitésimal", Hermann, 1968.
5. P.Henrici "Applied and computational complex analysis" v.1, Wiley, 1974.
6. B. Gleyse, "Number of zeros of real parameters polynomials in the unit disk : symbolic computation and graphic approach',' *Int. Workshop on computer algebra and applications*, Saint-Louis, 1992.
7. M.Marden, "Geometry of polynomials", *American Mathematical Society*, Providence, RI, 1966.
8. L.Pernebo "An algebraic theory for the design of controllers for linear multivariable systems", *IEEE Trans. Aut. Control*, **26**, n.1, 1981.
9. C.Schelin "Counting zeros of real polynomials within the unit disk", *SIAM Journ. Num. Anal.*, **20**, n.5, 1983.
10. A.Vieira and T.Kailath, "Another approach to the Schur-Cohn criterion", *IEEE Trans. Circuits and Systems*, **CAS-24**, p.218-220, 1977.

A COMPARISON OF ACCELERATION TECHNIQUES APPLIED TO THE SOR METHOD

RUDNEI DIAS DA CUNHA
Computing Laboratory, University of Kent at Canterbury, U.K.
Centro de Processamento de Dados, Universidade Federal do Rio Grande do Sul, Brasil

and

TIM HOPKINS
Computing Laboratory, University of Kent at Canterbury, U.K.

Abstract. In this paper we investigate the performance of four different SOR acceleration techniques on a variety of linear systems. Two of these techniques have been proposed by Dancis [1] who uses a polynomial acceleration together with a sub-optimal ω. The two other techniques discussed are vector accelerations; the ε algorithm proposed by Wynn [9] and a generalisation of Aitken's Δ^2 algorithm, proposed by Graves-Morris [3].

The experimental results show that these accelerations can reduce the amount of work required to obtain a solution and that their rates of convergence are generally less sensitive to the value of ω than the straightforward SOR method. However a poor choice of ω can result in particularly inefficient solutions and more work is required to enable cheap estimates of a effective parameter to be obtained.

Necessary conditions for the reduction in the computational work required for convergence are given for each of the accelerations, based on the number of floating-point operations.

It is shown experimentally that the reduction in the number of iterations is related to the separation between the two largest eigenvalues of the SOR iteration matrix for a given ω. This separation influences the convergence of all the acceleration techniques above.

Another important characteristic exhibited by these accelerations is that even if the number of iterations is not reduced significantly compared to the SOR method, they are competitive in terms of number of floating-point operations used and thus they reduce the overall computational workload.

Key words: SOR, Acceleration

1. Introduction

Consider the solution of the non-singular, symmetric, positive-definite system of n linear equations

$$Ax = b \tag{1}$$

by the SOR iteration

$$(I + \omega D^{-1} A_L)x^{(k+1)} = \left((1-\omega)I - \omega D^{-1} A_U\right) x^{(k)} + \omega D^{-1} b, \quad k = 0, 1, \ldots \tag{2}$$

where $D = diag(A)$, A_L and A_U are the strictly lower and upper triangular parts of A and ω is the relaxation parameter. Convergence of the method is guaranteed for $0 < \omega < 2$.

The SOR iterative method is commonly used for the solution of large, sparse, linear systems that arise from the approximation of partial differential equations. Its

A. Cuyt (ed.), Nonlinear Numerical Methods and Rational Approximation II, 247–260.

rate of convergence is dependent on the value chosen for the iteration parameter ω. Following Young [5], the minimum number of iterations required for convergence is obtained when this parameter has an optimal value, ω_b, which minimises the spectral radius of the SOR iteration matrix,

$$\mathcal{L}_\omega = (I + \omega D^{-1} A_L)^{-1} \left((1-\omega)I - \omega D^{-1} A_U\right).$$

The value of ω_b may be obtained from the largest eigenvalue, μ_1, of the Jacobi iteration matrix $B = -D^{-1}(A_L + A_U)$ using

$$\omega_b = \frac{2}{1 + \sqrt{1-\mu_1^2}} \tag{3}$$

However computing ω_b is relatively expensive in most cases. Adaptive procedures exist that can be used to update some initial approximation to ω_b, as in the ITPACK 2C package [4], but some of the initial iterations have large error vectors (when compared to SOR using some $\omega > 1$). In this case some of the initial estimates of the solution are "wasted" during the iterative process and this may be undesirable.

An alternative is to use an acceleration technique that may not be so sensitive to the choice of ω. The first two acceleration techniques, detailed in §2 were proposed by Dancis [1] and follow the usual approach of trying to select some ω for the SOR iteration. We show that for one of his techniques the selection of ω is not as sensitive as expected. In section 3 we look at an extension of the ε algorithm of Wynn applicable to vector and matrices iterations ([8][9]) and in §4 a generalisation of Aitken's Δ^2 algorithm as proposed by Graves-Morris [3] is described. These are *vector* accelerations and thus do not give a prescription for selecting ω.

Our investigation is aimed at establishing how sensitive these accelerations are with respect to the value chosen for ω. Section §6 describes the test problems used in the investigation and the results obtained from the experiments. In section §7 we summarise the results and draw some conclusions.

2. Dancis's Acceleration

Dancis proposes the use of a polynomial accelerator with SOR, ω being chosen such that the coordinate of the error vector, corresponding to the largest eigenvalue of $\mathcal{L}_\omega$, is annihilated. Dancis recommends that $\omega = \lambda_2 + 1$, where λ_2 is the second largest eigenvalue of $\mathcal{L}_\omega$. By computing the second largest eigenvalue, μ_2, of the Jacobi iteration matrix we can obtain the value of ω using Equation (3), with μ_1 replaced by μ_2.

Two different accelerations are proposed. The first, which we refer to as PSOR1, is as follows. Perform $r-1$ SOR iterations and then apply

$$x_{SOR}^{(r)} = \tilde{x} = \frac{1}{1-\lambda_1^r} x_{SOR}^{(r-1)} - \frac{\lambda_1^r}{1-\lambda_1^r} x^{(0)} \tag{4}$$

continuing with the SOR iterations thereafter.

The second acceleration (referred to as PSOR2) is obtained by

$$\hat{x}^{(i+1)} = \frac{1-a^i}{1-a^{i+1}} x_{SOR}^{(i+1)} + \frac{a^i(1-a)}{1-a^{i+1}} x^{(0)}, \quad i = 1, 2, \ldots, r \tag{5}$$

where $a = \omega - 1$. After r steps of the above iteration have been performed, we set $x_{SOR}^{(r+1)} = \hat{x}^{(r+1)}$ and the SOR iteration is used from then on.

Dancis propose that the value of r be chosen according to a spectral radius analysis (see [1, page 827]); its value is given by

$$\min_r = (\omega - 1)^{m-r}\left((\omega - 1)^r + \lambda_1^r\right)/(1 - \lambda_1^r)(\omega_b - 1)^m, \quad r = 1, 2, \ldots, \tag{6}$$

where m is some arbitrarily chosen number of iterations.

3. Wynn's ε Algorithm

In 1962 Wynn proposed an extension of the ε algorithm ([8]) for vector and matrix iterations ([9]). Consider a sequence $S = \{s_k\}_{k=0}^{\infty}$ which is slowly convergent. If we define the sequences $\varepsilon_{-1}^{(k)} = \{0\}_{k=1}^{\infty}$ and $\varepsilon_0^{(k)} = S$ then a new sequence $\varepsilon_{i+1}^{(k)}$ is generated by

$$\varepsilon_{i+1}^{(k)} = (\varepsilon_i^{(k+1)} - \varepsilon_i^{(k)})^{-1} + \varepsilon_{i-1}^{(k+1)}, \quad i = 0, 1, \ldots, \quad k = 0, 1, \ldots \tag{7}$$

In certain circumstances, the sequences $\varepsilon_{2i}^{(k)}$, $i = 1, 2, \ldots$ converge faster to the limit of S.

We investigate the behaviour of the $\varepsilon_2^{(k)}$ sequence obtained from vectors generated by SOR. We can express the new vector iterates, generated by two successive applications of (7) to three SOR vectors, $x_{SOR}^{(k-1)}$, $x_{SOR}^{(k)}$ and $x_{SOR}^{(k+1)}$, as

$$\varepsilon_2^{(k+1)} = \left((x_{SOR}^{(k+1)} - x_{SOR}^{(k)})^{-1} - (x_{SOR}^{(k)} - x_{SOR}^{(k-1)})^{-1}\right)^{-1} + x_{SOR}^{(k)} \tag{8}$$

where $u^{-1} = \bar{u}/(|\, u\,|^2)$ is the Moore-Penrose generalised inverse and $\bar{u}$ denotes the complex conjugate.

In [3] it is shown that the value of ω should be taken as $1 + \lambda_{2i}$ when using a sequence $\{\varepsilon_{2i}^{(k)}\}_i^{\infty}$. For the acceleration of SOR shown above, with $\varepsilon_2^{(k)}$ as the sequence to be used, then $\omega = 1 + \lambda_2$, which is the value proposed by Dancis.

4. Graves-Morris's Acceleration

Graves-Morris suggests, [3, page 25], that the sequence of vectors $x^{(k)}$ generated by (2) may be accelerated using

$$t^{(k)} = x^{(k-1)} - \frac{(\Delta x^{(k-2)})^2}{\Delta x^{(k-2)}\Delta^2 x^{(k-2)}}\Delta x^{(k-1)}, \quad k = 2, 3, \ldots \tag{9}$$

where $\Delta x^{(k)} = x^{(k+1)} - x^{(k)}$, which is a generalisation of Aitken's Δ^2 process [6]. We will refer to (9) as the G-M iteration.

Experimental results on a few model problems given in [3] show that, for the G-M iteration, using the value of ω which maximises the separation between the largest eigenvalue of the SOR iteration matrix and the other eigenvalues, a reduction in the number of iterations needed for convergence occurs. We investigate whether this reduction is also observed in larger, practical problems and, if so, how critical the eigenvalue separation is to the rate of convergence.

The value of ω is chosen in a similar way as for the ε algorithm, following [3].

5. Conditions for the Effectiveness of the Acceleration Techniques

Our aim is to discover whether we can reduce the amount of computation needed by SOR to solve a system of the form (1) by using one of the accelerations techniques.

For instance, using the G-M acceleration one might expect (intuitively) that if at least one iteration is saved, the computing workload is reduced by a factor of roughly n^2 multiplications compared to the SOR iteration. We present below necessary conditions for the techniques discussed to require a smaller number of floating point multiplications (*flops*) than SOR. These conditions are obtained in terms of the number of iterations (k) and the number of flops per iteration. For SOR and each of the accelerations we consider that the total number of flops is

$$flops_{SOR} = k_{SOR}(n^2 + n)O_*$$

$$flops_{PSOR1} = \left(k_{PSOR1}(n^2 + n) + 2n\right) O_*$$

$$flops_{PSOR2} = \left(k_{PSOR2}(n^2 + n) + 2rn\right) O_*$$

$$flops_{\varepsilon_2} = \left(k_{\varepsilon_2}(n^2 + 7n) - 6n\right) O_*$$

$$flops_{G\text{-}M} = k_{G\text{-}M}(n^2 + 3n)O_*$$

where O_* represents a floating-point multiplication.

It is easy to show that for $k_{SOR} > k_{accel}$ where *accel* denotes any of the acceleration techniques, we have

$$flops_{SOR} > flops_{accel}$$

if and only if the following conditions are satisfied

PSOR1: $n > (2 + k_{PSOR1} - k_{SOR})/(k_{SOR} - k_{PSOR1})$

PSOR2: $n > (2r + k_{PSOR2} - k_{SOR})/(k_{SOR} - k_{PSOR2})$

ε_2: $n > (7k_{\varepsilon_2} - k_{SOR} - 6)/(k_{SOR} - k_{\varepsilon_2})$

G-M: $n > (3k_{G\text{-}M} - k_{SOR})/(k_{SOR} - k_{G\text{-}M})$

6. Description of the Test Problems and Experiments

In this section we present the results obtained from solving a set of problems using MATLAB implementations of the above methods. Three of the problems are taken from the Boeing-Harwell library [2] and for these we used Fortran77 and BLAS routines to implement the methods and LAPACK subroutines to compute the eigenvalues.

The problems solved present different characteristics with respect to the distribution of eigenvalues and are of interest to the analysis of the acceleration techniques discussed.

In each experiment, we describe the system of linear equations used, the value of ω_b computed using (3) and the convergence criteria. The results are tabulated for each method in terms of number of iterations to achieve convergence and the flops counting of SOR and the ratios between the flops counting of each acceleration with respect to SOR.

For the analysis of the G-M iteration, we provide a graph showing the two largest eigenvalues of the SOR iteration matrix, λ_1 and λ_2, such that $|\lambda_1| > |\lambda_2|$. For these test problems, the value of r for the Dancis's accelerations was found to be 1.

6.1. Varga's Problem

This is a system of order $n = 16$ described in [7, Appendix B] derived from the five-point finite-difference discretisation of

$$-\frac{\partial}{\partial x}\frac{\partial D(x,y)u}{\partial x} - \frac{\partial}{\partial y}\frac{\partial D(x,y)u}{\partial y} + \sigma(x,y)u = S(x,y) \tag{10}$$

in the region $R = (0, 2.1) \times (0, 2.1)$ subject to the boundary condition $\partial n = 0$, where ∂n is the outward normal. The functions D, σ and S are as shown in Figure 1. The condition number is $\kappa(A) = 3.5888 \times 10^3$. The value of ω_b is 1.9177.

Fig. 1. Region for Varga's problem.

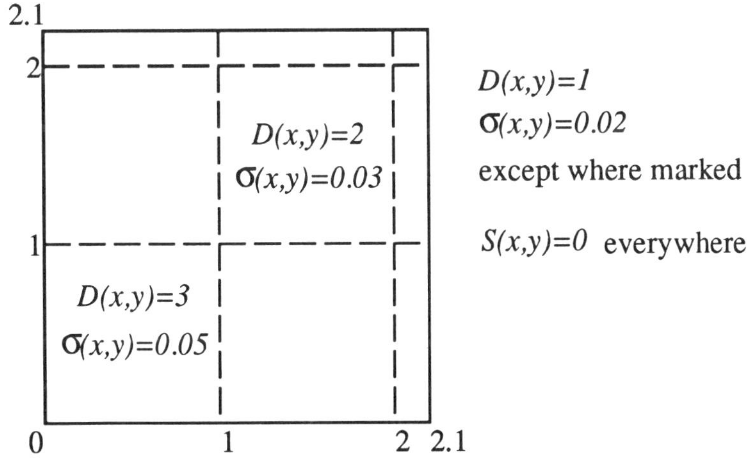

6.1.1. *Experiment*
In this problem, we are solving

$$Ax = 0 \tag{11}$$

which has the zero vector as the unique solution. The initial vector $x^{(0)}$ was set to $(1, 1, \ldots, 1)^T$ and we iterated until the ∞-norm of the solution vector was less than 10^{-4} (this stopping criterion was used in order to reproduce the behaviour of SOR presented in [7, Appendix B, page 304]). A maximum of 2000 iterations was allowed.

An impressive reduction in the number of iterations is achieved by the G-M, ε_2 and PSOR1 accelerations. Note that while the minimum number of iterations for SOR and PSOR2 is obtained when $\omega = \omega_b$, for G-M, ε_2 and PSOR1 this minimum occurs at some $\omega < \omega_b$.

TABLE I
Varga's problem: number of iterations.

ω	SOR	G-M	ε_2	PSOR1	PSOR2
1.1019	2000	60	90	60	2000
1.3059	1992	41	60	41	1992
1.5098	1282	23	31	23	1282
1.7137	708	37	52	37	708
1.9177	146	119	152	119	146

TABLE II
Varga's problem: flops counting.

	Mflops	ratio to SOR(%)			
ω	SOR	G-M	ε_2	PSOR1	PSOR2
1.1019	0.54	3.35	6.07	3.01	100.01
1.3059	0.54	2.30	4.06	2.06	100.01
1.5098	0.35	2.01	3.24	1.80	100.01
1.7137	0.19	5.84	9.89	5.24	100.02
1.9177	0.04	91.10	140.61	81.59	100.08
ω	ratio to minimum SOR flops(%)				
1.1019	1369.86	45.93	83.16	41.18	1369.94
1.3059	1364.38	31.39	55.36	28.16	1364.46
1.5098	878.08	17.61	28.49	15.83	878.16
1.7137	484.93	28.32	47.95	25.42	485.01
1.9177	100.00	91.10	140.61	81.59	100.08

Fig. 2. Varga's problem: Eigenvalue separation and number of iterations of the methods.

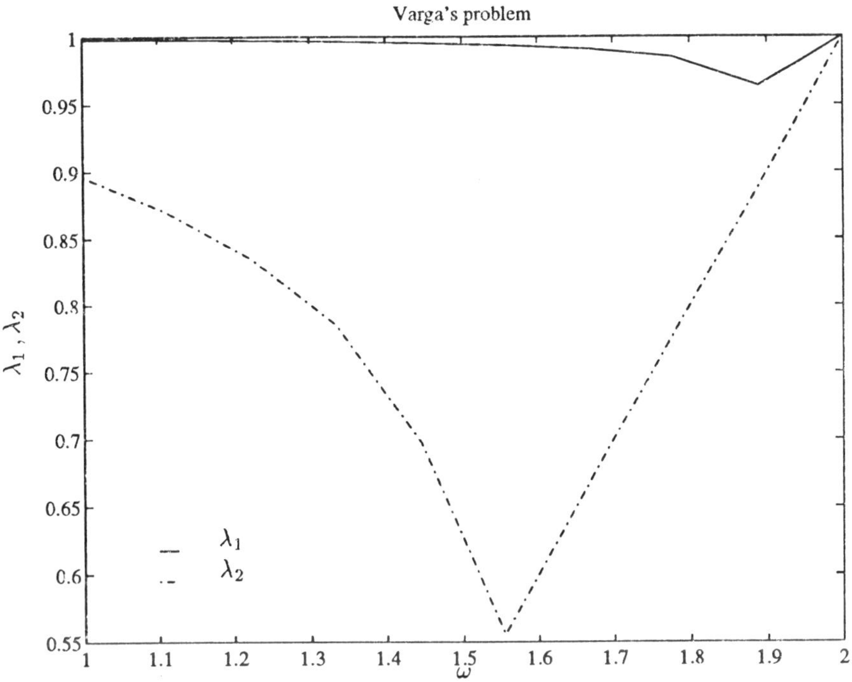

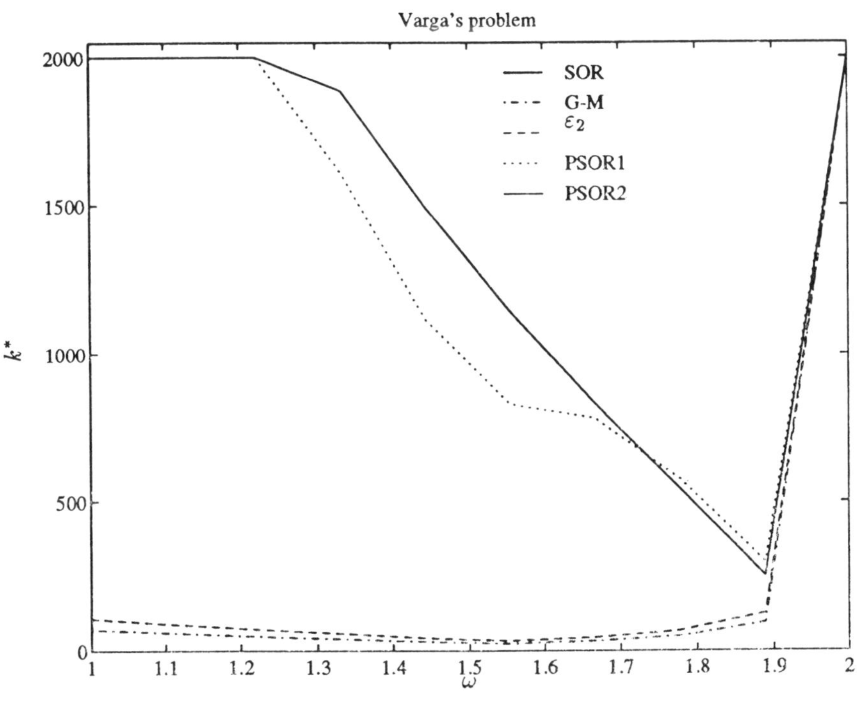

6.2. Problem TRIDIAG

This problem of order $n = 100$ is the system defined by

$$A = \begin{bmatrix} 10 & 3 & & & \\ 3 & 10 & 3 & & \\ & \ddots & \ddots & \ddots & \\ & & 3 & 10 & 3 \\ & & & 3 & 10 \end{bmatrix}, b = \begin{bmatrix} 1 \\ 1 \\ \vdots \\ 1 \\ 1 \end{bmatrix} \tag{12}$$

The condition number is $\kappa(A) = 3.9964$ and the value of ω_b is 1.1110.

6.2.1. *Experiment*

The starting vector used is $x^{(0)} = (0, 0, \ldots, 0)^T$ and the iterations proceed until the 2-norm of the residual of the solution vector is less than 10^{-10} or the number of iterations exceeded 200.

This example shows a situation where ω_b is close to 1. The experiment shows that in this case little, if any, gain is achieved by the accelerations. Nonetheless even if a single iteration is spared a reduction in the computational effort is verified.

TABLE III
Problem TRIDIAG: number of iterations.

ω	SOR	G-M	ε_2	PSOR1	PSOR2
1.0123	24	22	21	22	24
1.0369	23	21	22	22	23
1.0616	22	22	23	22	22
1.0863	22	21	21	23	22
1.1109	23	21	21	24	23

TABLE IV
Problem TRIDIAG: flops counting.

	Mflops	ratio to SOR(%)			
ω	SOR	G-M	ε_2	PSOR1	PSOR2
1.0123	0.24	93.48	92.45	91.75	100.08
1.0369	0.23	93.11	101.08	95.74	100.09
1.0616	0.22	101.98	110.49	100.09	100.09
1.0863	0.22	97.34	100.86	104.64	100.09
1.1109	0.23	93.11	96.47	104.43	100.09
ω	ratio to minimum SOR flops(%)				
1.0123	109.09	101.98	100.86	100.09	109.18
1.0369	104.55	97.34	105.67	100.09	104.64
1.0616	100.00	101.98	110.49	100.09	100.09
1.0863	100.00	97.34	100.86	104.64	100.09
1.1109	104.55	97.34	100.86	109.18	104.64

Fig. 3. Problem TRIDIAG: Eigenvalue separation and number of iterations of the methods.

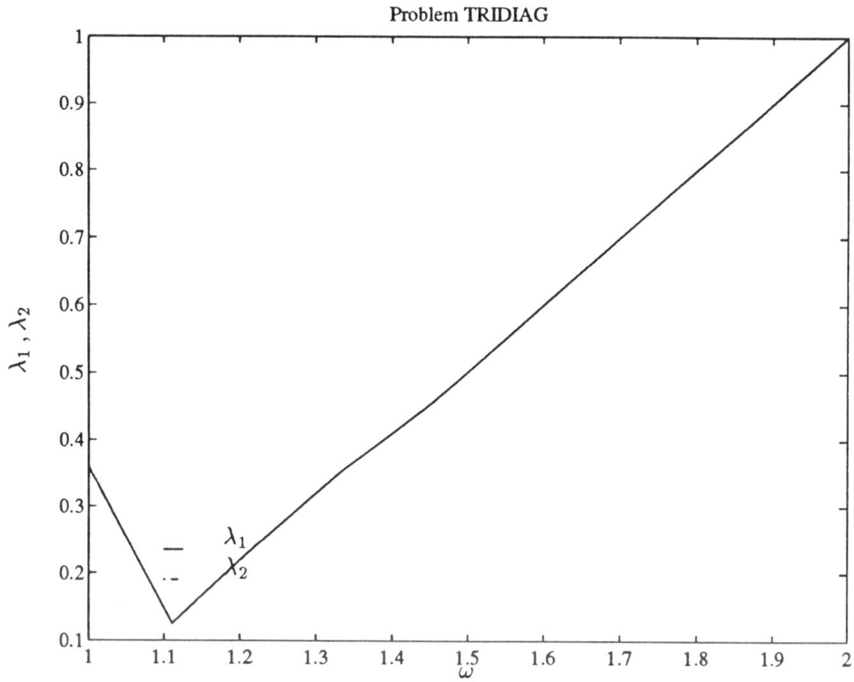

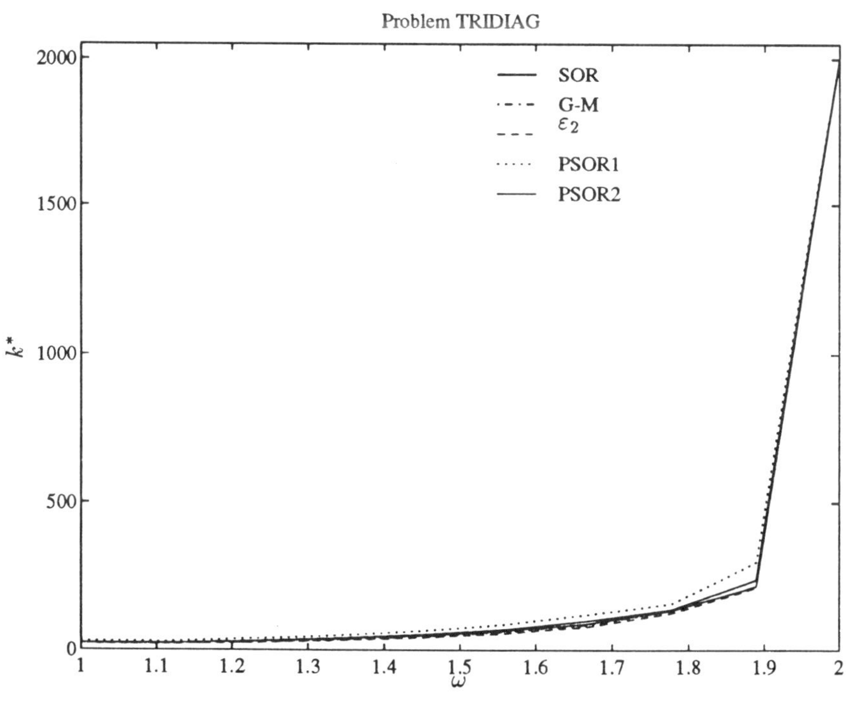

6.3. Problem NOS4

This problem is taken from the Harwell-Boeing sparse matrix collection [2, pp. 54-55]. It is derived from a finite-element approximation to a structural engineering problem. The system has order $n = 100$ and its condition number is $\kappa(A) = 1.5785 \times 10^3$. The RHS vector was chosen as $(1, 0, 0, \ldots, 0)^T$. The value of ω_b is 1.8810.

6.3.1. *Experiment*

In this problem we iterated until the 2-norm of the residual of the solution vector was less than 10^{-4} or the number of iterations exceeded 2000. The initial estimate of x was $(0, 0, \ldots, 0)^T$.

This example shows a similar behaviour to that of Varga's problem. However in this case the ε_2 acceleration is worse than SOR and PSOR1 and PSOR2 fail to produce any acceleration.

TABLE V
Problem NOS4: number of iterations.

ω	SOR	G-M	ε_2	PSOR1	PSOR2
1.0978	1099	699	1667	2000	2000
1.2933	748	212	1135	2000	2000
1.4889	489	126	729	2000	2000
1.6845	283	62	409	2000	2000
1.8810	95	103	105	2000	2000

TABLE VI
Problem NOS4: flops counting.

	Mflops		ratio to SOR(%)		
ω	SOR	G-M	ε_2	PSOR1	PSOR2
1.0978	11.10	64.86	160.69	181.99	181.99
1.2933	7.55	28.90	160.74	267.38	267.38
1.4889	4.94	26.28	157.92	409.00	409.00
1.6845	2.86	22.34	153.09	706.72	706.72
1.8810	0.96	110.57	117.03	2105.28	2105.28
ω	ratio to minimum SOR flops(%)				
1.0978	1156.84	750.36	1858.92	2105.28	2105.28
1.2933	787.37	227.58	1265.65	2105.28	2105.28
1.4889	514.74	135.26	812.89	2105.28	2105.28
1.6845	297.89	66.56	456.04	2105.28	2105.28
1.8810	100.00	110.57	117.03	2105.28	2105.28

Fig. 4. Problem NOS4: Eigenvalue separation and number of iterations of the methods.

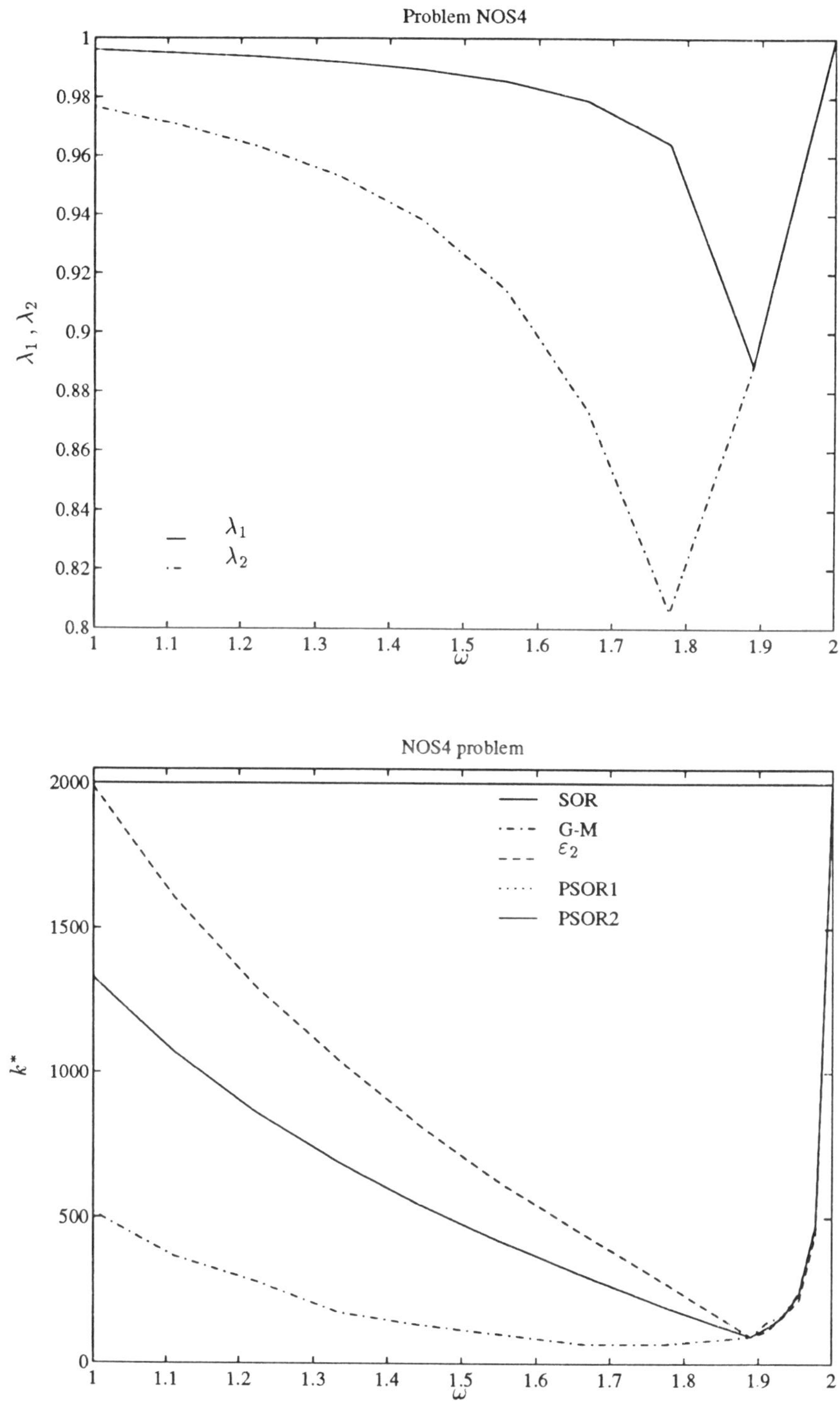

6.4. Problem 685 BUS

This problem is taken from the Harwell-Boeing sparse matrix collection [2, pp. 71]. The coefficient matrix is derived from the modelling of a power system network. The system has order $n = 685$ and its condition number is $\kappa(A) = 4.2305 \times 10^5$. The RHS vector was chosen as $(1, 0, 0, \ldots, 0)^T$. The value of ω_b is 1.9590.

6.4.1. *Experiment*

In this experiment we used the same stopping criteria as in problem NOS4. It shows a behaviour similar to that exhibited in Varga's and NOS4 problems except for the ε_2 acceleration which was always worse than SOR except at $\omega = \omega_b$.

TABLE VII
Problem 685 BUS: number of iterations.

ω	SOR	G-M	ε_2	PSOR1	PSOR2
1.1066	2000	614	2000	2000	2000
1.4762	1479	384	2000	2000	2000
1.6394	1064	251	2000	2000	2000
1.8525	547	117	952	2000	2000
1.9590	171	163	165	2000	2000

TABLE VIII
Problem 685 BUS: flops counting.

	Mflops	ratio to SOR(%)			
ω	SOR	G-M	ε_2	PSOR1	PSOR2
1.1066	939.82	30.79	100.87	100.00	100.00
1.4762	695.00	26.04	136.41	135.23	135.23
1.6394	499.98	23.66	189.61	187.97	187.97
1.8525	257.04	21.45	175.56	365.63	365.63
1.9590	80.35	95.60	97.33	1169.59	1169.59
ω	ratio to minimum SOR flops(%)				
1.1066	1169.59	360.11	1179.82	1169.59	1169.59
1.4762	864.91	225.22	1179.82	1169.59	1169.59
1.6394	622.22	147.21	1179.82	1169.59	1169.59
1.8525	319.88	68.62	561.59	1169.59	1169.59
1.9590	100.00	95.60	97.33	1169.59	1169.59

Fig. 5. Problem 685 BUS: Eigenvalue separation and number of iterations of the methods.

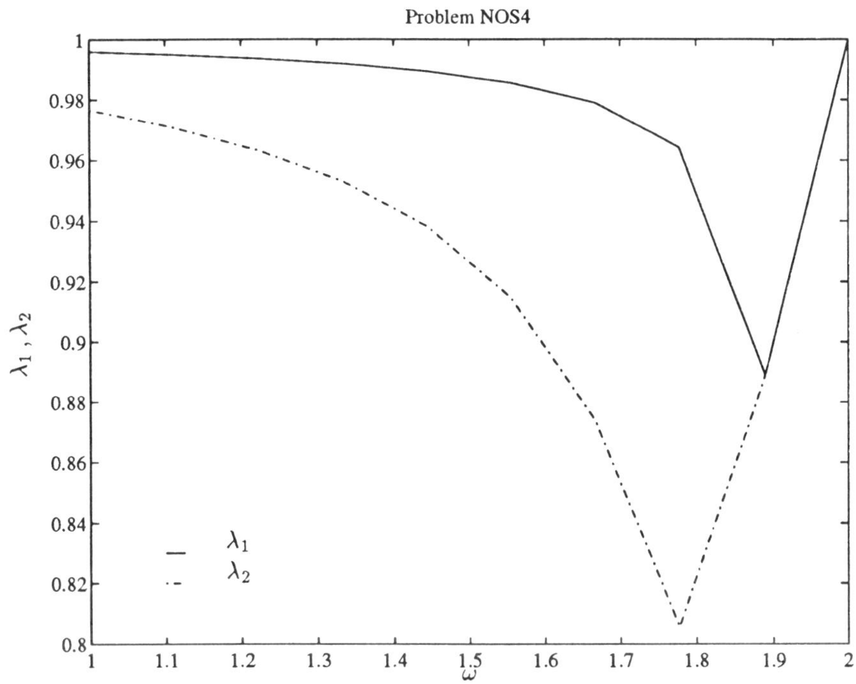

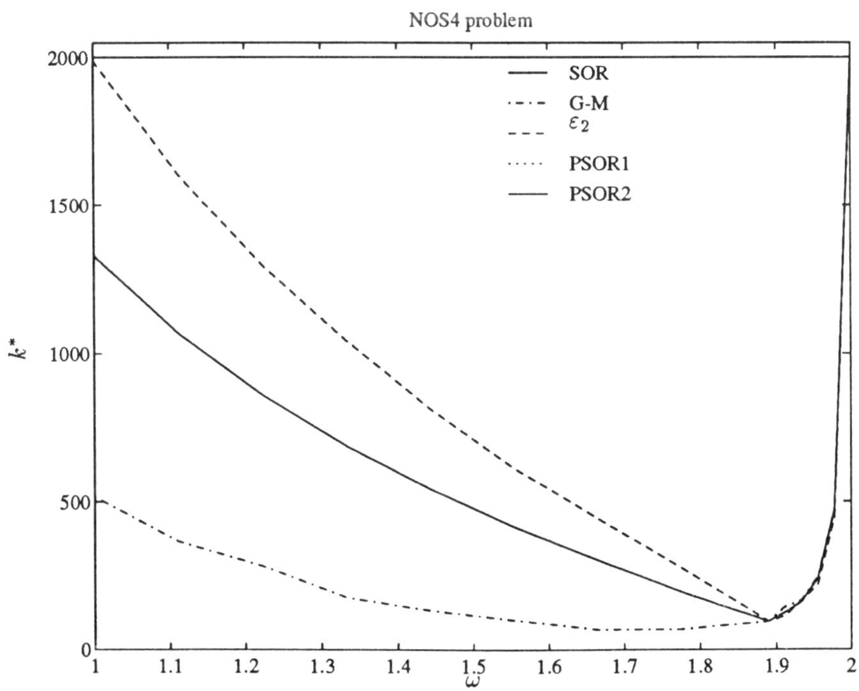

7. Summary

We summarise the results presented in §6. The main points are as follows

1. The PSOR2 iteration performed poorly in the test problems used in this paper. The other three methods generally outperformed the basic SOR method with the G-M acceleration showing the most consistent improvements.
2. The G-M, ε_2 and PSOR1 iterations almost invariably reduced the number of iterations required to obtain a specified accuracy when $\omega < \omega_b$. In some cases this reduction was observed for $\omega = \omega_b$.
3. As with the basic SOR method, the G-M, ε_2 and PSOR1 iterations are poor if ω chosen is greater than ω_b.
4. The reduction in the number of iterations using the G-M method is proportional to the separation of λ_1 and λ_2. Since the ε_2 and PSOR1 iterations produce similar behaviour to G-M, we believe this separation also has an influence on the convergence properties of these methods. Note that as the separation between λ_1 and λ_2 decreases all three iterations exhibit similar behaviour to SOR.

The experimental results presented show that the Graves-Morris's acceleration technique is the most attractive of the techniques discussed here, from the point of view both of the overall amount of computational work required and the range of the ω parameter for which the rate of convergence is improved. Though the number of experiments performed was small we believe the results indicate that these accelerations of the SOR method are effective and may be applicable to other systems.

Acknowledgements

We would like to thank Peter Graves-Morris for his valuable comments which helped to improve this paper.

References

1. J. Dancis. The optimal ω is not best for the SOR iteration method. *Linear Algebra and its Applications*, 154-156:819–845, 1991.
2. I.S. Duff, R.G. Grimes, and J.G. Lewis. Users' guide for the Harwell-Boeing sparse matrix collection. Report TR/PA/92/86, CERFACS, October 1992.
3. P.R. Graves-Morris. A review of Padé methods for the acceleration of convergence of a sequence of vectors. Report No. NA 92-31, Department of Mathematics, University of Bradford, October 1992.
4. R.G. Grimes, D.R. Kincaid, and D.M. Young. ITPACK 2.0 user's guide. Report No. CNA-150, Center for Numerical Analysis, University of Texas at Austin, August 1979.
5. L.A. Hageman and D.M. Young. *Applied Iterative Methods*. Academic Press, New York, 1981.
6. D.R. Kincaid and W. Cheney. *Numerical Analysis*. Brooks/Cole, Pacific Grove, 1991.
7. R.S. Varga. *Matrix Iterative Analysis*. Prentice-Hall, Englewood Cliffs, 1962.
8. P. Wynn. On a device for computing the $e_m(S_n)$ transformation. *Mathematical Tables and Aids of Computation*, 10:91–96, 1956.
9. P. Wynn. Acceleration techniques for iterated vector and matrix problems. *Mathematics of Computation*, 16:301–322, 1962.

HERMITE–PADÉ POLYNOMIALS AND APPROXIMATION PROPERTIES

K.A. DRIVER and D.S. LUBINSKY
Department of Mathematics, University of the Witwatersrand, P.O. Wits 2050, Johannesburg, Rep. of South Africa.

and

H. WALLIN
Department of Mathematics, University of Umeå, S–901 87 Umeå, Sweden

Abstract. Given a vector of functions $f = (f_0, \ldots, f_m)$, with each f_j analytic in an open set D containing the origin, $0 \leq j \leq m$, the relationship between Hermite–Padé polynomials of Type I and Type II is examined. A best simultaneous rational approximant for f on any compact subset K of D is defined, and we prove that when the error in best simultaneous rational approximation tends to zero faster than geometrically, certain sequences of Hermite–Padé Type I and Type II approximants to f converge in capacity.

AMS(MOS) Classification: Primary 41A21, Secondary30E10.

Key words: : Hermite–Padé approximant, best simultaneous rational approximant, Mahler's relations, Hermite–Padé Type I and Type II polynomials.

1. Introduction

Let $m \in \mathbb{N}$ and consider a vector $f = (f_0, \ldots, f_m)$ of $(m+1)$ formal power series in z and a multi-index $\rho = (\rho_0, \ldots, \rho_m) \in \mathbb{N}^{m+1}$, with $|\rho| := \rho_0 + \cdots + \rho_m$.

Associated with f and the multi-index ρ, Hermite [4] introduced two vectors of $(m+1)$ polynomials in z, Type I and Type II polynomials.

Definition 1.1. Hermite–Padé polynomials of Type I for f and ρ are polynomials $P_0, \ldots, P_m$, not all identically zero, such that

$$\deg(P_j) \leq \rho_j - 1, \quad 0 \leq j \leq m, \tag{1.1}$$

and

$$\sum_{j=0}^{m} P_j(z) f_j(z) = O(z^{|\rho|-1}) \text{ as } z \to 0. \tag{1.2}$$

The notation in (1.2) means that the order at zero of the power series in z is greater than or equal to $|\rho| - 1$ in the sense that all terms of order lower than $|\rho| - 1$ are zero. We observe that the existence of a non–trivial solution to (1.1) and (1.2) is assured since the number of unknown coefficients of the polynomials P_j, $0 \leq j \leq m$, is equal to $|\rho|$, while the number of homogeneous linear relations to be satisfied is $|\rho| - 1$.

A. Cuyt (ed.), Nonlinear Numerical Methods and Rational Approximation II, 261–268.

For simplicity in the formulation of Hermite–Padé polynomials of Type II for f and ρ, we make the assumption that $f_0 \equiv 1$, i.e., $f = (1, f_1, \ldots, f_m)$. (Note that this assumption, based on $f_0(0) \neq 0$, is not essential (cf. [5])).

Definition 1.2. Hermite–Padé polynomials of Type II for $f = (1, f_1, \ldots, f_m)$ and ρ are polynomials $Q_0, \ldots, Q_m$, not all identically zero, such that

$$\deg(Q_j) \leq |\rho| - \rho_j, \quad 0 \leq j \leq m, \tag{1.3}$$

and

$$(Q_0 f_j - Q_j)(z) = O(z^{|\rho|+1}) \text{ as } z \to 0, \quad 1 \leq j \leq m. \tag{1.4}$$

The number of unknown coefficients of the polynomials $Q_0, \ldots, Q_m$ is $m(|\rho|+1)+1$, while the number of homogeneous linear relations to be satisfied in (1.4) is $m(|\rho|+1)$. Such a system always has a non–trivial solution and, moreover, it is easy to verify that any non–trivial solution of (1.3) and (1.4) has $Q_0 \not\equiv 0$. It is natural to consider the vector of rational functions $(Q_1/Q_0, \ldots, Q_m/Q_0)$, with common denominator polynomial $Q_0 \not\equiv 0$, as an approximant to the vector of functions $(f_1, \ldots, f_m)$.

Definition 1.3. The vector of rational functions $(Q_1/Q_0, \ldots, Q_m/Q_0)$, where $Q_0, \ldots, Q_m$ satisfy (1.3) and (1.4), $Q_0 \not\equiv 0$, is called a Hermite–Padé Type II approximant for the vector of functions $(f_1, \ldots, f_m)$ and the multi–index $\rho \in \mathbb{N}^{m+1}$.

We remark that with $m = 1$, $\rho = (\rho_0, \rho_1) \in \mathbb{N}^2$, the Hermite–Padé Type II approximant is just the uniquely determined $[\rho_0/\rho_1]$ Padé approximant to the function f_1.

There are two basic aspects or directions of research in Hermite–Padé polynomial theory. The formal or algebraic aspect of the theory includes the Mahler relations (cf. [7]) which connect Hermite–Padé polynomials of Type I and Type II; while the analytic or asymptotic aspect of the theory is concerned with the asymptotic behaviour of the Hermite–Padé polynomials when $|\rho| \to \infty$ and has implications for and applications to convergence of the associated approximants. Excellent surveys of the historical development of the different aspects of Hermite–Padé polynomial theory can be found in [1] and [2].

This paper deals with both the algebraic and the analytic aspects of Hermite–Padé polynomials. In Section 2 we discuss the Mahler relations (cf. [7, 5]) for Type I and Type II polynomials. In Section 3 we study an approximation problem for Type II polynomials and use the result of this study, together with the Mahler relations, to prove an approximation theorem for Type I polynomials.

2. The Mahler Relations

We observe from (1.2) and (1.4) that the required order at zero is $|\rho| - 1$ and $|\rho| + 1$ respectively. In order to derive a relation between Type I and Type II polynomials, we need to work with the same order at zero. Let $\rho = (\rho_0, \ldots, \rho_m) \in \mathbb{N}^{m+1}$ be any given multi–index. We introduce the Kronecker symbol

$$\delta_{hk} := \begin{cases} 1, & h = k \\ 0, & h \neq k \end{cases},$$

and let

$$\delta_h := (\delta_{h0}, \dots, \delta_{hm})$$

be the vector with all components equal to zero, except at the hth place where the component is 1. Let

$$P_{kg}(z) := P_k(z; \rho + \delta_g), \quad 0 \leq k,\ g \leq m, \tag{2.1}$$

denote any Type I polynomials for $f = (1, f_1, \dots, f_m)$ and the multi–index $\rho + \delta_g = (\rho_0, \dots, \rho_g + 1, \dots, \rho_m)$. Then

$$\deg(P_{kg}) \leq \rho_k + \delta_{gk} - 1, \quad 0 \leq k,\ g \leq m, \tag{2.2}$$

and

$$\sum_{k=0}^{m} P_{kg}(z) f_k(z) = O(z^{|\rho|}) \text{ as } z \to 0, \quad 0 \leq g \leq m. \tag{2.3}$$

Similarly, let

$$Q_{kh}(z) := Q_k(z; \rho - \delta_h), \quad 0 \leq k,\ h \leq m, \tag{2.4}$$

denote any Type II polynomials for $f = (1, f_1, \dots, f_m)$ and the multi–index $\rho - \delta_h = (\rho_0, \dots, \rho_h - 1, \dots, \rho_m)$. Then

$$\deg(Q_{kh}) \leq |\rho| - 1 - (\rho_k - \delta_{kh}), \quad 0 \leq k,\ h \leq m, \tag{2.5}$$

and

$$(Q_{0h} f_k - Q_{kh})(z) = O(z^{|\rho|}) \text{ as } z \to 0, \quad 1 \leq k \leq m,\ 0 \leq h \leq m. \tag{2.6}$$

Using this notation, we introduce the polynomial

$$S_{gh}(z) := \sum_{k=0}^{m} P_{kg}(z) Q_{kh}(z), \quad 0 \leq g,\ h \leq m. \tag{2.7}$$

We record the fundamental identity due to Mahler (cf. [7]): For $0 \leq g,\ h,\ j \leq m$,

$$f_j S_{gh} = Q_{jh} \left(\sum_{k=0}^{m} P_{kg} f_k \right) + \sum_{k=0}^{m} P_{kg}(Q_{kh} f_j - Q_{jh} f_k). \tag{2.8}$$

Note that since $f_0 \equiv 1$, the identity (2.8) with $j = 0$ yields

$$S_{gh} = Q_{0h} \left(\sum_{k=0}^{m} P_{kg} f_k \right) + \sum_{k=0}^{m} P_{kg}(Q_{kh} - Q_{0h} f_k). \tag{2.9}$$

Note that $Q_{0h} \not\equiv 0$. The following theorem is due to Mahler (cf. [7]).

Theorem 2.1. For all z, $S_{gh}(z)$ defined in (2.7) satisfies

$$S_{gh}(z) = 0 \text{ if } g \neq h, \quad 0 \leq g,\ h \leq m, \tag{2.10}$$

and

$$S_{hh}(z) = c_h z^{|\rho|},\ c_h \text{ constant}, \quad 0 \leq h \leq m. \tag{2.11}$$

Proof. From the definition (2.7) of S_{gh} and equations (2.2) and (2.5) giving the maximum degrees of P_{kg} and Q_{kh} respectively, we have

$$\deg(S_{gh}) \leq \max_{0 \leq k \leq m} (|\rho| - 2 + \delta_{gk} + \delta_{kh}), \quad 0 \leq g,\ h \leq m. \tag{2.12}$$

However, applying the order relations (2.3) and (2.6) to the identity (2.9) yields

$$S_{gh}(z) = O(z^{|\rho|}) \text{ as } z \to 0, \quad 0 \leq g,\ h \leq m. \tag{2.13}$$

Combining (2.12) and (2.13), we obtain (2.10) and (2.11). ∎

3. Best simultaneous rational approximant and its connection to Type I and Type II approximants

It is well known (cf. [6]) that any function f, meromorphic in $\mathbb{C}$, admits fast rational approximation; i.e. if K is a compact set in which f is analytic, there exist rational functions R_n of type (n, n), $n \geq 1$, such that $\|f - R_n\|_{L_\infty(K)}^{\frac{1}{n}} \to 0$ as $n \to \infty$. More generally, any f defined on an open set U admitting such approximation on compact $K \subset U$ is said to belong to the Gonchar–Walsh class on U. In order to define a "best rational approximant" for a vector of functions $f = (f_1, \ldots, f_m)$ where f_j is analytic in an open set D, $1 \leq j \leq m$, that can be compared with a Hermite–Padé Type II approximant for $f = (f_1, \ldots, f_m)$, we require that the denominator polynomial should be common for all f_j, $1 \leq j \leq m$. We make the following definition.

Definition 3.1. Let $f = (f_1, \ldots, f_m)$ be a vector of functions with f_j analytic in an open set D, $1 \leq j \leq m$, and let $n \in \mathbb{N}$, $n \geq 1$. For every compact subset K of D, we define the error of best simultaneous rational approximation to f of type (n, n) by

$$r_n(f; K) := \inf \sum_{j=1}^{m} \|f_j - A_{jn}/B_n\|_K, \tag{3.1}$$

where the infimum is taken over all rational functions A_{jn}/B_n, $1 \leq j \leq m$, of type (n, n) with common denominator B_n. Here $\|\cdot\|_K$ denotes the sup or L_∞ norm on K.

For a vector of functions $f = (f_1, \ldots, f_m)$ of "fast" simultaneous rational approximability, we prove in Theorem 3.1 below that for quasi–diagonal multi–indices, the sequences of associated Type II approximants converge in capacity to f, while related diagonal sequences of Type I approximants for f converge to zero. The first part of Theorem 3.1 is a generalization of a result due to Gonchar (cf. [3]) on Padé approximants, which corresponds to the case $m = 1$. We denote the logarithmic capacity (cf. [6]) of a set A by cap (A).

Theorem 3.1. Let $f = (1, f_1, \ldots, f_m)$ be a vector of $(m+1)$ functions with each f_j analytic in an open set D containing the origin, $1 \le j \le m$. Assume that for every compact subset K_1 of D and $\tilde{f} = (f_1, \ldots, f_m)$,

$$[r_n(\tilde{f}; K_1)]^{\frac{1}{n}} \to 0 \text{ as } n \to \infty. \tag{3.2}$$

Let $\rho = (n, n+1, \ldots, n+1) \in \mathbb{N}^{m+1}$ be a multi–index.

(a) Let $Q_n, Q_{1n}, \ldots, Q_{mn}$ denote any Type II polynomials for f and $\rho - \delta_1 = (n, n, n+1, \ldots, n+1) \in \mathbb{N}^{m+1}$. Then for any $\delta > 0$ and any compact subset K of D, $0 \in K$, there exist sets E_{jn} with cap $E_{jn} < \delta$ such that

$$\left|(f_j - Q_{jn}/Q_n)(z)\right|^{\frac{1}{n}} \to 0 \text{ as } n \to \infty \text{ for } z \in K \backslash E_{jn},\ j = 1, \ldots, m. \tag{3.3}$$

(b) Let $P_{0n}, P_{1n} \ldots, P_{mn}$ denote any Type I polynomials for f and $\rho + \delta_0 = (n+1, n+1, \ldots, n+1) \in \mathbb{N}^{m+1}$. Then for any compact subset K of D with $0 \in K$, cap $(K) > 0$, we have

$$\left(\frac{\left\| \sum_{j=0}^{m} P_{jn} f_j \right\|_K}{\sum_{j=0}^{m} \|P_{jn}\|_K} \right)^{\frac{1}{n}} \to 0 \text{ as } n \to \infty. \tag{3.4}$$

Proof.

(a) For the multi–index $\rho - \delta_1 = (n, n, n+1, \ldots, n+1) \in \mathbb{N}^{m+1}$, it follows from the defining equations (1.3) and (1.4) for Type II polynomials that

$$\begin{aligned} \deg(Q_n) &\le m(n+1) - 1,\ \deg(Q_{1n}) \le m(n+1) - 1, \\ \deg(Q_{jn}) &\le m(n+1) - 2,\ j = 2, \ldots, m, \end{aligned} \tag{3.5}$$

and

$$(Q_n f_j - Q_{jn})(z) = O\big(z^{m(n+1)+n}\big) \text{ as } z \to 0,\ j = 1, \ldots, m. \tag{3.6}$$

Let $\epsilon > 0$, $\delta > 0$, $0 \in K \subset D$, K compact, be given. Take a contour Γ in D with $\operatorname{Ind}_\Gamma(a) = 1$, $a \in K$; $\operatorname{Ind}_\Gamma(a) = 0$, $a \notin D$. Assume that $r_n(\tilde{f};\Gamma) > 0$ and take rational functions A_{jn}/B_n of type (n,n), such that for $t \in \Gamma$,

$$\left|(f_j - A_{jn}/B_n)(t)\right| \le 2r_n(\tilde{f};\Gamma), \quad j = 1,\dots,m. \tag{3.7}$$

Now from (3.6), the functions $(Q_n f_j - Q_{jn})(z)/z^{m(n+1)+n}$ are analytic in D, $j = 1,\dots,m$, so that by Cauchy's theorem, for each j, $1 \le j \le m$, and $z \in K$, we have

$$\begin{aligned} \frac{B_n(z)(Q_n f_j - Q_{jn})(z)}{z^{m(n+1)+n}} &= \frac{1}{2i\pi}\int_\Gamma \frac{[B_n Q_n(f_j - A_{jn}/B_n)](t)}{t^{m(n+1)+n}(t-z)}dt \\ &+ \frac{1}{2i\pi}\int_\Gamma \frac{(Q_n A_{jn} - B_n Q_{jn})(t)}{t^{m(n+1)+n}(t-z)}dt. \end{aligned} \tag{3.8}$$

From (3.5), $\deg(Q_n A_{jn} - B_n Q_{jn})(t) \le m(n+1) - 1 + n$, $1 \le j \le m$, while $\deg\left(t^{m(n+1)+n}(t-z)\right) = m(n+1) + n + 1$. It follows that the second integral in (3.8) is zero and from (3.8), (3.7) we have the estimate, for $z \in K$,

$$\left|(f_j - Q_{jn}/Q_n)(z)\right| \le c r_n(\tilde{f};\Gamma) \max_{t\in\Gamma}\left|\frac{B_n(t)Q_n(t)z^{m(n+1)+n}}{B_n(z)Q_n(z)t^{m(n+1)+n}}\right|, \tag{3.9}$$

where $c = c(\tilde{f}, K, \Gamma)$ is independent of n. Using the method of estimation described by Wallin in [9], we obtain that for $n \ge n_0$, the maximum on the right hand side of (3.9) is less than $(\text{const})^n$ for $z \in K\backslash E_{jn}$, where cap $(E_{jn}) < \delta$. This estimate, together with assumption (3.2) then yields (3.3). We remark that we can choose E_{jn} independent of j.

(b) For the multi-index $\rho + \delta_0 = (n+1, n+1, \dots, n+1) \in \mathbb{N}^{m+1}$, we have by (1.1) and (1.2) that the Type I polynomials for $f = (1, f_1, \dots, f_m)$ satisfy

$$\deg(P_{jn}) \le n, \quad 0 \le j \le m, \tag{3.10}$$

and

$$\sum_{j=0}^{m} P_{jn}(z) f_j(z) = O(z^{m(n+1)+n}) \quad \text{as} \quad z \to 0. \tag{3.11}$$

Our choice of multi-indices, viz. $\rho - \delta_1$ in (a) and $\rho + \delta_0$ in (b), enable us to deduce from (2.9) and (2.10) that

$$0 = Q_n\left(\sum_{j=0}^{m} P_{jn} f_j\right) + \sum_{j=0}^{m} P_{jn}(Q_{jn} - Q_n f_j). \tag{3.12}$$

Furthermore, $Q_n = Q_{0n} \not\equiv 0$ and $f_0 \equiv 1$, so (3.12) yields

$$\sum_{j=0}^{m} P_{jn} f_j = \sum_{j=1}^{m} P_{jn}(f_j - Q_{jn}/Q_n). \tag{3.13}$$

Now let $\epsilon > 0$, $\delta > 0$, $0 \in K \subset D$, K compact, be given. By multiplying each P_{jn}, $0 \leq j \leq m$, by a constant depending on n but not on j, we see that we may assume

$$\sum_{j=0}^{m} \|P_{jn}\|_K = 1 \text{ forall } n \in \mathbb{N}. \tag{3.14}$$

By the Bernstein–Walsh lemma (cf. [8]), we deduce from (3.14) that for some compact subset K_2 of D, with K contained in Int (K_2), there exists a constant c such that

$$\sum_{j=0}^{m} \|P_{jn}\|_{K_2} \leq c^n. \tag{3.15}$$

Further, from (3.3), there exists an integer n_0 and a set E_n with cap $(E_n) < \delta$ such that

$$\left|(f_j - Q_{jn}/Q_n)(z)\right| < \epsilon^n \text{ for } n \geq n_0, \quad z \in K_2 \backslash E_n,\ 1 \leq j \leq m. \tag{3.16}$$

Hence, by (3.13), (3.15) and (3.16), we have

$$\left\| \sum_{j=0}^{m} P_{jn} f_j \right\|_{K_2 \backslash E_n} < c^n \epsilon^n \text{ for } n \geq n_0. \tag{3.17}$$

By the maximum principle, we can assume that each component of E_n intersects the boundary of K_2. Therefore, by choosing δ small enough, since $K \subseteq$ Int (K_2), we obtain $K \cap E_n = \phi$ and this observation together with (3.14) and (3.17) yields (3.4). ■

Remark. It has been observed by A. Ambroladze that, in Theorem 3.1, the condition $[r_n(\tilde{f}; K_1)]^{\frac{1}{n}} \to 0$ as $n \to \infty$ is equivalent to $[r_n(f_j; K_1)]^{\frac{1}{n}} \to 0$ as $n \to \infty$ for all j, $1 \leq j \leq m$.

References

1. Aptekarev, A.I. and Stahl, H.: 1992, 'Asymptotics of Hermite–Padé polynomials', in: Progress in Approximation Theory, eds. Gonchar, A.A. and Saff, E.B., Springer Verl., pp. 127–167.
2. de Bruin, M.G.: 1990, 'Some aspects of simultaneous rational approximation', in: Numerical Analysis and Mathematical Modelling, Banach Centre Publications, Vol. 24, PWN–Polish Scientific Publishers, Warsaw, pp. 51–84.
3. Gonchar, A.A.: 1978, 'On the speed of rational approximation of some analytic functions', *Math. USSR. Sbornik* Vol. no. **34**, pp. 131–145.
4. Hermite, Ch.: 1873, 'Sur la fonction exponentielle', Comptes rendus de l' Acad. des Sciences t. LXXVII, pp. 18–24, 74–79, 226–233, 285–293 = Oevres t. III, pp. 150–181.
5. Jäger, H.: 1964, 'A multi dimensional generalization of the Padé table', I–IV Nederl. Acad. Wetensch. Proc. Ser. A67 = Indag. Math. Vol. no. **26**, pp. 193–249.
6. Lubinsky, D.S.: 1992, 'Spurious poles in diagonal rational approximation', in: Progress in Approximation Theory, eds. Gonchar, A.A. and Saff, E.B., Springer Verl., pp. 191–214.

7. Mahler, K.: 1968, 'Perfect systems', *Compositio Math.* Vol. no. **19**, pp. 95–166.
8. Walsh, J.L.: 1969, 'Interpolation and approximation in the complex domain', 5th edn., Amer. Math. Soc. Colloq. Publns, Vol. 20, Amer. Math. Soc., Providence.
9. Wallin, H.: 1979, 'Potential theory and approximation of analytic functions by rational interpolation', in: Proc. of the Colloquium on Complex Analysis at Joensuu, Springer LNM, Vol. 747, Springer, Berlin, pp. 434–450.

ON THE EFFICIENCY OF LINEAR BUT NONREGULAR SEQUENCE TRANSFORMATIONS

ERNST JOACHIM WENIGER
Institut für Physikalische und Theoretische Chemie
Universität Regensburg
D-93040 Regensburg
Federal Republic of Germany

Abstract. Regular matrix transformations, which satisfy the conditions of the theorem of Toeplitz, have many advantageous theoretical properties, but they are at most moderately powerful. Theoretically, nonlinear sequence transformations, which are nonregular, are not yet very well understood, but they are frequently remarkably powerful. Consequently, in recent years the emphasis has been on nonlinear methods. However, the bad reputation of linear sequence transformations in general is not completely justified. By means of some examples it is demonstrated that linear sequence transformations can be constructed which are extremely powerful for special problems. The key to the success of these linear methods is that they are very well adapted to the problems, for which they were constructed, and that they are nonregular.

Key words: Regular matrix transformations, linear and nonlinear sequence transformations, acceleration of convergence, summation of divergent series.

1. Introduction

Principal tools for improving the numerical properties of slowly convergent or divergent sequences are *sequence transformations* which transform a given sequence $\{s_n\}_{n=0}^{\infty}$ into a new sequence $\{s'_n\}_{n=0}^{\infty}$ with hopefully better numerical properties. In practice, only finite subsets $s_n, s_{n+1}, \ldots, s_{n+\ell}$ of the input sequence are used to compute an element s'_n of the transformed sequence,

$$s'_n \;=\; \mathcal{T}_\ell(s_n, \ldots, s_{n+\ell}) \,. \tag{1}$$

Before the invention of electronic computers, mainly *linear* sequence transformations $\mathcal{T}_\ell$ were used which satisfy

$$\begin{aligned} &\mathcal{T}_\ell(\alpha s_n + \beta t_n, \ldots, \alpha s_{n+\ell} + \beta t_{n+\ell}) \\ &\quad = \; \alpha \mathcal{T}_\ell(s_n, \ldots, s_{n+\ell}) \; + \; \beta \mathcal{T}_\ell(t_n, \ldots, t_{n+\ell}) \end{aligned} \tag{2}$$

for arbitrary numbers α and β and for arbitrary sequences $\{s_n\}_{n=0}^{\infty}$ and $\{t_n\}_{n=0}^{\infty}$.

Linearity greatly facilitates a theoretical analysis. Accordingly, there has been a lot of research on linear transformations, and their theoretical properties are now well understood [1, 2, 3, 4, 5, 6]. The so-called *matrix transformations*, which compute the elements of the transformed sequence $\{s'_n\}_{n=0}^{\infty}$ as weighted averages

$$s'_n \;=\; \sum_{k=0}^{n} \mu_{nk}\, s_k \,, \tag{3}$$

A. Cuyt (ed.), Nonlinear Numerical Methods and Rational Approximation II, 269–282.

have particularly favorable theoretical properties. If the weights $\{\mu_{nk}\}$ in (3), which constitute an infinite triangular matrix, satisfy the conditions of the theorem of Toeplitz (see for instance [5, pp. 24 - 36] or [6, pp. 24 - 67]), then the transformation is *regular*. This means that the convergence of $\{s_n\}_{n=0}^{\infty}$ to some limit s implies the convergence of the transformed sequence $\{s'_n\}_{n=0}^{\infty}$ to the same limit.

Theoretically, regularity is a very desirable property since one never has to worry about convergence to a wrong limit. However, this undeniable advantage is also the major weakness of regular matrix transformations. Sequence transformations as *computational tools* are needed most in situations in which only the numerical values of a few elements of a slowly convergent or divergent sequence are known. Here, regular matrix transformations are not very helpful because they are at most moderately powerful. In Wimp's book [6, p. x] it is stated that *linear methods are distinctly limited in their usefulness primarily because the class of sequences for which the methods are regular is too large*, and *experience indicates that the size of the domain of regularity of a transformation and its efficiency seem to be inversely related*.

Before computers became generally available, nonlinear sequence transformations, which do not satisfy (2), were computationally not very attractive. However, it turned out that many nonlinear transformations could be programmed quite effectively. In addition, nonlinear sequence transformations as for instance Wynn's ϵ [7] and ρ [8] algorithm or Brezinski's θ algorithm [9] offer better chances of getting good transformation results. The theoretical properties of nonlinear transformations are not yet very well understood, and since they are in general nonregular, it is by no means guaranteed that the transformed sequence $\{s'_n\}_{n=0}^{\infty}$ will converge at all, let alone to the same limit as the original sequence $\{s_n\}_{n=0}^{\infty}$. Hence, using a nonlinear transformation often involves a certain amount of risk. However, this undeniable disadvantage is compensated by the empirical fact that if a nonlinear transformation works, it frequently does so in a spectacular fashion.

Thus, in recent years nonlinear sequence transformations have dominated research. This shift of interest is also seen in the books on sequence transformations. The older books [1, 2, 3, 4, 5] concentrate on linear transformations and almost completely neglect nonlinear transformations. In Wimp's book [6], linear transformations still play an important part, but nonlinear methods already dominate. The recent book by Brezinski and Redivo Zaglia [10] provides the currently most up-to-date and complete survey of nonlinear sequence transformations, but contains only relatively little material on linear transformations.

In my opinion, the current neglect of linear transformations is only justified in the case of regular matrix transformations. In this article some examples of *linear* but *nonregular* sequence transformations will be presented which are – at least for the problems for which they were constructed – as powerful as the most efficient nonlinear sequence transformations.

2. Strategies for the Construction of Sequence Transformations

Sequence transformations are based upon the assumption that the elements s_n of the sequence to be transformed can for all $n \geq 0$ be partitioned into a *limit* or *antilimit*

s and a *remainder* r_n according to $s_n = s + r_n$.

A sequence transformation maps $\{s_n\}_{n=0}^{\infty}$ into a new sequence $\{s'_n\}_{n=0}^{\infty}$ whose elements can also be partitioned into s and a transformed remainder according to $s'_n = s + r'_n$. In general, the transformed remainders $\{r'_n\}_{n=0}^{\infty}$ will be different from zero. However, convergence is accelerated if the transformed remainders $\{r'_n\}_{n=0}^{\infty}$ vanish more rapidly as $n \to \infty$ than the original remainders $\{r_n\}_{n=0}^{\infty}$, and a divergent sequence is summed if the transformed remainders vanish at all as $n \to \infty$.

Thus, a sequence transformation tries to obtain a better approximation to the limit or antilimit s by eliminating the remainder r_n as effectively as possible from s_n. Of course, there are some complications. The remainders $\{r_n\}_{n=0}^{\infty}$ of a sequence $\{s_n\}_{n=0}^{\infty}$ are in general unknown, and their determination is normally not easier than the determination of the limit or antilimit s of this sequence. In addition, there is an in principle unlimited variety of possible types of remainders. Thus, it would be too optimistic to expect that a *single* sequence transformation would be able to accelerate the convergence of *all* convergent sequences or to sum *all* divergent series.

In view of the large variety of possible types of remainders, a sequence transformation has to make either implicitly or explicitly some assumptions about the behaviour of the remainders of the sequence to be transformed. Consequently, a sequence transformation will only work well if the actual behaviour of the remainders of the sequence to be transformed is in close agreement with the assumptions made.

If, however, a sequence transformation, which makes some specific and detailed assumptions about the remainders, is applied to a sequence of a different type, it cannot be excluded that the transformed sequence either converges to a wrong limit or – more likely – that it does not converge at all. Thus, we cannot expect that a sequence transformation, which is able to accelerate the convergence of a certain class of sequences *efficiently*, will simultaneously be *regular* for all other sequences.

In realistic situations, the exact remainders of the sequence to be transformed are either unknown or not easily accessible. However, at least some *structural informations* about the dependence of the remainders r_n on n are usually available. For example, in the case of the geometric series, the truncation error satisfies

$$\frac{1}{1-z} - \sum_{m=0}^{n} z^m = \frac{z^{n+1}}{1-z} . \tag{4}$$

Similarly, in the case of the Euler series

$$\int_0^{\infty} \frac{\mathrm{e}^{-t}\,\mathrm{d}t}{1+zt} \sim \sum_{n=0}^{\infty} (-1)^n \, n!\, z^n = {}_2F_0(1,1;-z) , \qquad z \to 0 , \tag{5}$$

which diverges quite strongly for every argument $z \neq 0$, it can be shown that the truncation error satisfies (see for instance [11, Theorem 13-1])

$$\int_0^{\infty} \frac{\mathrm{e}^{-t}\,\mathrm{d}t}{1+zt} - \sum_{\nu=0}^{n} (-1)^{\nu} \, \nu!\, z^{\nu} = (-z)^{n+1} \int_0^{\infty} \frac{t^{n+1}\,\mathrm{e}^{-t}\,\mathrm{d}t}{1+zt} . \tag{6}$$

For $z \geq 0$, the remainder integral of the Euler series is bounded in magnitude by the first term not included in the partial sum:

$$(z)^{n+1} \int_0^\infty \frac{t^{n+1}\, e^{-t}\, dt}{1+zt} \; \leq \; (n+1)!\, (z)^{n+1} . \tag{7}$$

Thus, both in the case of the geometric series and the Euler series, the truncation error possesses the same sign as the first term not included in the partial sum, and is bounded in magnitude by this term. In general, the first term not included in the partial sum is a simple estimate for the truncation error of a convergent series with strictly alternating terms [2, p. 259]. Moreover, it is also the best simple estimate for the truncation error of a strictly alternating nonterminating hypergeometric series ${}_2F_0(\alpha, \beta; -z)$ with $\alpha, \beta, z > 0$, which diverges quite strongly for every $z \neq 0$ [12, Theorem 5.12-5].

In the case of the series expansion of the Riemann zeta function, which is notorious for its bad convergence, the first term not included in the partial sum would not be suited as an estimate for the truncation error since [6, p. 20]

$$\zeta(z) - \sum_{m=0}^{n} (m+1)^{-z} \; = \; O\bigl((n+1)^{1-z}\bigr) , \qquad n \to \infty . \tag{8}$$

Here, the first term not included in the partial sum multiplied by $(n+1)$ would be a simple estimate for the truncation error [13, p. 382]. Structural informations of that kind, which need not be very detailed, are often sufficient for the construction of powerful sequence transformations.

3. Sequence Transformations with Explicit Remainder Estimates

A very fruitful approach for the construction of sequence transformations, which are capable of incorporating such structural informations, is based upon *model sequences*. In this approach, a sequence transformation $T_k^{(n)}$ is constructed in such a way that it produces the exact limit or antilimit $\tilde{s}$ of a model sequence

$$\tilde{s}_n \; = \; \tilde{s} \, + \, \tilde{r}_n \, , \tag{9}$$

if it is applied to a finite set of $k+1$ elements of this model sequence:

$$T_k^{(n)}(\tilde{s}_n, \tilde{s}_{n+1}, \ldots, \tilde{s}_{n+k}) \; = \; \tilde{s} \, . \tag{10}$$

If the remainders $\{\tilde{r}_n\}_{n=0}^{\infty}$ of this model sequence are capable of providing sufficiently accurate approximations to the remainders $\{r_n\}_{n=0}^{\infty}$ of the sequence $\{s_n\}_{n=0}^{\infty}$ to be transformed, then it can be hoped that $T_k^{(n)}$ will be able either to accelerate the convergence of $\{s_n\}_{n=0}^{\infty}$ to its limit s, or to sum $\{s_n\}_{n=0}^{\infty}$ to its antilimit s in the case of divergence. More detailed treatments of this construction principle as well as many examples can be found in [10, 11, 14, 15, 16, 17].

In practical problems it frequently happens that the dependence of the remainder r_n upon n may be very complicated. In such a case, the elimination of r_n from s_n with the help of a sequence transformation would be very difficult. However, this problem can often be simplified considerably by means of a suitable reformulation. Let us consider the following model sequence [11, Section 3.2]:

$$\tilde{s}_n = \tilde{s} + \omega_n z_n . \tag{11}$$

Here, ω_n is a *remainder estimate*, which has to be chosen according to some rule and which may depend on n in a very complicated way, and z_n is a *correction term*, which has to be determined and eliminated by a sequence transformation.

Let us assume that a *linear* operator $\hat{T}$ can be found, which is defined on finite subsets $s_n, s_{n+1}, \ldots, s_{n+k}$ of sequences, and which annihilates the correction term z_n:

$$\hat{T}(z_n) = 0 . \tag{12}$$

If such a linear annihilation operator $\hat{T}$ is known, a sequence transformation, which is exact for the model sequence (11), can be constructed quite easily. Just apply $\hat{T}$ to the ratio $[\tilde{s}_n - \tilde{s}]/\omega_n = z_n$. Then, a sequence transformation $\mathcal{T}$ results which is *exact* for the model sequence (11) [11, Section 3.2]:

$$\tilde{s} = \mathcal{T}(\tilde{s}_n, \omega_n) = \frac{\hat{T}(\tilde{s}_n/\omega_n)}{\hat{T}(1/\omega_n)} . \tag{13}$$

If we want to apply the sequence transformation $\mathcal{T}(\tilde{s}_n, \omega_n)$ also to sequences, which are not of the form of the model sequence (11), we have to decide how to choose the remainder estimates $\{\omega_n\}_{n=0}^{\infty}$ in such a case.

The annihilation operator $\hat{T}$ is by assumption *linear*. Consequently, the effect of $\mathcal{T}$ on an arbitrary $\{s_n\}_{n=0}^{\infty}$ converging to s can be expressed as follows:

$$\mathcal{T}(s_n, \omega_n) = s + \frac{\hat{T}([s_n - s]/\omega_n)}{\hat{T}(1/\omega_n)} . \tag{14}$$

Obviously, $\mathcal{T}(s_n, \omega_n)$ converges to s if $\hat{T}$ annihilates $[s_n - s]/\omega_n$ more effectively than $1/\omega_n$. Consequently, an optimal choice of the remainder estimates $\{\omega_n\}_{n=0}^{\infty}$ does not only depend upon the sequence $\{s_n\}_{n=0}^{\infty}$ to be transformed, but also upon the annihilation operator $\hat{T}$ and therefore upon the correction terms $\{z_n\}_{n=0}^{\infty}$.

There are various possible ways of choosing the correction terms $\{z_n\}_{n=0}^{\infty}$ [11, 18]. Particularly simple results are obtained if the annihilation operators are based upon the finite difference operator Δ which is defined by $\Delta f(n) = f(n+1) - f(n)$ [11, Sections 7 - 9]. As is well known, the k-th power of Δ annihilates every polynomial $P_{k-1}(n)$ of degree $k-1$ in n. Let us now assume that the correction terms $\{z_n\}_{n=0}^{\infty}$ are chosen in such a way that multiplication of z_n by some suitable quantity $w_k(n)$ yields a polynomial $P_{k-1}(n)$ of degree $k-1$ in n:

$$w_k(n)\, z_n = P_{k-1}(n) . \tag{15}$$

Then, a suitable annihilation operator for z_n is the weighted difference operator

$$\hat{T} = \Delta^k\, w_k(n) , \tag{16}$$

and the corresponding sequence transformation (13) is given by the ratio

$$\mathcal{T}_k^{(n)}\left(w_k(n) \middle| s_n, \omega_n\right) = \frac{\Delta^k \{w_k(n) s_n / \omega_n\}}{\Delta^k \{w_k(n) / \omega_n\}} . \tag{17}$$

Different sequence transformations are obtained by specializing $w_k(n)$. For instance, if we choose $w_k(n) = (n+\zeta)^{k-1}$ with $\zeta > 0$, we obtain Levin's sequence transformation [13]:

$$\begin{aligned} \mathcal{L}_k^{(n)}(\zeta, s_n, \omega_n) &= \frac{\Delta^k \{(n+\zeta)^{k-1} s_n / \omega_n\}}{\Delta^k \{(n+\zeta)^{k-1} / \omega_n\}} \\ &= \frac{\displaystyle\sum_{j=0}^{k} (-1)^j \binom{k}{j} \frac{(\zeta+n+j)^{k-1}}{(\zeta+n+k)^{k-1}} \frac{s_{n+j}}{\omega_{n+j}}}{\displaystyle\sum_{j=0}^{k} (-1)^j \binom{k}{j} \frac{(\zeta+n+j)^{k-1}}{(\zeta+n+k)^{k-1}} \frac{1}{\omega_{n+j}}} . \end{aligned} \tag{18}$$

Here, the same notation as in [11] is used. $\mathcal{L}_k^{(n)}(\zeta, s_n, \omega_n)$ is by construction exact for the model sequence

$$s_n = s + \omega_n \sum_{j=0}^{k-1} c_j / (n+\zeta)^j , \tag{19}$$

which implies that the correction term z_n is by assumption a polynomial of degree $k-1$ in $1/(n+\zeta)$. The shift parameter ζ has to be positive in order to admit $n=0$ in (18) and (19). The most obvious choice would be $\zeta = 1$.

Similarly, we can choose $w_k(n) = (n+\zeta)_{k-1} = \Gamma(n+\zeta+k-1)/\Gamma(n+\zeta)$ with $\zeta > 0$. This yields the following sequence transformation [11, Eq. (8.2-7)],

$$\begin{aligned} \mathcal{S}_k^{(n)}(\zeta, s_n, \omega_n) &= \frac{\Delta^k \{(n+\zeta)_{k-1} s_n / \omega_n\}}{\Delta^k \{(n+\zeta)_{k-1} / \omega_n\}} \\ &= \frac{\displaystyle\sum_{j=0}^{k} (-1)^j \binom{k}{j} \frac{(\zeta+n+j)_{k-1}}{(\zeta+n+k)_{k-1}} \frac{s_{n+j}}{\omega_{n+j}}}{\displaystyle\sum_{j=0}^{k} (-1)^j \binom{k}{j} \frac{(\zeta+n+j)_{k-1}}{(\zeta+n+k)_{k-1}} \frac{1}{\omega_{n+j}}} , \end{aligned} \tag{20}$$

which is by construction exact for the model sequence [11, Eq. (8.2-1)]

$$s_n = s + \omega_n \sum_{j=0}^{k-1} c_j / (n+\zeta)_j . \tag{21}$$

In this case, the correction term z_n is by assumption a truncated factorial series. As in (18), the shift parameter ζ has to be positive, and $\zeta = 1$ is again the most obvious choice. The ratio (20) was originally derived by Sidi [19] for the construction of explicit expressions for Padé approximants of some special hypergeometric series, but it was apparently not used by Sidi as a sequence transformation. Later, (20) was used by Shelef [20] for the numerical inversion of Laplace transforms.

Further examples of sequence transformations, which are special cases of (17), can be found in sections 7 - 9 of [11]. A sequence transformation, which interpolates between Levin's transformation $\mathcal{L}_k^{(n)}(\zeta, s_n, \omega_n)$ and the closely related sequence transformation $\mathcal{S}_k^{(n)}(\zeta, s_n, \omega_n)$, was described in [21].

4. On the Choice of the Remainder Estimates

The sequence transformations (18) and (20) need as input data not only a sequence $\{s_n\}_{n=0}^{\infty}$, but also explicit remainder estimates $\{\omega_n\}_{n=0}^{\infty}$. The explicit incorporation of the information contained in the remainder estimates makes (18) and (20) both very versatile and very powerful[11, 21, 22, 23, 24, 25, 26, 27, 28].

The effect of a sequence transformation of the type of (17) on an arbitrary sequence $\{s_n\}_{n=0}^{\infty}$ can be expressed as follows:

$$\mathcal{T}_k^{(n)}\bigl(w_k(n)\big|s_n, \omega_n\bigr) \;=\; s + \frac{\Delta^k\{w_k(n)(s_n - s)/\omega_n\}}{\Delta^k\{w_k(n)/\omega_n\}} \,. \tag{22}$$

Obviously, $\mathcal{T}_k^{(n)}\bigl(w_k(n)\big|s_n, \omega_n\bigr)$ converges to s if the remainder estimates $\{\omega_n\}_{n=0}^{\infty}$ are chosen in such a way that the ratio $(s_n - s)/\omega_n$ depends less strongly on n than $1/\omega_n$. Repeated differencing will then annihilate the numerator more rapidly than the denominator. Hence, $\mathcal{T}_k^{(n)}\bigl(w_k(n)\big|s_n, \omega_n\bigr)$ will be successful if the remainder estimates can be chosen such that

$$(s_n - s)/\omega_n \;=\; c + O(1/n)\,, \qquad c \neq 0\,, \qquad n \to \infty\,. \tag{23}$$

The asymptotic condition (23) does not determine the remainder estimates $\{\omega_n\}_{n=0}^{\infty}$ uniquely. Hence, given a sequence $\{s_n\}_{n=0}^{\infty}$ which converges to some limit s, it is at least in principle possible to find an unlimited variety of different sequences $\{\omega_n\}_{n=0}^{\infty}$ of remainder estimates which all satisfy (23).

On the basis of heuristic and asymptotic arguments Levin [13] suggested for sequences of partial sums

$$s_n \;=\; \sum_{\nu=0}^{n} a_\nu \,, \tag{24}$$

the following simple explicit remainder estimates:

$$\omega_n \;=\; (\zeta + n)\, a_n \;=\; (\zeta + n)\, \Delta s_{n-1}\,, \qquad \zeta > 0\,, \tag{25}$$

$$\omega_n \;=\; a_n \;=\; \Delta s_{n-1}\,, \tag{26}$$

$$\omega_n \;=\; \frac{a_n a_{n+1}}{a_n - a_{n+1}} \;=\; \frac{\Delta s_{n-1} \Delta s_n}{\Delta s_{n-1} - \Delta s_n}\,. \tag{27}$$

The use of these remainder estimates in (18) yields Levin's u, t, and v transformation, respectively. Later, Smith and Ford [22] suggested the remainder estimate

$$\omega_n \;=\; a_{n+1} \;=\; \Delta s_n\,, \tag{28}$$

which is – as already remarked at the end of section 2 – actually the best simple truncation error estimate for convergent and divergent alternating series. A more detailed discussion of the properties of these remainder estimates, some generalizations, an extended heuristic motivation, and a description of the types of sequences, for which these estimates should be effective, can be found in [11, Section 7.3].

The remainder estimates (25) - (28) can also be used in (20) [11, Section 8.4]. In spite of their simplicity, their use in (18) or (20) leads to powerful sequence transformations, which are obviously nonlinear and nonregular [11, Sections 7.3 and 8.4]. The theoretical convergence properties of (18) and (20) were analyzed in articles by Sidi [23, 24, 29], in [11, Sections 12 - 14], and also in [28, Section 4].

The main advantage of the simple remainder estimates (25) - (28) is that they can be used in situations in which only the numerical values of a few terms of a slowly convergent or divergent series are known. If, however, explicit expressions for the terms of the series to be transformed are known, a different approach may be more profitable. For instance, it may be possible to construct explicit analytical approximations to the truncation errors, which satisfy the asymptotic condition (23) and which can therefore be used as remainder estimates. If remainder estimates $\{\omega_n\}_{n=0}^{\infty}$, which do not depend explicitly on the input data $\{s_n\}_{n=0}^{\infty}$, are used in (18) or (20), the resulting sequence transformation will are *linear* and normally also *nonregular*.

Remainder estimates of that kind, which do not depend explicitly on the terms of series to be transformed, were first proposed by Sidi in connection with moment series [24, p. 249] and used for the construction of quadrature rules [30, 31, 32]. However, these ideas were used at least implicitly much earlier, for instance in [33]. Then, a linear and nonregular variant of Levin's transformation (18) had already been used long before the discovery of Levin's transformation. If we set in (18) $\omega_n = 1/(n+\zeta)$ with $\zeta > 0$, the denominator of (18) can be expressed in closed form yielding a linear sequence transformation [11, Eq. (7.3-20)] which is nonregular [6, p. 36]. In Wimp's book [6, Section 2.3.4], this sequence transformation is attributed to Salzer [34]. However, it is actually older since it is a special case of the Richardson extrapolation process [35] (see [11, p. 244]). The practical usefulness of this sequence transformations is limited. Nevertheless, it is one of the best accelerators for certain logarithmically convergent sequences [6, Section 2.3.4].

Similarly, if we set in (20) $\omega_n = 1/(n+\zeta-1)$ with $\zeta > 1$, the denominator in (20) can be expressed in closed form yielding a linear sequence transformation [11, Eq. (8.4-11)] which is also nonregular [6, Section 2.3.5]. It is a pwerful accelerator for certain logarithmically convergent sequences, but otherwise, its practical usefulness is limited.

5. An Expansion in Terms of Reduced Bessel Functions

In this section, the acceleration of convergence of the expansion [36, Eq. (6.5)]

$$1/z \;=\; \sum_{m=0}^{\infty} \hat{k}_{m-1/2}(z) \,/\, [2^m\, m!] \,, \qquad z > 0 \,, \tag{29}$$

by means of a linear variant of Levin's transformation (18) will be discussed. The reduced Bessel function of order ν is defined by [37, p. 275]

$$\hat{k}_\nu(z) \; = \; (2/\pi)^{1/2} \, z^\nu \, K_\nu(z) \, , \tag{30}$$

with $K_\nu(z)$ being a modified Bessel function of the second kind:

$$K_\nu(z) \; = \; \frac{\pi}{2\sin(\pi\nu)} \left\{ I_{-\nu}(z) \, - \, I_\nu(z) \right\} , \tag{31}$$

$$I_\nu(z) \; = \; \sum_{m=0}^{\infty} \frac{(z/2)^{\nu+2m}}{m! \, \Gamma(\nu+m+1)} \, . \tag{32}$$

If the order of $\hat{k}_\nu(z)$ is half integral, $\nu = n + 1/2$ with n being a nonnegative integer, it can be represented as an exponential multiplied by a terminating confluent hypergeometric series ${}_1F_1$:

$$\hat{k}_{n+1/2}(z) \; = \; 2^n \, (1/2)_n \, \mathrm{e}^{-z} \, {}_1F_1(-n; -2n; 2z) \, . \tag{33}$$

The reduced Bessel functions can be computed with the help of the following three-term recursion which is stable in forward direction:

$$\hat{k}_{\nu+1}(z) \; = \; 2\nu \hat{k}_\nu(z) \, + \, z^2 \hat{k}_{\nu-1}(z) \, . \tag{34}$$

Suitable starting values are $\hat{k}_{-1/2}(z) = \mathrm{e}^{-z}/z$ and $\hat{k}_{1/2}(z) = \mathrm{e}^{-z}$.

Because of its extremely slow convergence, the series expansion (29) is a very challenging test problem. With the help of some known asymptotic properties of reduced Bessel functions [11, Section 14.3] it can be shown that

$$1/z \, - \, \sum_{m=0}^{n} \hat{k}_{m-1/2}(z) \, / \, [2^m \, m!] \; = \; O(n^{-1/2}) \, , \qquad n \to \infty \, . \tag{35}$$

This asymptotic estimate implies that increasing the number of terms of the partial sum by a factor of 100 leads only to a gain of a single decimal digit. Consequently, the series (29) is a very demanding problem.

In view of (35), it is an obvious idea to try to use Levin's sequence transformation (18) in combination with the remainder estimate

$$\omega_n \; = \; (n+\zeta)^{-1/2} \, , \qquad \zeta > 0 \, . \tag{36}$$

This yields the sequence transformation

$$\mathcal{L}_k^{(n)}(\zeta, s_n(z), (n+\zeta)^{-1/2}) \; = \; \frac{\displaystyle\sum_{j=0}^{k} (-1)^j \binom{k}{j} \frac{(\zeta+n+j)^{k-1/2}}{(\zeta+n+k)^{k-1}} \, s_{n+j}(z)}{\displaystyle\sum_{j=0}^{k} (-1)^j \binom{k}{j} \frac{(\zeta+n+j)^{k-1/2}}{(\zeta+n+k)^{k-1}}} \, , \tag{37}$$

Table I: Acceleration of the Convergence of the Series (29) for $z = 1/2$.

n	Partial sum $s_n(z)$ Eq. (38)	$u_n^{(0)}(1/2, s_0(z))$ $\omega_n = (n+1/2)\Delta s_{n-1}$	$\mathcal{L}_n^{(0)}(1/2, s_0(z), \omega_0)$ $\omega_n = (n+1/2)^{-1/2}$
5	1.7550395054422	2.00738324942506	2.00024590012830
6	1.7752650093680	1.99057913476461	2.00016604510045
7	1.7911964491931	2.00215775399122	1.99995687998898
8	1.8041633171580	2.00043327863322	1.99999950221378
9	1.8149828698370	1.99962798058347	2.00000233598712
10	1.8241884943835	2.00008193424340	1.99999953857509
11	1.8321449539650	2.00000542454101	1.99999998438564
12	1.8391112148225	1.99999184482571	2.00000002531206
13	1.8452768640345	2.00000213245876	1.99999999511376
14	1.8507843002336	1.99999990763032	1.99999999995522
15	1.8557428278275	1.99999989059952	2.00000000022561
16	1.8602379276877	2.00000003775919	1.99999999995024
17	1.8643375404571	1.99999999543626	2.00000000000196
18	1.8680964369459	1.99999999919590	2.00000000000160
19	1.8715593276470	2.00000000047323	1.99999999999955
20	1.8747631197850	1.99999999990910	2.00000000000004
21	1.8777385849386	2.00000000000144	2.00000000000001
22	1.8805116108815	2.00000000000405	2.00000000000000

which is obviously a *linear* weighted average of the partial sums

$$s_n(z) \;=\; \sum_{m=0}^{n} \widehat{k}_{m-1/2}(z)/[2^m\, m!]\,. \tag{38}$$

However, the linear transformation (37) is nonregular since the conditions of the theorem of Toeplitz are not satisfied.

The use of the remainder estimate (25) in (18) yields Levin's u transformation [13]

$$u_k^{(n)}(\zeta, s_n) \;=\; \mathcal{L}_k^{(n)}(\zeta, s_n, \omega_n)\,, \tag{39}$$

which is one of the best accelerators for logarithmically convergent series of the type of (29).

Table I shows the effect of Levin's u transformation (39) and of the linear transformation (37) on the partials sums (38) of (29). The results in Table I show that the remainder estimate $\omega_n = (n+1/2)^{-1/2}$, which is the basis of (37), leads to a more rapid convergence than Levin's simple remainder estimate $\omega_n = (\zeta + n)\Delta s_{n-1}$ with $\zeta = 1/2$, which is the basis of the u transformation (39).

The results produced by the linear transformation (37) are very good. The only sequence transformations, which transform the infinite series (29) more efficiently, are the modified Aitken process of Bjørstad, Dahlquist, and Grosse [38] and Osada's [39] variant of Wynn's ρ algorithm (compare [40, Table III]).

Table II: Summation of the Divergent Asymptotic Series
${}_2F_0\left(1/2+\nu, 1/2-\nu; -1/(2z)\right) = (2z/\pi)^{1/2}\, e^z\, K_\nu(z)$ for $z = 5/2$ and $\nu = 1/3$.

n	$\mathcal{L}_n^{(0)}\left(1, s_0(z), \omega_0\right)$ $\omega_n = \Delta s_n(z)$	$\mathcal{L}_n^{(0)}\left(1, s_0(z), \omega_0\right)$ $\omega_n = \frac{n!}{(-2z)^{n+1}}$	$\mathcal{S}_n^{(0)}\left(1, s_0(z), \omega_0\right)$ $\omega_n = \Delta s_n(z)$	$\mathcal{S}_n^{(0)}\left(1, s_0(z), \omega_0\right)$ $\omega_n = \frac{n!}{(-2z)^{n+1}}$
5	0.976586962847258	0.976586976053676	0.976586751881455	0.976586753512561
6	0.976586714703266	0.976586713038724	0.976586743113963	0.976586743231855
7	0.976586745119273	0.976586745311558	0.976586742166994	0.976586742175300
8	0.976586742028637	0.976586742010056	0.976586742223496	0.976586742225109
9	0.976586742220275	0.976586742221389	0.976586742227861	0.976586742228148
10	0.976586742232954	0.976586742233019	0.976586742227523	0.976586742227570
11	0.976586742226434	0.976586742226400	0.976586742227483	0.976586742227491
12	0.976586742227589	0.976586742227595	0.976586742227485	0.976586742227487
13	0.976586742227489	0.976586742227489	0.976586742227485	0.976586742227486
14	0.976586742227482	0.976586742227482	0.976586742227485	0.976586742227485
15	0.976586742227486	0.976586742227486	0.976586742227485	0.976586742227485
16	0.976586742227485	0.976586742227485	0.976586742227485	0.976586742227485

6. The Summation of the Divergent Asymptotic Series for the Modified Bessel Function of the Second Kind

The ideas, which were described in the last section, can also be used for the summation of strongly divergent series.

The modified Bessel function $K_\nu(z)$ possesses the following asymptotic expansion as $|z| \to \infty$:

$$K_\nu(z) \sim [\pi/(2z)]^{1/2}\, e^{-z}\, {}_2F_0\big(1/2+\nu, 1/2-\nu; -1/(2z)\big)\,. \tag{40}$$

This series diverges quite strongly for every $|z| < \infty$, but is Stieltjes summable and Borel summable [26, Section 2]. In addition, the divergent sequence of partial sums

$$s_n(z) = \sum_{m=0}^{n} \frac{(1/2+\nu)_m\, (1/2-\nu)_m}{m!} \frac{(-1)^m}{(2z)^m} \tag{41}$$

of the hypergeometric series ${}_2F_0$ in (40) can be summed quite efficiently by (18) and (20) in combination with the remainder estimate (28) [26, Table 2 and 3].

The coefficients of the asymptotic series ${}_2F_0$ satisfy as $n \to \infty$ [26, Eq. (3.2)]:

$$\frac{(1/2+\nu)_n\, (1/2-\nu)_n}{n!} = \frac{(n-1)!}{\Gamma(1/2+\nu)\Gamma(1/2-\nu)}\big\{1 + O(1/n)\big\}\,. \tag{42}$$

The remainder estimate (28) corresponds to the first term of the divergent hypergeometric series ${}_2F_0$ not included in the partial sum (41). Consequently, it is an obvious idea to use in (18) and (20) the remainder estimate

$$\omega_n = \frac{n!}{(-2z)^{n+1}} \tag{43}$$

which is just the asymptotic ($n \to \infty$) and ν-independent form of (28). This yields the following sequence transformations acting on the partial sums (41):

$$\mathcal{L}_k^{(n)}(\zeta, s_n(z), n!/(-2z)^{n+1}) = \frac{\sum_{j=0}^{k} (-1)^j \binom{k}{j} \frac{(\zeta+n+j)^{k-1}}{(\zeta+n+k)^{k-1}} \frac{(-2z)^{n+j+1} s_{n+j}(z)}{(n+j)!}}{\sum_{j=0}^{k} (-1)^j \binom{k}{j} \frac{(\zeta+n+j)^{k-1}}{(\zeta+n+k)^{k-1}} \frac{(-2z)^{n+j+1}}{(n+j)!}}, \tag{44}$$

$$\mathcal{S}_k^{(n)}(\zeta, s_n(z), n!/(-2z)^{n+1}) = \frac{\sum_{j=0}^{k} (-1)^j \binom{k}{j} \frac{(\zeta+n+j)_{k-1}}{(\zeta+n+k)_{k-1}} \frac{(-2z)^{n+j+1} s_{n+j}(z)}{(n+j)!}}{\sum_{j=0}^{k} (-1)^j \binom{k}{j} \frac{(\zeta+n+j)_{k-1}}{(\zeta+n+k)_{k-1}} \frac{(-2z)^{n+j+1}}{(n+j)!}}. \tag{45}$$

Although these two transformations are *rational* in z, they are *linear* with respect to the partial sums (41).

Table II shows the effect of (18) and (20) in combination with the remainder estimates (28) and (43), respectively, on the partial sums (41).

Although the best results in Table II are achieved by (20) in combination with the remainder estimate (28), the results obtained by (18) and (20) in combination with the remainder estimate (43) are only slightly less good. In fact, they are much better than the summation results which can be obtained by Padé approximants. For instance, the application of Wynn's ϵ algorithm [7] to the partial sums (41) with $z = 5/2$ and $\nu = 1/3$ yielded

$$\epsilon_{14}^{(1)} = 0.976\,586\,741\,439\,565\,, \tag{46}$$

$$\epsilon_{16}^{(0)} = 0.976\,586\,742\,559\,173\,. \tag{47}$$

7. Conclusions

There can be no reasonable doubt that *regular* matrix transformations are at most moderately powerful, and that nonlinear sequence transformations will continue to dominate practical applications since they offer better chances of getting good transformation results. Nevertheless, the numerical results in Tables I and II show that suitably constructed linear variants of the sequence transformations (18) and (20) can be extremely powerful.

The results in Tables I and II seem to indicate that the power of (18) and (20) depends mainly on quality of the remainder estimates $\{\omega_n\}_{n=0}^{\infty}$ being used. Whether the resulting sequence transformation is a linear or nonlinear function of the elements of the sequence $\{s_n\}_{n=0}^{\infty}$ to be transformed seems to be only of secondary importance.

The sequence transformations (18) and (20) would only need two elements s_n and s_{n+1} to produce the exact limit or antilimit s of a sequence $\{s_n\}_{n=0}^{\infty}$, if remainder estimates $\{\omega_n\}_{n=0}^{\infty}$ could be found such that the ratios $(s_n - s)/\omega_n$ are for all $n \geq 0$ a constant $c \neq 0$. In practice, this is of course impossible, and one has to be content with remainder estimates satisfying the asymptotic condition (23). Nevertheless, it is always the ultimate goal to make the ratio $(s_n - s)/\omega_n$ as n-independent as possible. Whether this can be accomplished more effectively with the help of the simple remainder estimates (25) - (28), which are computed from the elements of the input sequence $\{s_n\}_{n=0}^{\infty}$, or with the help of alternative remainder estimates, which do not depend not explicitly on the input data, depends of course on the problem under consideration and cannot be answered in general.

The approach described in this article completely ignores the question of regularity. The linear transformations are all constructed in such a way that they give good results for a special problem, and there is no intention of using them otherwise. Whether these transformations are regular also for other classes of sequences is therefore quite irrelevant from a practical point of view.

References

1. Hardy, G.H.: 1949, *Divergent series* (Clarendon Press, Oxford).
2. Knopp, K.: 1964, *Theorie und Anwendung der unendlichen Reihen* (Springer-Verlag, Berlin)
3. Petersen, G.M.: 1966, *Regular Matrix Transformations* (McGraw-Hill, London).
4. Peyerimhoff, A.: 1969, *Lectures on Summability* (Springer-Verlag, Berlin).
5. Powell, R.E. and Shah, S.M.: 1988, *Summability Theory and Its Applications* (Prentice-Hall of India, New Delhi).
6. Wimp, J.: 1981, *Sequence Transformations and Their Applications* (Academic Press, New York).
7. Wynn, P.: 1956, 'On a Device for Computing the $e_m(S_n)$ Transformation', *Math. Tables Aids Comput.* Vol. no. **10**, pp. 91 - 96.
8. Wynn, P.: 1956, 'On a Procrustean Technique for the Numerical Transformation of Slowly Convergent Sequences and Series', *Proc. Camb. Phil. Soc.* Vol. no. **52**, pp. 663 - 671.
9. Brezinski, C.: 1971, 'Accélération de Suites à Convergence Logarithmique', *C. R. Acad. Sc. Paris* Vol. no. **273**, pp. 727 - 730.
10. Brezinski, C. and Redivo Zaglia M.: 1991, *Extrapolation Methods* (North-Holland, Amsterdam).
11. Weniger, E.J.: 1989, 'Nonlinear Sequence Transformations for the Acceleration of Convergence and the Summation of Divergent Series', *Comput. Phys. Rep.* Vol. no. **10**, pp. 189 - 371.
12. Carlson, B.C.: 1977, *Special Functions of Applied Mathematics* (Academic Press, New York).
13. Levin, D.: 1973, 'Development of Non-Linear Transformations for Improving Convergence of Sequences', *Int. J. Comput. Math.* Vol. no. B **3**, 371 - 388.
14. Brezinski, C.: 1988, 'A New Approach to Convergence Acceleration Methods', in Cuyt. A., *Nonlinear Numerical Methods and Rational Approximation* (Reidel, Dordrecht), pp. 373 - 405.
15. Matos, A.: 1990, 'Acceleration Methods Based upon Convergence Tests', *Numer. Math.* Vol. no. **58**, pp. 329 - 340.
16. Matos, A.: 1990, 'A Convergence Acceleration Method Based on a Good Estimation of the Absolute Error', *IMA J. Numer. Anal.* Vol. no. **10**, pp. 243 - 251.
17. Brezinski, C. and Redivo Zaglia M.: 1991, 'Construction of Extrapolation Processes', *Appl. Numer. Math.* Vol. no. **8**, pp. 11 - 23.
18. Sidi, A.: 1982, 'An Algorithm for a Special Case of a Generalization of the Richardson Extrapolation Process', *Numer. Math.* Vol. no. **38**, pp. 299 - 307.
19. Sidi, A.: 1981, 'A New Method for Deriving Padé Approximants for Some Hypergeometric Functions', *J. Comput. Appl. Math.* Vol. no. **7**, pp. 37 - 40.
20. Shelef, R.: 1987, *New Numerical Quadrature Formulas for Laplace Transform Inversion by Bromwich's Integral* (in Hebrew), (Master Thesis, Technion, Israel Institute of Technology, Haifa).

21. Weniger, E.J.: 1992, 'Interpolation between Sequence Transformations', *Numer. Algor.* Vol. no. **3**, pp. 477 - 486.
22. Smith, D.A. and Ford, W.F.: 1979, 'Acceleration of linear and logarithmic convergence', *SIAM J. Numer. Anal.* Vol. no. **16**, pp. 223 - 240.
23. Sidi, A.: 1979, 'Convergence Properties of Some Nonlinear Sequence Transformations', *Math. Comput.* Vol. no. **33**, pp. 315 - 326.
24. Sidi, A.: 1980, 'Analysis of Convergence of the T-Transformation for Power Series', *Math. Comput.* Vol. no. **35**, pp. 833 - 850.
25. Weniger, E.J.: 1990, 'On the summation of some divergent hypergeometric series and related perturbation expansions', *J. Comput. Appl. Math.* Vol. no. **32**, pp. 291 - 300.
26. Weniger, E.J. and Čížek, J.: 1990, 'Rational Approximations for the Modified Bessel Function of the Second Kind', *Comput. Phys. Commun.* Vol. no. **59**, pp. 471 - 493.
27. Weniger, E.J., Čížek, J. and Vinette F.: 1993, 'Very accurate summation for the infinite coupling limit of the perturbation series expansions of anharmonic oscillators', *Phys. Lett. A* Vol. no. **156**, pp. 169 - 174.
28. Weniger, E.J., Čížek, J. and Vinette F.: 1993, 'The Summation of the Ordinary and Renormalized Perturbation Series for the Ground State Energy of the Quartic, Sextic, and Octic Anharmonic Oscillators Using Nonlinear Sequence Transformations', *J. Math. Phys.* Vol. no. **34**, pp. 571 - 609.
29. Sidi, A.: 1986, 'Borel Summability and Converging Factors for Some Everywhere Divergent Series', *SIAM J. Math. Anal.* Vol. no. **17**, pp. 1222 - 1231.
30. Sidi, A.: 1980, 'Numerical Quadrature and Nonlinear Sequence Transformations; Unified Rules for Efficient Computation of Integrals With Algebraic and Logarithmic Endpoint Singularities', *Math. Comput.* Vol. no. **35**, pp. 851 - 874.
31. Sidi, A.: 1982, 'Numerical Quadrature Rules for Some Infinite Range Integrals', *Math. Comput.* Vol. no. **38**, pp. 127 - 142.
32. Sidi, A.: 1982, 'Converging Factors for Some Asymptotic Moment Series that Arise in Numerical Quadrature', *J. Austral. Math. Soc. (Series B)* Vol. no. **24**, pp. 223 - 233.
33. Wimp, J.: 1972, 'Some Transformations of Monotone Sequences', *Math. Comput.* Vol. no. **26**, pp. 251 - 254.
34. Salzer, H.E.: 1954, 'A Simple Method for Summing Certain Slowly Convergent Series', *J. Math. and Phys. (Cambridge, Mass.)* Vol. no. **33**, pp. 356 - 359.
35. Richardson, L.F.: 1927, 'The Deferred Approach to the Limit. I. Single Lattice', *Phil. Trans. Roy. Soc. London A* Vol. no. **226**, pp. 229 - 349.
36. Filter, E. and Steinborn, E.O.: 1978, 'The Three-Dimensional Convolution of Reduced Bessel Functions and Other Functions of Physical Interest', *J. Math. Phys.* Vol. no. **19**, pp. 79 - 84.
37. Steinborn, E.O. and Filter, E.: 1975, 'Translations of Fields Represented by Spherical-Harmonic Expansions for Molecular Calculations. III. Translations of Reduced Bessel Functions, Slater-Type *s*-Orbitals, and Other Functions', *Theor. Chim. Acta* Vol. no. **38**, pp. 273 - 281.
38. Bjørstad, P., Dahlquist, G. and Grosse, E.: 1981, 'Extrapolations of Asymptotic Expansions by a Modified Aitken δ^2-Formula', *BIT* Vol. no. **21**, pp. 56 - 65.
39. Osada, N.: 1990, 'A Convergence Acceleration Method for Some Logarithmically Convergent Sequences', *SIAM J. Numer. Anal.* Vol. no. **27**, pp. 178 - 189.
40. Weniger, E.J.: 1991, 'On the Derivation of Iterated Sequence Transformations for the Acceleration of Convergence and the Summation of Divergent Series', *Comput. Phys. Commun.* Vol. no. **64**, pp. 19 - 45.

THE ACCURACY-THROUGH-ORDER AND THE EQUIVALENCE PROPERTIES IN THE ALGEBRAIC APPROXIMANT *

GEORGE A. BAKER, JR.
Theoretical Division, Los Alamos National Laboratory
University of California, Los Alamos, N. M. 87545 USA

Abstract. In addition to the accuracy-through-order requirement that the defining polynomials not all be divisible by z, as required for Padé and integral approximants, there is the further problem of deficiency as pointed out by McInnes. I prove a finite bound on the deficiency and also prove the accuracy-through-order property for algebraic approximants. In addition I prove the equivalence property for algebraic approximants.

Key words: Padé approximant, Algebraic approximant, Accuracy-through-order property, Equivalence property

In the study of Padé approximants, there are a number of basic properties, which are fundamental, and are so well known that they are often taken for granted. The theory of algebraic approximants is not so well developed and some of these properties are not yet established. The theory of both types of approximants is a special case of the theory of Hermite-Padé approximants and starts with a linear, polynomial-defining equation,

$$\sum_{j=0}^{k} P_j(z) f_j(z) = O(z^{s+1}),$$

for the $P_j(z)$'s plus some initial conditions. The approximant involves the solution of

$$\sum_{j=0}^{k} P_j(z) y_j(z) = 0,$$

where the $y_j(z)$'s are related to each other in a manner akin the the way the $f_j(z)$'s are related to each other. The first basic property is to show that as the degree of contact at the origin in the polynomial defining equation increase indefinitely, then so too does the degree of contact between the $f_j(z)$'s and the $y_j(z)$'s. This property is called the accuracy-through-order property. The next basic property is uniqueness. It consists of two parts. First the uniqueness of the approximant polynomials and second the uniqueness of the solution, given the polynomials, for the approximant. Finally there is the problem of equivalence. It is, in the case of

* Work supported in part by the US Department of Energy (contract W-7405-ENG-36) and in part by the NSF (grant INT-90 13977)

A. Cuyt (ed.), *Nonlinear Numerical Methods and Rational Approximation II*, 283–289.

the Padé approximant stated as, for a function $g(z)$ which is analytic at the origin,

$$g(z) = \frac{\sum_{j=0}^{l} c_j z^j}{1+\sum_{j=1}^{m} e_j z^j}$$

if and only if,

$$[L/M]_g = [l/m]_g \quad \forall\, L \geq l,\ M \geq m,$$

where $[\lambda/\mu]_g$ is the Padé approximant to $g(z)$ with the degree of the numerator polynomial equal to at most λ and the degree of the denominator polynomial equal to at most μ. The Padé polynomials satisfy a defining equation with $s = \lambda + \mu$. The problem of uniqueness of the polynomials has been addressed by Baker and Graves-Morris (1990) and of the solution by McInnes (1992). The Padé approximants also have very useful invariance properties with respect to linear fractional transformations of both their arguments and their values. This property is shared by the algebraic approximants (Baker, 1984). So far however, the accuracy-through-order property and the equivalence property for algebraic approximants have not been adequately treated. We will give results on these properties.

The algebraic approximants were introduce by Padé (1892) and an investigation of the quadratic case was made by Shafer (1974).

DEFINITION 1. *Let a function $f(z)$ be given in terms of its Maclaurin expansion. We define the algebraic polynomials Q_{j,m_j} by the accuracy-through-order principle by means of*

$$\sum_{j=-1}^{k} Q_{j,m_j}(z)[f(z)]^{j+1} = O(z^{s+1}), \tag{1}$$

where the $Q_{j,m_j}(z)$ are polynomials of degree at most m_j and s is given by

$$M = \sum_{j=0}^{k}(m_j+1) - 1, \quad s = M + m_{-1}. \tag{2}$$

We will use the convention that if $m_j = -1$, then $Q_{j,-1}(z) \equiv 0$. The algebraic approximant is denoted by $\langle m_{-1}/m_0, \ldots, m_k\rangle$ and is defined by the solution $y(z)$ of

$$\sum_{j=-1}^{k} Q_{j,m_j}(z)[y(z)]^{j+1} = 0, \quad \text{where} \quad y(0) = f(0). \tag{3}$$

We will always assume that $f(0) \neq 0$. Of course, we can impose additional boundary conditions at $z = 0$, if necessary to break the degeneracy of the possible solutions. See McInnes (1992) for an additional discussion of this point.

In order to complete the definition of these approximants, attention must be paid to the possibility that (1) does not uniquely define the polynomials. The points where $Q_{k,m_k}(z)$ vanishes correspond to singularities. Consequently, in the case where the polynomials are not uniquely determined it would be desirable, in my view, to eliminate any arbitrary or spurious singularities that are introduced by this lack of uniqueness. We complete these definitions by using the minimal polynomials of Baker and Graves-Morris (1990)

DEFINITION 2. *A solution for the algebraic polynomials of type $(m_{-1}, \ldots, m_k)$ to $f(z)$ is called minimal if it is of the lowest degree in the following sense. First there exists no other solution of type $(m_{-1}, \ldots, m_k)$ for which the actual degree of Q_{k,m_k} is smaller. If there exist solutions of type $(m_{-1}, \ldots, m_k)$ for which $Q_{k,m_k} \equiv 0$, then we minimize the degree of $Q_{k-1,m_{k-1}}$, $Q_{k-2,m_{k-2}}$, etc. to find the minimal solution.*

Uniqueness is insured by their arguments. In addition, we will impose the condition that $\vec{Q}_{\vec{m}}(0) \neq \vec{0}$. Baker and Graves-Morris (1994) have proven that we may do so and, although some types $\vec{m}$ may then fail to exist, the table of approximants is as complete, in so far as the derived approximants are concerned, as was the table of approximants derived from the minimal polynomials. Some of the repeats in the table may however have been dropped by this restriction.

A very important property of Padé approximants is that of accuracy-through order. In that case, the requirement that the denominator polynomial not vanish at the origin was sufficient to show that the Padé approximant agreed with the defining series to the same order in z, as the accuracy of the polynomial defining equations. For integral approximants the same results obtain, but for algebraic approximants, as we shall see, the results are roughly true as well, but the situation is a bit more complex. This question has been addressed by Baker and Graves-Morris (1990) for the case of integral approximants, and by McInnes (1992) in the algebraic case.

In order to analyze the problem of accuracy-through-order for the algebraic approximants, we need, following McInnes (1992), to introduce the concept of *deficiency.* While our minimal definition for the algebraic polynomials and the results of Baker and Graves-Morris (1994) insure that we need only look at cases for which $\vec{Q}_{\vec{m}}(0) \neq \vec{0}$, there is another quantity which is important to consider in the algebraic case, *i.e.*, the coefficient of dy/dz in (5) below.

DEFINITION 3. *The deficiency d of the algebraic polynomials to $f(z)$ is given by*

$$\sum_{j=0}^{k} (j+1) Q_{j,m_j}(z)[f(z)]^j \underset{z\to 0}{\propto} z^d. \tag{4}$$

THEOREM 4. *If the error in the algebraic polynomial defining equation is $O(z^{\mathcal{M}})$ and the deficiency of the algebraic polynomials is d, then the Maclaurin series for the algebraic approximant differs from the defining series with an error at worst $O(z^{\mathcal{M}-d})$ when we specify that $y^{(j)}(0) = f^{(j)}(0)$ for $j = 0, \ldots, d$.*

PROOF: First, if we differentiate the approximant defining equation (with error displayed for convenience), we get,

$$\left(\sum_{j=0}^{k} (j+1) Q_{j,m_j}(z)[y(z)]^j\right) \frac{dy(z)}{dz} = -\sum_{j=-1}^{k} Q'_{j,m_j}(z)[y(z)]^{j+1} + O(z^{s+t}), \tag{5}$$

where s is defined by (2) and t is the degree of oversatisfaction defined by,

$$\sum_{j=-1}^{k} Q_{j,m_j}(z)[f(z)]^{j+1} = O(z^{s+t+1}), \tag{6}$$

and where the right-hand side is to be of true stated order. Suppose that the deficiency is $d = 0$, then by letting $z \to 0$ in (5) we can compute y_1 from $y_0 = f_0$, where $y(z) = \sum_{j=0}^{\infty} y_j z^j$. If we now compute the coefficient of z^j in (5) we find that it involves y_{j+1} linearly with the coefficient proportional to the limit in (4) which is non-zero in this case. Thus we can solve in a unique recursive manner for all the coefficients y_j. By the defining equation, the f_j are also a solution within the error, so we get $y_j = f_j$ for $j = 0, \ldots s+t$ which establishes the accuracy-through-order results for this case. If the deficiency $d > 0$, then if we examine (5), since it is also satisfied by $f(z)$ as well as $y(z)$, and $f'(0)$ is finite, the left-hand side vanishes like z^d and so too must the right-hand side. Here we have used the initial conditions $y^{(j)}(0) = f^{(j)}(0)$ for $j = 0, \ldots, d$. We next observe that in the defining equation (3) the coefficients of z^j to z^{2j-1} are linear in y_j, since y_j^2 can not appear at lower order that z^{2j}, as every y_j appears only in the combination $y_j z^j$ so that y_j^2 carries at least a factor of z^{2j}, *etc.* Consider the case of deficiency d. The coefficient of z^{2d+1} must be linear in $y_{d+1}, \ldots, y_{2d+1}$. The coefficient in (3) of $z^{2d+1} y_{d+n}$ for $n = 1, \ldots, d+1$ can be seen to be the coefficient of z^{d+1-n} in the coefficient of dy/dz on the left-hand side of (5). This result follows by noting that this coefficient is the partial derivative of (3) with respect to $y(z)$ and so will give the coefficient of the linear terms. Thus because the deficiency is d, only y_{d+1} has a non-zero coefficient. This coefficient necessarily involves only $y_0, \ldots, y_d$, which we have set equal to the corresponding f_j's by the initial conditions. As we examine the higher orders we find that the coefficient of z^{d+n} is linear in y_n with a non-zero coefficient for $n \geq d+1$, and that, as in the case $n = d+1$ just discussed, the coefficients in this term of y_{n+l} all vanish for any $l > 0$. Thus we may, in a unique and recursive manner, compute all the y_j for $j = d+1, d+2, \ldots, s+t-d$. Since as we have remarked, the f_j's satisfy the equation, we have proven the theorem, unless $d > s+t$. But it can not be that $d > s$, since we have a minimal solution. For if it were so, then the quantity in (4) multiplied by $f(z)$, would be a different solution set (of the same order and the same polynomial degrees) to the polynomial defining equations. A linear combination of these two polynomial sets could then be formed which would give a new solution set of order $k-1$ instead of k, which is a contradiction to our selection of a minimal set of algebraic polynomials. ∎

In order for algebraic approximants to be generally useful, we need to show that their degree of contact with the defining series goes to infinity with $\mathcal{M}$.

THEOREM 5. *Given an essentially unique, minimal algebraic approximant whose polynomials satisfy $\vec{Q}_{\vec{m}}(0) \neq \vec{0}$ and for which the error in the polynomial defining equations is $O(z^{\mathcal{M}})$, the degree of contact c of the algebraic approximant with the defining series is bounded by*

$$c \geq \left[\frac{\mathcal{M}-1}{\mu}\right], \tag{7}$$

where $[x]$ denotes the greatest interger less than or equal to x, and $1 \leq \mu \leq k+1$ is the smallest value of ν for which the coefficient of $[f(z) - f(0)]^\nu$ in (8) does not vanish for $z = 0$. The value of (7) for $\mu = k+1$ is always a bound.

PROOF: Let us expand the polynomial defining equation about the given value of $f(0)$. We get,

$$\sum_{\nu=0}^{k+1} \frac{[f(z)-f(0)]^\nu}{\nu!} \left(\sum_{j=\nu-1}^{k} \frac{Q_{j,m_j}(z)(j+1)![f(0)]^{j+1-\nu}}{(j+1-\nu)!} \right) = O(z^{s+t+1}). \quad (8)$$

Now we note that the $\nu = 0$ term vanishes at $z = 0$ as is required by the polynomial defining equations. If the second term does not vanish at $z = 0$, then the deficiency is zero. It is the case, when we denote the coefficient of $[f(z)-f(0)]^\nu$ by $\mathcal{Q}_\nu(z)$, that for at least one $\nu = \mu$ that $\mathcal{Q}_\mu(0) \neq 0$. This result can be seen as follows. First consider the case $\nu = k+1$. If $\mathcal{Q}_{k+1}(0) \neq 0$, we have the desired result. Otherwise, consider the case, $\nu = k$. In this circumstance, $\mathcal{Q}_k(0) = 0$ if and only if $Q_{k-1,m_{k-1}}(0) = 0$. We can continue in this manner through all $(k+1 \geq \nu > 0)$ the $\mathcal{Q}_\nu(0)$'s. We must eventually find one which does not vanish, since by hypothesis, $\vec{Q}_{\vec{m}}(0) \neq \vec{0}$. In order to focus on the computation of c, it is simplest to consider the determination of f_1. Let μ be the smallest value of ν for which $\mathcal{Q}_\nu(0) \neq 0$. From (8) there will be a non-vanishing term, proportional to f_1^μ, and potentially other terms of order z^μ involving f_1 but no other series coefficients except f_0. Let us now examine the series in z term by term. The first term (beyond the $\nu = 0$ term) is linear in f_1 and if the coefficient does not vanish, we can solve for f_1 directly. If it does vanish, then the second term will be a quadratic equation in f_1 and there will be no mixed terms involving f_1 or terms involving f_j with $j > 1$ because their coefficients vanish as a consequence of the vanishing of the first term. The examination continues in this manner so that we either find an equation for f_1, or the form of the next order term in z involves f_1 alone. We are however guaranteed that we will find an equation for f_1 with at least one non-zero coefficient by the z^μ term. It will be of the μth order in f_1 and corresponds to μ roots which are coincident at the origin. If the appropriate root of this equation for f_1 is not a multiple root, then we can solve for all the rest of the coefficients, f_j, $j > 1$ because they will first appear linearly and if $R(f) = 0$ is the equation for f_1 then their coefficient will always be $\left.\frac{\partial R(f)}{\partial f}\right|_{f_1}$ which is non-zero for this case. By analysis of this case through some rather tedious but straightforward algebra, one can see that there will always be an equation with at least one non-vanishing coefficients to determine f_j which includes the term $\mathcal{Q}_\mu(0) f_j^\mu z^{j\mu}$. Thus the accuracy-through-order equations are always sufficient to determine at least $[(\mathcal{M}-1)/\mu]$ coefficients. ∎

An illustration of these theorems is given by the function

$$f(z) = -1 + z - \frac{1}{2}z^2 + \frac{3}{8}z^3 - \frac{5}{16}z^4 + \frac{35}{128}z^5 - \frac{63}{256}z^6 + O(z^7), \quad (9)$$

which satisfies,

$$(1+z)[f(z)]^2 + 2(1+z)f(z) + 1 + z - z^2 = O(z^7). \quad (10)$$

Condition (5) becomes in this case,

$$2(1+z)(y(z)+1)\frac{dy(z)}{dz} = -[y(z)]^2 - 2y(z) - 1 + 2z + O(z^6). \quad (11)$$

We see by direct substitution of the defining series in (11) that, as z divides (11), the deficiency is $d = 1$ thereby reducing the degree of contact by unity. This deficiency is associated with the the coincidence of two solutions at $z = 0$, and the corresponding vanishing of the first two terms in (8). Thus the solution for y(z) is,

$$y(z) = -1 \pm \sqrt{\frac{z^2}{1+z}} + O(z^6). \tag{12}$$

We now consider the equivalence properties. In this regard it is useful to define the Beckermann (1990) minimal polynomials. First, we need an ordering relation. If $\vec{m} = (m_{-1}, \ldots, m_k)$ is a vector in the index space labeling the algebraic polynomials, then the partial ordering relation $\vec{a} \leq \vec{c}$ means that every component of $\vec{a}$ is less than or equal to the corresponding component of $\vec{c}$. The relation $\vec{a} = \vec{c}$ means that every component of the two vectors is equal, and $\vec{a} < \vec{c}$ means that $\vec{a} \leq \vec{c}$ holds but that $\vec{a} = \vec{c}$ fails. Notice that if $a_i > c_i$ and $a_j < c_j$ for $i \neq j$, then $\vec{a}$ and $\vec{c}$ are incomparable by these partial ordering relations.

DEFINITION 6. *A nontrivial solution of (1) for the algebraic polynomials is called a Beckermann minimal solution if, among all the solutions of (1), there is no other nontrivial solution whose degree is less according to the above given partial ordering relation.*

Notice that a minimal solution according to Definition 2, is also a Beckermann minimal solution, but that there may be Beckermann minimal solutions which are not minimal according to Definition 2. We now give the following results.

THEOREM 7. *The statement (i) $f(z)$ is a functional element at $z = 0$ which satisfies*

$$\sum_{j=-1}^{k} Q_{j,m_j}(z)[f(z)]^{j+1} = 0, \tag{13}$$

where $\vec{Q}_{\vec{m}}$ is of true nominal degree and essentially unique, is equivalent to (ii) there exists a functional element $g(z)$ at $z = 0$ for which,

$$\vec{\mathcal{Q}}_{\vec{M}} = \vec{Q}_{\vec{m}}, \quad \forall \vec{M} \geq \vec{m}, \tag{14}$$

where $\vec{Q}_{\vec{m}}$ is minimal and $\vec{\mathcal{Q}}_{\vec{M}}$ is a Beckermann minimal solution of type $\vec{M}$ for $g(z)$, the inequality is in the sense of Definition 6, and $g(z) = f(z)$.

PROOF: First, (i) implies (ii) as, if we pick $g(z) = f(z)$, then $\vec{Q}$ is a solution of type $\vec{M}$ for any $\vec{M} \geq \vec{m}$ and as by hypothesis, as $\vec{Q}_{\vec{m}}$ is minimal, it is at least a Beckermann minimal solution of type $\vec{M}$.

Second we must consider, whether (ii) implies (i). Suppose that (ii) holds, therefore we must have

$$Q_{-1,m_{-1}}(z) + \sum_{j=0}^{k} Q_{j,m_j}(z)[g(z)]^{j+1} = O(z^t), \quad \forall t < \infty. \tag{15}$$

By the minimality of $\vec{Q}_{\vec{m}}(z)$, there are no common factors of the type $1 + az$ and the division by any factor of z^j leaves (15) unchanged except that $m_i \to m_i - j$, as it holds for all $t < \infty$. As we have seen in the proof of Theorem 4 that we can divide both sides of (5) by z^d, the integration of this divided equation shows by the implicit function theorem that $g(z)$ is a functional element. Thus the left-hand side of (15) is regular at $z = 0$ and so by (15) must be identically zero. The remaining problem to complete the proof is to show that $g(z) = f(z)$. This result follows from the accuracy-through-order Theorem 4 or 5 and the principle of analytic continuation. The point is that either of these theorems show that the degree of contact at the origin between the algebraic approximant and $g(z)$ is infinite, provided that at most $d+1$, where d is the deficiency, of the initial conditions $y^{(j)}(0) = g^{(j)}(0)$ hold. This result, by the principle of analytic continuation, proves that any $y(z) = g(z)$ and so $f(z) = g(z)$. ∎

Acknowledgements

I wish to thank P. R. Graves-Morris for helpful discussions. I also wish to thank the referee for pointing out an incompleteness in the original version.

References

Baker, George A. Jr.: 1984, 'Invariance properties in Hermite–Padé approximation theory', *J. Comp. & Appl. Math.* Vol. no. **11**, pp. 49–55

Baker, George A. Jr. and Graves-Morris, P. R.: 1990, 'Definition and uniqueness of integral approximants', *J. Comp. & Appl. Math.* Vol. no. **31**, pp. 357–372

Baker, George A. Jr. and Graves-Morris, P. R.: 1994, 'Existence of certain sequences of Hermite-Padé approximants', *J. Comp. & Appl. Math.* to be published

Beckermann, B.: 1990, 'The structure of the singular solution table of the M-Padé approximation problem', *J. Comp. & Appl. Math.* Vol. no. **32**, pp. 3–15

McInnes, A. W.: 1992, 'Existence and uniqueness of algebraic function approximations', *Constructive Approximation* Vol. no. **8**, pp. 1–21

Padé, H.: 1892, 'Sur la répresentation approchée d'une fonction par des fractions rationelles', *Ann. de l'Ecole Normale Sup.* 3^{iéme} *Série* Vol. no. **9**, Suppl., pp. 3–93

Shafer, R. E.: 1974, 'On quadratic approximation', *SIAM J. Num. Anal.* Vol. no. **11**, pp. 447-460

ON THE VECTOR-VALUED PADÉ APPROXIMANTS AND THE VECTOR ε-ALGORITHM

A.SALAM
Université des Sciences et Technologies de Lille,
Laboratoire d'analyse numérique et d'optimisation,
UFR IEEA-M3, 59655 Villeneuve d'Ascq- France.

Abstract. In a recent work, thanks to the use of Clifford algebras and designants, we have shown that Wynn's vector ε-algorithm can be written as a ratio of two designants. The present work, according to these new results, has the aim of
1. Finding easily and differently the results of some authors in this scope, particularly those of Wynn, D.E. Roberts and P.R. Graves-Morris.
2. Computing explicitly the expressions of the vector-valued Padé approximants as a ratio of two designants, which generalizes the results given by the authors above, and establishing the connection between these approximants and the vector ε-algorithm.

Key words: Designant, Clifford algebra, vector ε-algorithm, vector-valued Padé approximants.

1. Introduction

let $\mathbb{K}$ denote a non-commutative field (or ring), $\mathbb{M}_n(\mathbb{K})$ the set of matrices whose coefficients are in $\mathbb{K}$. By a determinant we mean an application, denoted by *det*

$$
\begin{aligned}
det : \mathbb{M}_n(\mathbb{K}) &\longrightarrow \mathbb{K} \\
A &\longrightarrow det(A)
\end{aligned}
$$

and satisfying fhe following conditions

$$
\begin{aligned}
&1)\ det \text{ is multilinear with respect to the rows of } A \\
&2)\ det(A) = 0 \Longleftrightarrow A \text{ singular} \\
&3)\ det(A \times B) = det(A).det(B).
\end{aligned}
$$

Dyson's theorem ([10, 19]) states that determinants do not exist when $\mathbb{K}$ is a non-commutative ring. The study of many extrapolation methods use the theory of determinants for building up recursive algorithms corresponding to these extrapolation methods (E-algorithm [4], [16], ε-algorithm [28]). In [26], it was shown that extrapolation methods can be extended to the non-commutative case. Determinants are replaced by an old notion which is more general: Designants ([17]). Using designants and Clifford algebra, the conjecture about Wynn's vector ε-algorithm is

A. Cuyt (ed.), Nonlinear Numerical Methods and Rational Approximation II, 291–301.

solved: The ε-algorithm can be written as a 'ratio' of two designants. Other results are found. In this paper, we will apply these results to the vector-valued Padé approximants ([23, 24, 15]). Now, we give briefly the definition of designants. For more informations, see [25, 26, 17].

2. Designants

Let $\mathbb{K}$ be a non-commutative ring. Consider the system of homogeneous linear equations of two unknowns $x_1, x_2 \in \mathbb{K}$, with coefficients on the right

$$\begin{cases} x_1 a_{11} + x_2 a_{12} = 0 \\ x_1 a_{21} + x_2 a_{22} = 0 \end{cases} \quad a_{ij} \in \mathbb{K} \;\; i,j = 1,2. \tag{1}$$

Suppose that a_{11} is invertible, then, by elimininting the unknown x_1 in the second equation of the system, we get :

$$x_2(a_{22} - a_{12}a_{11}^{-1}a_{21}) = 0.$$

Set

$$\Delta_r = \begin{vmatrix} a_{11} & a_{12} \\ a_{21} & a_{22} \end{vmatrix}_r = a_{22} - a_{12}a_{11}^{-1}a_{21}.$$

Δ_r is called the right designant of the system (1)
The suffix r indicates that the designant in question is a designant of the right system (1). It indicates also the direction of the computation.

In the same way, consider the system of homogeneous linear equations of two unknowns x_1, x_2, with coefficients on the left

$$\begin{cases} a_{11} x_1 + a_{12} x_2 = 0 \\ a_{21} x_1 + a_{22} x_2 = 0 \end{cases} \quad a_{ij} \in \mathbb{K} \;\; i,j = 1,2. \tag{2}$$

Suppose that a_{11} invertible, then, by elimininting the unknown x_1 in the second equation of the system, we get :

$$(a_{22} - a_{21}a_{11}^{-1}a_{12})x_2 = 0.$$

Set

$$\Delta_l = \begin{vmatrix} a_{11} & a_{12} \\ a_{21} & a_{22} \end{vmatrix}_l = (a_{22} - a_{21}a_{11}^{-1}a_{12}).$$

Δ_l is called the left designant of the system (2)
The suffix l indicates that the designant in question is a designant of the left system (2). It indicates also the direction of the computation.

If $\Delta_r \neq 0$ (resp $\Delta_l \neq 0$) the system (1) (resp the system (2)) has uniquely the trivial solution.
If $\Delta_r = 0$ (resp $\Delta_l = 0$) the system (1) (resp the system (2)) has more than one

solution.

In the general case, we proceed in the same way : consider the system of homogeneous linear equations of n unknowns $x_1, x_2, ..., x_n$, with coefficients on the right

$$\begin{cases} x_1 a_{11} + & x_2 a_{12} + & \cdots & + x_n a_{1n} = 0 \\ x_1 a_{21} + & x_2 a_{22} + & \cdots & + x_n a_{2n} = 0 \\ \vdots & \vdots & \ddots & \vdots \\ x_1 a_{n1} + & x_2 a_{n2} + & \cdots & + x_n a_{nn} = 0 \end{cases} \tag{3}$$

where $a_{ij} \in I\!K \;\; i, j = 1, \cdots, n$.

By eliminating x_1 from the $(n-1)$ last equations, then x_2 from the $(n-2)$ last equations, and so on, we obtain $x_n \Delta_r = 0$.
Δ_r is called the right designant of the system (3) and it is denoted by

$$\Delta_r = \begin{vmatrix} a_{11} & \cdots & a_{1n} \\ \vdots & \ddots & \vdots \\ a_{n1} & \cdots & a_{nn} \end{vmatrix}_r .$$

Δ_r has a meaning only if its principal minor

$$\begin{vmatrix} a_{11} & \cdots & a_{1,n-1} \\ \vdots & \ddots & \vdots \\ a_{n-1,1} & \cdots & a_{n-1,n-1} \end{vmatrix}_r$$

is invertible. In its turn, this r.designant has a meaning only if its principal minor

$$\begin{vmatrix} a_{11} & \cdots & a_{1,n-2} \\ \vdots & \ddots & \vdots \\ a_{n-2,1} & \cdots & a_{n-2,n-2} \end{vmatrix}_r$$

is invertible, and so on. Finally Δ_r has a meaning only if

$$a_{11}, \begin{vmatrix} a_{11} & a_{12} \\ a_{21} & a_{22} \end{vmatrix}_r, \ldots, \begin{vmatrix} a_{11} & \cdots & a_{1,n-1} \\ \vdots & \ddots & \vdots \\ a_{n-1,1} & \cdots & a_{n-1,n-1} \end{vmatrix}_r$$

are all invertible. In the same way, we obtain

$$\Delta_l = \begin{vmatrix} a_{11} & \cdots & a_{1n} \\ \vdots & \ddots & \vdots \\ a_{n1} & \cdots & a_{nn} \end{vmatrix}_l$$

from the system above but we take $a_{ij} x_j$ instead of $x_j a_{ij}$.

Property 1 *Let A^p_{qr} (resp.B^p_{qr}) be the r.designant (resp. the l.designant) of order $p+1$ obtained from Δ_r (resp. Δ_l) by keeping the rows $1,2,...,p,q$ and the columns $1,2,...,p,r$. We have*

$$\Delta_r = \begin{vmatrix} a_{11} & \cdots & a_{1n} \\ \vdots & \ddots & \vdots \\ a_{n1} & \cdots & a_{nn} \end{vmatrix}_r = \begin{vmatrix} A^p_{p+1,p+1} & \cdots & A^p_{p+1,n} \\ \vdots & \ddots & \vdots \\ A^p_{n,p+1} & \cdots & A^p_{n,n} \end{vmatrix}_r .$$

Taking $p = n-2$, we obtain Sylvester's identity ([1]) for designants

$$\Delta_r = \begin{vmatrix} a_{11} & \cdots & a_{1n} \\ \vdots & \cdots & \vdots \\ a_{n1} & \cdots & a_{nn} \end{vmatrix}_r = \begin{vmatrix} A^{n-2}_{n-1,n-1} & A^{n-2}_{n-1,n} \\ A^{n-2}_{n,n-1} & A^{n-2}_{n,n} \end{vmatrix}_r .$$

We have exactly the same equalities for Δ_l but we should take B^p_{qr} instead of A^p_{qr} and the suffix l instead of r.

We cite also the following property which connects the r.designants with the l.desigants.

Property 2

$$\begin{vmatrix} a_{11} & \cdots & a_{1n} \\ \vdots & \ddots & \vdots \\ a_{n1} & \cdots & a_{nn} \end{vmatrix}_r = \begin{vmatrix} a_{11} & \cdots & a_{n1} \\ \vdots & \ddots & \vdots \\ a_{1n} & \cdots & a_{nn} \end{vmatrix}_l .$$

and finally

Property 3

$$\begin{vmatrix} a_{11} & \cdots & a_{1n} & 0 \\ \vdots & \ddots & \vdots & \vdots \\ a_{n+11} & \cdots & a_{n+1n} & 1 \end{vmatrix}_r = 1.$$

all proofs of these properties are in [17, 25].

3. Real and Complex Clifford algebra

Let V be a real linear space, q its quadratic form. There exists ([9]) an algebra $(C(V), +, \times, .)$ called Clifford algebra, which satisfies the following fundamental property:

There exists a linear transformation $\varphi : V \longrightarrow C(V)$ such that $\forall x \in V \ \ (\varphi(x))^2 = q(x)$.

From now on, we take $V = \mathbb{R}^d$ and $q(x) = (x.x)$ where (x.y) denotes the usual scalar product. Let $e_1, ..., e_d$ be an orthonormal basis of V and set $E_i = \varphi(e_i), i = 1, ..., d$. The Clifford algebra associated to V is generated by d elements $E_i, i = 1, ..., d$ (see [2, 22]) which satisfy the anti-commutation relations

$$E_i \times E_j + E_j \times E_i = 2\delta_{ij} E_0, \tag{4}$$

E_0 being the identity element, δ_{ij} the Kronecker symbol for $1 \leq i, j \leq d$. The real linear space spanned by the products $E_{j_1} \times E_{j_2} \times ... \times E_{j_r}$, $0 \leq j_1 < j_2 < ... < j_r \leq d$ i.e. $E_0, E_1, E_2, ..., E_d, E_1 \times E_2, ..., E_{d-1} \times E_d, ..., E_1 \times E_2 \times ... \times E_d$ forms the associative but non-commutative ($d > 1$) algebra $C(V)$. If the elements $E_{j_1} \times E_{j_2} \times ... \times E_{j_r}$ are linearly independent then $C(V)$ is a universal Clifford algebra of dimension 2^d. However, it is not a division algebra ($d \geq 1$) as is proved by the relation $(E_0 + E_1)(E_0 - E_1) = 0$.

There are various matrix representations of E_i e.g. [18, 20]. Furthermore, we may identify each $x = \sum_{i=1}^{d} x_i e_i \in V$, with $X = \sum_{i=1}^{d} x_i E_i \in C(V)$ and each $\lambda \in \mathbb{R}$ with $\lambda E_0 \in C(V)$ (see [2]). According to this identification, one can consider V and $\mathbb{R}$ as subsets of $C(V)$.

From the equation (4), we can easily establish

$$\forall X \in V, \forall Y \in V \quad X \times Y + Y \times X = 2(X.Y) \tag{5}$$

By setting $X = Y$, relation (5) admits the following particular rule

$$X \times X = (X.X). \tag{6}$$

This last relation allows us to conclude that, if $X \neq 0$, then

$$X^{-1} = \frac{X}{(X.X)}. \tag{7}$$

Let us notice that this inverse is the same as the Moore-Penrose pseudo-inverse (see [21]). The complex Clifford algebra is the linear space spanned by taking linear combinations, over $\mathbb{C}$, of the products $E_{j_1} \times E_{j_2} \times ... \times E_{j_r}$. We remark in this case that a vector $x \in \mathbb{C}^d$ different from 0 is not necessarily invertible. Other defintions of complex Clifford algebra exist (see [18]), but for our need, we use the definition above.

4. The vector ε-algorithm

The vector ε-algorithm is a quite powerful method for accelerating the convergence of vector sequences. It has many other applications : solution of systems of linear equations [5], [11], systems of nonlinear equations [7] and the computation of eigenvalues of a matrix [6]. It occurs likewise in approximation theory (vector continued fractions [29], vector interpolation and Padé approximants [12],[13], [14], [15]).

In [26], it was shown that the vector ε-algorithm, like other extrapolation algorithms, realizes some extrapolation in $C(V)$, solves some system of linear equations in $C(V)$ and can be written as a "ratio" of two designants.

In fact, set $\mathbb{K} = C(V)$ where $V = \mathbb{R}^d$ or $\mathbb{C}^d$. The vector ε-algorithm of Wynn can be written as follows

$$\varepsilon_{-1}^{(n)} = 0 \qquad n = 0, 1, ...$$
$$\varepsilon_{0}^{(n)} = S_n \in V \qquad n = 0, 1, ...$$

$$\varepsilon_{k+1}^{(n)} = \varepsilon_{k-1}^{(n+1)} + \left(\varepsilon_{k}^{(n+1)} - \varepsilon_{k}^{(n)}\right)^{-1} \quad k = 0, 1, ...; n = 0, 1, ...$$

The inverse considered here is the above-mentioned inverse. The $\varepsilon_k^{(n)}$'s belong always to V, this is because the sum and the difference are closed operations in V and the inverse of an element of V also belong to V.

Theorem 1 *Setting*

$$\varepsilon_{2k}(S_n) = \begin{vmatrix} \Delta S_n & \cdots & \Delta S_{n+k-1} & S_n \\ \vdots & \ddots & \vdots & \vdots \\ \Delta S_{n+k} & \cdots & \Delta S_{n+2k-1} & S_{n+k} \end{vmatrix}_r \begin{vmatrix} \Delta S_n & \cdots & \Delta S_{n+k-1} & 1 \\ \vdots & \ddots & \vdots & \vdots \\ \Delta S_{n+k} & \cdots & \Delta S_{n+2k-1} & 1 \end{vmatrix}_r^{-1}$$

The quantities $\varepsilon_k^{(n)}$ computed by the vector ε-algorithm can be expressed as a "ratio" of two designants as follows

$$\varepsilon_{2k}^{(n)} = \varepsilon_{2k}(S_n) \text{ and } \varepsilon_{2k+1}^{(n)} = (\varepsilon_{2k}(\Delta S_n))^{-1}.$$

We also have

Theorem 2 $\varepsilon_{2k}^{n} = S \Longleftrightarrow \exists A_1, ..., A_k \in C(V)$ *such that* $S_n = S + A_1 \Delta S_n + ... + A_k \Delta S_{n+k-1}$.

Remark
The fundamental algebraic result of McLeod ([20]) is a particular case of the theorem 2 by imposing that A_i's belong to $\mathbb{R}$ (if $V = \mathbb{R}^d \subset C(\mathbb{R}^d)$) or $\mathbb{C}$ (if $V = \mathbb{C}^d \subset C(\mathbb{C}^d)$). From theorem 1, we can also deduce some other interesting results and obtain some other well known results as particular cases:

Theorem 3 *Let $\varepsilon_k^{(n)}$ be the vectors obtained by applying the ε-algorithm to the vector sequence (S_n) of elements of V.*
Let A, B be two invertible elements of $C(V)$, C an element of V.
Let $\tilde{\varepsilon}_k^{(n)}$ be the elements of $C(V)$ obtained by applying the ε-algorithm to the sequence of the elements $(A \times S_n \times B + C)$ of $(C(V)$ then

$$\tilde{\varepsilon}_{2k}^{(n)} = A \times \varepsilon_{2k}^{(n)} \times B + C$$

$$\tilde{\varepsilon}_{2k+1}^{(n)} = B^{-1} \times \varepsilon_{2k+1}^{(n)} \times A^{-1}.$$

A particular case of this theorem is the following known theorem ([3])

Theorem 4 *Let M be a square orthogonal matrix of dimension $d \times d$ ($M^T M = I$). Let $\tilde{\varepsilon}_k^{(n)}$ be the vector obtained by applying the vector ε-algorithm to the vector sequence $MS_n + C$ where C is a vector of V and MS_n the product of the matrix M by the vector S_n. then*

$$\tilde{\varepsilon}_{2k}^{(n)} = M\varepsilon_{2k}^{(n)} + C$$

$$\tilde{\varepsilon}_{2k+1}^{(n)} = M\varepsilon_{2k+1}^{(n)}.$$

5. Vector-valued Padé approximants

Vector-valued Padé approximants is a subject of interest to many authors. Among them, I cite for exemple D.E. Roberts and P.R. Graves-Morris. However, there exist different approachs from those above. I cite Brezinski [8] and J.Van Iseghem [27]. In this paper we take an interest in the work of D.E. Roberts ([23]) and we will apply the theorems above for obtaining new results and finding differently some known resullts.

Consider a vector-valued function $f(z)$, $z \in \mathbb{C}$ possessing the following expansion: $f(z) = c_0 + c_1 z + c_2 z^2 + \ldots + c_n z^n + \ldots$ in which $c_i \in \mathbb{C}^d$, $i = 0, 1, 2, \ldots, n$.

Roberts defines the Right-handed vector-valued Padé approximants $[M/N]$V.P.A (if it exists) by $p_M^R(z)\{q_N^R(z)\}^{-1} - f(z) = O(z^{M+N+1})$ where $p_M^R(z)$ and $q_N^R(z)$ are polynomials of degrees M and N respectively whose coefficients belong to $C(\mathbb{C}^d)$. The left-handed $[M/N]$V.P.A. is defined in a similar manner $\{q_N^L(z)\}^{-1} p_M^L(z) - f(z) = O(z^{M+N+1})$.

We will show how we can express explicitly $p_M^R(z), q_M^R(z), p_M^L(z)$ and $q_M^L(z)$ in terms of $c_i z^i$ of $f(z)$. This explicit form is a designant which allows us the connection between V.P.A. and vector ε-algorithm. In fact set $S_n = \sum_{i=0}^{n} c_i z^i$. We have the

Theorem 5

$$\varepsilon_{2k}^{(M)} = \begin{vmatrix} c_{M+1} & \cdots & c_{M+k} & z^k S_M \\ c_{M+2} & \cdots & c_{M+k+1} & z^{k-1} S_{M+1} \\ \vdots & \ddots & \vdots & \vdots \\ c_{M+k+1} & \cdots & c_{M+2k} & S_{M+k} \end{vmatrix}_r \begin{vmatrix} c_{M+1} & \cdots & c_{M+k} & z^k \\ c_{M+2} & \cdots & c_{M+k+1} & z^{k-1} \\ \vdots & \ddots & \vdots & \vdots \\ c_{M+k+1} & \cdots & c_{M+2k} & 1 \end{vmatrix}_r^{-1} = [M+k/k]\,V.P.A.$$

More precisely

$$p_{M+k}^R(z) = \begin{vmatrix} c_{M+1} & \cdots & c_{M+k} & z^k S_M \\ c_{M+2} & \cdots & c_{M+k+1} & z^{k-1} S_{M+1} \\ \vdots & \ddots & \vdots & \vdots \\ c_{M+k+1} & \cdots & c_{M+2k} & S_{M+k} \end{vmatrix}_r$$

$$q_k^R(z) = \begin{vmatrix} c_{M+1} & \cdots & c_{M+k} & z^k \\ c_{M+2} & \cdots & c_{M+k+1} & z^{k-1} \\ \vdots & \ddots & \vdots & \vdots \\ c_{M+k+1} & \cdots & c_{M+2k} & 1 \end{vmatrix}_r$$

$$p_{M+k}^L(z) = \begin{vmatrix} c_{M+1} & \cdots & c_{M+k} & z^k S_M \\ c_{M+2} & \cdots & c_{M+k+1} & z^{k-1} S_{M+1} \\ \vdots & \ddots & \vdots & \vdots \\ c_{M+k+1} & \cdots & c_{M+2k} & S_{M+k} \end{vmatrix}_l$$

$$q_k^L(z) = \begin{vmatrix} c_{M+1} & \cdots & c_{M+k} & z^k \\ c_{M+2} & \cdots & c_{M+k+1} & z^{k-1} \\ \vdots & \ddots & \vdots & \vdots \\ c_{M+k+1} & \cdots & c_{M+2k} & 1 \end{vmatrix}_l$$

Proof:
Theorem 1 allows us to write

$$\varepsilon_{2k}^{(M)} = \begin{vmatrix} \Delta S_M & \cdots & \Delta S_{M+k-1} & S_M \\ \vdots & \ddots & \vdots & \vdots \\ \Delta S_{M+k} & \cdots & \Delta S_{M+2k-1} & S_{M+k} \end{vmatrix}_r \begin{vmatrix} \Delta S_M & \cdots & \Delta S_{M+k-1} & 1 \\ \vdots & \ddots & \vdots & \vdots \\ \Delta S_{M+k} & \cdots & \Delta S_{M+2k-1} & 1 \end{vmatrix}_r^{-1} .$$

Since $S_M = \sum_{i=0}^{M} c_i z^i$, we have $\Delta S_M = c_{M+1} z^{M+1}$. So we can write

$$\varepsilon_{2k}^{(M)} = \begin{vmatrix} c_{M+1} z^{M+1} & \cdots & c_{M+k} z^{M+k} & S_M \\ c_{M+2} z^{M+2} & \cdots & c_{M+k+1} z^{M+k+1} & S_{M+1} \\ \vdots & \ddots & \vdots & \vdots \\ c_{M+k+1} z^{M+k+1} & \cdots & c_{M+2k} z^{M+2k} & S_{M+k} \end{vmatrix}_r \begin{vmatrix} c_{M+1} z^{M+1} & \cdots & c_{M+k} z^{M+k} & 1 \\ c_{M+2} z^{M+2} & \cdots & c_{M+k+1} z^{M+k+1} & 1 \\ \vdots & \ddots & \vdots & \vdots \\ c_{M+k+1} z^{M+k+1} & \cdots & c_{M+2k} z^{M+2k} & 1 \end{vmatrix}_r^{-1} .$$

Multiplying the first row of each designant by z^k, the second row by z^{k-1} and so on, these designants do not change ([25]). So

$$\varepsilon_{2k}^{(M)} = \begin{vmatrix} c_{M+1} z^{M+k+1} & \cdots & c_{M+k} z^{M+2k} & z^k S_M \\ c_{M+2} z^{M+k+1} & \cdots & c_{M+k+1} z^{M+2k} & z^{k-1} S_{M+1} \\ \vdots & \ddots & \vdots & \vdots \\ c_{M+k+1} z^{M+k+1} & \cdots & c_{M+2k} z^{M+2k} & S_{M+k} \end{vmatrix}_r \times$$

$$\begin{vmatrix} c_{M+1} z^{M+k+1} & \cdots & c_{M+k} z^{M+2k} & z^k \\ c_{M+2} z^{M+k+1} & \cdots & c_{M+k+1} z^{M+2k} & z^{k-1} \\ \vdots & \ddots & \vdots & \vdots \\ c_{M+k+1} z^{M+k+1} & \cdots & c_{M+2k} z^{M+2k} & 1 \end{vmatrix}_r^{-1} .$$

Simplifying the first column by dividing top and bottom by z^{M+k+1}, the second column by z^{M+k+2} and so on, the last but one by the z^{M+2k}, we have

$$\varepsilon_{2k}^{(M)} = \begin{vmatrix} c_{M+1} & \cdots & c_{M+k} & z^k S_M \\ c_{M+2} & \cdots & c_{M+k+1} & z^{k-1} S_{M+1} \\ \vdots & \ddots & \vdots & \vdots \\ c_{M+k+1} & \cdots & c_{M+2k} & S_{M+k} \end{vmatrix}_r \begin{vmatrix} c_{M+1} & \cdots & c_{M+k} & z^k \\ c_{M+2} & \cdots & c_{M+k+1} & z^{k-1} \\ \vdots & \ddots & \vdots & \vdots \\ c_{M+k+1} & \cdots & c_{M+2k} & 1 \end{vmatrix}_r^{-1} .$$

This expression is formally similar to this of scalar $[M+k/k]$ approximant, but we have designants instead of determinants and vectors instead of scalars. Now using property 3,

$$\begin{vmatrix} c_{M+1} & \cdots & c_{M+k} & 0 \\ c_{M+2} & \cdots & c_{M+k+1} & 0 \\ \vdots & \ddots & \vdots & \vdots \\ c_{M+k+1} & \cdots & c_{M+2k} & 1 \end{vmatrix}_r = 1 \qquad (8)$$

we will show that

$$\begin{vmatrix} c_{M+1} & \cdots & c_{M+k} & z^k S_M \\ c_{M+2} & \cdots & c_{M+k+1} & z^{k-1} S_{M+1} \\ \vdots & \ddots & \vdots & \vdots \\ c_{M+k+1} & \cdots & c_{M+2k} & S_{M+k} \end{vmatrix}_r \begin{vmatrix} c_{M+1} & \cdots & c_{M+k} & z^k \\ c_{M+2} & \cdots & c_{M+k+1} & z^{k-1} \\ \vdots & \ddots & \vdots & \vdots \\ c_{M+k+1} & \cdots & c_{M+2k} & 1 \end{vmatrix}_r^{-1} - f(z) = O(z^{M+2k+1}).$$

$$\Longleftrightarrow \begin{vmatrix} c_{M+1} & \cdots & c_{M+k} & z^k S_M \\ c_{M+2} & \cdots & c_{M+k+1} & z^{k-1} S_{M+1} \\ \vdots & \ddots & \vdots & \vdots \\ c_{M+k+1} & \cdots & c_{M+2k} & S_{M+k} \end{vmatrix}_r - f(z) \begin{vmatrix} c_{M+1} & \cdots & c_{M+k} & z^k \\ c_{M+2} & \cdots & c_{M+k+1} & z^{k-1} \\ \vdots & \ddots & \vdots & \vdots \\ c_{M+k+1} & \cdots & c_{M+2k} & 1 \end{vmatrix}_r = O(z^{M+2k+1}).$$

Since designant is linear with respect to the last column,

$$\Longleftrightarrow \begin{vmatrix} c_{M+1} & \cdots & c_{M+k} & z^k S_M \\ c_{M+2} & \cdots & c_{M+k+1} & z^{k-1} S_{M+1} \\ \vdots & \ddots & \vdots & \vdots \\ c_{M+k+1} & \cdots & c_{M+2k} & S_{M+k} \end{vmatrix}_r - \begin{vmatrix} c_{M+1} & \cdots & c_{M+k} & z^k f(z) \\ c_{M+2} & \cdots & c_{M+k+1} & z^{k-1} f(z) \\ \vdots & \ddots & \vdots & \vdots \\ c_{M+k+1} & \cdots & c_{M+2k} & f(z) \end{vmatrix}_r = O(z^{M+2k+1})$$

$$\Longleftrightarrow \begin{vmatrix} c_{M+1} & \cdots & c_{M+k} & z^k (S_M - f(z)) \\ c_{M+2} & \cdots & c_{M+k+1} & z^{k-1} (S_{M+1} - f(z)) \\ \vdots & \ddots & \vdots & \vdots \\ c_{M+k+1} & \cdots & c_{M+2k} & S_{M+k} - f(z) \end{vmatrix}_r = O(z^{M+2k+1}) \text{ which is true since}$$

$z^{k-i}(S_{M+i} - f(z)) = O(z^{M+k+1})$ for $i = 0..k$. then, replacing in the above numerator, we see that $\varepsilon_{2k}^{(M)} = f(z) + O(z^{M+2k+1})$.
So we can deduce that

$$p_M^R(z)[q_k^R(z)]^{-1} = \begin{vmatrix} c_{M+1} & \cdots & c_{M+k} & z^k S_M \\ c_{M+2} & \cdots & c_{M+k+1} & z^{k-1} S_{M+1} \\ \vdots & \ddots & \vdots & \vdots \\ c_{M+k+1} & \cdots & c_{M+2k} & S_{M+k} \end{vmatrix}_r \begin{vmatrix} c_{M+1} & \cdots & c_{M+k} & z^k \\ c_{M+2} & \cdots & c_{M+k+1} & z^{k-1} \\ \vdots & \ddots & \vdots & \vdots \\ c_{M+k+1} & \cdots & c_{M+2k} & 1 \end{vmatrix}_r^{-1}.$$

Since we have $q_k^R(0) = 1$ ([23]) and relation (8), we can deduce the results.
We have immediately

Corollary 1 $p_M^R(z)$ *exists if and only if*

$$c_{M+1}, \dots, \begin{vmatrix} c_{M+1} & \cdots & c_{M+k} \\ c_{M+2} & \cdots & c_{M+k+1} \\ \vdots & \ddots & \vdots \\ c_{M+k} & \cdots & c_{M+2k-1} \end{vmatrix}_r \quad \textit{are invertible.}$$

Corollary 2 $q_k^R(z)$ *exists and is invertible if and only if*

$$c_{M+1}, \dots, \begin{vmatrix} c_{M+1} & \cdots & c_{M+k} \\ c_{M+2} & \cdots & c_{M+k+1} \\ \vdots & \ddots & \vdots \\ c_{M+k} & \cdots & c_{M+2k-1} \end{vmatrix}_r, \begin{vmatrix} c_{M+1} & \cdots & c_{M+k} & z^k \\ c_{M+2} & \cdots & c_{M+k+1} & z^{k-1} \\ \vdots & \ddots & \vdots & \vdots \\ c_{M+k+1} & \cdots & c_{M+2k} & 1 \end{vmatrix}_r$$

are invertible.

Now we will give an example of proof using only designants. D.E.Roberts has shown ([23]) that

$$p_M^R(z) = \widetilde{p_M^L}(z) \tag{9}$$

where $\tilde{}$ is the anti-automorphism

$$\begin{aligned} \tilde{} : C(V) &\longrightarrow C(V) \\ u &\longrightarrow \tilde{u}, \end{aligned}$$

satisfying $\tilde{E_i} = E_i$ and $\forall u \in C(V)$, $\forall v \in C(V)$, $\widetilde{(uv)} = \tilde{v}\tilde{u}$. We need only to prove that $\tilde{\Delta_r} = \Delta_l$ where

$$\Delta_r = \begin{vmatrix} a_{11} & \cdots & a_{1n} \\ \vdots & \ddots & \vdots \\ a_{n1} & \cdots & a_{nn} \end{vmatrix}_r \text{ and } \Delta_l = \begin{vmatrix} a_{11} & \cdots & a_{1n} \\ \vdots & \ddots & \vdots \\ a_{n1} & \cdots & a_{nn} \end{vmatrix}_l.$$

Proof: by induction.
For n=1 the property is true because $a_{11} \in V$.
Now suppose the property true for all designants of order n-1.
For n, using property 1, we have: $\Delta_r = A_{n,n}^{n-2} - A_{n,n-1}^{n-2}\{A_{n-1,n-1}^{n-2}\}^{-1}A_{n-1,n}^{n-2}$. So $\tilde{\Delta_r} = \widetilde{A_{n,n}^{n-2}} - \widetilde{A_{n-1,n}^{n-2}}\{\widetilde{A_{n-1,n-1}^{n-2}}\}^{-1}\widetilde{A_{n,n-1}^{n-2}}$. By the induction assumption, we have $\tilde{\Delta_r} = B_{n,n}^{n-2} - B_{n-1,n}^{n-2}\{B_{n-1,n-1}^{n-2}\}^{-1}B_{n,n-1}^{n-2}$.
Using property 1, we obtain $\tilde{\Delta_r} = \Delta l$.

Remark

1) Let Δ_r^T be the transpose of Δ_r i.e. $\Delta_r^T = \begin{vmatrix} a_{11} & \cdots & a_{n1} \\ \vdots & \ddots & \vdots \\ a_{1n} & \cdots & a_{nn} \end{vmatrix}_r$. The relation (9) is directly derived from the property 2. In fact, since $\tilde{\Delta_r} = \Delta_r^T$ and $\Delta_r^T = \Delta_l$, the relation (9) follows immediately.
2) Taking the vector continued fractions proposed by Wynn in [29], we obtain exactly the same results.

Acknowledgements

I would like thank Professor C.Brezinski for helpful discussions and encouragements.

References

1. A.C.Aitken, *Determinants and Matrices*, Oliver Boyd(Edinburgh),1965.
2. A.Artin, *Geometric algebra*, Interscience(NewYork)1966.
3. C.Brezinski, M.Redivo Zaglia, *Extrapolation Methods. Theory and Practice*,North-Holland(Amesterdam),1991.
4. C.Brezinski, A general extrapolation algorithm, Numer.Math, 35(1980)175-187.
5. C.Brezinski, Some results in the theory of the vector ε-algorithm., Linear Algebra Appl., 8(1974)77-86.
6. C.Brezinski, Computation of the eigenelements of a matrix by the ε-algorithm, Linear Algebra Appl., 11(1975)7-20.
7. C.Brezinski, Application de l'ε-algorithme à la résolution des systèmes non linéaires, C.R.Acad.Sci.Paris, 271A(1970)1174-1177.

8. C.Brezinski, *Padé type approximants and general orthogonal polynomials*, Birkauser Verlag, Basel (1980).
9. R.Deheuvels, *formes quadratiques et groupes classiques*, Presses Universitaires de France,1981.
10. F.J.Dyson, Quaternion determinants,Helv.Phys.Acta,45(1972)289-302.
11. E.Gekeler, On the solution of systems of equations by the epsilon algorithm of Wynn,Math.Comput.,26(1972)427-436.
12. P.R.Graves-Morris, Vector-valued Rational Interpolants I, Numer.Math., 42(1983)331-348.
13. P.R.Graves-Morris, Vector-valued Rational Interpolants II,I.M.A.J.Num.Anal, 4(1984)209-224.
14. P.R.Graves-Morris and CD Jenkins, Vector-valued Rational Interpolants III,Constr.Approx, 2(1986)263-289.
15. P.R.Graves-Morris and DE Roberts,From Matrix to Vector Padé Approximants, J.Comp.Appld.Math., to appear.
16. T.Håvie, Generalized Neville type extrapolation schemes, BIT, 19(1979)204-213.
17. A.Heyting, Die theorie der linearen Gleichungen in einer Zahlenspezies mit nichtkommutativer Multiplikation,Math.Ann,98(1927)465-490.
18. G.N. Hile and P.Lounesto, Matrix representations of Clifford algebras,Linear Algebra Appl.,128(1990)51-63.
19. M.L.Mehta, *Matrix theory. Selected Topics and Useful Results*, Les éditions de Physique, Les Ulis(1989).
20. J.B.McLeod, A note on the ε-algorithm, Computing, 7(1971)17-24.
21. R.Penrose , A generalised inverse for matrices, Pro.Cambridge Phil.Soc, 51(1955),406-413.
22. IR.Porteous, *Topological Geometry,2nd ed*,Cambridge University Press,1981.
23. DE.Roberts, Clifford algebras and vector-valued rational forms I,Proc.Roy.Soc,A431(1990)285-300.
24. DE.Roberts, Clifford algebras and vector-valued rational forms II, to appear in the proceedings of the International Congres on Extrapolation and Rational Approximation, to be published in Numerical Algorithms.
25. A.Salam, *Extrapolation: Extension et Nouveaux Résultats*,Thesis Université des Sciences et Technologies de Lille,1993.
26. A.Salam, Non-commutative extrapolation algorithms, to appear.
27. J.Van Iseghem, Vector Padé approximants, 11th IMACS Worl Congres Numer. Math. and Appl. 1 ed. Vichnevetsky and Vignes (1986), North Holland, Amsterdam.
28. P.Wynn, On a device for computing the $e_m(S_n)$ transformation,MTAC,10(1956)91-96.
29. P.Wynn, Vector continued fractions,Linear Algebra Appl.,1(1968)357-395.

QUADRATURE FORMULAS ON THE UNIT CIRCLE AND TWO-POINT PADÉ APPROXIMATION

ADHEMAR BULTHEEL
Dept. Computer Science, K.U. Leuven, Belgium

PABLO GONZÁLEZ-VERA
Dept. Mathematical Analysis, Univ. of La Laguna, Tenerife, Spain

ERIK HENDRIKSEN
Dept. Mathematics, Univ. of Amsterdam, The Netherlands

and

OLAV NJÅSTAD
Dept. Mathematical Sciences, Univ. of Trontheim-NTH, Norway

Abstract. In this paper our aim is to estimate integrals of the form $I_\mu\{f\} = \int_{-\pi}^{\pi} f(e^{i\theta})d\mu(\theta)$ where μ is, in general a complex measure. We consider quadrature formulas like $I_n\{f\} = \sum_{j=1}^{n} A_{j,n} f(x_{j,n})$ with n distinct nodes $x_{j,n}$ on the unit circle and so that $I_\mu\{f\} = I_n\{f\}$ for any $f \in \mathcal{R}_n$ (a certain subspace of Laurent polynomials with dimension n). Under appropriate assumptions on the function f we show that such quadratures are immediately related to the estimation of the so-called Riesz-Herglotz transform of the measure μ. Results concerning the rate of convergence of these quadratures are also given.

Key words: Numerical quadrature, two-point Padé approximation

1. Introduction

In the approximate calculation of definite integrals,

$$\int_a^b f(x)\omega(x)dx \tag{1}$$

where $[a, b]$ is a finite interval and ω a complex function, by a linear quadrature formule with coefficients or weights A_j and nodes x_j, $j = 1, 2, \ldots, n$, it is fundamental that f can be uniformly approximated by polynomials on $[a, b]$. Thus the coefficients and nodes will be chosen so that the corresponding quadrature $\sum_{j=1}^{n} A_j f(x_j)$ integrates exactly polynomials up to a certain degree as large as possible. When $\omega(x) \geq 0$, the well known Gauss-Christoffel formulas arise. See e.g. [13] for a survey. When ω is in general a complex function, not so many results are known. Orthogonal polynomials with respect to complex measures play now a crucial role. A recent contribution in this matter is the paper by González-Vera et al [16]. (See also, [25] and [1]).

On the other hand, the estimation of integrals not in the form (1) but now on

A. Cuyt (ed.), Nonlinear Numerical Methods and Rational Approximation II, 303–317.

the unit circle, that is

$$I_\mu\{f\} = \int_{-\pi}^{\pi} f(e^{i\theta})d\mu(\theta) \tag{2}$$

and related topics such as Szegő polynomials and the trigonometric moment problem have recently received much attention as a result of their application in the rapidly growing field of digital signal processing ([3, 11]) and in operator theory [15, 22]. In such cases μ in (2) used to be a finite positive Borel measure on $[-\pi, \pi]$ and the concept of "Szegő quadrature formula" appears as an anologue on the unit circle to a Gaussian qaudrature formula on an interval. It was formally introduced by W.B. Jones, O. Njåstad and W. Thron in [21], (see also [26]). Results concerning the error and convergence for such formulas were given in [5]. As for numerical results see [18].
Finally, the extension of these quadratures to certain spaces of rational functions with prescribed poles not on the unit circle were first introduced in [4]. (See also [6, 7] with an application to the Poisson integral).
Next, some remarks on notation are given. For every pair (p, q) of integers, where $p \leq q$, we denote by $\Delta_{p,q}$ the linear space of all Laurent polynomials (L-polynomials)

$$L(z) = \sum_{j=p}^{q} c_j z^j; \qquad c_j \in \mathbf{C}.$$

We write Δ for the linear space of all L-polynomials, Π for the space of all polynomials and $\Pi_n (= \Delta_{0,n})$ for the space of all polynomials of degree at most n. We shall also write $\mathbf{T} = \{w \in \mathbf{C} : |z| = 1\}$, $\mathbf{D} = \{w \in \mathbf{C} : |z| < 1\}$ and $\mathbf{E} = \{z \in \mathbf{C} : |z| > 1\}$ for the unit circle, the open unit disk and the exterior of the unit circle. Finally, let us remark that as for quadratures on $\mathbf{T}$, L-polynomials play the same role as the usual polynomials when quadratures for integrals (1) over an interval are considered. This fact is basically motivated by the following

Theorem 1 *([27, p. 39]) Let C be an arbitrary Jordan curve of the finite z-plane. Then, any function $f(z)$ continuous on C can be uniformly approximated on C by the sum of a polynomial in z and a polynomial in $\bar{z}$.*

As an immediate consequence, we have

Corollary 1 *If f is any continuous function on $\mathbf{T}$, then f can be uniformly approximated on $\mathbf{T}$ by L-polynomials.*

2. Preliminary results

Let $\mathbf{X} = \{x_{j,n}\}$, $n \in \mathbf{N}$, $j = 1, 2, \ldots, n$ be a triangular table of nodes, $\mathbf{X} \subset \mathbf{T}$. In order to construct quadrature formulas not involving values of derivatives, we shall assume that $x_{j,n} \neq x_{k,n}$ if $j \neq k$. Thus, for a given integral $I_\mu\{f\}$ on the unit circle (2), where μ is always allowed to be a complex measure, we shall be concerned with its estimation by using a formula of the kind,

$$I_n\{f\} = \sum_{j=1}^{n} A_{j,n} f(x_{j,n}) \tag{3}$$

such that $I_\mu\{f\} = I_n\{f\}$ for any $f \in \mathcal{R}_n$, which, as will be established later on, represents a certain subspace of Δ.

Let G be domain of the complex z-plane (i.e., a closed and connected point set), which contains the unit circle $\mathbf{T}$ but not the origin. Let Γ be the boundary of G which is assumed to be a union of Jordan curves and let $f(z)$ be a function which is analytic in G. Then, from Cauchy's Theorem

$$\frac{1}{2\pi i}\int_\Gamma \frac{z_0+z}{z_0-z} f(z)dz = -2z_0 f(z_0) \tag{4}$$

whenever z_0 is in the interior of G. Now, setting $g(z) = -f(z)/(2z)$ (which is also analytic in G), it follows that

$$f(z_0) = \frac{1}{2\pi i}\int_\Gamma \frac{z_0+z}{z_0-z} g(z)dz. \tag{5}$$

From (5) and Fubini's Theorem, the following holds,

$$I_\mu\{f\} = \frac{1}{2\pi i}\int_\Gamma F_\mu(z)g(z)dz \tag{6}$$

where $F_\mu(z) = \int_{-\pi}^{\pi} \frac{e^{i\theta}+z}{e^{i\theta}-z} d\mu(\theta)$ represents the well-known Herglotz-Reisz transform of the measure μ, which is an analytic function outside $\mathbf{T}$.
Let $f_n(z)$ denote a certain rational approximant to $F_\mu(z)$ with n distinct poles $x_{j,n}$, $j = 1, \ldots, n$ on $\mathbf{T}$. That is,

$$f_n(z) = Q_n(z)/P_n(z); \qquad \deg(Q_n) \le \deg(P_n), \qquad P_n(z) = \prod_{j=1}^{n}(z - x_{j,n})$$

For such rational functions, one has the decomposition,

$$f_n(z) = \lambda_n + \sum_{j=1}^{n} A_{j,n}\frac{x_{j,n}+z}{x_{j,n}-z} \tag{7}$$

and it can be written (again by (5)) as

$$\frac{1}{2\pi i}\int_\Gamma f_n(z)g(z)dz = \frac{1}{2\pi i}\sum_{j=1}^{n} A_{j,n}\int_\Gamma \frac{x_{j,n}+z}{x_{j,n}-z} g(z)dz = \sum_{j=1}^{n} A_{j,n} f(x_{j,n}) = I_n\{f\}. \tag{8}$$

In other words, when replacing in (6) $F_\mu(z)$ by a certain rational approximation, say $f_n(z)$, a quadrature formula $I_n\{f\}$ of the desired form (3) is obtained.

In order to provide an explicit representation of coefficients $\{A_{j,n}\}$ in such a formula, we can write

$$f_n(z) = \frac{Q_n(z)}{P_n(z)} = \frac{a_0 z^n + a_1 z^{n-1} + \cdots + a_n}{z^n + b_1 z^{n-1} + \cdots + b_n} = a_0 + \frac{\alpha_1 z^{n-1} + \cdots + \alpha_n}{z^n + b_1 z^{n-1} + \cdots + b_n}$$

with $\alpha_i = a_i - a_0 b_i$; $i = 1, 2, \ldots, n$. Thus, $f_n(z) = a_0 + R_{n-1}(z)/P_n(z)$, where $R_{n-1} \in \Pi_{n-1}$. Setting, (by simple partial fraction decomposition)

$$\frac{R_{n-1}(z)}{P_n(z)} = \sum_{j=1}^{n} \frac{B_j}{z - x_{j,n}}, \quad B_j = \frac{R_{n-1}(x_{j,n})}{P'_n(x_{j,n})}, \quad j = 1, \ldots, n \tag{9}$$

yields

$$f_n(z) = \lambda_n + \sum_{j=1}^{n} A_{j,n} \frac{x_{j,n} + z}{x_{j,n} - z},$$

where $\lambda_n = a_0 - \sum_{j=1}^{n} B_j / 2x_{j,n}$ and

$$A_{j,n} = -B_j / 2x_{j,n}, \quad j = 1, \ldots, n \tag{10}$$

since

$$\frac{1}{z - x_{j,n}} = \frac{1}{2x_{j,n}} \left[\frac{z + x_{j,n}}{z - x_{j,n}} - 1 \right].$$

As for the error in the numerical quadrature (8), from (6) and (8) one immediately gets,

$$E_n\{f\} = I_\mu\{f\} - I_n\{f\} = -\frac{1}{2\pi i} \int_\Gamma (F_\mu(z) - f_n(z)) \frac{f(z)}{2z} dz \tag{11}$$

and it follows trivially that:

$$|E_n\{f\}| \leq \frac{1}{4\pi} \max\{|f(\zeta)/\zeta| : \zeta \in \Gamma\} \int_\Gamma |F_\mu(z) - f_n(z)||dz|. \tag{12}$$

Equation (12) clearly implies that rational approximants $\{f_n(z)\}$, converging locally uniformly on $\mathbf{C} - \mathbf{T}$ to $F_\mu(z)$, can be "good candidates" to provide "suitable quadrature formulas".

In the sequel, we shall assume that, in order to compute (2) with μ a complex measure, the integrals

$$\mu_k = \int_{-\pi}^{\pi} e^{ik\theta} d\mu(\theta) \tag{13}$$

exist and are known for $k = 0, \pm 1, \pm 2, \ldots$ (Actually, it is sufficient to assume that $\int_{-\pi}^{\pi} |d\mu(\theta)| \leq M < +\infty$.) Furthermore, it can be easily checked that $F_\mu(z)$ admits the expansion (around $z = 0$ and $z = \infty$):

$$\begin{aligned} L_0 &= \mu_0 + 2\sum_{k=1}^{\infty} \mu_k z^k; \quad |z| < 1 \\ L_\infty &= -\mu_0 - 2\sum_{k=1}^{\infty} \mu_{-k} z^{-k}; \quad |z| > 1. \end{aligned} \tag{14}$$

Untill now, for a fixed n, nothing has been said about the approximation of $F_\mu(z)$ by $f_n(z)$. Thus, let us assume that $f_n(z)$ satisfies

$$L_0 - f_n(z) = \sum_{j=p+1}^{\infty} \mu_j^* z^j = O(z^{p+1}), \quad p \geq 0 \tag{15}$$

$$L_\infty - f_n(z) = \sum_{j=q+1}^{\infty} \mu^*_{-j} z^{-j} = O(z^{q+1}), \quad q \geq 0. \tag{16}$$

Now, if we write (from (7))

$$f_n(z) = \lambda_n + \sum_{j=1}^{n} A_{j,n} + 2\sum_{k=1}^{\infty}\left(\sum_{j=1}^{n} A_{j,n}/x_{j,n}^k\right) z^k; \quad |z| \to 0$$

and

$$f_n(z) = \lambda_n - \sum_{j=1}^{n} A_{j,n} - 2\sum_{k=1}^{\infty}\left(\sum_{j=1}^{n} A_{j,n}x_{j,n}^k\right) z^{-k}; \quad |z| \to \infty,$$

then, from (15) and (15), one first gets

$$\lambda_n + \sum_{j=1}^{n} A_{j,n} = \mu_0 \quad \text{and} \quad \lambda_n - \sum_{j=1}^{n} A_{j,n}^n = -\mu_0,$$

which implies $\lambda_n = 0$. On one hand, we have from (15)

$$I_n\{z^{-k}\} = \sum_{j=1}^{n} A_{j,n}/x_{j,n}^k = \mu_k = I_\mu\{z^{-k}\}, \quad (z = e^{i\theta}), \quad 0 \leq k \leq p$$

and, on the other, from (15),

$$I_n\{z^k\} = \sum_{j=1}^{n} A_{j,n}x_{j,n}^k = \mu_{-k} = I_\mu\{z^k\}, \quad 0 \leq k \leq q.$$

As a result, the following Proposition has been proved.

Proposition 1 *Let n, p and q be nonnegative integers ($n \geq 1$) and $f_n(z)$ a rational function (7) satisfying (15-15). Then, $\lambda_n = 0$ and the formula (3) with weights given by (10) and as nodes the poles of $f_n(z)$, is valid in $\Delta_{-p,q}$, that is, $I_n\{f\} = I_\mu\{f\}$, $\forall f \in \Delta_{-p,q}$.*

We see that when formulas (3) valid in subspaces of Δ (see Corollary 1) are required, so that conditions (15-15) are satisfied, then this clearly leads to the concept of two-point Padé-type approximant ([12, 17]) (As for two-point Padé approximants see, e.g. [23, 20].) Thus, in general, let L and L^* be two formal power series of the form

$$L = \sum_{j=0}^{\infty} c_j z^j, \quad \text{and} \quad L^* = \sum_{j=0}^{\infty} c^*_{-j} z^{-j},$$

k and n nonnegative integers such that $0 \leq k \leq n$. Let $P_n(z)$ be an arbitrary polynomial of degree n, satisfying $P_n(0) \neq 0$. Under these conditions a unique polynomial $Q_n \in \Pi_n$ exists so that (see [12])

$$L - Q_n/P_n = O(z^k) \quad \text{and} \quad L^* - Q_n/P_n = O((z^{-1})^{n-k+1})$$

The rational function Q_n/P_n is said to be a "two-point Padé-type approximant" (2PTA) of order $(k, n-k+1)$ to the pair (L, L^*). According to our purposes (see (15) and (15)) 2PTA's with $k > 0$ will be used on the sequel. This means that we shall deal with 2PTA's to the pair (14) and order $(p+1, q+1)$ with $p+q = n-1$. Such approximant will be denoted, in short, by

$$(p/n)_{L_0,L_\infty} = Q_n(z)/P_n(z); \quad 0 \le p \le n-1, \quad (q = n-1-p)$$

Remark 1 A reciprocal result to Proposition 1 also holds. Indeed, let $I_n\{f\} = \sum_{j=1}^n A_{j,n} f(x_{j,n})$ be a quadrature formula with distinct nodes and valid in $\Delta_{-p,q}$ $(p+q = n-1)$. If $P_n(z) = \prod_{j=1}^n (z - x_{j,n})$ is the nodal polynomial, then the rational function given by

$$\sum_{j=1}^n A_{j,n} \frac{x_{j,n} + z}{x_{j,n} - z}$$

represents a 2PTA to (L_0, L_∞) of order $(p+1, q+1)$ with denominator $P_n(z)$. For further references concerning connections between rational funcions and quadrature formulas in a rather general framework, see the paper by Njåstad and Thron [24].

Remark 2 In order to give a more compact formula for such 2PTA, it will be convenient to recall (see [5]) that a formula (3) is valid in $\Delta_{-p,q}$ with $p+q = n-1$, if and only if $I_n\{f\} = I_\mu\{R_{p,q}\}$, where $R_{p,q}$ is unique function in $\Delta_{-p,q}$ satisfying the interpolation conditions $R_{p,q}(x_{j,n}) = f(x_{j,n})$, $j = 1, 2, \ldots, n$.

On the other hand, one has also,

Proposition 2 *Let us consider $g(x,z) = \frac{x+z}{x-z}$ as a function in the variable x, z being a parameter. Let $R_{p,q}$ be the L-polynomial in $\Delta_{-p,q}$, $(p+q = n-1)$ interpolating $g(x,z)$ at distinct gives nodes $x_{j,n}$, $j = 1, \ldots, n$ $(x_{j,n} \neq 0)$, then*

$$R_{p,q}(x) = 1 + \frac{2z}{x-z}\left(1 - \frac{z^p P_n(x)}{x^p P_n(z)}\right), \quad P_n(z) = \prod_{j=1}^n (z - x_{j,n}) \tag{17}$$

Proof. Since $x_{j,n} \neq 0$ and $P_n(x_{j,n}) = 0$, $R_{p,q}(x)$ given by (17) clearly fullfils the interpolation conditions. Furthermore, if we write,

$$R_{p,q}(x) = 1 + \frac{2z}{P_n(z)} x^{-p} \left[\frac{x^p P_n(z) - z^p P_n(x)}{x - z}\right]$$

we see that the expression in brackets is a polynomial in the variable x of degree $n-1$ at most. Therefore, factor x^{-p} makes the resulting function belong to $\Delta_{-p,q}$ and the proof follows by virtue of the unicity property for such interpolating Laurent polynomials. □

Remark 3 The above result is also valid when not necessarily distinct nodes are used. In that case, Hermite interpolation should be considered.

Thus from Proposition 2 and remarks 2-3, one has

Corollary 2 *Let $P_n(z)$ be a polynomial of degree n such that $P_n(0) \neq 0$, p and q nonnegative integers with $p+q=n-1$. Then, the 2PTA with denominator P_n to $F_\mu(z)$ (or equivalently to the pair (L_0, L_∞)) and order $(p+1, q+1)$ is given by*

$$(p/n)_{(L_0,L_\infty)} = \frac{Q_n(z)}{P_n(x)} = I_\mu\left\{1 + \frac{2z}{x-z}\left(1 - \frac{z^p P_n(x)}{x^p P_n(z)}\right)\right\}. \tag{18}$$

Furthermore, the following formula holds for the error

$$F_\mu(x) - \frac{Q_n(x)}{P_n(z)} = \frac{2z^{p+1}}{P_n(z)}\int_{-\pi}^{\pi}\frac{e^{-ip\theta}P_n(e^{i\theta})}{e^{i\theta}-z}d\mu(\theta). \tag{19}$$

It results from (18) that the numerator Q_n in this approximant, equals

$$Q_n(z) = I_\mu\left\{P_n(z) + \frac{2z}{x-z}\left(P_n(x) - \frac{z^p}{x^p}P_n(x)\right)\right\}$$

which yields

$$Q_n(z) = I_\mu\left\{\frac{x+z}{x-z}\left(P_n(z) - \frac{z^p}{x^p}P_n(x)\right)\right\} + I_\mu\left\{\frac{z^p P_n(x)}{x^p}\right\}, \quad 0 \le p \le n-1. \tag{20}$$

Finally, the following integral expression for coefficients $A_{j,n}$ will be used in Section 4.

Proposition 3 *If $I_n\{f\} = \sum_{j=1}^n A_{j,n} f(x_{j,n})$ is a quadrature formula valid in $\Delta_{-p,q}$, $(p+q=n-1)$, then*

$$A_{j,n} = x_{j,n}^p I_\mu\left\{\frac{z^{-p}P_n(z)}{(z-x_{j,n})P_n'(x_{j,n})}\right\}, \quad j=1,\ldots,n \tag{21}$$

where, as usual, $P_n(z) = \prod_{j=1}^n (z - x_{j,n})$.

Proof. Let $R_{p,q}(z)$ be the interpolating L-polynomial in $\Delta_{-p,q}$ satisfying $R_{p,q}(x_{j,n}) = f(x_{j,n})$, $j=1,\ldots,n$, f being a given function on $\mathbf{T}$. Set

$$L_j(z) = x_{j,n}^p z^{-p}\frac{P_n(z)}{(z-x_{j,n})P_n'(x_{j,n})}, \quad j=1,\ldots,n.$$

Then clearly $L_j \in \Delta_{-p,q}$ and $L_j(x_{k,n}) = \delta_{jk}$; $1 \le j,k \le n$. Therefore,

$$R_{p,q}(z) = \sum_{j=1}^n L_j(x) f(x_{j,n})$$

Now, since $I_n\{f\} = I_\mu\{R_{p,q}\}$, the proof follows. □

Untill now our starting point has been a triangular table $\mathbf{X}$ of distinct nodes on $\mathbf{T}$ such that for each fixed n a formula (3) valid in $\Delta_{-p,q}$ $(p+q=n-1)$ can be constructed. In the next sections we are concerned with choices of nodes $\{x_{j,n}\}$ which will give rise to easy computations and which result in sequences of quadrature formulas that are convergent for a class of functions as large as possible.

3. Positive measures: Szegő formulas

As an illustration, we consider in this section the usual case of a finite positive Borel measure μ. By introducing the inner product,

$$\langle f, g\rangle_\mu = \int_{-\pi}^{\pi} f(e^{i\theta})\overline{g(e^{i\theta})}d\mu(\theta) \tag{22}$$

a monic sequence of polynomials $\{\rho_n\}$, $(\deg(\rho_n) = n)$ can be uniquely determined so that $\langle \rho_n, z^m\rangle_\mu = 0$ for $0 \le m \le n-1$ and $\langle \rho_n, \rho_n\rangle_\mu > 0$. $\{\rho_n\}$ is called the sequence of Szegő polynomials with respect to the measure μ. Such polynomials have been extensively studied. (See e.g. [21] for a survey and references found there.) Furthermore, as is well known [14, 19], the zeros of the Szegő polynomials all lie inside the unit disc $\mathbf{D}$. Hence, such zeros can not be used as nodes in our quadratures. In order to overcome such drawback, W.B. Jones et al. introduced in [21] certain polynomials with orthogonality properties w.r.t. μ whose zeros lie on $\mathbf{T}$ and are distinct. Indeed, set $\rho_n^*(z) = z^n\overline{\rho_n(1/\bar{z})}$ and define $X_n(z) = \rho_n(x) + \omega_n\rho_n^*(z)$ with $\omega_n \in \mathbf{T}$, then the following holds (see Th. 6.1 and Th. 6.2 in [21].

Proposition 4 *Let $\{\omega_n\}$ be a given sequence of complex numbers satisfying $|\omega_n| = 1$ for $n \ge 1$ and let $X_n(z)$ be as defined above. Then:*
(i) $\langle X_n, 1\rangle_\mu \ne 0$, $\langle X_n, z^m\rangle_\mu = 0$, $1 \le m \le n-1$, and $\langle X_n, z^n\rangle_\mu \ne 0$
(ii) The n zeros of $X_n(z)$ are simple and lie on the unit circle $\mathbf{T}$.

(Polynomials satisfying conditions in (i) are called para-orthogonal.)

For a fixed sequence $\{\omega_n\} \subset \mathbf{T}$, let $x_{j,n}$, $j = 1, \ldots, n$, be the zeros of $X_n(z)$, $n = 1, 2, \ldots$. We now investigate, from our approach, the corresponding quadrature formulas based on such nodes or equivalently the 2PTA's with denominator $X_n(z)$. Thus, set

$$R_n^p(z) = Q_n^p(z)/X_n(z) = (p/n)_{L_0, L_\infty}, \quad 0 \le p \le n-1.$$

Formula (19) and orthogonality properties for $X_n(z)$ allow one to show easily that, independently of p,

$$F_\mu(z) - R_n^p(z) = O(z^n) \quad \text{and} \quad F_\mu(z) - R_n^p(z) = O((1/z)^n). \tag{23}$$

From (23) and Proposition 1, we have

Corollary 3 *Let μ be a positive measure on $[-\pi, \pi]$. Let $\{x_{j,n}\}_{j=1}^n$ be the zeros of $X_n(z)$ as defined above and $I_n\{f\} = \sum_{j=1}^n A_{j,n}f(x_{j,n})$ the corresponding quadrature formula valid in $\Delta_{-p,q}$, $(p+q = n-1)$. Then, it holds*

$$I_n\{f\} = I_\mu\{f\}, \quad \forall f \in \Delta_{1-n,n-1}. \tag{24}$$

Thus, we see from (24) that the so-called Szegő formulas arise, [21]. As for the weights $A_{j,n}$ in those formulas, from (21) and taking $p = 0$ (recall that p can be any integer with $0 \le p \le n-1$), it follows that

$$A_{j,n} = I_\mu\left\{\frac{X_n(z)}{(z - x_{j,n})X_n'(x_{j,n})}\right\} = I_\mu\left\{\left|\frac{X_n(z)}{(z - x_{j,n})X_n'(x_{j,n})}\right|^2\right\} > 0. \tag{25}$$

The last equality in (25) is a consequence of the orthogonality properties for $X_n(z)$. (For more details, see [21, pp. 134–135].)

Remark 4 For the numerator $Q_n^p(z)$ of the 2PTA $R_n^p(z)$, from (20) and Proposition 4, (*i*) (take into account that $I_\mu\{z^p x^{-p} X_n(x)\} = z^p \langle X_n, x^p \rangle_\mu = 0$, $1 \le p \le n-1$ with $x = e^{i\theta}$), the following holds

$$Q_n^p(z) = I_\mu \left\{ \frac{x+z}{x-z} \left(X_n(z) - \frac{z^p}{x^p} X_n(x) \right) \right\}, \quad 1 \le p \le n-1.$$

(Compare with relation (5.6) in [21, p. 127].) Furthermore, if we take $p = 0$ in (20), we can also write

$$Q_n^p(z) = I_\mu \left\{ \frac{x+z}{x-z} (X_n(z) - X_n(x)) \right\} + \omega_n \langle z^n, \rho_n \rangle_\mu .$$

Thus, by taking $\varphi_n(z)$ as the corresponding orthonormal polynomials of degree n and with leading coefficient $\alpha_n > 0$, then

$$Q_n^p(z) - \frac{\omega_n}{\alpha_n} = I_\mu \left\{ \frac{x+z}{x-z} (X_n(z) - X_n(x)) \right\}$$

where $X_n(z) = \varphi_n(z) + \omega_n \varphi_n^*(z)$ and $\varphi_n(z)$ orthonormal.

To end this Section, a result concerning convergence is stated.

Theorem 2 (Convergence of Szegő formulas) *Let $\{\omega_n\}$ be a given sequence of complex numbers satisfying $|\omega_n| = 1$, for $n \ge 1$. Let $I_n\{f\}$ be the sequence of quadratures (Szegő formulas) based on the zeros of $X_n(z) = \rho_n(z) + \omega_n \rho_n^*(z)$, $n = 1, 2, \ldots$. (The $\{\rho_n\}$ are the monic Szegő polynomials.) Then,* $\lim_{n\to\infty} I_n\{f\} = I_\mu\{f\}$ *for any integrable function f on $\mathbf{T}$ w.r.t. the measure μ.*

Proof. By (11) and Th. 7.3 in [21, pp. 135–136] convergence is first guaranted for f being analytic in $\mathbf{D}$. Corollary 1 and (25) allow to extend convergence to the class of the continuous functions on $\mathbf{T}$. Finally, paralleling rather closely the arguments given in [10, pp. 127–129], one can show that Theorem 2 is also valid for integrable functions, because of the fact that any sequence of integration rules with positive weights, which converges for all continuous functions, converges for all integrable functions w.r.t. a finite Borel measure μ on $[-\pi, \pi]$. □

Remark 5 For an alternative proof of this Theorem see [5]. However, it should be clear that boundedness (12) and relation (19) are valid only for analytic functions f.

4. Complex measures. Choice of nodes and convergence

Let μ be a complex measure on $[-\pi, \pi]$, so that nothing, can be said in general, about the location of the zeros of the corresponding Szegő polynomials w.r.t. an inner product $\langle \cdot, \cdot \rangle_\mu$ (22) which is not Hermitian. As a result, for a given sequence of para-orthogonal polynomials (in case they exist), their zeros do not necessarily lie on $\mathbf{T}$ and cannot be used as nodes in a quadrature formula (3). Therefore we describe a method by which easily computable nodes can be selected such that, convergence of the quadrature process is assured. Let us first consider the "natural" selection of equally distributed nodes $\{x_{j,n}\}$ on $\mathbf{T}$; that is,

$$P_n(z) = \prod_{j=1}^{n}(z - x_{j,n}) = z^n + \omega_n, \quad |\omega_n| = 1, \quad n = 1, 2, \ldots \tag{26}$$

Let $\{p(n)\}_1^\infty$ and $\{q(n)\}_1^\infty$ be two arbitrary sequences of nonnegative integers with $p(n) + q(n) = n - 1$ and $\lim_{n\to\infty} \frac{p(n)}{n} = r$, $0 < r < 1$. We consider the sequence of quadrature formulas (3) with nodes given by (26), so that for $n \geq 1$, $I_n\{f\}$ is exact in $\Delta_{-p(n),q(n)}$. One has,

Theorem 3 *Let $I_n\{f\}$ and μ be as in the preceding paragraph. Let f be an analytic function in a domain G such that $\mathbf{T} \subset G$ and $0 \notin G$. Then, $\lim_{n\to\infty} I_n\{f\} = I_\mu\{f\}$. Furthermore,*

$$\lim_{n\to\infty} |E_n\{f\}|^{1/n} = \lim_{n\to\infty} |I_n\{f\} - I_\mu\{f\}|^{1/n} \leq \gamma < 1, \tag{27}$$

where $\gamma = \max\{\gamma_1, \gamma_2\}$ with $\gamma_1 = \max\{|z|^r : z \in \Gamma \cap \mathbf{D}\}$ and $\gamma_2 = \max\{|z|^{1-r} : z \in \Gamma \cap \mathbf{E}\}$, Γ being the boundary of G.

Proof. By (12) it suffices to prove that the sequence of 2PTA's of order $(p(n) + 1, q(n) + 1)$ with denominators (26) converge locally uniformly to $F_\mu(z)$ outside $\mathbf{T}$. Indeed, for any $z \notin \mathbf{T}$, (19) yields,

$$\left|F_\mu(z) - \frac{Q_n(z)}{P_n(z)}\right| \leq \frac{2|z|^{p(n)+1}}{|P_n(z)|} \max_{x\in\mathbf{T}} |P_n(x)| \int_{-\pi}^{\pi} \frac{|d\mu(\theta)|}{|e^{i\theta} - z|}. \tag{28}$$

Now, if $z \in \mathbf{D}$ then $\lim_{n\to\infty} |P_n(z)|^{1/n} = 1$. Therefore, from (28)

$$\lim_{n\to\infty} |F_\mu(z) - Q_n(z)/P_n(z)|^{1/n} \leq |z|^r < 1. \tag{29}$$

Next, let us assume that $z \in \mathbf{E}$, that is $|z| > 1$. Since

$$|P_n(z)| = |z^n + \omega_n| = |z^n||z^{-n} + \bar{\omega}_n|, \quad (\omega_n \in \mathbf{T}),$$

then clearly $\lim_{n\to\infty} |P_n(z)|^{1/n} = |z|$ for any $z \in \mathbf{E}$, which implies

$$\lim_{n\to\infty} |F_\mu(z) - Q_n(z)/P_n(z)|^{1/n} \leq |z|^{r-1} < 1. \tag{30}$$

Hence, by (29), (30) and (11) the proof follows. □

In order to study other selections of nodes $\{x_{j,n}\} \subset \mathbf{T}$, it is helpful to recall that for integrals (1), interpolatory quadrature formulas based on zeros of orthogonal polynomials w.r.t. a positive measure on $[a, b]$ have been largely considered in recent years (see e.g. [2]). Accordingly, for a given positive measure α on $[-\pi, \pi]$, we let $\{X_n^\alpha\}$ be a sequence of para-orthogonal polynomials, that is

$$X_n^\alpha(z) = \phi_n(x, \alpha) + \omega_n \phi_n^*(z, \alpha)$$

$\{\phi_n(z, \alpha)\}$ being the monic Szegő polynomials w.r.t. α. Thus quadrature formulas (3) based on the zeros of $X_n^\alpha(z)$ to estimate $I_\mu\{f\}$ could provide good results. Immediately, a question arises: Which measure α should be chosen?
By taking $d\alpha(\theta) = d\theta/(2\pi)$ (Lebesgue measure), then $\phi_n(z, \alpha) = z^n$ and $X_n^\alpha(z) = z^n + \omega_n$. Thus, the case of equally distributed nodes on $\mathbf{T}$ is now contemplated. We can also think of measures induced by the Poisson kernel, that is

$$d\alpha(\theta) = \frac{1}{2\pi} \frac{1 - |r|^2}{|z - r|^2} d\theta, \quad z = e^{i\theta} \quad \text{and} \quad r \in \mathbf{D}. \tag{31}$$

(Observe that when $r = 0$, the Lebesgue measure is recovered.) We now find $\phi_0(z, \alpha) = \phi_0^*(z, \alpha) = 1$, $\phi_n(z, \alpha) = z^n - rz^{n-1}$, $\phi_n^*(z, \alpha) = 1 - \bar{r}z$, $n \geq 1$ and for the para-orthogonal polynomials [26],

$$X_n^\alpha(z) = z^n - rz^{n-1} - \omega_n \bar{r} z + \omega_n. \tag{32}$$

From (32) $z = 1$ is a zero iff $r \in \mathbf{R}$ and $\omega_n = -1$, and $z = -1$ iff $\omega_n = (-1)^{n-1}$. When, for simplicity, we choose $\omega_n = 1$, $n \geq 1$, all other zeros come in conjugate pairs whenever r is real, $0 < r < 1$. (Take into account that the more general case $r \in \mathbf{D}$ can be easily reduced to this one.) In any case, for large n-values, the zeros of (32) must be computed numerically. As for the convergence of the corresponding sequence of quadrature formulas based on polynomials (32), one has: for any given sequence $\{\omega_n\} \subset \mathbf{T}$ and $z \in \mathbf{D}$,

$$\lim_{n\to\infty} |X_n^\alpha(z)|^{1/n} = 1. \tag{33}$$

Furthermore, $X_n^\alpha(z) = \omega_n z^n T_n^\alpha(1/z)$, where $T_n^\alpha(z) = \phi_n(z, \alpha) + \bar{\omega}_n \phi_n^*(z, \alpha)$. Hence, for any $z \in \mathbf{E}$, it results

$$\lim_{n\to\infty} |X_n^\alpha(z)|^{1/n} = |z|. \tag{34}$$

Thus, from (33-34), Theorem 3 remains still valid when replacing polynomials (26) ($r = 0$) by polynomials (32) ($r \in \mathbf{D}$). Hence, from the point of view of the rate of convergence for the sequence $I_n\{f\}$, measures (31) do not seem to provide relevant contributions. Certainly the problem of choosing the most appropriate measure is an open question and much more work could be done on this in the future. To end, the extension of Theorem 3 to the class of continuous functions on $\mathbf{T}$ is studied. By Corollary 1, this extension would be a simple matter, provided that the sequence $S_n = \sum_{j=1}^m |A_{j,n}|$ is bounded. As usual $A_{j,n}$ are the weights in formula (3); that is, (recall (21)),

$$A_{j,n} = x_{j,n}^{p(n)} \int_{-\pi}^{\pi} \frac{z^{-p(n)} P_n(z)}{(z - x_{j,n}) P_n'(x_{j,n})} d\mu(\theta), \quad z = e^{i\theta}.$$

On the other hand, if P_n is para-orthogonal w.r.t. a positive measure α supported in $[-\pi, \pi]$, the weights $\lambda_{j,n}$ in the Szegő formula (Section 3) are given by (25):

$$\lambda_{j,n} = \int_{-\pi}^{\pi} \left| \frac{P_n(z)}{(z - x_{j,n})P_n'(x_{j,n})} \right|^2 d\alpha(\theta), \quad z = e^{i\theta}; \quad j = 1, 2, \ldots, n.$$

The measure α is also assumed to be normalized, i.e. $\int d\alpha(\theta) = 1$. So, we have

$$\sum_{j=1}^{n} \lambda_{j,n} = 1. \tag{35}$$

Now, let us assume that $d\alpha(\theta) = \alpha'(\theta)d\theta$, $\alpha'(\theta) > 0$, a.e. in $[-\pi, \pi]$, $d\mu(\theta) = w(\theta)d\theta$ and,

$$\int_{-\pi}^{\pi} \frac{|w(\theta)|^2}{\alpha'(\theta)} d\theta = K < +\infty. \tag{36}$$

Then, making use of the Cauchy-Schwartz inequality, we can write ($z = e^{i\theta}$):

$$\begin{aligned} |A_{j,n}| &= \left| \int_{-\pi}^{\pi} \left[\frac{P_n(z)}{(z - x_{j,n})P_n'(x_{j,n})} \sqrt{\alpha'(\theta)} \right] \frac{w(\theta) z^{-p(n)}}{\sqrt{\alpha'(\theta)}} d\theta \right| \leq \\ &\leq \left\{ \int_{-\pi}^{\pi} \left| \frac{P_n(z)}{(z - x_{j,n})P_n'(x_{j,n})} \right|^2 \alpha'(\theta) d\theta \right\}^{1/2} \left\{ \int_{-\pi}^{\pi} \frac{|w(\theta)|^2}{\alpha'(\theta)} d\theta \right\}^{1/2} = K_1 \lambda_{j,n}^{1/2}, \end{aligned}$$

with $K_1 = K^{1/2}$. Therefore, by using again the Cauchy-Schwartz inequality, we get

$$S_n = \sum_{j=1}^{n} |A_{j,n}| \leq K_1 \sum_{j=1}^{n} [\lambda_{j,n}]^{1/2} \leq K_1 \sqrt{n} \left\{ \sum_{j=1}^{n} \lambda_{j,n} \right\}^{1/2} = K_1 \sqrt{n}. \tag{37}$$

Remark 6 Under the above conditions, for a positive measure α, we have in general the inequality $\sum_{j=1}^{n} [\lambda_{j,n}]^{1/2} \leq \sqrt{n}$ which turns out to be an equality when taking the Lebesgue measure, i.e. $d\alpha(\theta) = d\theta/2\pi$, because then $\lambda_{j,n} = 1/n$, $j = 1, 2, \ldots, n$. (See, [26] and [8].)

As is seen from (37), we cannot in general conclude that the sequence $\{S_n\}$ is bounded; thus additional assumptions on the function f are required. For a continuous function f on $\mathbf{T}$, one can write

$$f(e^{i\theta}) = f_1(\theta) + i f_2(\theta), \quad \theta \in [-\pi, \pi] \tag{38}$$

where $f_1(\theta) = \operatorname{Re}\left(f(e^{i\theta})\right)$ and $f_2(\theta) = \operatorname{Im}\left(f(e^{i\theta})\right)$ are 2π-periodic continuous functions on $[-\pi, \pi]$. Let us introduce the modulus of continuity for the function f,

$$\omega(f, \delta) = \max\{|f(e^{is}) - f(e^{it})| : |s - t| < \delta\}, \quad -\pi < s, t < \pi.$$

Let $C_{2\pi}$ stand for the space of continuous 2π-periodic functions with supremum norm. For $g \in C_{2\pi}$, we write

$$\gamma_n(g) = \min_{a_k,b_k} \max_{\theta} |g(\theta) - \sum_{k=1}^{n}(a_k \cos k\theta + b_k \sin k\theta)|, \quad a_k, b_k \in \mathbf{R}.$$

For our purposes, the following result will be needed,

Theorem 4 (Jackson's Theorem III, [9] p. 144)

$$\forall g \in C_{2\pi}: \quad \gamma_n(g) \leq \omega\left(g, \frac{\pi}{n+1}\right).$$

Now, we are in a situation to prove the following:

Theorem 5 *If f is a continuous function on $\mathbf{T}$, then there exists $R_n \in \Delta_{-n,n}$ such that,*

$$\max_{z\in\mathbf{T}} |f(z) - R_n(z)| \leq 2\omega\left(f, \frac{\pi}{n+1}\right).$$

Proof. Let $R_{n,j}(\theta)$, $j = 1, 2$ be the real trigonometric polynomials satisfying

$$\gamma_n(f_1) = \max_{\theta} |f_1(\theta) - R_{n,1}(\theta)| \quad \text{and} \quad \gamma_n(f_2) = \max_{\theta} |f_2(\theta) - R_{n,2}(\theta)|$$

with $f_j(\theta)$, $j = 1, 2$ given by (38). Set $R_{n,1}(\theta) = \text{Re}R_n$ and $R_{n,2}(\theta) = \text{Im}R_n$, i.e.,

$$R_n(z) = R_{n,1}(\theta) + iR_{n,2}(\theta); \quad z = e^{i\theta}. \tag{39}$$

Then, $R_n \in \Delta_{-n,n}$. Furthermore, since

$$|f_j(s) - f_j(t)| \leq |f(e^{is}) - f(e^{it})|, \quad j = 1, 2$$

it follows that,

$$\omega(f_j, \delta) \leq \omega(f, \delta); \quad j = 1, 2. \tag{40}$$

Thus, by Theorem 4 and (38),

$$\max_{z\in\mathbf{T}} |f(z) - R_n(z)| \leq \max_{\theta} |f_1(\theta) - R_{n,1}(\theta)| + \max_{\theta} |f_2(\theta) - R_{n,2}(\theta)| \leq 2\omega\left(f, \frac{\pi}{n+1}\right)$$

□

It follows that, if f is a continuous function on $\mathbf{T}$ with $\omega(f, \delta) = O(\delta^p)$ with $p > 1/2$, then by Theorem 5 and (37), constants A, B and C indenpendent of n exist so that for sufficiently large n, one gets

$$|I_n\{f\} - I_\mu\{f\}| \leq \frac{A + B\sqrt{n}}{(C + n)^p}.$$

In other words, the following Theorem has been proved.

Theorem 6 *Let f be a continuous function on* $\mathbf{T}$ *with modulus of continuity* $\omega(f,\delta) = O(\delta^p)$ $(p > 1/2)$ *and, for* $n = 1, 2, \ldots$, *let* $\{x_{j,n}\}$ *be the zeros of the para-orthogonal polynomials* $X_n^\alpha(z)$ *w.r.t. a positive measure* α *supported in* $[-\pi, \pi]$. *Then, if the measure* μ *satisfies (36), the sequence of quadrature formulas for* μ *with nodes* $\{x_{j,n}\}$ *converges to* $I_\mu\{f\}$.

Corollary 4 *Under the conditions of Theorem 6, the sequence* $I_n\{f\}$ *of quadrature formulas (3) with equally distributed nodes on* $\mathbf{T}$, *converges to* $I_\mu\{f\}$.

Remark 7 Note that Theorem 6 holds for any function f satisfying a Lipschitz condition of the form:

$$|f(z_1) - f(z_2)| \leq \lambda |z_1 - z_2|^p, \quad (p > 1/2); \quad z_1, z_2 \in \mathbf{T}.$$

References

1. D. Barrios, G. López, and E. Torrano. Location of zeros and asymptotics of polynomials satisfying three-term recurrence relations. *Math. Sbornik*, 1993. To appear.
2. T. Bloom, D.S. Lubinski, and H. Stahl. Interpolatory integration rules and orthogonal polynomials with varying weights. *Numerical Algorithms*, 3:55–66, 1992.
3. A. Bultheel. Algorithms to compute the reflection coefficients of digital filters. In L. Collatz, G. Meinardus, and H. Werner, editors, *Numerische Methoden der Approximationstheorie 7*, volume 67 of *ISNM*, pages 33–50. Birkhäuser Verlag, 1984.
4. A. Bultheel, P. González-Vera, E. Hendriksen, and O. Njåstad. A Szegő theory for rational functions. Technical Report TW131, Department of Computer Science, K.U. Leuven, May 1990.
5. A. Bultheel, P. González-Vera, E. Hendriksen, and O. Njåstad. Orthogonality and quadrature on the unit circle. In C. Brezinski, L. Gori, and A. Ronveaux, editors, *Orthogonal polynomials and their applications*, volume 9 of *IMACS annals on computing and applied mathematics*, pages 205–210, Basel, 1991. J.C. Baltzer AG.
6. A. Bultheel, P. González-Vera, E. Hendriksen, and O. Njåstad. Orthogonal rational functions and quadrature on the unit circle. *Numer. Alg.*, 3:105–116, 1992.
7. A. Bultheel, P. González-Vera, E. Hendriksen, and O. Njåstad. Quadrature formulas on the unit circle based on rational functions. *J. Comput. Appl. Math.*, 1992. To appear.
8. M. Camacho and P. González-Vera. A note on para-orthogonality and biorthogonality. *Det Kongelige Norske Videnskaber Selskab*, Skrifter 3:1–16, 1992.
9. E.W. Cheney. *Introduction to approximation theory*. McGraw Hill, 1966.
10. P.J. Davis and P. Rabinowitz. *Methods of numerical integration*. Academic Press, 2nd edition, 1984.
11. Ph. Delsarte, Y. Genin, Y. Kamp, and P. Van Dooren. Speech modelling and the trigonometric moment problem. *Philips J. of Research*, 37:277–292, 1982.
12. A. Draux. Approximation de type-Padé en deux points. Technical Report ANO-110, Université des Sciences et Techniques de Lille, Lab. d'analyse numérique et d'optimisation, 1983.
13. W. Gautchi. A survey of Gauss-Christoffel quadrature formulae. In P.L. Butzer and F. Feher, editors, *E.B. Christoffel. The influence of his work on mathematical and physical sciences*, pages 72–147, Basel, 1081. Birkhäuser Verlag.
14. Ya. Geronimus. *Polynomials orthogonal on a circle and their applications*, volume 3 of *Trans. Math. Monographs, Ser. 1*, pages 1–78. AMS, 1954.
15. I. Gohberg, editor. *I. Schur methods in operator theory and signal processing*, volume 18 of *Operator Theory: Advances and Applications*. Birkhäuser Verlag, Basel, 1986.
16. P. González-Vera, G. López, R. Orive, and J.C. Santos-Leon. On the convergence of quadrature formulas for complex weight functions. Preprint., 1993.

17. P. González-Vera and R. Orive. Optimization of two-point Padé-type approximants. *J. Comp. Appl. Math.*, 1993. To appear.
18. P. González-Vera, J.C. Santos-Leon, and O. Njåstad. Some numerical results about quadrature formulas on the unit circle. Preprint., 1993.
19. U. Grenander and G. Szegő. *Toeplitz forms and their applications*. University of California Press, Berkley, 1958.
20. W.B. Jones, O. Njåstad, and W.J. Thron. Two-point Padé expansions for a family of analytic functions. *J. Comp. Appl. Math.*, 9:105–124, 1983.
21. W.B. Jones, O. Njåstad, and W.J. Thron. Moment theory, orthogonal polynomials, quadrature and continued fractions associated with the unit circle. *Bull. London Math. Soc.*, 21:113–152, 1989.
22. H.J. Landau. Maximum entropy and the moment problem. *Bull. Am. Math. Soc. (New Series)*, 16(1):47–77, 1987.
23. A. Magnus. On the structure of the two-point Padé table. In *Analytic theory of continued fractions*, volume 932 of *Lecture Notes Math.*, pages 362–372, Berlin, 1982. Springer.
24. O. Njåstad and W.J. Thron. Rational functions and quadrature formulas. *Analysis*, 8:347–366, 1988.
25. J. Nuttal and C.J. Wherry. Gauss integration for complex weight functions. *J. Inst. Math. Appl.*, 21:165–170, 1987.
26. H. Waadeland. A Szegő quadrature formula for the Poisson integral. In *Proceedings IMACS'91 (Dublin)*, pages 12–13, 1991.
27. J. L. Walsh. *Interpolation and approximation*, volume 20 of *AMS Colloquium publications*. American Mathematical Society, Providence, Rhode Island, 3rd edition, 1960. First edition 1935.

CONTINUED

FRACTIONS

A SURVEY OF TRUNCATION ERROR ANALYSIS FOR PADÉ AND CONTINUED FRACTION APPROXIMANTS

CATHLEEN CRAVIOTTO, WILLIAM B. JONES and W. J. THRON*
Department of Mathematics, University of Colorado, Boulder, CO 80309-0395, U.S.A.

Contents

1. Introduction

Many important functions $f(z)$ of mathematical physics, chemistry, engineering, and statistics are represented by convergent sequences $\{f_n(z)\}$ of rational functions that are entries of a (1-point or multipoint) Padé table for $f(z)$. In most cases of practical interest $\{f_n(z)\}$ is the sequence of approximants of a continued fraction (see, e.g., [1],

* Research supported in part by the U.S. National Science Foundation under Grants INT-9113400 and DMS-9302584.

A. Cuyt (ed.), Nonlinear Numerical Methods and Rational Approximation II, 321–373.

[37], [45] and references contained therein). One reason for the importance of Padé tables and related continued fractions is that sequences of their approximants may converge in larger regions of the complex plane $\mathbb{C}$ than the power series expansion, which may not converge at all. Also the algorithmic character of continued fractions and Padé approximants provides efficient methods for the computation of special functions.

To compute the value $f(z) = \lim_{n\to\infty} f_n(z)$ at a point $z \in \mathbb{C}$ (using $\{f_n(z)\}$), it is essential to have realistic upper bounds for the truncation error $|f(z) - f_n(z)|$ that results from replacing the true value $f(z)$ by an approximant $f_n(z)$. Following the advent of high-speed computers, an extensive literature on truncation error analysis for Padé and continued fraction approximants has developed. This paper is a survey of constructive methods for obtaining such truncation error bounds. References to a large number of the original research publications on this subject are contained in the bibliography.

Three types of truncation error estimates are considered. *A posteriori bounds* for the truncation error $|f(z)-f_n(z)|$ are determined after calculating the approximants $f_0(z), f_1(z), \ldots, f_n(z)$ and related expressions. *A priori bounds* are expressed in terms of z and parameters defining $f_n(z)$; they can be used to appraise the truncation error at the start of the computations. A third type of error estimation describes asymptotically the speed of convergence of $\{f_n(z)\}$. This paper contains examples of all three types of error bounds. However, emphasis is given to *a posteriori* bounds, since they generally give the sharpest error estimates.

In some cases dealt with in this paper the approximant sequence $\{f_n(z)\}$ is not the sequence of approximants of a continued fraction (cf., sections 3.2.3 and 3.2.4). In order to treat all of the approximant sequences $\{f_n(z)\}$ with a uniform framework, we introduce *linear fractional approximant sequences* **(LFASs)**. An LFAS F is an ordered pair

$$F = \langle\langle\{\langle a_j, b_j, c_j, d_j\rangle\}, \{w_j\}\rangle, \{f_n\}\rangle, \tag{1.1a}$$

where the *elements* a_j, b_j, c_j, d_j and converging factors w_j are complex numbers (possibly functions of a complex variable z) satisfying

$$a_j d_j - b_j c_j \neq 0, \qquad j = 0, 1, 2, \ldots. \tag{1.1b}$$

The *n-th approximant* $f_n = v_n(F)$ of F is given by

$$f_n := v_n(F) := T_n(F, w_n), \qquad n = 0, 1, 2, 3, \ldots, \tag{1.1c}$$

where $\{T_n(F, w)\}$ and the *generating sequence* $\{t_n^F(w)\}$ are defined by

$$t_j^F(w) := \frac{a_j + c_j w}{b_j + d_j w}, \qquad j = 0, 1, 2, \ldots, \tag{1.1d}$$

and

$$T_0(F, w) := t_o^F(w), \quad T_n(F, w) := T_{n-1}(F, t_n^F(w)), \quad n = 1, 2, 3, \ldots. \tag{1.1e}$$

An LFAS F is said *to converge to a value* $v(F) \in \widehat{\mathbb{C}} = \mathbb{C} \cup [\infty]$, if its sequence of approximants $\{v_n(F)\}$ converges to $v(F)$; i.e., $v(F) = \lim_{n\to\infty} v_n(F)$. To indicate explicitly the association with F of its elements and converging factors we write $a_j(F), b_j(F), c_j(F), d_j(F)$ and $w_j(F)$. For convenience we sometimes use the abbreviated notation

$$\Gamma_j(F) := \langle a_j(F), b_j(F), c_j(F), d_j(F)\rangle, \qquad j = 0, 1, 2, \ldots . \tag{1.2}$$

The *LFAS algorithm* $\mathcal{A}(F)$ is the mapping of the ordered pair $\langle\{\Gamma_j(F)\}, \{w_j(F)\}\rangle$ to $\{f_n\} = \{v_n(F)\}$. If F depends on a complex variable z, we may write $F(z), v(F(z))$ and $v_n(F(z))$.

To obtain upper bounds for the truncation error $|v(F) - v_n(F)|$, it is useful to work with special families $\mathcal{F}$ of LFASs that contain F. For that purpose we consider *sequences of element regions* $\Omega = \{\Omega_j\}$ and *converging factors* $W = \{w_j\}$ satisfying

$$\phi \neq \Omega_j \in \boldsymbol{C}^4, \qquad j = 0, 1, 2, \ldots, \tag{1.3a}$$

where

$$\Gamma_j = \langle a_j, b_j, c_j, d_j\rangle \in \Omega_j \text{ implies } a_j d_j - b_j c_j \neq 0, \tag{1.3b}$$

and

$$w_j \in \boldsymbol{C}, \qquad j = 0, 1, 2, \ldots . \tag{1.3c}$$

For each such pair of sequences (Ω, W), we define the family of LFASs

$$\mathcal{F} = \mathcal{F}(\Omega, W) := [\text{LFASs } F : \Gamma_j(F) \in \Omega_j \text{ and } w_j(F) = w_j,\ j \geq 0]. \tag{1.4}$$

For brevity we write $\mathcal{F}$ instead of $\mathcal{F}(\Omega, W)$ if the dependence of $\mathcal{F}$ on (Ω, W) is clearly understood. We also use subfamilies of $\mathcal{F}$ defined, for each $F \in \mathcal{F}$ and $n = 0, 1, 2, 3, \ldots$, by

$$\mathcal{F}_n(F) := [G \in \mathcal{F} : \Gamma_j(G) = \Gamma_j(F), \quad j = 0, 1, 2, \ldots, n]. \tag{1.5}$$

If $G \in \mathcal{F}_n(F)$, we say that G is *n-equivalent to F*, and we call $\mathcal{F}_n(F)$ the *n-th equivalence class of F in $\mathcal{F}$*. For each $f \in \mathcal{F}$ and integer $n \geq 0$, we define the *n-th limit region $L_n(F, \mathcal{F})$ for $\mathcal{F}_n(F)$* by

$$L_n(F, \mathcal{F}) := c(\ell_n(F, \mathcal{F})), \quad (c(S) \text{ denotes closure of } S) \tag{1.6a}$$

where

$$\ell_n(F, \mathcal{F}) := [\lambda \in \boldsymbol{C} : \lambda = \lim_{j\to\infty} v_{m_j}(G) \text{ for } G \in \mathcal{F}_n(F) \text{ and subsequence } \{m_j\}]. \tag{1.6b}$$

If F converges to a finite value $v(F)$, then $L_n(F, \mathcal{F})$ is not empty, since $v(F) \in \ell_n(F, \mathcal{F}) \subseteq L_n(F, \mathcal{F})$. The concept of limit region was first used in the context of truncation error analysis by L. Lorentzen, M. Overholt, W. J. Thron and H. Waadeland (see, e.g., [48]). Our definition (1.6) differs from their's in that we allow $\ell_n(F, \mathcal{F})$ to contain finite limits of subsequences $\{f_{m_j}(G)\}$ with $G \in \mathcal{F}_n(F)$ even if $\{v_m(G)\}$ diverges. For a given family $\mathcal{F} = \mathcal{F}(\Omega, W)$ and for a finitely convergent $F \in \mathcal{F}$, we define the *best bound $\beta_n(F, \mathcal{F})$ of the truncation error $|v(F) - v_n(F)|$ for $v_n(F)$ with respect to $\mathcal{F}$* by

$$\beta_n(F, \mathcal{F}) := \sup[|\lambda - v_n(F)| : \lambda \in L_n(F, \mathcal{F})]. \tag{1.7}$$

Clearly $|v(F) - v_n(F)| \leq \beta_n(F, \mathcal{F})$, since $v(F) \in L_n(F, \mathcal{F})$. The term "best" for $\beta_n(F, \mathcal{F})$ is based on the fact that the values $\lambda \in \ell_n(F, \mathcal{F})$ are all possible candidates for $v(F)$, if we assume that our knowledge about F is limited to the following: (a) $F \in \mathcal{F}$, (b) F is finitely convergent, and (c) the only known elements of F are $\Gamma_j(F)$, $j = 0, 1, 2, \ldots, n$. One can readily see that a given LFAS F can belong to many families $\mathcal{F}^{(\alpha)}$, $\alpha \in A$. If F is finitely convergent and $F \in \mathcal{F}^{(1)} \subseteq \mathcal{F}^{(2)}$, then

$$\beta_n(F, \mathcal{F}^{(1)}) \leq \beta_n(F, \mathcal{F}^{(2)}). \tag{1.8}$$

Thus an LFAS F may have different "best" error bounds $\beta_n(F, \mathcal{F}^{(\alpha)})$ corresponding to different families $\mathcal{F}^{(\alpha)}$. It is therefore advantageous to use the smallest family $\mathcal{F}$ that is feasible.

For a given LFAS F, the linear fractional transformations $T_n(F, w)$ defined by (1.1) can be expressed in the form

$$T_n(F, w) = \frac{A_n + wC_n}{B_n + wD_n}, \qquad n = 0, 1, 2, \ldots, \tag{1.9}$$

where the $A_n = A_n(F)$, $B_n = B_n(F)$, $C_n = C_n(F)$ and $D_n = D_n(F)$ are defined by the *difference equations*

$$\begin{array}{lll} \text{a)} & A_0 := a_0,\ b_0 := b_0,\ C_0 := c_0, & D_0 := d_0 \\ \text{b)} & A_n := a_n C_{n-1} + b_n A_{n-1},\ C_n := c_n C_{n-1} + d_n A_{n-1},\ n = 1, 2, 3, \ldots, \\ \text{c)} & B_n := a_n D_{n-1} + b_n B_{n-1},\ D_n := c_n D_{n-1} + d_n B_{n-1},\ n = 1, 2, 3, \ldots . \end{array} \tag{1.10}$$

They satisfy the *determinant formulas*

$$A_n D_n - B_n C_n = (-1)^n \prod_{j=0}^{n} (a_j d_j - b_j c_j) \neq 0, \qquad n = 0, 1, 2, \ldots \tag{1.11}$$

(see, e.g., [37], Section 2.2).

An LFAS F in (1.1) reduces to a *continued fraction* (**CF**)

$$F = a_0 + \mathop{\mathrm{K}}_{j=1}^{\infty} \left(\frac{a_j}{b_j}\right) = a_0 + \frac{a_1}{b_1} + \frac{a_2}{b_2} + \frac{a_3}{b_3 + \cdots}, \tag{1.12a}$$

if the elements a_j, b_j, c_j, d_j in (1.12) and converging factors w_j satisfy

$$a_0 \in \boldsymbol{C}, \quad b_0 = 1,\ c_0 = 1,\ d_0 = 0, \tag{1.12b}$$

$$a_j \neq 0, \quad b_j \in \boldsymbol{C},\ c_j = 0, d_j = 1,\ j = 1, 2, 3, \ldots, \tag{1.12c}$$

and

$$w_j = 0, \qquad j = 0, 1, 2, \ldots . \tag{1.12d}$$

The *n-th approximant of a CF* (1.12) is then

$$v_n(F) := T_n(F, 0) =: a_0 + \frac{a_1}{b_1} + \frac{a_2}{b_2} + \cdots + \frac{a_n}{b_n}. \tag{1.13}$$

An LFAS F in (1.1) reduces to a *modified continued fraction* (MCF)

$$F = a_0 + \overset{\infty}{\underset{j=1}{\mathrm{K}}} (a_j, b_j; w_j) \tag{1.14}$$

if the elements satisfy (1.12 b,c). The *n-th approximant of a MCF* (1.14) is given by

$$v_n(F) := T_n(F, w_n) =: a_0 + \frac{a_1}{b_1} + \frac{a_2}{b_2} + \cdots + \frac{a_{n-1}}{b_{n-1}} + \frac{a_n}{b_n + w_n}. \tag{1.15}$$

For CFs (1.12) and MCFs (1.14), the difference equations (1.10) reduce to

$$\begin{aligned} A_{-1} &:= 1, \quad A_0 := a_0, \quad B_{-1} := 0, \quad B_0 := 1, \\ A_n &= b_n A_{n-1} + a_n A_{n-2}, \quad n = 1, 2, 3, \ldots, \\ B_n &= b_n B_{n-1} + a_n B_{n-2}, \quad n = 1, 2, 3, \ldots . \end{aligned} \tag{1.16}$$

Here $C_n = A_{n-1}$ and $D_n = B_{n-1}$, $n \geq 1$. Throughout this paper, when referring to CFs and MCFs, we make use of the familiar notation in (1.12) and (1.14), respectively.

Special classes of LFASs that are dealt with here (Section 3), which are neither CFs nor MCFs, are those associated with *normalized Carathéodory functions* ($\mathcal{C}$-*functions*)

$$\mathcal{C} := [f : f \text{ is analytic and } \operatorname{Re} f(z) > 0 \text{ for } |z| < 1, \ f(0) > 0] \tag{1.17}$$

and *normalized Schur functions* ($\mathcal{S}$-*functions*)

$$\mathcal{S} := [f : f \text{ is analytic and } |f(z)| < 1 \text{ for } |z| < 1, \ -1 < f(0) < 1]. \tag{1.18}$$

Associated with $\mathcal{C}$-functions are the LFASs $F = C[\{\delta_j\}z]$ with generating sequences $\{t_j^F(w)\}$ of the form

$$t_0^F(w) := \delta_0 \frac{1-w}{1+w}, \quad t_j^F(w); = z\frac{\bar{\delta}_j + w}{1 + \delta_j w}, \quad j = 1, 2, 3, \ldots \tag{1.19a}$$

where

$$\delta_0 > 0 \quad \text{and} \quad \delta_j \in \boldsymbol{C}, \quad 0 \leq |\delta_j| < 1, \quad j = 1, 2, 3, \ldots, \tag{1.19b}$$

and with converging factors

$$w_j = 0, \qquad j = 0, 1, 2, \ldots . \tag{1.19c}$$

Associated with $\mathcal{S}$-functions are LFASs $F = S[\{\gamma_j\}, z]$ with generating sequences of the form

$$t_j^F(w) := \frac{\gamma_j + zw}{1 + \bar{\gamma}_j zw}, \quad \gamma_0 \in I\!R, \ |\gamma_0| < 1, \ \gamma_j \in \boldsymbol{C}, \ |\gamma_j| < 1, \ j = 1, 2, 3, \ldots, \tag{1.20a}$$

and converging factors

$$w_j = 0, \qquad j = 0, 1, 2, \ldots . \tag{1.20b}$$

Sequences of value regions $V = \{V_n\}$ corresponding to sequences of element regions $\Omega = \{\Omega_j\}$ and converging factors $W = \{w_j\}$ are discussed in Section 2. Methods for obtaining truncation error bounds based on sequences of value regions $\{V_n\}$ are developed (Theorems 2.5 and 2.6). For many special families $\mathcal{F}(\Omega, W)$ of LFASs, we are able to determine best truncation error bounds $\beta_n(F, \mathcal{F})$ by using best sequences of value regions. Applications of the method are described in Sections 3 and 4. In Section 3 the method is applied to the following 7 special families of LFASs:

$$\mathcal{F}^{W(\rho)} := [K(a_j/1) : 0 \neq |a_j| \leq \rho(1-\rho),\ a_j \in \mathbb{C} \text{ for } j \geq 1],\ 0 < \rho \leq \tfrac{1}{2}, \text{ (Worpitzky)} \tag{1.21}$$

$$\mathcal{F}^{SP(p)} := [K(1/b_j) : |b_j| \geq \rho + 1/\rho,\ b_j \in \mathbb{C} \text{ for } j \geq 1],\ 0 < \rho \leq 1, \text{ (Śleszyński–Pringsheim)} \tag{1.22}$$

$$\mathcal{F}^{St(z)} := [K(a_j z/1) : a_j > 0 \text{ for } j \geq 1,\ 0 \neq z \in \mathbb{C},\ |\arg z| < \pi], \text{ (Stieltjes)} \tag{1.23}$$

$$\mathcal{F}^{T(z)} := [K(F_j z/(1 + G_j z)) : F_j,\ G_j > 0,\ 0 \neq z \in \mathbb{C},\ |arg z| < \pi], \text{ (Thron)} \tag{1.24}$$

$$\mathcal{F}^{J(z)} := [K(-\alpha_j^2/(\beta_j + z)) : -\alpha_1^2 = 1, \beta_1 \in \mathbb{R}; 0 \neq \alpha_j \in \mathbb{R}, \beta_j \in \mathbb{R} \text{ for } j \geq 2; \operatorname{Im} z \neq 0], \text{ (real J-fractions)} \tag{1.25}$$

$$\mathcal{F}^{PPC(z)} := [C[\{\delta_j\}, z] : \{t_j^F(w)\} \text{ and } w_j \text{ in (1.19)},\ |z| < 1], \text{ (Carathéodory)} \tag{1.26}$$

$$\mathcal{F}^{Sh(z)} := [S[\{\gamma_j\}, z] : \{t_j^F(w)\} \text{ and } w_j \text{ in (1.20)},\ |z| < 1], \text{ (Schur)} \tag{1.27}$$

In Section 4 the value region method is applied to the following 4 special families of LFASs that are limit k-periodic CFs or MCFs:

$$K(a_j, 1, x_1),\ a_j \to a \in \mathbb{C} - (-\infty, -1/4],\ \text{as } j \to \infty, \tag{1.28a}$$

$$K(a_j/1),\ a_j \to 0,\ \text{as } j \to \infty, \tag{1.28b}$$

$$K(a_j, 1, w_j),\ a_j \to \infty,\ \text{as } j \to \infty, \tag{1.29}$$

$$K(1/b_j),\ b_j \to \infty,\ \text{as } j \to \infty, \tag{1.30}$$

$$K(1, b_j, w_j),\ b_{4i+j} \to \beta_j,\ j = 1, 2, 3, 4,\ \text{as } i \to \infty. \tag{1.31}$$

Section 5 deals with asymptotically best truncation error bounds for limit periodic LFASs (including limit periodic MCFs). Due to constraints of space and time we have had to omit some topics and results on the subject of this paper. Among the omissions is a formal discussion about simple sequences developed in [35]. Some examples of simple sequences are given in (3.18), (3.35) and Section 3.3. We have also omitted applications of truncation error bounds to particular special functions and results from computational experiments. Examples of such applications and experiments can be found in many of the papers given in the references. Before continuing with Section 2 we summarize some additional definitions and notation that are subsequently used.

For $m = 0, 1, 2, \ldots$, the *m-th tail of an LFAS F* (see (1.1)) is the LFAS, denoted by $F^{(m)}$, with elements $a_j^{(m)}, b_j^{(m)}, c_j^{(m)}, d_j^{(m)}$ and converging factors $w_j^{(m)}$ defined by

$$\Gamma_j^{(m)} := \langle a_j^{(m)}, b_j^{(m)}, c_j^{(m)}, d_j^{(m)} \rangle := \langle a_{m+j}, b_{m+j}, c_{m+j}, d_{m+j} \rangle \tag{1.32a}$$

and

$$w_j^{(m)} ;= w_{m+j}. \tag{1.32b}$$

We note that $F^{(0)} = F$,

$$t_j^{F^{(m)}}(w) = t_{m+j}^F(w), \quad m = 0, 1, 2, \ldots \text{ and } j = 0, 1, 2, \ldots, \tag{1.33a}$$

and

$$T_1(F^{(m)}, w) = t_{m+1}^F(w), \; T_n(F^{(m)}, w) = T_{n-1}(F^{(m)}, t_{m+n}^F(w)), \; m = 0, 1, 2, \ldots, \tag{1.33b}$$

It follows that, for $m = 0, 1, 2, \ldots$ and $n = 1, 2, 3, \ldots$,

$$T_n(F^{(m)}, w) = t_{m+1}^F \circ t_{m+2}^F \circ \cdots \circ t_{m+n}^F(w), \tag{1.34}$$

$$T_{m+n}(F, w) = T_m(F, T_n(F^{(m)}, w), \tag{1.35}$$

$$v_n(F^{(m)}) := T_n(F^{(m)}, w_n^{(m)}) = T_n(F^{(m)}, w_{m+n}), \tag{1.36}$$

$$v_{m+n}(F) := T_{m+n}(F, w_{m+n}) = T_m(F, T_n(F^{(m)}, w_{m+n}) = T_m(F, v_n(F^{(m)}), \tag{1.37}$$

and

$$v(F) := \lim_{n \to \infty} v_n(F) = T_m(F, v(F^{(m)})), \; m = 0, 1, 2, \ldots \; . \tag{1.38}$$

A sequence $\{\tau_n\}$, where $\tau_n \in \widehat{\mathbb{C}}$, is called a *tail sequence of an LFAS F* if, for some $\tau \in \widehat{\mathbb{C}}$,

$$\tau_m = T_m^{-1}(F, \tau), \qquad m = 0, 1, 2, \ldots \; . \tag{1.39}$$

An example of a tail sequence of a LFAS F is given by

$$\tau_m := v(F^{(m)}), \qquad m = 0, 1, 2, \ldots, \tag{1.40}$$

provided, of course, that the tails $F^{(m)}$ are convergent. The sequence $\{v(F^{(m)}\}$ is called the *right* (i.e., correct) *tail sequence of F* since the τ in (1.39) is given by $\tau = v(F)$ (see (1.38)). Another important tail sequence, called the *critical tail sequence*, is defined by

$$\tau_m := -h_m(F) := T_m^{-1}(F, \infty), \; m = 0, 1, 2, \ldots \; . \tag{1.41}$$

It follows from (1.9) and (1.41) that

$$h_m(F) = B_m(F)/D_m(F), \quad m = 0, 1, 2, \ldots \; . \tag{1.42}$$

2. Truncation Error Bounds from Value Regions

Let $\mathcal{F} = \mathcal{F}(\Omega, W)$ denote a given non-empty family (1.4) of LFASs. Let $\{U_n(\mathcal{F})\}_{n=-1}^{\infty}$ be defined by

$$U_n(\mathcal{F}) := [t_{n+1}^F \circ t_{n+2}^F \circ \cdots \circ t_{n+m}^F(w_{n+m}) : F \in \mathcal{F} \text{ and } m = 1, 2, 3, \ldots]. \tag{2.1}$$

We begin with the following

Theorem 2.1 *Let $\mathcal{F} = \mathcal{F}(\Omega, W)$ be a given non-empty family* (1.4) *of LFASs. Then*

$$[t_n^F(w_n) : F \in \mathcal{F}] \subseteq U_{n-1}(\mathcal{F}), \quad n = 0, 1, 2, 3, \ldots, \tag{2.2a}$$

and

$$[t_n^F(U_n(\mathcal{F})) : F \in \mathcal{F}] \subseteq U_{n-1}(\mathcal{F}), \quad n = 0, 1, 2, 3, \ldots . \tag{2.2b}$$

Proof. Condition (2.2a) is an immediate consequence of (2.1). To prove (2.2b) let $F \in \mathcal{F}$, $n \in [0, 1, 2, \ldots]$ and $u \in U_n(\mathcal{F})$ be given. Then by (2.1) there exists a $G \in \mathcal{F}_n(F)$ and an integer $m \geq 1$ such that

$$u = t_{n+1}^G \circ t_{n+2}^G \circ \cdots \circ t_{n+m}^G(w_{n+m}). \tag{2.3}$$

It follows from this, (2.1) and $t_n^F(w) = t_n^G(w)$ that

$$t_n^F(u) = t_n^G \circ t_{n+1}^G \circ \cdots \circ t_{n+m}^G(w_{n+m}) \in U_{n-1}(\mathcal{F}). \Box \tag{2.4}$$

A sequence $\{V_n\}_{n=-1}^{\infty}$ of non-empty subsets of $\widehat{\mathbb{C}}$ is called *a sequence of value regions with respect to* $\mathcal{F} = \mathcal{F}(\Omega, W)$ if the following conditions are satisfied:

$$[t_n^F(w_n) : F \in \mathcal{F}] \subseteq V_{n-1}, \quad n = 0, 1, 2, 3, \ldots, \tag{2.5a}$$

$$[t_n^F(V_n) : F \in \mathcal{F}] \subseteq V_{n-1}, \quad n = 0, 1, 2, 3, \ldots . \tag{2.5b}$$

The family of all sequences of value regions $\{V_n\}$ with respect to $\mathcal{F}$ is denoted by $\mathcal{V}(\mathcal{F})$. It is clear that

$$\{U_n(\mathcal{F})\}_{n=-1}^{\infty} \in \mathcal{V}(\mathcal{F}). \tag{2.6}$$

From our next result (Theorem 2.2) we see that $\{U_n(\mathcal{F})\}$ is the "smallest" sequence in $\mathcal{V}(\mathcal{F})$. We therefore call $\{U_n(F)\}$ the *best sequence of value regions with respect to* $\mathcal{F}$.

Theorem 2.2 *Let $\mathcal{F} = \mathcal{F}(\Omega, W)$ be a non-empty family of LFASs. If $\{V_n\} \in \mathcal{V}(\mathcal{F})$, then*

$$U_n(\mathcal{F}) \subseteq V_n, \quad n = -1, 0, 1, 2, \ldots . \tag{2.7}$$

Proof. Let $\{V_j\} \in \mathcal{V}(\mathcal{F})$ and $n \in [-1, 0, 1, 2, \ldots]$ and $u \in U_n(\mathcal{F})$ be given. Then there exists a $G \in \mathcal{F}$ and an integer $m \geq 1$ such that u can be expressed by (2.3). If $m = 1$, then $u = t_{n+1}^G(w_{n+1}) \in V_n$ by (2.5a). If $m = 2$, then $u = t_{n+1}^G \circ t_{n+2}^G(w_{n+2}) \in t_{n+1}^G(V_{n+1})$ by (2.5a) and hence $u \in t_{n+1}^G(V_{n+1}) \subseteq V_n$, by (2.5b). Continuing in this manner one can show (by induction) that all expressions of the form (2.3) are in V_n. This proves (2.7). $\Box$

Some elementary but useful properties of value regions are summarized in our next result (Theorem 2.3). A proof is an immediate consequence of the above definitions and hence is omitted.

Theorem 2.3 *Let $\mathcal{F} = \mathcal{F}(\Omega, W)$ be a non-empty family of LFASs. Then*

(A) *If $\{V_n\}_{n=-1}^{\infty}$ is a family of non-empty subsets of $\widehat{\mathbb{C}}$ such that*

$$w_n \in V_n, \quad n = 0, 1, 2, \ldots, \tag{2.8}$$

and (2.2b) *holds, then* $\{V_n\} \in \mathcal{V}(\mathcal{F})$.

(B) *If* $\{V_n\} \in \mathcal{V}(\mathcal{F})$, *then* $\{c(V_n)\} \in \mathcal{V}(\mathcal{F})$, *where* $c(V_n)$ *denotes the closure of* V_n.

(C) *If* $\{V_n^{(\alpha)}\} \in \mathcal{V}(\mathcal{F})$ *for all* α *in an index set* A, *then*

$$\left\{ \bigcap_{\alpha \in A} V_n^{(\alpha)} \right\} \in \mathcal{V}(\mathcal{F}). \tag{2.9}$$

An approach for obtaining truncation error bounds by use of value regions is based on the following:

Theorem 2.4 *Let* $\mathcal{F} = \mathcal{F}(\Omega, W)$ *be a given non-empty family of LFASs. Let* $F \in \mathcal{F}$ *converge to a finite value* $v(F) = \lim v_n(F)$. *Let* $\{V_n\} \in \mathcal{V}(\mathcal{F})$ *and let* k *be a non-negative integer such that*

$$w_n \in c(V_n), \qquad n = k, k+1, k+2, \dots . \tag{2.10}$$

Then

$$|v(F) - v_n(F)| \le diam\, T_n(F, c(V_n)), \quad n = k, k+1, k+2, \dots . \tag{2.11}$$

Remarks on Theorem 2.4. Determination of truncation error bounds by use of Theorem 2.4 involves the following steps: (a) First we obtain a sequence $\{V_n\} \in \mathcal{V}(\mathcal{F})$ such that (2.10) holds for some $k \ge 0$. (b) Next we find a description of the set $T_n(F, c(V_n))$ such that its diameter $(\operatorname{diam} T_n(F, c(V_n)))$ can be computed. Many examples that illustrate these steps are described in Sections 3 and 4.

Proof of Theorem 2.4. By Theorem 2.3(B), $\{c(V_n)\} \in \mathcal{V}(\mathcal{F})$. Thus an application of (2.5b) yields

$$T_n(F, c(V_n)) = T_{n-1}(F, t_n^F(c(V_n))) \subseteq T_{n-1}(F, c(V_{n-1})), \; n = 1, 2, 3, \dots . \tag{2.12}$$

Hence $\{T_n(F, c(V_n))\}$ is a nested sequence of non-empty closed subsets of $\widehat{\mathbb{C}}$. From this, (1.1c) and (2.10) we obtain, for all $n \ge k$ and $m \ge 0$,

$$v_{n+m}(F) := T_{n+m}(F, w_{n+m}) \in T_{n+m}(F, c(V_{n+m})) \subseteq T_n(F, c(V_n)), \tag{2.13}$$

and hence

$$|v_{n+m}(F) - v_n(F)| \le \operatorname{diam} T_n(F, c(V_n)), \quad m = 0, 1, 2, \dots . \tag{2.14}$$

Assertion (2.11) follows from (2.14). □

We note in passing that many important convergence theorems for LFASs have been proved by first establishing (2.14) and then showing that $\lim_{n\to\infty} \operatorname{diam} T_n(F, c(V_n)) = 0$ (see, e.g., [37] and [45]). Every closed set that contains the set

$$[v_{n+m}(G) : G \in \mathcal{F}_n(F), \; m \ge 0] \tag{2.15}$$

is called an *n-th inclusion region for* F *with respect to* $\mathcal{F}$. We denote the family of all such regions by $I_n(F, \mathcal{F})$. Clearly $T_n(F, c(V_n)) \in I_n(F, \mathcal{F})$ for all $\{V_n\} \in \mathcal{V}(\mathcal{F})$ and $F \in \mathcal{F}$. Since

$$T_n(F, c(U_n(\mathcal{F}))) = c[v_{n+m}(G) : G \in \mathcal{F}_n(F), \; m \ge 0], \tag{2.16}$$

$T_n(F, c(U_n(\mathcal{F})))$ is called the *best n-th inclusion region for F with respect to* $\mathcal{F}$. Henrici and Pfluger [21] were the first to use inclusion regions in their development of truncation error bounds for S-fractions (see (1.23) and Section 3). In our next result (Theorem 2.5) we show that, subject to stated sufficient conditions, the best truncation error bound $\beta_n(F, \mathcal{F})$ (see (1.7)) can be expressed in terms of $T_n(F, c(U_n(\mathcal{F})))$.

Theorem 2.5 *Let $\mathcal{F} = \mathcal{F}(\Omega, W)$ be a non-empty family of LFASs. Let $F \in \mathcal{F}$ be convergent to a finite value $v(F)$ and let n be a non-negative integer such that $T_n(F, c(U_n(\mathcal{F})))$ is bounded. Then*

$$\beta_n(F, \mathcal{F}) = \sup[|\lambda - v_n(F)| : \lambda \in T_n(F, c(U_n(\mathcal{F})))], \tag{2.17}$$

provided that at least one of the following conditions holds:

$$\text{For} \quad m \geq n+1, \quad w_m \in U_m(\mathcal{F}). \tag{2.18}$$

$$\text{For} \quad m \geq n+1, \quad w_m \in c(U_m(\mathcal{F})) \tag{2.19a}$$

and for every $k \geq 1$, there exists a sequence $\{G_j\}$ of finitely converging LFASs such that

$$G_j \in \mathcal{F}_n(F) \text{ for } j \geq 1 \text{ and } w_{n+k} = \lim_{j\to\infty} v(G_j^{(n+k)}). \tag{2.19b}$$

Proof. In view of the definition of $\beta_n(F, \mathcal{F})$ in (1.7) and of (2.17) it suffices to show that, if the conditions of Theorem 2.5 hold, then

$$L_n(F, \mathcal{F}) = T_n(F, c(U_n(\mathcal{F}))), \tag{2.20}$$

where the n-th limit region $L_n(F, \mathcal{F})$ for $\mathcal{F}_n(F)$ is defined by (1.6).

First we suppose that condition (a) *holds.* Let $\lambda \in T_n(F, U_n(\mathcal{F}))$ be given. Then by the definition of $U_n(\mathcal{F})$ in (2.1), there exists a $G_1 \in \mathcal{F}_n(F)$ and an integer $m_1 \geq n+1$ such that

$$\lambda = T_n(F, t_{n+1}^{G_1} \circ t_{n+2}^{G_1} \circ \cdots \circ t_{m_1}^{G_1}(w_{m_1})) = v_{m_1}(G_1).$$

Since by (2.18) $w_{m_1} \in U_{m_1}(\mathcal{F})$, there exists a $G_2 \in \mathcal{F}_{m_1}(G_1)$ and an $m_2 \geq m_1 + 1$ such that

$$w_{m_1} = t_{m_1+1}^{G_2} \circ t_{m_1+2}^{G_2} \circ \cdots \circ t_{m_2}^{G_2}(w_{m_2})$$

and hence $\lambda = v_{m_2}(G_2)$ and $w_{m_2} \in U_{m_2}(\mathcal{F})$. Continuing in this manner, we obtain a sequence $\{G_j\}$ of LFASs and a sequence of integers $\{m_j\}$ such that, for each $j \geq 1$,

$$G_{j+1} \in \mathcal{F}_{m_j}(G_j), \; m_{j+1} \geq m_j + 1, \; w_{m_j} = t_{m_j+1}^{G_{j+1}} \circ t_{m_j+2}^{G_{j+1}} \circ \cdots \circ t_{m_{j+1}}^{G_{j+1}}(w_{m_{j+1}}), \tag{2.21a}$$

and hence

$$\lambda = v_{m_j}(G_j) \quad \text{for} \quad j = 1, 2, 3, \ldots . \tag{2.21b}$$

From the definition of $\mathcal{F} = \mathcal{F}(\Omega, W)$ in (1.4) and from (2.21a) it follows that there exists a $G \in \mathcal{F}$ such that

$$G \in \mathcal{F}_{m_j}(G_j) \text{ for } j = 1, 2, 3, \ldots, \text{ and } G \in \mathcal{F}_n(F). \tag{2.22}$$

Therefore by (2.21) and (2.22), $\lambda = v_{m_j}(G)$ for $j \geq 1$ so that

$$\lambda = \lim_{j\to\infty} v_{m_j}(G) \in L_n(F, \mathcal{F}).$$

We have shown that $T_n(F, U_n(\mathcal{F})) \subseteq L_n(F, \mathcal{F})$ and since $L_n(F, \mathcal{F})$ is a closed set we have

$$T_n(F, c(U_n(\mathcal{F}))) \subseteq L_n(F, \mathcal{F}). \tag{2.23}$$

To prove that the inclusion in (2.23) holds in the opposite direction, we let $\lambda \in \ell_n(F, \mathcal{F})$ be given. Then by the definition of $\ell_n(F, \mathcal{F})$ in (1.6a), there exists a $G \in \mathcal{F}_n(F)$ and a subsequence of natural numbers $\{m_j\}$ (with $m_1 \geq n+1$) such that

$$\lambda = \lim_{j\to\infty} v_{m_j}(G). \tag{2.24}$$

Therefore, for all $j \geq 1$,

$$v_{m_j}(G) := T_{m_j}(G, w_{m_j}) = T_n(F, t^G_{n+1} \circ \cdots \circ t^G_{m_j}(w_{m_j})) \in T_n(F, U_n(\mathcal{F})).$$

Hence

$$\lambda = \lim_{j\to\infty} v_{m_j}(G) \in T_n(F, c(U_n(\mathcal{F}))),$$

which shows that $\ell_n(F, \mathcal{F}) \subseteq T_n(F, c(U_n(\mathcal{F})))$ and so

$$L_n(F, \mathcal{F}) \subseteq T_n(F, c(U_n(\mathcal{F}))), \tag{2.25}$$

since the right side of (2.25) is a closed set. The equation (2.20) follows from (2.23) and (2.25).

Next we suppose that condition (b) *holds.* Let $\lambda \in T_n(F, U_n(\mathcal{F}))$ be given. Then by the definition of $U_n(\mathcal{F})$ in (2.1) and of $\mathcal{F}_n(F)$ in (1.5), there exists a $G \in \mathcal{F}_n(F)$ and a positive integer k such that

$$\lambda = T_n(F, t^G_{n+1} \circ t^G_{n+2} \circ \cdots \circ t^G_{n+k}(w_{n+k})). \tag{2.26}$$

By condition (b) there exists a sequence $\{G_j\}$ (depending upon n and k) such that for all $j = 1, 2, 3, \ldots$,

$$G_j \in \mathcal{F}_{n+k}(G) \text{ and } w_{n+k} = \lim_{j\to\infty} v(G_j^{(n+k)}). \tag{2.27}$$

To verify the first relation in (2.27) we note that condition (b) of the hypothesis places no restrictions on the elements $\mathcal{C}_m(G_j)$ for $n+1 \leq m \leq n+k$. Hence we can set $\mathcal{C}_m(G_j) = \mathcal{C}_m(G)$ for $n+1 \leq m \leq n+k$ and $j \geq 1$. Therefore by (2.26) and (2.27)

$$\begin{aligned}\lambda &= T_{n+k}(G, w_{n+k}) = \lim_{j\to\infty} T_{n+k}(G, v(G_j^{(n+k)})) \\ &= \lim_{j\to\infty} T_{n+k}(G_j, v(G_j^{(n+k)})), \text{ (since } G_j \in \mathcal{F}_{n+k}(G) \subseteq \mathcal{F}_n(F)) \\ &= \lim_{j\to\infty} v(G_j), \quad \text{by (1.38).}\end{aligned}$$

It follows that $\lambda = \lim_{j\to\infty} v(G_j) \in L_n(F, \mathcal{F})$ and hence $T_n(F, U_n(\mathcal{F})) \subseteq L_n(F, \mathcal{F})$. Since the right side of the last inclusion is a closed set, we obtain

$$T_n(F, c(U_n(\mathcal{F}))) \subseteq L_n(F, \mathcal{F}). \tag{2.28}$$

Finally, we note that $v(F) \in \ell_n(F,\mathcal{F})$ so that $\ell_n(F,\mathcal{F})$ is not empty. Let $\lambda \in \ell_n(F,\mathcal{F})$ be given. Then by definition of $\ell_n(F,\mathcal{F})$ in (1.6a), there exists a $G \in \mathcal{F}_n(F)$ and a subsequence $\{m_j\}$ of the natural numbers such that

$$\lambda = \lim_{j\to\infty} v_{m_j}(G). \tag{2.29}$$

Without loss of generality we may assume that $m_1 > n$ and let $k_j := m_j - n$, $j \geq 1$. From this and (2.29) it follows that

$$\lambda = \lim_{j\to\infty} T_n(F, t^G_{n+1} \circ t^G_{n+2} \circ \cdots \circ t^G_{n_{k_j}}(w_{n_{k_j}})) \in T_n(F, c(U_n(\mathcal{F}))),$$

since $t^G_{n+1} \circ \cdots \circ t^G_{n+k_j}(w_{n+k_j}) \in U_n(\mathcal{F})$ for all $j \geq 1$. Therefore $\ell_n(F,\mathcal{F}) \subseteq T_n(F, c(U_n(\mathcal{F})))$ and, since the right side of this inclusion is a closed set, we obtain

$$L_n(F,\mathcal{F}) \subseteq T_n(F, c(U_n(\mathcal{F}))). \tag{2.30}$$

The relations (2.28) and (2.30) imply (2.20) and this completes our proof. □

Our next result (Theorem 2.6) provides explicit and easily computable bounds for the truncation error $|v(F)-v_n(F)|$ when one has value regions V_n that are closed circular disks centered at the corresponding converging factors w_n. If in addition the hypotheses of Theorem 2.5 hold, then $V_n = c(U_n(\mathcal{F}))$ and hence the explicit error bound is the best bound $\beta_n(F,\mathcal{F})$.

Theorem 2.6 *Let $\mathcal{F} = \mathcal{F}(\Omega, W)$ be a non-empty family of LFASs. Let $\{V_m\}$ be a sequence of value regions corresponding to $\mathcal{F}$ such that for some integer $k \geq 0$ and some sequence of positive numbers $\{\rho_m\}_{m=k}^{\infty}$,*

$$V_m := [u \in \boldsymbol{C} : |u - w_m| \leq \rho_m], \quad m = k, k+1, k+2, \dots . \tag{2.31}$$

Let $F \in \mathcal{F}$ have a finite value $v(F)$ and let n be an integer such that $n \geq k$ and the n-th inclusion region $T_n(F, V_n)$ is a closed circular disk (and hence bounded). Let $D_n(F)$ and $h_n(F)$ be defined as in (1.10) and (1.41), respectively. Then:

(A)

$$\begin{aligned} |v(F) - v_n(F)| &\leq \sup[|\lambda - v_n(V)| : \lambda \in T_n(F, V_n)] \\ &= \frac{\rho_n \prod_{j=0}^{n} |a_j(F)d_j(F) - b_j(F)c_j(F)|}{|D_n(F)|^2 \cdot |w_n + h_n(F)|(|w_n + h_n(F)| - \rho_n)}. \end{aligned} \tag{2.32}$$

(B) *If, in addition*

$$V_n = c(U_n(\mathcal{F})) \tag{2.33}$$

and the hypotheses of Theorem 2.5 hold, then the expression on the right side of (2.32) is the best truncation error bound $\beta_n(F,\mathcal{F})$ for F with respect to $\mathcal{F}$.

Proof. (A): By (1.41) $T_n(F, -h_n(F)) = \infty$. Therefore, since $T_n(F, V_n)$ is a bounded closed circular disk, we obtain

$$-h_n(F) \notin V_n.$$

Let $u_n \in \boldsymbol{C}$ denote the point of intersection of the circular boundary $\partial T_n(F, V_n)$ and the line segment $[w_n, -h_n(F)]$. From the defining relations for value regions (2.5) (see also the proof of Theorem 2.4) we see that $\{T_m(F, V_m)\}_{m=n}^{\infty}$ is a nested sequence of closed circular disks and, for $m = 0, 1, 2, \ldots$,

$$v_{n+m}(F) := T_{n+m}(F, w_{n+m}) \in T_{n+m}(F, V_{n+m}) \subseteq T_n(F, V_n).$$

Hence

$$v(F) = \lim_{m \to \infty} v_{n+m}(F) \in T_n(F, V_n).$$

Let $\lambda \in T_n(F, V_n)$ be given and let $u \in V_n$ be chosen so that $\lambda = T_n(F, u)$. Then by (1.9) we obtain

$$\begin{aligned} |\lambda - v_n(F)| &= |T_n(F, u) - T_n(F, w_n)| \\ &= \left| \frac{A_n(F) + C_n(F)u}{B_n(F) + D_n(F)u} - \frac{A_n(F) + C_n(F)w_n}{B_n(F) + D_n(F)w_n} \right| \\ &= \left| \frac{(w_n - u)(A_n(F)D_n(F) - B_n(F)C_n(F))}{(B_n(F) + D_n(F)u)(B_n(F) + D_n(F)w_n)} \right|. \end{aligned}$$

Using this with the determinant formulas (1.11) and $B_n(F) = h_n(F)D_n(F)$ from (1.42) yields

$$|\lambda - v_n(F)| = \frac{|w_n - u| \prod_{j=0}^{n} |a_j(F)d_j(F) - b_j(F)c_j(F)|}{|D_n(F)|^2 \cdot |u + h_n(F)| \cdot |w_n + h_n(F)|}. \tag{2.34}$$

It is readily seen that

$$\max_{u \in V_n} |w_n - u| = \rho_n \text{ and } \min_{u \in V_n} |u + h_n(F)| = |w_n + h_n(F)| - \rho_n > 0, \tag{2.35}$$

where the extremum in both cases is attained with $u = u_n$. An application of (2.35) to (2.34) gives (2.32).

(B) follows immediately from part (A) proved above and Theorem 2.5. □

One can use Theorems 2.5 and 2.6 to obtain best truncation error bounds $\beta_n(F, \mathcal{F})$ by determining a simple explicit (geometrical or analytical) description of $c(U_n(\mathcal{F}))$ and of its image $T_n(F, c(U_n(\mathcal{F})))$. Applications of that kind are given in Sections 3 and 4 for a number of important special families $\mathcal{F}$ of LFASs (see (1.21) to (1.31)). For some of these applications, $c(U_n(\mathcal{F}))$ can be determined by direct elementary methods. For other families we have made use of the following:

Theorem 2.7 *Let $\mathcal{F} = \mathcal{F}(\Omega, W)$ be a given non-empty family of LFASs, such that every $F \in \mathcal{F}$ and its tails $F^{(m)}$ converge to finite values $v(F^{(m)})$, $m \geq 0$. Let $\{V_n\}$ be a sequence of value regions with respect to $\mathcal{F}$ such that, for some integer $k \geq 0$,*

$$[t_n^F(V_n) : F \in \mathcal{F}] = V_{n-1}, \quad n = k+1, k+2, \ldots, \tag{2.36}$$

and such that

$$\lim_{n \to \infty} \left\{ \sup_{F \in \mathcal{F}} [diam\, T_n(F^{(m)}, V_{n+m})] \right\} = 0, \quad m = k+1, k+2, \ldots . \tag{2.37}$$

Then

$$c(V_n) = c(U_n(\mathcal{F})), \qquad n = k, k+1, k+2, \dots . \tag{2.38}$$

Proof. By Theorem 2.2, $c(U_n(\mathcal{F})) \subseteq c(V_n)$ for $n \geq 0$. Thus it suffices to show that

$$c(V_n) \subseteq c(U_n(\mathcal{F})), \quad n = k, k+1, k+2, \dots . \tag{2.39}$$

Let $n \geq k$ and $u_n \in V_n$ be given. We show that $u_n \in c(U_n(\mathcal{F}))$. From (2.36) there exists an $F_1 \in \mathcal{F}$ and a $u_{n+1} \in V_{n+1}$ such that $u_n = t^{F_1}_{n+1}(u_{n+1})$. Again by (2.36) there exists an $F_2 \in \mathcal{F}$ and a $u_{n+2} \in V_{n+2}$ such that $u_{n+1} = t^{F_2}_{n+2}(u_{n+2})$. Continuing in this manner we see that there exist sequences $\{u_j\}$ and $\{F_j\}$ such that, for $j = 1, 2, 3, \dots$,

$$F_j \in \mathcal{F}, \; u_{n+j-1} \in V_{n+j-1}, \text{ and } u_{n+j-1} = t^{F_j}_{n+j}(u_{n+j}). \tag{2.40}$$

Let $F \in \mathcal{F}$ be defined by $\Gamma_{n+j}(F) := \Gamma_{n+j}(F_j)$ for $j \geq 1$. From this and (2.40) we have

$$u_{n+j-1} = t^F_{n+j}(u_{n+j}), \quad j = 1, 2, 3, \dots,$$

and hence by (1.34), for all $m = 1, 2, 3, \dots$,

$$u_n = t^F_{n+1} \circ t^F_{n+2} \circ \cdots \circ t^F_{n+m}(u_{n+m}) = T_m(F^{(n)}, u_{n+m}). \tag{2.41}$$

By (2.5) and (1.33)

$$T_m(F^{(n)}, V_{n+m}) = T_{m-1}(F^{(n)}, t^F_{m+n}(V_{n+m})) \subseteq T_{m-1}(F^{(n)}, V_{n+m-1}), \tag{2.42}$$

and hence $\{T_m(F^{(n)}, V_{m+n})\}_{m=1}^{\infty}$ is a nested sequence of non-empty subsets of $\widehat{\mathbf{C}}$. For $m \geq 1$ and $j \geq 1$, we obtain by (1.33) and (2.5)

$$\begin{aligned} v_{m+j}(F^{(n)}) &:= T_{m+j}(F^{(n)}, w_{n+m+j}) = T_{m+j-1}(F^{(n)}, t^F_{n+m+j}(w_{n+m+j})) \\ &\in T_{m+j-1}(F^{(n)}, V_{n+m+j-1}) \subseteq \cdots \subseteq T_m(F^{(n)}, V_{n+m}). \end{aligned} \tag{2.43}$$

Therefore

$$v(F^{(n)}) := \lim_{j \to \infty} v_{m+j}(F^{(n)}) \in T_n(F^{(n)}, c(V_{n+m})), \; m = 1, 2, 3, \dots . \tag{2.44a}$$

By (2.41) and the fact that $u_{n+m} \in V_{n+m}$ for $m \geq 0$, we have

$$u_n = T_n(F^{(n)}, u_{n+m}) \in T_m(F^{(n)}, V_{n+m}), \; m = 1, 2, 3, \dots . \tag{2.44b}$$

Thus we conclude from (2.43), (2.44) and the hypothesis (2.37) that

$$u_n = v(F^{(n)}). \tag{2.45}$$

We also have from the definition of $U_n(\mathcal{F})$ in (2.1) that

$$v_m(F^{(n)}) := T_m(F^{(n)}, w_{n+m}) \in U_n(\mathcal{F}) \tag{2.46}$$

so that

$$v(F^{(n)}) = \lim_{m \to \infty} v_m(F^{(n)}) \in c(U_n(\mathcal{F})). \tag{2.47}$$

Combining (2.45) with (2.47) yields

$$u_n \in c(U_n(\mathcal{F})).$$

We have shown that $V_n \subseteq c(U_n(\mathcal{F}))$, from which (2.38) follows. □

3. Special Families of LFASs with Simple Value Regions

In Sections 3 and 4 we apply Theorem 2.5 to obtain best truncation error bounds $\beta_n(F, \mathcal{F})$ for a number of important special families $\mathcal{F}$ of LFASs. Other truncation error bounds are included which, though not best, are sharp enough to be useful and are easy to compute. Sequences of value regions $V = \{V_j\}$ with respect to families $\mathcal{F}(\Omega, W)$ play an essential role in these two sections. The procedure used to determine families $\mathcal{F}(\Omega, W)$ and associated value regions $\{V_j\}$ is a generalization of an approach developed for continued fractions. It rests on the observation first made in [44] that, starting in a "natural" way with element regions $\{\Omega_j\}$ and converging factors $\{w_j\}$, it may be very difficult to find corresponding value regions $\{V_j\}$ (or $\{U_j(\mathcal{F})\}$ in (2.1)). A simpler approach is to start with sequences $\{V_j\}$ and $\{w_j\}$ and determine a corresponding sequence $\{\Omega_j\}$, which may lead to null sets $\Omega_j = \emptyset$, $j \geq 1$. One way of doing this for continued fractions (with $w_j = 0$, $j \geq 0$) was used in [49] and [52] for special cases, and then was formalized by Lane [42] for circular disks V_n and arbitrary continued fractions $K(a_n/b_n)$. We refer to the generalization of this procedure, described in Section 3.1, as the $VW\Omega$-method.

3.1. The $VW\Omega$-Method

Starting with a sequence $V = \{V_j\}$ of non-empty subsets of $\widehat{\boldsymbol{C}}$ and a sequence of complex numbers $W = \{w_j\}$ satisfying

$$w_j \in V_j, \qquad j = 0, 1, 2, \ldots, \tag{3.1}$$

we determine a sequence $\Omega = \{\Omega_j\}$ of subsets of $\boldsymbol{C}^4$ by

$$\Omega_j := [\Gamma_j := \langle a_j, b_j, c_j, d_j \rangle \in \boldsymbol{C}^4 : \frac{a_j + c_j V_j}{b_j + d_j V_j} \subseteq V_{j-1}],\ j = 0, 1, 2, \ldots, \tag{3.2a}$$

with the restriction that

$$a_j d_j - b_j c_j \neq 0 \text{ for all } \Gamma_j = \langle a_j, b_j, c_j, d_j \rangle \in \Omega_j \tag{3.2b}$$

We call this procedure the *$VW\Omega$-method.* It follows from (1.1) and (2.5) that $\{V_j\}$ is a sequence of value regions with respect to the family $\mathcal{F}(\Omega, W)$ of LFASs (1.4) provided

$$\Omega_j \neq \emptyset, \qquad j = 0, 1, 2, \ldots . \tag{3.3}$$

In practice, conditions (3.2b) are ensured by imposing special conditions for the generating sequence

$$t_j^F(w) := \frac{a_j + c_j w}{b_j + d_j w}, \qquad j = 0, 1, 2, \ldots . \tag{3.4}$$

As an illustration of the above we start with

$$V_j := V_0 := [u \in \boldsymbol{C} : 0 \leq |u| \leq 1/2],\ w_j := 0,\ j = -1, 0, 1, \ldots, \tag{3.5a}$$

and generating functions of the form

$$t_0^F(w) := w,\ t_j^F(w) := \frac{a_j}{1 + w},\ a_j \neq 0,\ j = 0, 1, 2, \ldots . \tag{3.5b}$$

Then (3.1) and (3.2b) are satisfied and (3.2a) reduces to

$$\Omega_0 := [\Gamma_0 = \langle 0,1,1,0\rangle \in \boldsymbol{C}^4] \tag{3.6a}$$

and

$$\Omega_j := \left[\Gamma_j = \langle a_j,1,0,1\rangle \in \boldsymbol{C}^4 : \frac{a_j}{1+V_0} \subseteq V_0\right],\ j=1,2,3,\dots . \tag{3.6b}$$

It is readily shown from (3.5a) and (3.6b) that

$$\Omega_j = \Omega_1 = [\Gamma_j = \langle a_j,1,0,1\rangle \in \boldsymbol{C}^4 : 0 < |a_j| \le 1/4],\ j=1,2,3,\dots . \tag{3.7}$$

In the example described above $\mathcal{F} = \mathcal{F}(\Omega, W)$ is the family of all continued fractions (CFs)

$$F = \operatorname*{K}_{j=1}^{\infty}\left(\frac{a_j}{1}\right) \text{ such that } a_j \in E = [a \in \boldsymbol{C} : 0 < |a_j| \le 1/4]. \tag{3.8}$$

In 1865 Julius Worpitzky [66] proved that all CFs (3.8) converge to finite values. The set E is therefore called a *simple convergence region* for CFs of the form $K(a_j/1)$ and this set E is the first known example of a convergence region for CFs (see [28] for a discussion of Worpitzky's contributions to CF theory and his times). Best truncation error bounds $\beta_n(F, \mathcal{F})$ for the family of CFs (3.8) are given in Section 3.2.

In most (but not all) of the special families $\mathcal{F}(\Omega, W)$ of LFASs that have been studied extensively, the determination of $\{\Omega_j\}$ defined by (3.2) is simplified (as in the preceding example) by holding constant all but one of the components in $\Gamma_j = \langle a_j, b_j, c_j, d_j\rangle \in \Omega_j$. The determination of $\{\Omega_j\}$ in (3.2) can be (and usually is) simplified further by choosing regions V_j whose boundaries are circles or lines in $\boldsymbol{C}$, or else intersections of such regions. An additional simplification is attained when $\{V_j\}$, $\{w_j\}$ and $\{\Omega_j\}$ are all constant sequences; that is,

$$V_j = V_0,\ w_j = w_1,\ \Omega_j = \Omega_1 \quad \text{for all} \quad j \ge 1. \tag{3.9}$$

When this occurs we use the terms *simple value region* V_0 and *simple element region* Ω_1. In the following Section 3.2 we consider families $\mathcal{F}$ with simple value regions V_0 such that $c(V_0)$ is a closed circular disk and the corresponding Ω_1 is a simple element region.

3.2. Families of LFASs with Simple Circular Disk Value Regions

We begin this section with a result that is an immediate consequence of Theorem 2.6.

Theorem 3.1 *Let $\mathcal{F} = \mathcal{F}(\Omega, W)$ be a family of LFAS continued fractions (CFs)*

$$\operatorname*{K}_{j=1}^{\infty}\left(\frac{a_j}{b_j}\right) = \frac{a_1}{b_1} \genfrac{}{}{0pt}{}{}{+} \frac{a_2}{b_2} \genfrac{}{}{0pt}{}{}{+} \frac{a_3}{b_3} \genfrac{}{}{0pt}{}{}{+} \cdots,\ (c_j = 0, d_j = 1, w_j = 0 \text{ see } (1.12)). \tag{3.10}$$

Let $\{V_j\}$ be a sequence of value regions with respect to $\mathcal{F}$ such that for some integer $k \ge 0$ and sequence of positive numbers $\{\rho_j\}_{j=k}^{\infty}$,

$$V_j = [u \in \boldsymbol{C} : |u| \le \rho_j],\quad j = k, k+1, k+2, \dots .$$

Let $F \in \mathcal{F}$ have a finite value $v(F)$ and let n be a given positive integer with $n > k$. Let $v_n(F)$ and $h_n(F)$ be defined by (1.15) and $h_n(F) := B_n(F)/B_{n-1}(F)$. If the n-th inclusion region $T_n(F, V_n)$ is a bounded circular disk, then:

(A)

$$\begin{aligned} |v(F) - v_n(F)| &\leq \sup_\lambda[|\lambda - v_n(F)| : \lambda \in T_n(F, V_n)] \\ &= \frac{\rho_n \prod_{j=1}^n |a_j(F)|}{(|h_n(F)| - \rho_n) \cdot |B_n(F) B_{n-1}(F)|} \\ &= \frac{\rho_n}{(|h_n(F)| - \rho_n)} |v_n(F) - v_{n-1}(F)|. \end{aligned} \tag{3.11}$$

(B) *If, in addition,*

$$V_n = c(U_n(\mathcal{F})) \tag{3.12}$$

and the hypotheses of Theorem 2.5 hold, then the expressions on the right side of (3.11) give the best truncation error bound $\beta_n(F, \mathcal{F})$ for F with respect to $\mathcal{F}$.

A number of special families $\mathcal{F} = \mathcal{F}(\Omega, W)$ have best value regions $\{U_m(\mathcal{F})\}$ and converging factors $\{w_m\}$ satisfying

$$U_m(\mathcal{F}) = U_0(\mathcal{F}), \qquad m = 0, 1, 2, \ldots, \tag{3.13a}$$

and

$$w_m = 0 \in c(U_0(\mathcal{F})) := [u \in \mathbf{C} : |u| \leq \rho], \ \rho > 0, \ m \geq 0. \tag{3.13b}$$

We give results for four such families in this section.

3.2.1. *Worpitzky Family $\mathcal{F}^{W(\rho)}$*

For $0 < \rho \leq \frac{1}{2}$, we call

$$\mathcal{F}^{W(\rho)} := \left[\mathop{\mathrm{K}}_{j=1}^{\infty} (a_j/1) : a_j \in \mathbf{C}, \ 0 < |a_j| \leq \rho(1-\rho), \ j \geq 1 \right] \tag{3.14}$$

the *ρ-Worpitzky family of* LFASs. Since $0 < \rho(1-\rho) \leq 1/4$, it follows from Worpitzky's convergence region result (see (3.8)) that every $F \in \mathcal{F}^{W(\rho)}$ has a finite value $v(F)$. The family $\mathcal{F}^{W(\rho)}(\Omega, W)$ has element regions

$$\Omega_0 := \langle 0, 1, 1, 0 \rangle \text{ so that } t_0^F(w) \equiv w \text{ for all } F \in \mathcal{F}^{W(\rho)}, \tag{3.15a}$$

and

$$\Omega_j = \Omega_1 = \langle E_a(\rho), 1, 0, 1 \rangle, \text{ where } E_a(\rho) := [u \in \mathbf{C} : 0 < |u| \leq \rho(1-\rho)], \tag{3.15b}$$

and converging factors $w_j = 0$, $j \geq 0$. Our results for $\mathcal{F}^{W(\rho)}$ are summarized by the following:

Theorem 3.2 ($\mathcal{F}^{W(\rho)}$). *Let ρ satisfying $0 < \rho \leq 1/2$ be given. Then:*

(A) *$\mathcal{F}^{W(\rho)}$ has a simple best value region*

$$U_m(\mathcal{F}^{W(\rho)}) = U_0(\mathcal{F}^{W(\rho)}) = [u \in \mathbf{C} : 0 < |u| < \rho], \ m = -1, 0, 1, 2, \ldots . \tag{3.16}$$

(B) *For each $F \in \mathcal{F}^{W(\rho)}$ and each positive integer n, the best truncation error bound $\beta_n(F, \mathcal{F}^{W(\rho)})$ for $v_n((F)$ with respect to $\mathcal{F}^{W(\rho)}$ is given by*

$$\beta_n(F, \mathcal{F}^{W(\rho)}) = \frac{\rho \prod_{j=1}^{n} |a_j(F)|}{(|h_n(F)| - \rho) \cdot |B_n(F) B_{n-1}(F)|} = \frac{\rho}{(|h_n(F)| - \rho)} |v_n(F) - v_{n-1}(F)|. \tag{3.17}$$

(C) *If, in addition, $0 < \rho < 1/2$, then*

$$|v(F) - v_n(F)| \leq \left(\frac{2\rho}{1-2\rho}\right) |v_n(F) - v_{n-1}(F)|, \; n = 2, 3, 4, \ldots . \tag{3.18}$$

Remarks: It follows from (3.16) and the definition of $U_m(\mathcal{F}^{W(\rho)})$ in (2.1) that, if $0 < \rho \leq 1/2$, then

$$|v(F)| \leq \rho \quad \text{for all} \quad F \in \mathcal{F}^{W(\rho)}.$$

Proof of Theorem 3.2. (A): Let $\{V_j(\rho)\}$ be defined by

$$V_j(\rho) := V_0(\rho) := [u \in \mathbf{C} : 0 < |u| < \rho], \; j = -1, 0, 1, 2, \ldots . \tag{3.19}$$

To prove (A) it suffices to show that

$$\{V_j(\rho)\} \in \mathcal{V}(\mathcal{F}^{W(\rho)}) \text{ and } V_0(\rho) \subseteq U_0(\mathcal{F}^{W(\rho)}). \tag{3.20}$$

We prove $\{V_j(\rho)\} \in \mathcal{V}(\mathcal{F}^{W(\rho)})$ by verifying that conditions (2.5) hold. Condition (2.5a) is an immediate consequence of (3.14), (3.19) and $\rho(1-\rho) < \rho$. Condition (2.5b) (with $n \geq 1$) is equivalent to

$$\frac{1 + V_0(\rho)}{a} \subseteq \frac{1}{V_0(\rho)} \text{ for all } a \in E_a(\rho) := [z \in E : 0 < |z| \leq \rho(1-\rho)],$$

which can be readily proven. To show that $V_0(\rho) \subseteq U_0(\mathcal{F}^{W(\rho)})$ we let u denote an arbitrary point in V_0. For each $n \geq 0$, let g_n denote the n-th approximant of the CF

$$1 + K\left(\frac{-\rho(1-\rho)}{1}\right) = 1 - \frac{\rho(1-\rho)}{1} + \frac{-\rho(1-\rho)}{1} + \frac{-\rho(1-\rho)}{1} + \cdots, \tag{3.21}$$

so that

$$g_0 := 1 \text{ and } g_n := 1 - \frac{\rho(1-\rho)}{g_{n-1}}, \text{ for } n = 1, 2, 3, \ldots . \tag{3.22}$$

We now prove (by induction) that

$$\frac{1}{2} \leq 1 - \rho < g_n < g_{n-1} \leq 1, \quad n = 1, 2, 3, \ldots . \tag{3.23}$$

Since $g_0 := 1$ and $g_1 = 1 - \rho(1-\rho)$, one can see that (3.23) holds for $n = 1$. As our induction hypothesis we assume that

$$1 - \rho < g_k < g_{k-1} < 1, \quad k = 2, 3, \ldots, n-1, \tag{3.24}$$

for some positive integer n. Then $1-\rho < g_{n-1}$ implies

$$g_n = 1 - \frac{\rho(1-\rho)}{g_{n-1}} > 1 - \frac{\rho(1-\rho)}{1-\rho} = 1-\rho. \tag{3.25}$$

Furthermore,

$$g_{n-1} - g_n = g_{n-1} - \left(1 - \frac{\rho(1-\rho)}{g_{n-1}}\right) = \frac{\rho(1-\rho) - g_{n-1}(1-g_{n-1})}{g_{n-1}} > 0 \tag{3.26}$$

iff

$$g_{n-1}(1-g_{n-1}) < \rho(1-\rho). \tag{3.27}$$

This inequality holds since $\frac{1}{2} \le (1-\rho) < g_{n-1} < 1$ and $f(x) := x(1-x)$ is decreasing on the interval $\frac{1}{2} \le x \le 1$. We have established (3.23). Worpitzky's theorem ensures that the CF (3.21) converges to a finite value $g = \lim_{n\to\infty} g_n$. Therefore from the recurrence relations (3.22) we see that g satisfies the quadratic equation

$$g = 1 - \frac{\rho(1-\rho)}{g},$$

whose roots are ρ and $(1-\rho)$. From this and (3.23) we conclude that $\{g_n\}_{n=0}^{\infty}$ decreases monotonically to the limit g, with

$$\frac{1}{2} \le g = 1-\rho < 1. \tag{3.28}$$

Let $\varepsilon := \rho - |u|$, $\varepsilon_n := g_n - (1-\rho)$, $n \ge 0$, and let $n_0 \ge 1$ be chosen so that

$$\varepsilon_{n_0} \le \frac{\varepsilon(1-\rho)}{\rho-\varepsilon}. \tag{3.29}$$

We then define a by

$$|a| := (\rho-\varepsilon)[(1-\rho)+\varepsilon_{n_0}] \text{ and } \arg a := \arg u. \tag{3.30}$$

It follows from (3.30) that

$$\frac{|a|}{g_{n_0}} = |u| \text{ and } |a| = \rho(1-\rho) - [\varepsilon(1-\rho) - \varepsilon_n(\rho-\varepsilon)] \le \rho(1-\rho),$$

and hence by (2.1)

$$u = \frac{a}{g_{n_0}} \in U_0(\mathcal{F}^{W(\rho)}).$$

This completes the proof of (A).

(B): It follows from conditions (2.2), that $\{T_n(F, U_0(\mathcal{F}^{W(\rho)}))\}$ is a nested sequence of subsets of $\mathbb{C}$ if $F \in \mathcal{F}^{W(\rho)}$. Therefore since for $n \ge 1$,

$$T_n(F, c(U_0(\mathcal{F}^{W(\rho)}))) \subseteq \cdots \subseteq T_1(F, c(U_0(\mathcal{F}^{W(\rho)}))) \subseteq c(U_o(\mathcal{F}^{W(\rho)})),$$

we see that $T_n(F, c(U_0(\mathcal{F}^{W(\rho)})))$ is a bounded, closed circular disk. We wish to apply Theorem 3.1(B). For that purpose it suffices to verify that condition (b) of Theorem

2.5 holds. Let $n \geq 1$ and $k \geq 1$ be given. Then for each $j \geq 1$, we define a CF $G_j \in \mathcal{F}^{W(\rho)}$ as follows:

$$a_m(G_j) := a_m(F) \text{ for } m = 1, 2, \ldots, n+k,$$

$$a_{n+k+1}(G_j) := \frac{1}{j} \text{ and } a_m(G_j) = -\rho(1-\rho) \text{ for } m \geq n+k+2.$$

Condition (b) of Theorem 2.5 follows from the fact that the CF (3.21) has value $1-\rho$, and

$$\lim_{j\to\infty} v(G_j^{(n+k)}) = \lim_{j\to\infty} \left(\frac{1/j}{1-\rho}\right) = 0 =: w_{n+k}.$$

Assertion (B) follows then from Theorem 3.1(B).

(C) follows immediately from (B) and the fact that $T_n(F, -h_n(F)) = \infty$, so that $-h_n(F) \notin c(V_0(1/4)) = c(U_0(\mathcal{F}^{W(1/4)}))$ and hence $|h_n(F)| > 1/2$. □

3.2.2. *Pringsheim–Śleszyński Family* $\mathcal{F}^{PS(\rho)}$

For $0 < \rho \leq 1$, we call

$$\mathcal{F}^{PS(\rho)} := \left[\mathop{\mathrm{K}}_{j=1}^{\infty} \left(\frac{1}{b_j}\right) : b_j \in \boldsymbol{C},\ \rho + \frac{1}{\rho} \leq |b_j| < \infty\right] \tag{3.31}$$

the ρ-Pringsheim–Śleszyński family of LFASs. Since $\rho + \dfrac{1}{\rho} \geq 2$, it follows from the Pringsheim–Śleszyński criterion (see, e.g., [37], Theorem 4.35 and [59]) that every $F \in \mathcal{F}^{PS(\rho)}$ has a finite value $v(F)$. The family $\mathcal{F}^{PS(\rho)}$ has element regions

$$\Omega_0 := \langle 0, 1, 1, 0\rangle \text{ so that } t_0^F(w) \equiv w \text{ for all } F \in \mathcal{F}^{PS(\rho)} \tag{3.32a}$$

and, for $j \geq 1$,

$$\Omega_j := \Omega_1 := \langle 1, E_b(\rho), 0, 1\rangle, \text{ where } E_b(\rho) := \left[u \in \boldsymbol{C} : \rho + \frac{1}{\rho} \leq |u| < \infty\right], \tag{3.32b}$$

and converging factors $w_j = 0$, $j \geq 0$. Our results for $\mathcal{F}^{PS(\rho)}$ are summarized in the following:

Theorem 3.3 ($\mathcal{F}^{PS(\rho)}$). *Let ρ satisfying $0 < \rho \leq 1$ be given. Then:*

(A) *$\mathcal{F}^{PS(\rho)}$ has a simple best value region $U_0(\mathcal{F}^{PS(\rho)})$ satisfying*

$$c(U_0(\mathcal{F}^{PS(\rho)})) = [u \in \boldsymbol{C} : 0 \leq |u| \leq \rho]. \tag{3.33}$$

(B) *For each $F \in \mathcal{F}^{PS(\rho)}$ and each positive integer n, the best truncation error bound $\beta_n(F, \mathcal{F}^{PS(\rho)})$ for $v_n(F)$ with respect to $\mathcal{F}^{PS(\rho)}$ is given by*

$$\beta_n(F, \mathcal{F}^{PS(\rho)}) = \frac{\rho}{(|h_n(F)| - \rho) \cdot |B_n(F) B_{n-1}(F)|} = \frac{\rho}{(|h_n(F)| - \rho)} |v_n(F) - v_{n-1}(F)|. \tag{3.34}$$

(C) *If, in addition, $0 < \rho < 1$, then*

$$|v(F) - v_n(F)| \leq \frac{\rho}{1-\rho} |v_n(F) - v_{n-1}(F)|,\ n = 2, 3, 4, \ldots . \tag{3.35}$$

Remark: It follows from (3.33) and the definition of $U_0(\mathcal{F}^{PS(\rho)})$ in (2.1) that, if $0 < \rho \leq 1$, then

$$|v(F)| \leq \rho \quad \text{for} \quad F \in \mathcal{F}^{PS(\rho)}. \tag{3.36}$$

Proof. (A): Let $\{V_j(\rho)\}$ be defined by

$$V_j(\rho) := V_0(\rho) := [u \in \boldsymbol{C} : 0 \leq |u| \leq \rho], \; j = -1, 0, 1, 2, \dots . \tag{3.37}$$

To prove (A) it suffices to show that

$$\{V_j(\rho)\} \in \mathcal{V}(\mathcal{F}^{PS(\rho)}) \quad \text{and} \quad V_0(\rho) = c(U_0(\mathcal{F}^{PS(\rho)})). \tag{3.38}$$

We prove $\{V_j(\rho)\} \in \mathcal{V}(\mathcal{F}^{PS(\rho)})$ by verifying conditions (2.5). Condition (2.5a) follows directly from (3.31), (3.37) and $\rho + 1/\rho > \rho$. Condition (2.5b) is equivalent to

$$b + V_0(\rho) \subseteq \frac{1}{V_0(\rho)} \text{ for all } b \in E_b(\rho) := \left[u \in \boldsymbol{C} : \rho + \frac{1}{\rho} \leq |u| < \infty\right], \tag{3.39}$$

which can be readily shown. To prove that

$$V_0(\rho) = c(U_0(\mathcal{F}^{PS(\rho)})) \tag{3.40}$$

we make use of Theorem 2.7. First we show that

$$[t_n^F(V_0(\rho)) : F \in \mathcal{F}^{PS(\rho)}] = V_0(\rho). \tag{3.41}$$

In view of (2.5b) it suffices to verify

$$V_0(\rho) \subseteq [t_n^F(V_0(\rho)) : F \in \mathcal{F}^{PS(\rho)}] \tag{3.42}$$

or, equivalently, $V_0(\rho) \subseteq 1/(E_b(\rho) + V_0(\rho))$; that is,

$$\frac{1}{V_0(\rho)} \subseteq E_b(\rho) + V_0(\rho). \tag{3.43}$$

Let $v \in 1/V_0(\rho)$ be given and let $\varphi := \arg v$, so that $\dfrac{1}{\rho} \leq |v| < \infty$ and $0 \leq \varphi < 2\pi$.

Let b and u be defined by

$$|b| := |v| + \rho, \quad \arg b := \varphi, \quad u := -\rho e^{i\varphi}.$$

It follows from this that

$$b + u = (|v| + \rho)e^{i\varphi} - \rho e^{i\varphi} = v, \; b \in E_b(\rho) \text{ and } u \in V_0(\rho).$$

This proves (3.43) and hence also (3.41). Condition (2.37) can be written

$$\lim_{n\to\infty} \left\{ \sup_{F \in \mathcal{F}^{PS(\rho)}} [\operatorname{diam} T_n(F^{(m)}, V_0(\rho))] \right\} = 0. \tag{3.44}$$

This is an immediate consequence of a theorem due to Hillam (see, e.g., [12], Theorem 2.7; [22]). Thus (A) follows from Theorem 2.7.

(B): We apply Theorem 3.1(B). For that purpose we note that $T_n(F, V_0(\rho))$ is a bounded circular disk, since by (2.2) $\{T_n(F, V_0(\rho))\}$ is a nested sequence of closed disks and $T_n(F, V_0(\rho)) \subseteq T_{n-1}(F, V_0(\rho)) \subseteq \cdots \subseteq V_0(\rho)$. It suffices to verify that condition (b) of Theorem 2.5 is satisfied. Let $n \geq 1$ and $k \geq 1$ be given. Then for each $j \geq 1$, we define a CF $G_j \in \mathcal{F}_{n+k}^{PS(\rho)}(F)$ as follows:

$$\begin{aligned} b_m(G_j) &:= b_m(F), \quad m = 1, 2, \ldots, n+k, \\ b_{n+k+1}(G_j) &:= j, \\ b_m(G_j) &:= \rho + 1/\rho, \quad m = n+k+2,\ n+k+3, \ldots. \end{aligned}$$

Then

$$\lim_{j\to\infty} v(G_j^{(n+k)}) = \lim_{j\to\infty} \frac{1}{j + v(G_j^{(n+k+1)})} = O =: w_{n+k},$$

since $v(G_j^{(n+k+1)})$ is the value of the periodic CF $K\left(\frac{1}{\rho + \frac{1}{\rho}}\right)$ and hence, by (3.36), $|v(G_j^{(n+k+1)})| \leq \rho$.

Therefore (B) follows from Theorem 3.1(B).

(C) follows from Theorem 3.1(A) and the fact that $T_n(F, -h_n(F)) = \infty$, so that $-h_n(F) \notin V_0\left(\frac{1}{2}\right) = [u \in \boldsymbol{C} : 0 \leq |u| \leq 1]$; hence $|h_n(F)| > 1$. □

3.2.3. *Positive Perron–Carathéodory Family* $\mathcal{F}^{PPC(z)}$.

Let

$$z \in D := [u \in \boldsymbol{C} : 0 \leq |u| < 1] \tag{3.45}$$

be given. We define the family $\mathcal{F}^{PPC(z)}$ of LFASs $F = C[\{\delta_j\}, z]$, called *positive Perron–Carathéodory approximant sequences*, as follows:

$$\mathcal{F}^{PPC(z)} := \left[\text{LFASs } F : t_j^F(w) = \frac{a_j + c_j w}{b_j + d_j w},\ j = 0, 1, 2, \ldots\right], \tag{3.46a}$$

where the generating sequences $\{t_j^F(w)\}_{j=0}^{\infty}$ have the form

$$t_0^F(w) := \delta_0 \frac{1 - w}{1 + w},\ t_j^F(w) := z \frac{\bar{\delta}_j + w}{1 + \delta_j w},\ j = 1, 2, 3, \ldots, \tag{3.46b}$$

where

$$\delta_0 > 0 \quad \text{and} \quad \delta_j \in D,\ j = 1, 2, 3, \ldots, \tag{3.46c}$$

and the converging factors $w_j = 0$, for $j \geq 0$. To emphasize the dependence of the δ_j on F we may write $\delta_j(F)$. Each $F \in \mathcal{F}^{PPC(z)}$ is related to the *positive Perron–Carathéodory CF (PPC-fraction)*

$$\delta_0 - \frac{2\delta_0}{1} \, \frac{1}{+ \bar{\delta}_1 z +} \, \frac{(1 - |\delta_1|^2) z}{\delta_1} \, \frac{1}{+ \bar{\delta}_2 z +} \, \frac{(1 - |\delta_2|^2) z}{\delta_2} + \cdots \tag{3.47}$$

in the following way: We define sequences $\{s_n^F(w)\}$ and $\{S_n(F,w)\}$ by

$$s_0^F(w) := \delta_0 + w, \; s_{2j}^F(w) := \frac{1}{\delta_j z + w}, \quad j = 1,2,3,\ldots \tag{3.48a}$$

$$s_1^F(w) := \frac{-2\delta_0}{1+w}, \; s_{2j+1}^F(w) := \frac{(1-|\delta_j|^2)z}{\delta_j + w}, \; j = 1,2,3,\ldots, \tag{3.48b}$$

$$S_0(F,w) := s_0^F(w), \; S_n(F,w) := S_{n-1}(F, s_n^F(w)), \; n = 1,2,3,\ldots, \tag{3.46c}$$

and we let $P_n(F,z)$ and $Q_n(F,z)$ denote the n-th numerator and denominator, respectively, of the CF (3.47). It follows that

$$S_n(F,w) = \frac{P_n(F,z) + wP_{n-1}(F,z)}{Q_n(F,z) + wQ_{n-1}(F,z)}, \quad n = 1,2,3,\ldots, \tag{3.49}$$

$$t_0^F(w) = s_0^F \circ s_1^F(w^{-1}), \; t_j^F(w) = [s_{2j}^F \circ s_{2j+1}^F(w^{-1})]^{-1}, \; j = 1,2,3,\ldots, \tag{3.50a}$$

$$T_n(F,w) = S_{2n+1}(F,w^{-1}) = \frac{P_{2n+1}(F,z)w + P_{2n}(F,z)}{Q_{2n+1}(F,z)w + Q_{2n}(F,z)}, \; n = 0,1,2,\ldots. \tag{3.50b}$$

Therefore, for $n = 0,1,2,\ldots$,

$$v_n(G) := T_n(F,0) = \frac{P_{2n}(F,z)}{Q_{2n}(F,z)}, \tag{3.51}$$

and

$$\begin{aligned} A_n(F) &= P_{2n}(F,z), \; B_n(F) = Q_{2n}(F,z), \; C_n(F) = P_{2n+1}(F,z), \\ D_n(F) &= Q_{2n+1}(F,z), \end{aligned}$$

where A_n, B_n, C_n, D_n are defined by the difference equations (1.10).

The class $\mathcal{C}$ of *normalized Carathéodory functions* is defined by

$$\mathcal{C} := [f : f \text{ is analytic and } \operatorname{Re} f(z) > 0 \text{ for } |z| < 1, \; f(0) > 0]. \tag{3.52}$$

It can be seen that all functions of the form

$$f(z) = \sum_{j=1}^{n} \lambda_j \frac{e^{i\theta_j} + z}{e^{i\theta_j} - z}, \; \lambda_j > 0 \text{ for } 1 \le j \le n, \; -\pi < \theta_1 < \theta_2 < \cdots < \theta_n = \pi \tag{3.53}$$

are in $\mathcal{C}$. We consider the subclass $\mathcal{C}_c$ of $\mathcal{C}$ defined by

$$\mathcal{C}_c := [f \in \mathcal{C} : f \text{ is not of the form (3.53) and } f \text{ is not constant}]. \tag{3.54}$$

For each $F \in \mathcal{F}^{PPC(z)}$, we let $v(F(z))$ denote the value of F considered as a function of z. In [33], Theorem 10.2, it was shown that,

$$F(z) \in \mathcal{F}^{PPC(z)} \Rightarrow f(z) := v(F(z)) \in \mathcal{C}_c,$$

and, conversely, $f(z) \in \mathcal{C}_c$ implies that there exists a unique $F(z) \in \mathcal{F}^{PPC(z)}$ such that $f(z) = v(F(z))$. The following result (Theorem 3.4) gives best truncation error bounds for $v_n(F(z))$.

Theorem 3.4 ($\mathcal{F}^{PPC(z)}$). *Let $z \in \boldsymbol{C}$, satisfying $0 < |z| < 1$ be given. Then:*

(A) *The family of LFASs $\mathcal{F}^{PPC(z)}$ has a sequence of best value regions given by*

$$U_{-1}(\mathcal{F}^{PPC(z)}) = \bigcup_{\delta_0 > 0} \Delta(\delta_0), \tag{3.55a}$$

where

$$\begin{aligned} \Delta(\delta_0) &:= [u \in \boldsymbol{C} : |u - \Gamma(\delta_0)| < R(\delta_0)],\ \Gamma(\delta_0) := \delta_0 \tfrac{1+|z|^2}{1-|z|^2}, \\ R(\delta_0) &:= \tfrac{2\delta_0 |z|}{1-|z|^2}, \end{aligned} \tag{3.55b}$$

and

$$U_m(\mathcal{F}^{PPC(z)}) := [u \in \boldsymbol{C} : 0 \le |u| < |z|],\ m = 0, 1, 2, \ldots . \tag{3.55c}$$

(B) *For each $F \in \mathcal{F}^{PPC(z)}$ and each integer $n \ge 1$, the best truncation error bound for $v_n(F)$ with respect to $\mathcal{F}^{PPC(z)}$ is given by*

$$\beta_n(F, \mathcal{F}^{PPC(z)}) = \frac{2\delta_0 \prod_{j=1}^{n}(1 - |\delta_j|^2) z^{n+1}}{|Q_{2n}(F,z)|(|Q_{2n}(F,z)| - |zQ_{2n+1}(F,z)|)}, \tag{3.56}$$

where Q_{2n} and Q_{2n+1} are defined by (3.49) and (3.52).

(C) *For each $F \in \mathcal{F}^{PPC(z)}$ and integer $n \ge 1$*

$$|v(F) - v_n(F)| \le \frac{4\delta_0 |z|^{n+1}}{1 - |z|^2}. \tag{3.57}$$

Remarks. (1) We have omitted the point $z = 0$ in Theorem 3.4, since, if $z = 0$, $v_n(F(0)) = T_n(F(0), 0) = \delta_0$ for all $n \ge 0$, and hence $v(F(0)) = \delta_0$. (2) $\delta_0 \in \Delta(\delta_0)$, since $\delta_0 > \delta_0(1 - |z|) = \Gamma(\delta_0) - R(\delta_0) > 0$.

Proof. (A): Let $\{V_j\}$ be defined by

$$V_{-1} := \bigcup_{\delta_0 > 0} \Delta(\delta_0),\ \Delta(\delta_0) \text{ defined by (3.55b)}, \tag{3.58a}$$

$$V_j := [u \in \boldsymbol{C} : 0 \le |u| < |z|], \quad j = 0, 1, 2, \ldots . \tag{3.58b}$$

By (3.55) and (3.58)

$$t_0^F(0) = \delta_0(F) \in \Delta(\delta_0) = t_0^F(V_0) \subseteq V_{-1}, \text{ for all } \delta_0 > 0. \tag{3.59a}$$

Therefore (2.5) holds for $n = 0$. For all $n \ge 1$,

$$t_n^F(0) = \bar{\delta}_n(F) z \in V_{n-1} \text{ for all } 0 \le |\delta_n(F)| < 1, \tag{3.60a}$$

and

$$t_n^F(V_n) = [u \in \boldsymbol{C} : |u - \Gamma_n| < R_n] \subseteq V_{n-1} \text{ for all } F \in \mathcal{F}^{PPC(z)}, \tag{3.60b}$$

since

$$|\Gamma_n| + R_n < |z| < 1 \tag{3.60c}$$

where

$$\Gamma_n := \frac{z\bar{\delta}_n(1-|z|^2)}{1-|z|^2|\delta_n|^2}, \; R_n := \frac{|z|^2(1-|\delta_n|^2)}{1-|z|^2|\delta_n|^2} \tag{3.60d}$$

(see, e.g., [39], Lemma 3.2, for more details on proof of (3.60c)).

It follows from (3.60) that (2.5) holds for all $n \geq 1$. Therefore

$$\{V_j\} \in \mathcal{V}(\mathcal{F}^{PPC(z)}). \tag{3.61}$$

We now show that

$$V_j \subseteq U_j(\mathcal{F}^{PPC(z)}), \quad j = -1, 0, 1, 2, \ldots . \tag{3.62}$$

In fact, for $j \geq 0$,

$$\begin{aligned} V_j = V_0 &:= [u \in \boldsymbol{C} : 0 \leq |u| < |z|] = [\bar{\delta}_1 z : 0 \leq |\delta_1| < 1] \\ &= [t_1^F(0) : F \in \mathcal{F}^{PPC(z)}] \subseteq U_0(\mathcal{F}^{PPC(z)}). \end{aligned}$$

and, since $t_0^F(V_0) = \Delta(\delta_0(F))$, we have

$$\begin{aligned} V_{-1} &:= [\Delta(\delta_0(F)) : F \in \mathcal{F}^{PPC(z)}] = [t_0^F(V_0) : F \in \mathcal{F}^{PPC(z)}] \\ &\subseteq U_{-1}(\mathcal{F}^{PPC(z)}). \end{aligned}$$

This proves (3.62) and hence (A).

(B): Since the conditions of Theorem 2.6(B) hold (with condition (a) of Theorem 2.5), we have for $n \geq 1$

$$\begin{aligned} \beta_n(F, \mathcal{F}^{PPC(z)}) &= \frac{|z| \prod_{j=0}^{n} |a_j(F)d_j(F) - b_j(F)c_j(F)|}{|D_n(F)|^2 |h_n(F)|(|h_n(F)| - |z|)} \\ &= \frac{2\delta_0 \prod_{j=1}^{n}(1-|\delta_j|^2)|z|^{n+1}}{|B_n(F)|(|B_n(F)| - |zD_n(F)|)} \end{aligned}$$

which gives (3.56), using (3.52) and

$$\begin{aligned} a_0(F) &= \delta_0, \; b_0(F) = 1, \; c_0(F) = -\delta_0, \; d_0(F) = 1, \\ a_j(F) &= \bar{\delta}_j z, \; b_j(F) = 1, \; c_j(F) = z, \; d_j(F) = \delta_j, \; j = 1, 2, 3, \ldots . \end{aligned}$$

This proves (B).

(C): Our proof of (3.57) makes use of Theorem 2.4. In [39], Lemma 3.3, we obtain

$$\operatorname{diam} T_n(F, V_n) = \frac{4\delta_0 \prod_{j=1}^{n}(1-|\delta_j|^2)|z|^{n+1}}{|Q_{2n}(F,z)|^2 - |zQ_{2n+1}(F,z)|^2}, \; n = 1, 2, 3, \ldots . \tag{3.63}$$

Using Christoffel–Darboux formulas derived in [40, Section 2], we obtain the inequality

$$|Q_{2n}(F,z)|^2 - |zQ_{2n+1}(F,z)|^2 \geq (1-|z|^2) \prod_{j=1}^{n}(1-|\delta_j|^2), \; n = 1, 2, 3, \ldots . \tag{3.64}$$

Combining (3.63) and (3.64) with Theorem 2.4 yields

$$|v(F) - v_n(F)| \leq \frac{4\delta_0 \prod_{j=1}^{n}(1-|\delta_j|^2)|z|^{n+1}}{|Q_{2n}(F,z)|^2 - |zQ_{2n+1}(F,z)|^2} \tag{3.65}$$

and hence (3.57). □

Remark. One can readily show that $\beta_n(F, \mathcal{F}^{PPC(z)})$ is at least as small as the bound given by (3.65). In fact, that statement holds iff

$$|Q_{2n}(F,z)|^2 - |zQ_{2n+1}(F,z)|^2 \leq 2|Q_{2n}(F,z)|(|Q_{2n}(F,z)| - |zQ_{2n+1}(F,z)|). \tag{3.66}$$

Dividing both sides of (3.66) by $|Q_{2n}(F,z)|^2$ and rearranging terms, we obtain the following inequality that is equivalent to (3.66):

$$\left(1 - \left|z\frac{Q_{2n+1}(F,z)}{Q_{2n}(F,z)}\right|\right)^2 \geq 0. \tag{3.67}$$

3.2.4. *Positive Schur Family* $\mathcal{F}^{Sh(z)}$.

Let

$$z \in D := [u \in \mathbb{C} : 0 \leq |u| < 1] \tag{3.68}$$

be given. We define the family $\mathcal{F}^{Sh(z)}$ of LFASs $F = S[\{\gamma_j\}, z]$, called *positive Schur approximant* sequences, as follows:

$$\mathcal{F}^{Sh(z)} := \left[\text{LFASs } F : t_j^F(w) = \frac{\gamma_j + zw}{1 + \bar{\gamma}_j zw},\ j = 0,1,2,\ldots\right], \tag{3.69a}$$

where

$$\gamma_0 \in \mathbb{R},\ |\gamma_0| < 1 \text{ and } \gamma_j \in \mathbb{C},\ |\gamma_j| < 1,\ j = 1,2,3,\ldots, \tag{3.69b}$$

and converging factors $w_j := 0$, $j = 0,1,2,\ldots$. To emphasize dependence of γ_j on F we may write $\gamma_j(F)$. Each $F \in \mathcal{F}^{Sh(z)}$ is related to the *positive Schur CF*

$$\gamma_0 + \frac{(1-|\gamma_0|^2)z}{\bar{\gamma}_0 z} \frac{1}{+\gamma_1 +} \frac{(1-|\gamma_1|^2)z}{\bar{\gamma}_1 z} \frac{1}{+\gamma_2 + \cdots}, \tag{3.70}$$

in the following way: We define sequences $\{s_n^F(w)\}$ and $\{S_n(F,w)\}$ by

$$s_0^F(w) := \gamma_0 + w,\ s_{2j}^F(w) := \frac{1}{\gamma_j + w},\ j = 1,2,3,\ldots, \tag{3.71a}$$

$$s_{2j+1}^F(w) := \frac{(1-|\gamma_j|^2)z}{\bar{\gamma}_j z + w},\quad j = 0,1,2,\ldots, \tag{3.71b}$$

$$S_0(F,w) := s_0^F(w),\ S_n(F,w) := S_{n-1}(F, s_n^F(w)),\ n = 1,2,3,\ldots, \tag{3.71c}$$

and let $P_n(F,z)$ and $Q_n(F,z)$ denote the n-th numerator and denominator, respectively, of the CF (3.70). It follows that

$$S_n(F,w) = \frac{P_n(F,z) + wP_{n-1}(F,z)}{Q_n(F,z) + wQ_{n-1}(F,z)},\quad n = 0,1,2,\ldots, \tag{3.72}$$

$$t_0^F(w) := s_0^F s_1^F(w^{-1}),\ t_j^F(w) := [s_{2j}^F \circ s_{2j+1}^F(w^{-1})]^{-1},\ j = 1,2,3,\dots, \tag{3.73a}$$

$$T_n(F,w) = S_{2n+1}(F,w^{-1}) = \frac{P_{2n+1}(F,z)w + P_{2n}(F,z)}{Q_{2n+1}(F,z)w + Q_{2m}(F,z)},\ n = 0,1,2,\dots. \tag{3.73b}$$

Therefore, for $n = 0,1,2,\dots$,

$$v_n(F) := T_n(F,0) = S_{2n+1}(F,\infty) = \frac{P_{2n}(F,z)}{Q_{2n}(F,z)}, \tag{3.74}$$

and

$$\begin{aligned} A_n(F) &= P_{2n}(F,z),\ B_n(F) = Q_{2n}(F,z),\ C_n(F) = P_{2n+1}(F,z), \\ D_n(F) &= Q_{2n+1}(F,z), \end{aligned} \tag{3.75}$$

where A_n, B_n, C_n, D_n are defined by the difference equations (1.10). The class $\mathcal{S}$ of *normalized Schur functions* is defined by

$$\mathcal{S} := [f : f \text{ is analytic and } |f(z)| \le 1 \text{ for } |z| < 1,\ -1 < f(0) < 1]. \tag{3.76}$$

It can be seen that all functions of the form

$$f(z) = \varepsilon \prod_{j=1}^{n} \frac{z+\omega_j}{1+\bar{\omega}_j z}\quad |\omega_j| < 1,\ j = 1,2,\dots,n,\ |\varepsilon| = 1,\ \varepsilon \prod_{j=1}^{n} \omega_j \in \mathbb{R} \tag{3.77}$$

are members of $\mathcal{S}$. We consider the subclass $\mathcal{S}_c$ of $\mathcal{S}$ defined by

$$\mathcal{S}_c := [f \in \mathcal{S} : f \text{ is not of the form (3.77) and } f \text{ is not constant}]. \tag{3.78}$$

For each $F \in \mathcal{F}^{Sh(z)}$, we let $v(F(z))$ denote the value of F considered as a function of z. In [50] and [32] it is shown that

$$F \in \mathcal{F}^{Sh(z)} \Rightarrow v(F(z)) \in \mathcal{S}_c$$

and, conversely, $f(z) \in \mathcal{S}_c$ implies that there exists a unique $F \in \mathcal{F}^{Sh(z)}$ such that $f(z) = v(F(z))$. The following result (Theorem 3.5) gives best truncation error bounds for $v_n(F(z))$ with respect to $\mathcal{F}^{Sh(z)}$.

Theorem 3.5 ($\mathcal{F}^{Sh(z)}$). *Let $z \in D := [u \in \mathbb{C} : 0 \le |u| < 1]$ be given. Then*

(A) *The family of LFASs $\mathcal{F}^{Sh(z)}$ has a sequence of best value regions given by*

$$U_{-1}(\mathcal{F}^{Sh(z)}) = \bigcup_{-1<\gamma_0<1} t_0^F(D) = \bigcup_{-1<\gamma_0<1} [u \in \mathbb{C} : |u - c(\gamma_0)| < r(\gamma_0)] \subseteq D \tag{3.79a}$$

where

$$c(\gamma_0) := \frac{\gamma_0(1-|z|^2)}{1-\gamma_0^2|z|^2},\quad r(\gamma_0) := \frac{(1-\gamma_0^2)|z|}{1-\gamma_0^2|z|^2}, \tag{3.79b}$$

and

$$U_m(\mathcal{F}^{Sh(z)}) = D,\quad m = 0,1,2,\dots. \tag{3.79c}$$

(B) *For each $F \in \mathcal{F}^{Sh(z)}$ and each integer $n \ge 0$, the best truncation error bound for $v_n(F)$ with respect to $\mathcal{F}^{Sh(z)}$ is given by*

$$\beta_n(F, \mathcal{F}^{Sh(z)}) = \frac{|z|^{n+1}\prod_{j=0}^{n}(1-|\gamma_j|^2)}{|Q_{2n}(F,z)| \cdot (|Q_{2n}(F,z)| - |Q_{2n+1}(F,z)|)}. \tag{3.80}$$

Proof. (A): A proof of (3.79c) can be found in [31], Lemma 7. The first equality in (3.79a) follows from the definition of U_{-1} in (2.1) and from (3.79c). The second equality in (3.79a) follows from elementary conformal mapping of D by the linear fractional transformation $t_0^F(w)$.

(B) follows immediately from Theorem 2.6(B), since $w_m = 0 \in U_m(\mathcal{F}^{Sh(z)})$, $a_j(F) = \gamma_j$. $b_j(F) = 1$, $c_j(F) = z$, $d_j(F) = \bar{\gamma}_j z$, $h_n(F) = B_n(F)/D_n(F)$, $B_n(F) = Q_{2n}(F, z)$ and $D_n(F) = Q_{2n+1}(F, z)$. □

Remark. A proof of Theorem 3.5 was given in [31], Theorem 10, using essentially the same methods as employed in Theorem 2.6(B).

3.3. Families of LFASs with Other Simple Value Regions

In this section we obtain best truncation error bounds for Real J-Fractions, Stieltjes Fractions, Modified Stieltjes Fractions, and Positive T-Fractions. For each of these families of LFASs (CFs), the best value regions are simple and they are half-planes or intersections of half-planes with part or none of the boundaries included. We make use of the following:

Theorem 3.6 *Let $\mathcal{F} = \mathcal{F}(\Omega, \{0\})$ be a family of LFASs of continued fractions (CFs). Let $\{U_j(\mathcal{F})\}$ denote the best sequence of value regions corresponding to $\mathcal{F}$. Let $F = \mathrm{K}_{j=1}^{\infty}(a_j/b_j) \in \mathcal{F}$ be convergent to a finite value $v(F)$, let n be a positive integer such that $T_n(F, c(U_n(\mathcal{F})))$ is bounded and let condition (a) or (b) of Theorem 2.5 hold. Then*

$$\beta_n(F, \mathcal{F}) = \frac{|v_n(F) - v_{n-1}(F)|}{|h_n(F)| \inf\left[\left|\frac{-1}{h_n(F)} - \frac{1}{u}\right| : u \in c(U_n(\mathcal{F}))\right]}. \tag{3.81}$$

Proof. From Theorem 2.5,

$$\beta_n(F, \mathcal{F}) = \sup[|T_n(F, u) - v_n(F)| : u \in c(U_n(\mathcal{F}))]. \tag{3.82}$$

From (1.1c), (1.9) and (1.11) we have

$$\begin{aligned} |T_n(F, u) - v_n(F)| &= |T_n(F, u) - T_n(F, 0)| \\ &= \left|\frac{A_n + uA_{n-1}}{B_n + uB_{n-1}} - \frac{A_n}{B_n}\right| = \frac{|v_n(F) - v_{n-1}(F)|}{|h_n|\left|\frac{1}{-h_n}\right.} - \left.\frac{1}{u}\right|. \end{aligned} \tag{3.83}$$

From (3.82) and (3.83) we obtain (3.81). □

To apply Theorem 3.6 to a CF $K(a_j/b_j)$, we make use of specific information on the location of $h_n(F)$. From (1.42) and (1.10c), we have

$$h_n(F) = b_n + \frac{a_n}{h_{n-1}(F)} = b_n + \frac{a_n}{b_{n-1}} + \frac{a_{n-1}}{b_{n-2}} + \cdots + \frac{a_2}{b_1}, \quad n = 1, 2, 3, \ldots. \tag{3.84}$$

For each of the following special families of LFASs (CFs), we use information about the value regions to obtain information about the location of $\{h_n(F)\}$, and then we apply geometric arguments to determine $\inf\left[\left|\frac{-1}{h_n(F)} - \frac{1}{u}\right| : u \in c(U_n(\mathcal{F}))\right]$.

3.3.1. *Real J-Fractions $\mathcal{F}^{J(z)}$.*
Let $z \in \mathbb{C} - \mathbb{R}$ be given. The family $\mathcal{F}^{J(z)}$ of *real J-fractions* is defined by

$$\begin{aligned}\mathcal{F}^{J(z)} &:= \mathcal{F}(\Omega, W)\\ &= \left[F : F = \frac{1}{\beta_1 + z +} \frac{-\alpha_1^2}{\beta_2 + z +} \frac{-\alpha_2^2}{\beta_3 + z + \cdots}, \ 0 \neq \alpha_j \in \mathbb{R}, \beta_j \in \mathbb{R}, j \geq 1\right],\end{aligned} \tag{3.85}$$

where $\Omega = \{\Omega_j\}$ and $W = \{w_j\} = \{0\}$,

$$\Omega_1 := \langle 1, [\beta_1 + z : \beta_1 \in \mathbb{R}], 0, 1\rangle \tag{3.86a}$$

$$\Omega_j := \langle [-\alpha^2 : 0 \neq \alpha \in \mathbb{R}], [\beta + z : \beta \in \mathbb{R}], 0, 1\rangle, \ j = 2, 3, 4, \dots . \tag{3.86b}$$

The generating sequence $\{t_j^F(w)\}$ for $F \in \mathcal{F}^{J(z)}$ is given by

$$t_0^F(w) := w, \ t_1^F(w) := \frac{1}{\beta_1 + z + w}, \ t_j^F(w) := \frac{-\alpha_{j-1}^2}{\beta_j + z + w}, \ j = 2, 3, 4, \dots . \tag{3.87}$$

Theorem 3.7 *Let $z \in \mathbb{C} - \mathbb{R}$ be given and let $\mathcal{F}^{J(z)}$ denote the family of real J-fractions (3.85). Then:*

(A) *The best sequence of value regions $\{U_n(\mathcal{F}^{J(z)})\}_{n=0}^{\infty}$ with respect to $\mathcal{F}^{J(z)}$ is given by*

$$U_0(\mathcal{F}^{J(z)}) = \left[u \in \mathbb{C} : \left|u + \frac{i}{2\mathrm{Im}\,(z)}\right| \leq \frac{1}{2|\mathrm{Im}\,z|}, \ u \neq 0\right], \tag{3.88a}$$

$$U_n(\mathcal{F}^{J(z)}) = V(z) := \begin{cases} H^+ := [u \in \mathbb{C} : \mathrm{Im}\,u > 0], & \text{if } z \in H^+, \\ H^- := [u \in \mathbb{C} : \mathrm{Im}\,u < 0], & \text{if } z \in H^-, \end{cases} \quad n = 1, 2, 3, \dots . \tag{3.88b}$$

(B) *If $F = F(z) \in \mathcal{F}^{J(z)}$ converges to a finite value $v(F)$ and if n is a positive integer, then the best truncation error bound $\beta_n(F, \mathcal{F}^{J(z)})$ for $v_n(F)$ with respect to $\mathcal{F}^{J(z)}$ is given by*

$$\beta_n(F, \mathcal{F}^{J(z)}) = \frac{|h_n(F(z))|}{|\mathrm{Im} h(F(z))|} |v_n(F(z)) - v_{n-1}(F(z))|. \tag{3.89}$$

Proof. (A) follows directly from the definition of $\{U_n(\mathcal{F}^{J(z)})\}$ in (2.1) (see also Theorem 9 in [37]).

(B): We show that condition (b) of Theorem 2.5 holds and then apply Theorem 3.6. From (3.88) it is clear that

$$w_m := 0 \in c(U_m(\mathcal{F}^{J(z)}), \text{ for } m = 0, 1, 2, \dots .$$

Let k be a given positive integer. We then define $\{G_j\}$, for $j \geq 1$, by

$$G_j := \frac{1}{\beta_1(F) + z +} \frac{-\alpha_1^2(F)}{\beta_2(F) + z + \cdots +} \frac{-\alpha_{n+k-1}^2(F)}{\beta_{n+k}(F) + z +} \frac{-1}{j + z +} \frac{-1}{z} \frac{-1}{+ z} \frac{-1}{+ z} + \cdots . \tag{3.90}$$

The periodic CF

$$H := \frac{-1}{z} \underset{+}{} \frac{-1}{z} \underset{+}{} \frac{-1}{z} + \cdots \tag{3.91}$$

converges to a finite value $v(H)$ satisfying

$$\left| v(H) - \frac{i}{2\mathrm{Im}\, z} \right| \leq \frac{1}{2|\,\mathrm{Im}\, z|}. \tag{3.92}$$

It follows that each G_j converges to a value $v(G_j) \in \boldsymbol{C}$ and $|z + v(H)| > |\mathrm{Im}\, z| > 0$. Hence

$$|v(G_j^{(n+k)}| = \left| \frac{-1}{j + z + v(H)} \right| \leq \frac{1}{j - |z + v(H)|} \leq \frac{1}{j - |\mathrm{Im}\, z|}, \text{ for sufficiently large } j.$$

Therefore $\lim_{j \to \infty} v(G_j^{(n+k)}) = 0 = w_m$ for all $m \geq 1$, and so condition (b) of Theorem 2.5 holds. Next we show that $T_n(F, U_n(\mathcal{F}^{J(z)})) = T_n(F, V(z))$ is bounded. By (2.2)

$$T_n(F, V(z)) = T_{n-1}(F, t_n^F(V(z))) \subseteq T_{n-1}(F, V(z)) \subseteq \cdots \subseteq T_1(F, V(z)) \subseteq U_0(F^{J(z)}).$$

From (3.88a), $U_0(\mathcal{F}^{J(z)})$ is a bounded set. By (3.84) and (2.56) we obtain

$$-\frac{1}{h_n(F)} \in \frac{1}{-\beta_n - z - V(z)}$$

and hence

$$-\frac{1}{h_n(F)} \in \left[0 \neq w \in \boldsymbol{C} : \left| w - \frac{i}{2\mathrm{Im}\, z} \right| \leq \frac{1}{2|\mathrm{Im}\, z|} \right].$$

It follows from this and (3.81) that (3.89) holds. □

3.3.2. *Stieltjes Fractions* $\mathcal{F}^{St(z)}$.
Let $z \in S_\pi := [u \in \boldsymbol{C} : 0 \leq |\arg u| < \pi]$ be given. We define the *family* $\mathcal{F}^{St(z)}$ *of Stieltjes Fractions* by

$$\mathcal{F}^{St(z))} := \mathcal{F}(\Omega, W) = \left[F(z) : F(z) = K\left(\frac{a_j z}{1}\right),\ a_j > 0,\ j \geq 1 \right], \tag{3.93a}$$

where

$$\Omega = \{\Omega_j\}_{j=0}^{\infty}, \quad W = \{w_j\}_{j=0}^{\infty} = \{0\}, \tag{3.93b}$$

$$\Omega_0 := \langle 0, 1, 1, 0 \rangle,\ \Omega_j := \langle [a_j z : a_j > 0], 1, 0, 1 \rangle,\ j = 1, 2, 3, \ldots. \tag{3.93c}$$

The generating sequence $\{t_j^F(w)\}$ for $F = F(z) = K(a_j z/1) \in \mathcal{F}^{St(z)}$ is given by

$$t_0^F(w) := w,\ t_j^F(w) := \frac{a_j(F) z}{1 + w}, \quad j = 1, 2, 3, \ldots. \tag{3.94}$$

Theorem 3.8 *Let* $\mathcal{F}^{St(z)}$ *be the family of Stieltjes fractions* (3.93) *for a given* $z \in S_\pi$. *Then:*

(A) *The best sequence of value regions* $\{U_n(\mathcal{F}^{St(z)})\}_{m=0}^{\infty}$ *with respect to* $\mathcal{F}^{St(z)}$ *is given, for* $m \geq 0$, *by*

$$U_m(\mathcal{F}^{St(z)}) = U(z) := \begin{cases} [u \in \boldsymbol{C} : 0 < \arg u \leq \arg z], & \textit{if } 0 < \arg z < \pi, \\ [u \in \boldsymbol{C} : \arg z \leq \arg u < 0], & \textit{if } -\pi < \arg z < 0, \\ [u \in \boldsymbol{C} : \arg u = 0], & \textit{if } \arg z = 0. \end{cases} \quad (3.95)$$

(B) *Let* $F = F(z) := K(a_j z/1) \in \mathcal{F}^{St(z)}$ *converge to a finite value* $v(F)$ *and let* n *be a given positive integer. Then the best truncation error bound* $\beta_n(F, \mathcal{F}^{St(z)})$ *for* $v_n(F)$ *with respect to* $\mathcal{F}^{St(z)}$ *is given by*

$$\beta_n(F, \mathcal{F}^{St(z)}) = K_n(F(z))|v_n(F(z)) - v_{n-1}(F(z))|, \quad (3.96a)$$

where $K_n(F(z))$ *is defined as follows:*

Case a. $(0 \leq |\arg z| \leq \pi/2)$

$$K_n(F(z)) = 1, \quad \textit{if} \quad 0 \leq |\arg z| \leq \frac{\pi}{2}. \quad (3.96b)$$

Case b. *Suppose that* $\pi/2 < |\arg z| < \pi$.

$\mathbf{b}_1$*:*

$$K_n(F(z)) = \frac{|h_n(F)|}{|\mathrm{Im}\,(h_n(F))|}, \ \textit{if } 0 < \left|\arg\left(\frac{-1}{h_n(F)}\right)\right| \leq \frac{\pi}{2}. \quad (3.96c)$$

$\mathbf{b}_2$*:*

$$K_n(F(z)) = 1, \ \textit{if } \frac{\pi}{2} < \left|\arg\left(-\frac{1}{h_n(F)}\right)\right| \leq \frac{3\pi}{2} - |\arg z|. \quad (3.96d)$$

$\mathbf{b}_3$*:*

$$K_n(F(z)) = |\csc\beta| \ \textit{if } \frac{3\pi}{2} - |\arg z| < \left|\arg\left(-\frac{1}{h_n(F)}\right)\right| < \pi, \quad (3.96e)$$

where

$$\beta := 2\pi - |\arg z| - \left|\arg\left(-\frac{1}{h_n(F)}\right)\right|. \quad (3.96f)$$

Proof. (A): Suppose $0 < \arg z < \pi$. Let $\{V_j\}$ be defined by (see (3.95))

$$V_j := V := V(z) := [u \in \boldsymbol{C} : 0 < \arg u \leq \arg z], \ j = 0, 1, 2, \ldots .$$

Let

$$H_1 := [u \in \boldsymbol{C} : \mathrm{Im}\, u > 0], \ H_2 := [u \in \boldsymbol{C} : \arg z - \pi \leq \arg u \leq \arg z],$$

so that $V = H_1 \cap H_2$. It follows that

$$t_j^G(H_1) \subseteq H_2 \text{ and } t_j^G(H_2) \subseteq H_1 \text{ for all } G \in \mathcal{F}^{St(z)}, \ j = 1, 2, 3, \ldots,$$

$$t_j^G(V) = t_j^G(H_1) \cap t_j^G(H_2) \subseteq H_1 \cap H_2 = V, \ j = 0, 1, 3, \ldots,$$

and

$$t_j^G(0) = a_j(G) z \in V, \quad j = 1, 2, 3, \ldots .$$

We have shown that $\{V_j\}_{j=0}^{\infty}$ is a sequence of value regions with respect to $\mathcal{F}^{St(z)}$. By an elementary geometrical argument one can show that

$$V = \left[\frac{a_1 z}{1 + a_2 z} : a_1 > 0,\ a_2 > 0\right] = [t_0^G \circ t_1^G \circ t_2^G(0) : G \in \mathcal{F}^{St(z)}].$$

Therefore $V = V_j = U_j(\mathcal{F}^{St(z)})$, $j = 0, 1, 2, \ldots$. A similar argument holds for $-\pi < \arg z < 0$ and for $\arg z = 0$. This proves (A).

(B) To apply Theorem 3.6, we verify that condition (b) of Theorem 2.5 holds. Let k be a given positive integer. Let $\{G_j\}_{j=1}^{\infty}$ be defined by

$$G_j := \frac{a_1(F)z}{1} + \frac{a_2(F)z}{1} + \cdots + \frac{a_{n+k}(F)z}{1} + \frac{(1/j)z}{1} + \frac{z}{1} + \frac{z}{1} + \frac{z}{1} + \cdots.$$

It follows that

$$G_j \in \mathcal{F}_{n+k}(F) \text{ for } j \geq 1 \text{ and } w_{n+k} = 0 = \lim_{j \to \infty} v(G_j^{(n+k)}).$$

Moreover, since $w_m = 0 \in c(U_m(\mathcal{F}^{St(z)}))$ for $m \geq n+1$, condition (b) of Theorem 2.5 holds. We also note that $T_n(F, U(z))$ is a bounded set, since

$$T_n(F, U(z)) \subseteq T_{n-1}(F, U(z)) \subseteq \cdots \subseteq T_1(F, U(z)) = t_1^F(U(z)),$$

and the set $t_1^F(U(z))$ is the intersection of a circular disk and a half-plane, provided $0 < |\arg z| < \pi$.

If $\arg z = 0$, then $z > 0$ and

$$t_1^F(U(z)) = \left[\frac{a_1 z}{1+u} : 0 < u < \infty\right] = [x \in \mathbb{R}^+ : 0 < x < a_1 z]$$

is bounded. Therefore by Theorem 3.6, $\beta_n(F, \mathcal{F}^{St(z)})$ is given by (3.81). It remains to find estimates for $\inf\left[\left(\frac{-1}{h_n(F)} - \frac{1}{u}\right) : u \in U(z)\right]$. By (3.84)

$$h_n(F) = 1 + \frac{a_n z}{1} + \frac{a_{n-1} z}{1} + \cdots + \frac{a_2 z}{1}.$$

It follows from this and $t_j^G(U(z)) \subseteq U(z)$ for $j \geq 1$ and $G \in \mathcal{F}^{St(z)}$, that

$$h_n(F) \in 1 + U(z) \text{ and so } \frac{-1}{h_n(F)} \in \frac{-1}{1 + U(z)}.$$

We consider cases for which $0 < \arg z < \pi$. (Similar arguments hold for $-\pi < \arg z < 0$ and $\arg z = 0$ and hence they are omitted.) One can readily show that

$$\frac{1}{U(z)} = [u \in \mathbb{C} : \arg z \leq \arg u < 0]$$

and that $-1/(1 + U(z))$ is a region in $\mathbb{C}$ bounded by the interval $-1 < u < 0$ and by the circular arc passing through -1 and 0, tangent at $u = -1$ to the line with angle of inclination equal to $\arg z$.

Case a. If $0 < \arg z \le \pi/2$, then

$$\inf\left[\left|\left(\frac{-1}{h_n(F)}\right) - \frac{1}{u}\right| : u \in U(z)\right] = \frac{1}{|h_n(F)|}$$

and hence (3.96b) follows from (3.81).

Case b. Suppose that $\pi/2 < \arg z < \pi$.

$\mathbf{b}_1$: If $0 < \left|\arg\left(\frac{-1}{h_n(F)}\right)\right| \le \frac{\pi}{2}$, then

$$\inf\left[\left|\left(\frac{-1}{h_n(F)}\right) - \frac{1}{u}\right| : u \in U(z)\right] = \left|\mathrm{Im}\left(\frac{-1}{h_n(F)}\right)\right| = \frac{|\mathrm{Im}\, h_n(F)|}{|h_n(F)|^2}.$$

Hence (3.96c) follows from (3.81).

$\mathbf{b}_2$: If $\frac{\pi}{2} < \left|\arg\left(\frac{-1}{h_n(F)}\right)\right| < \frac{3\pi}{2} - \arg z$, then

$$\inf\left[\left|\left(\frac{-1}{h_n(F)}\right) - \frac{1}{u}\right| : u \in U(z)\right] = \frac{1}{|h_n((F)|},$$

and so (3.96d) follows from (3.81).

$\mathbf{b}_3$: If $\frac{3\pi}{2} - \arg z < \arg\left(\frac{-1}{h_n(F)}\right) < \pi$, then

$$\inf\left[\left|\left(\frac{-1}{h_n(F)}\right) - \frac{1}{u}\right| : u \in U(z)\right] = \frac{\left|\sin\left[2\pi - \arg z - \arg\left(\frac{-1}{h_n(F)}\right)\right]\right|}{|h_n(F)|}$$

and hence (3.96e) follows from (3.81). For the computations used to obtain b_2 and b_3, we have used the fact that the ray $\arg u = \frac{3\pi}{2} - \arg z$ is perpendicular to the line passing through the ray $\arg u = -\arg z$. □

We state without proof the following useful result originally given by Henrici and Pfluger [21] (see also Theorem 4.4 in [39]).

Theorem 3.9 *([21]; Theorem 4.4 in [39]). If $F(z) = K(a_j z/1)$, $a_j > 0$, $j \ge 1$ is an S-fraction converging to a finite value $v(F(z))$, then, for $n \ge 2$,*

$$|v(F(z)) - v_n(F(z))| \le \begin{cases} |v_n(F(z)) - v_{n-1}(F(z))|, & \text{if } |\arg z| \le \pi/2, \\ \csc(|\arg z|)|v_n(F(z)) - v_{n-1}(F(z))|, & \text{if } \pi/2 < |\arg z| < \pi. \end{cases}$$

3.3.3. *Adjusted Stieltjes Fractions* $\mathcal{F}^{ASt(z)}$.

Let $z \in \boldsymbol{C}$ be given with $|\mathrm{Arg} z| < \pi/2$. We define

$$\mathcal{F}^{ASt(z)} := (\Omega, \{0\}_{n=1}^{\infty}), \tag{3.97}$$

where

$$\Omega = \{\Omega_n\}_{n=1}^{\infty} := \{\langle 1, [u \in \boldsymbol{C} : \arg u = \arg z], 0, 1\rangle\}_{n=1}^{\infty}. \tag{3.98}$$

Then

$$\mathcal{F}^{ASt(z)} = \left[\mathop{\mathrm{K}}_{j=1}^{\infty} \left(\frac{1}{b_j z} \right) : b_j > 0 \right] \tag{3.99}$$

is called the family of *Adjusted Stieltjes fractions.*

Theorem 3.10 *Let $\mathcal{F}^{ASt(z)}$ be the family of LFASs defined by* (3.97)–(3.99). *Then*
(A) *The best sequence of value regions $\{U_j(\mathcal{F}^{ASt(z)})\}_{n=0}^{\infty}$ with respect to $\mathcal{F}^{ASt(z)}$ satisfies*

$$c(U_j(\mathcal{F}^{ASt(z)})) := c(U(\mathcal{F}^{ASt(z)})) = V := [u \in \boldsymbol{C} : 0 < |Arg\, u| \leq |Arg\, z|], \quad j = 0, 1, 2, \ldots . \tag{3.100}$$

(B) *If $F = F(z) \in \mathcal{F}^{ASt(z)}$ converges to a finite value $v(F(z)) \in \boldsymbol{C}$ and if n is a positive integer, then the best truncation error bound for $v_n(F(z))$ with respect to $\mathcal{F}^{ASt(z)}$ is given by*

$$\beta_n(F, \mathcal{F}^{ASt(z)}) = K_n(F(z))|v_n(F) - v_{n-1}(F)|, \tag{3.101a}$$

where

$$K_n(F(z)) = \begin{cases} 1, \ \text{if} \ |\arg z| \leq \left|\arg\left(\frac{-1}{h_n(F)}\right)\right| - \frac{\pi}{2}, \\ \csc\left(\left|\arg\left(\frac{-1}{h_n(F)}\right)\right| - |\arg z|\right), \ \text{if} \ \left|\arg\left(\frac{-1}{h_n(F)}\right)\right| - \frac{\pi}{2} < |\arg z|. \end{cases} \tag{3.101b}$$

A proof of this theorem can be given that is very similar to that of Theorem 3.8 and hence it is omitted.

3.3.4. *Positive T-Fractions:* $\mathcal{F}^{T(z)}$.
Let $z \in S_\pi := [u \in \boldsymbol{C} : 0 \leq |\arg z| < \pi]$ be given. We define

$$\mathcal{F}^{T(z)} := \mathcal{F}(\Omega, \{0\}_{n=0}^{\infty}), \tag{3.102}$$

where

$$\Omega = \{\Omega_n\}_{n=1}^{\infty}, \quad \Omega_n := \langle [F_n z : F_n > 0], [1 + G_n z : G_n > 0], 0, 1, \rangle \tag{3.103}$$

for $n = 1, 2, 3, \ldots$. Then

$$\mathcal{F}^{T(z)} = \left[F : F = \frac{F_1 z}{1 + G_1 z} + \frac{F_2 z}{1 + G_2 z} + \cdots, F_n, G_n > 0, \ n = 1, 2, 3, \ldots \right] \tag{3.104}$$

is called the family of *Positive T-fractions.*

Theorem 3.11 *Let $\mathcal{F}^{T(z)}$ be the family of LFASs defined by* (3.102)–(3.104). *Then:*
(A) *The best sequence of value regions $\{U_n(\mathcal{F}^{T(z)})\}_{n=0}^{\infty}$ with respect to $\mathcal{F}^{F(z)}$ satisfies*

$$U_n(\mathcal{F}^{T(z)}) := U(\mathcal{F}^{T(z)}) := \begin{cases} [u : 0 < \arg u < \arg z] \ \text{if} \ \arg z > 0, \\ [u : \arg z < \arg u < 0] \ \text{if} \ \arg z < 0, \\ [u : \arg u = 0], \ \text{if} \ \arg z = 0. \end{cases} \tag{3.105}$$

(B) *If* $F(z) \in \mathcal{F}^{T(z)}$ *converges to a finite value* $v(F(z)) \in \mathbb{C}$ *and if* n *is a positive integer, then the best truncation error bound for* $v_n(F(z))$ *with respect to* $\mathcal{F}^{T(z)}$ *is given by*

$$\beta_n(F, \mathcal{F}^{T(z)}) = K_n(F(z))|v_n(F) - v_{n-1}(F)|, \tag{3.106a}$$

where $K_n(F(z))$ *is defined as follows:*
Case a. $(0 \leq |\arg z| \leq \pi/2)$

$$K_n(F(z)) := 1, \quad \text{if} \quad 0 \leq |\arg z| \leq \frac{\pi}{2}. \tag{3.106b}$$

Case b. *Suppose that* $\pi/2 < |\arg z| < \pi$.
$\mathbf{b}_1$*:*

$$K_n(F(z)) := \frac{|h_n(z)|}{|\operatorname{Im} h_n(F)|}, \ \text{if}\, 0 < \left|\arg\left(\frac{-1}{h_n(F)}\right)\right| \leq \frac{\pi}{2}. \tag{3.106c}$$

$\mathbf{b}_2$*:*

$$K_n(F(z)) := 1, \ \text{if}\, \frac{\pi}{2} < \left|\arg\left(\frac{-1}{h_n(F)}\right)\right| \leq \frac{3\pi}{2} - |\arg z|. \tag{3.106d}$$

$\mathbf{B}_3$*:*

$$K_n(F(z)) := |\csc \beta|, \ \text{if}\, \frac{3\pi}{2} - |\arg z| < \left|\arg\left(\frac{-1}{h_n(F)}\right)\right|, \tag{3.106e}$$

where

$$\beta := 2\pi - |\arg z| - \left|\arg\left(\frac{-1}{h_n(F)}\right)\right|. \tag{3.106f}$$

Proof. (A): Suppose $0 < \arg z < \pi$. Let $\{V_n\}$ denote the (constant) sequence of sets

$$V_n := V := [u : 0 < \arg u < \arg z], \ n = 0, 1, 2, \dots .$$

Suppose $F(z) \in \mathcal{F}^{T(z)}$. Then $t_n^{F(z)}(V) = \dfrac{F_n z}{1 + G_1 z + V}$ for some $F_n > 0$. Let $H_1 := [u : \operatorname{Im} u > 0]$ and $H_2 := [u : \arg z - \pi < \arg u < \arg z]$. One can show that $t_n^{F(z)}(H_1) \subseteq H_1$ and $t_n^{F(z)}(H_2) \subseteq H_2$. It follows that $t_n^{F(z)}(V) = t_n^{F(z)}(H_1 \cap H_2) = T_n^{F(z)}(H_1) \cap T_n^{F(z)}(H_2) \subseteq H_1 \cap H_2 = V$. Furthermore, every point in V can be expressed as $\dfrac{F_1 z}{1 + G_1 z}$ for some $F_1, G_1 > 0$. Therefore $V \subseteq c(U_n(\mathcal{F}^{T(z)})$ and by (2.4) $U_n(\mathcal{F}^{T(z)}) \subseteq V$. A similar argument holds when $-\pi < \arg z < 0$.

(B) Let $k \geq 1$ be given. Define $\{G_j\}_{j=1}^{\infty}$ by

$$G_j := \frac{F_1 z}{1 + G_1 z} + \frac{F_2 z}{1 + G_2 z} + \dots + \frac{F_{n+k} z}{1 + G_{n+k} z} + \frac{z}{(1/j) + z} + \frac{z}{1 + z} + \frac{z}{1 + z} + \frac{z}{1 + z} + \cdots .$$

An argument analogous to one in the proof of Theorem 3.7 shows $\{G_j\}_{j=1}^{\infty}$ satisfies the hypothesis of Theorem 2.5. By (3.84)

$$h_n(F(z)) = 1 + G_n z + \frac{F_n z}{1 + G_{n-1} z} + \dots + \frac{F_2 z}{1 + G_1 z}, \ n = 2, 3, 4, \dots .$$

By (2.5b), $\frac{F_n z}{1+G_{n-1}z+U(\mathcal{F}^{T(z)})} \subseteq U(\mathcal{F}^{T(z)})$, whenever $F_n, G_{n-1} > 0$ and thus $h_n(F(z)) \in 1+G_n z+U(\mathcal{F}^{T(z)})$ and hence $\frac{-1}{h_n(F(z))} \in \frac{1}{-1-G_n z-U(\mathcal{F}^{T(z)})}$. The remainder of our proof is similar to that given for Theorem 3.8(B) and hence is omitted. □

We conclude this section by stating the following result. Jefferson [30] gave this result for the special case with $G_n = 1$, $n \geq 1$. The general form given here was proved by Gragg [18].

Theorem 3.12 *If a positive T-fraction*

$$F = \mathop{\mathrm{K}}_{n=1}^{\infty} \left(\frac{F_n z}{1+G_n z}\right) = \frac{F_1 z}{1+G_1 z} + \frac{F_2 z}{1+G_2 z} + \frac{F_3 z}{1+G_3 z} + \cdots, \quad F_n, G_n > 0$$

converges to a finite value $v(F)$, then for $n \geq 2$,

$$|v(F) - v_n(F)| \leq K_n(F(z))|v_n(F) - v_{n-1}(F)|,$$

where

$$K_n(F(z)) = \begin{cases} 1, & \text{if } 0 \leq |\arg z| \leq \pi/2, \\ |\csc(\arg z)|, & \text{if } \pi/2 < |\arg z| < \pi. \end{cases}$$

4. Best Truncation Error Bounds for Limit k-Periodic MCFs

Many modified continued fraction (MCF) expansions of special functions have the form

$$\mathop{\mathrm{K}}_{j=1}^{\infty} (a_j(z), b_j(z), w_j(z)),$$

where the elements $a_j(z), b_j(z)$ and converging factors $w_j(z)$ are complex-valued functions of a complex variable z. The MCF is called *periodic with period k* if $a_{rk+m}(z) = a_m(z)$, $b_{rk+m}(z) = b_m(z)$, $w_{rk+m}(z) = w_m(z)$ for $m \geq 1$ and $r \geq 0$. The MCF is called *limit k-periodic* if, for $m = 1, 2, 3, \ldots$,

$$\lim_{r\to\infty} a_{rk+m}(z) = \alpha_m(z), \quad \lim_{r\to\infty} b_{rk+m}(z) = \beta_m(z) \text{ and } \lim_{r\to\infty} w_{rk+m}(z) = \omega_m(z).$$

An MCF is called *limit periodic* if it is limit 1-periodic. Section 4.1 deals with limit periodic MCFs

$$\mathop{\mathrm{K}}_{j=1}^{\infty} (a_j(z), 1, w_j(z)), \text{ where } \lim_{j\to\infty} a_j(z) = \alpha(z), \ \lim_{j\to\infty} w_j(z) = \omega(z),$$

where $\alpha(z) \in \mathbb{C} - (-\infty, -1/4]$ in Section 4.1.1 and $\alpha(z) = \infty$ in Section 4.1.2. Section 4.2 deals with MCFs

$$\mathop{\mathrm{K}}_{j=1}^{\infty} (1, b_j(z), w_j(z)),$$

where $\lim_{j\to\infty} b_j(z) = \infty$ in Section 4.2.1 and where the MCF is limit 4-periodic in Section 4.2.2.

4.1. Limit Periodic CFs $K(a_j/1)$ and MCFs $K(a_j, 1, w_j)$

Our interest in this section is in best truncation error bounds for continued fractions (CFs)

$$\operatorname*{K}_{j=1}^{\infty}\left(\frac{a_j}{1}\right) = \frac{a_1}{1} \, {}_{+} \, \frac{a_2}{1} \, {}_{+} \, \frac{a_3}{1} \, {}_{+} \cdots \tag{4.1}$$

and modified continued fractions (MCFs) $K(a_j, 1, w_j)$ whose elements a_j satisfy a limit-periodic condition of the form

$$\lim_{j\to\infty} a_j = a \in \widehat{\boldsymbol{C}} = \boldsymbol{C} \cup [\infty]. \tag{4.2}$$

Most of the results in this section apply to the case in which

$$\lim_{j\to\infty} a_j = a \in \boldsymbol{C} - (-\infty, -1/4]. \tag{4.3}$$

By using the parabola theorem of [53] one can readily prove:

Theorem 4.1 *If the elements a_j of a CF $F = K(a_j/1)$ satisfy (4.3), then F converges to a value*

$$v(F) = \lim_{n\to\infty} v_n(F) \in \widehat{\boldsymbol{C}} = \boldsymbol{C} \cup [\infty]. \tag{4.4}$$

If a number a satisfies (4.3), then the fixed points x_1 and x_2 of the transformation

$$t(w) = \frac{a}{1+w} \tag{4.5}$$

are given by

$$x_1 = \sqrt{a + \frac{1}{4}} - \frac{1}{2}, \; x_2 = -\sqrt{a + \frac{1}{4}} - \frac{1}{2}, \; (\mathrm{Re}\sqrt{} > 0), \tag{4.6a}$$

and they satisfy

$$|x_1| < |x_2|, \; x_2 = -(x_1 + 1), \; a = -x_1 x_2. \tag{4.6b}$$

The periodic CF $K(a/1)$ converges to the attractive fixed point

$$x_1 = v(K(a/1)), \text{ where } K\left(\frac{a}{1}\right) = \frac{a}{1} \, {}_{+} \, \frac{a}{1} \, {}_{+} \, \frac{a}{1} \, {}_{+} \cdots, \tag{4.7}$$

(see, e.g., [37], Theorem 3.2). If $F = K(a_j/1)$ is a limit-periodic CF satisfying (4.3), then for the n-th tail $F^{(n)}$ of F, we have

$$\lim_{n\to\infty} v(F^{(n)}) = v(K(a/1)) = v(F) = x_1. \tag{4.8}$$

(see, e.g., [37], pp. 113–114). With $F = K(a_j/1)$ and $\{T_n(F, w)\}$ defined by (1.1e) and with generating sequence $\{t_j^F(w)\}$ given by

$$t_0^F(w) := w, \; t_j^F(w) := \frac{a_j}{1+w}, \quad j = 1, 2, 3, \ldots, \tag{4.9}$$

we obtain (see (1.38))

$$v(F) = T_n(F, v(F^{(n)})), \quad n = 1, 2, 3, \dots . \tag{4.10}$$

Equations (4.8) and (4.10) provide motivation for considering MCFs $K(a_j, 1, w_j)$ with converging factors

$$w_j = x_1 = v(K(a/1)), \quad j = 1, 2, 3, \dots . \tag{4.11}$$

Theorem 4.2 *Lemma 2.1 in [5]. If the elements a_j of $F = K(a_j/1)$ satisfy (4.3), then the critical tail sequence $\{-h_n(F)\}$ (see (1.42)) satisfies*

$$\lim_{n\to\infty} (-h_n(F)) = x_1 \textit{ if } v(F) = \infty \tag{4.12a}$$

and

$$\lim_{n\to\infty} (-h_n(F)) = x_2 = -(x_1 + 1), \textit{ if } v(F) \in \boldsymbol{C}. \tag{4.12b}$$

We now consider families $\mathcal{F} = \mathcal{F}(\Omega, W)$ of LFASs defined in the following:

4.1.1. $K(a_j, 1, x_1)$, $a_j \to a \in \boldsymbol{C} - (-\infty, -1/4]$.
Let $a, k, \{a_j\}_{j=1}^k$ and $\{\rho_j\}_{j=k}^\infty$ satisfy

$$a \in \boldsymbol{C} - (-\infty, -1/4],\ 0 \leq k \in \boldsymbol{Z},\ 0 \neq a_j \in \boldsymbol{C},\ j = 1, 2, \dots, k, \tag{4.13a}$$

$$0 < \rho_j < |x_2|, \text{ for } j = k, k+1, k+2, \dots, \tag{4.13b}$$

and

$$\rho_j \rho_{j-1} < \rho_{j-1}|x_2| - \rho_j |x_1|, \text{ for } j = k+1, k+2, k+3, \dots, \tag{4.13c}$$

where x_1 and x_2 are defined by (4.6a). Let $\Omega = \{\Omega_j\}$ and $W = \{w_j\}$ be defined by

$$\Omega_0 := [\langle 0, 1, 1, 0 \rangle], \tag{4.14a}$$

$$\Omega_j := \langle E_j, 1, 0, 1 \rangle \text{ and } w_j = x_1,\ j = 1, 2, 3, \dots, \tag{4.14b}$$

where

$$E_j := \begin{cases} [a_j], \quad j = 1, 2, \dots, k, \\ [u \in \boldsymbol{C} : |u(1 + \bar{x}_1) - x_1(|x_2|^2 - \rho_j^2)| + \rho_j |u| \\ \leq \rho_{j-1}(|x_2|^2 - \rho_j^2)],\ j \geq k+1. \end{cases} \tag{4.14c}$$

We define a family $\mathcal{F} := \mathcal{F}(\Omega, W)$ of LFASs by

$$\begin{aligned} \mathcal{F} &:= \mathcal{F}(\Omega, W) := \mathcal{F}(a, 1, k, \{a_j\}_1^k, \{\rho_j\}_k^\infty) \\ &:= [F = K(a_j, 1, x_1) : \{a_j\} \text{ satisfies (4.3) and } 0 \neq a_j \in E_j,\ j \geq 1]. \end{aligned} \tag{4.15}$$

Condition (4.13c) implies that

$$a \in E_j \text{ and hence } x_1 = \frac{a}{1 + x_1} \in U_j(\mathcal{F}) \text{ for } j \geq k+1.$$

If $a \neq 0$, the element set E_j, for $j \geq k+1$, in (4.14c) is a closed, bounded, convex subset of $\boldsymbol{C}$ with an axis of symmetry given by the line passing through the ray $\arg u = \arg a$. The boundary ∂E_j of E_j (for $j \geq k+1$) is called a *Cartesian oval.*

If $a = 0$, then $x_1 = 0$, $x_2 = -1$ and E_j (for $j \geq k+1$) in (4.14c) reduces to the circular region

$$E_j = [u \in \boldsymbol{C} : |u| \leq \rho_{j-1}(1-\rho_j)], \quad j \geq k+1. \tag{4.16}$$

A sequence of value regions $\{V_n\}$ with respect to $\mathcal{F}(\Omega, W)$ is given by

$$V_n := \begin{cases} [u \in \boldsymbol{C} : |u - x_1| \leq \rho_n], & n = k, k+1, k+2, \ldots, \\ \frac{a_{n+1}}{1+(V_{n+1} \cup [x_1])}, & n = k-1, k-2, \ldots, 1, 0. \end{cases} \tag{4.17}$$

For $n \geq k$, the value-region-defining conditions (2.5) can be verified by using the $VW\Omega$-method described in Section 3.1. For $0 \leq n \leq k-1$, conditions (2.5) follow directly from (4.17). The following result was proven in [5], Theorems 2.2 and 2.4.

Theorem 4.3 *Let $\mathcal{F} = \mathcal{F}(\Omega, W)$ be a family of LFASs of the form* (4.15) *and let $F = K(a_j, 1, x_1) \in \mathcal{F}$ be given. Then:*

(A) *$F = K(a_j, 1, x_1)$ and $K(a_j/1)$ converge to the same value $v(F) \in \widehat{\boldsymbol{C}} = \boldsymbol{C} \cup [\infty]$.*

(B) *If there exists an integer $k_0 \geq k$ such that*

$$|h_{k_0}(F) + x_1| > \rho_{k_0}, \tag{4.18}$$

then

$$|h_n(F) + x_1| > \rho_n, \quad n = k_0+1, k_0+2, k_0+3, \ldots, \tag{4.19}$$

and $F = K(a_j, 1, x_1)$ and $K(a_j/1)$ converge to the same finite value $v(F) \in \boldsymbol{C}$.

(C) *If $\lim_{j\to\infty} \rho_j = 0$ and $K(a_j/1)$ converges to a finite value f, then there exists an integer $k_0 \geq k$ such that* (4.18) *holds. Hence $f = v(F) \in \boldsymbol{C}$.*

The results in our next theorem were proven in [5], Theorems 3.1 and 4.1.

Theorem 4.4 *Let $\mathcal{F} = \mathcal{F}(\Omega, W) = \mathcal{F}(a, 1, k, \{a_j\}_1^k, \{\rho_j\}_k^\infty)$ be a family of LFASs* (4.15) *and let $F = K(a_j, 1, x_1) \in \mathcal{F}$ be given. Then:*

(A) *If there exists an integer $k_0 \geq k$ such that*

$$|h_{k_0}(F) + x_1| > \rho_{k_0} \quad (i.e., \ -h_{k_0}(F) \notin V_{k_0}), \tag{4.20}$$

then F converges to a finite value $v(F) \in \boldsymbol{C}$ and, for all $n \geq k_0+1$,

$$\begin{aligned} |v(F) - v_n(F)| &\leq \frac{\rho_n \prod_{j=1}^n |a_j(F)|}{|B_{n-1}(F)|^2 |h_n(F) + x_1| (|h_n(F) + x_1| - \rho_n)} \\ &= \frac{\rho_n |h_n(F)|}{|h_n(F) + x_1| (|h_n(F) + x_1| - \rho_n)} \cdot |v_n(F) - v_{n-1}(F)|. \end{aligned} \tag{4.21}$$

(B) *If $K(a_j/1)$ converges to a finite value $v(K(a_j/1))$ and if $\lim_{j\to\infty}\rho_j = 0$, then there exists an integer $k_0 \geq k$ such that* (4.20) *holds and hence* (4.21) *holds for $n \geq k_0+1$:*

(C) *Let $\{\rho_j\}$ satisfy the following additional conditions for all $j \geq k+1$:*

(a) *If $a \in \mathbb{C} - (-\infty, 0]$ and $\alpha := \arg a$, then*

$$\rho_{j-1}|x_2| - \rho_j|x_1| < \sqrt{|a|}\cos(\alpha/2), \tag{4.22a}$$

and

$$\rho_{j-1} \leq \frac{1}{2}\cos\left(\frac{\alpha}{2}\right) + \mathrm{Re}\,(x_1 e^{-i\alpha/2}) = \mathrm{Re}\,(\sqrt{a+\frac{1}{4}}\; e^{-i\alpha/2}). \tag{4.22b}$$

(b) *If $-\frac{1}{4} < a \leq 0$, then*

$$(\rho_{j-1} + |x_1|)(|x_2| - \rho_j) \leq \frac{1}{4} \text{ and } \rho_{j-1} \leq \sqrt{a+\frac{1}{4}} = x_1 + \frac{1}{2}. \tag{4.23}$$

If there exists an integer $k_0 \geq k$ such that (4.20) *holds, then the truncation error bound in* (4.21) *is the best bound $\beta_n(F, \mathcal{F})$ for $v_n(F)$ with respect to $\mathcal{F}$ for $n \geq k_0+1$.*

Proof. (A): We make use of Theorem 2.6 with $w_m = x_1$. Condition (4.20) implies that $-h_{k_0}(F) \notin V_{k_0}$. Therefore since $T_{k_0}(F, -h_{k_0}(F)) = \infty$, the set $T_{k_0}(F, V_{k_0})$ is a closed, bounded circular disk. Hence the nestedness of the sequence $\{T_n(F, V_n)\}_{k_0}^{\infty}$ implies that $T_n(F, V_n)$ is a closed, bounded disk for all $n \geq k_0$. By Theorem 4.1, F converges to a finite value $v(F)$. Assertion (4.21) is then an immediate consequence of Theorem 2.6(A).

(B): If $K(a_j/1)$ converges to a finite value $v(K(a_j/1))$, it follows from Lemma 4.2 that

$$\lim_{n\to\infty} h_n(F) = -x_2 = x_1 + 1.$$

Thus if $\lim_{j\to\infty}\rho_j = 0$, there exists an integer $k_0 \geq k$ such that (4.20) holds, and hence by (A), (4.21) holds for $n \geq k_0+1$.

(C): It was shown in [4], Theorem 3.1. that, subject to the additional conditions (4.22) and/or (4.23),

$$c(U_n(\mathcal{F})) = V_n := [u \in \mathbb{C} : |u| \leq \rho_n], \; n = k, k+1, k+2, \dots . \tag{4.24}$$

Now suppose $a \neq 0$. Then $a \in E_n$ for all $n \geq k_0+1$ and so

$$x_1 = \frac{a}{1+x_1} \in U_n(\mathcal{F}), \text{ for } n = k_0+1, k_0+2, k_0+3, \dots . \tag{4.25}$$

Hence assertion (C) follows from Theorem 2.6(B) since (4.25) implies condition (a) of Theorem 2.5. *On the other hand, if* $a = 0$, then assertion (C) follows from Theorem 2.6(B), since condition (b) of Theorem 2.5 holds. □

Remark. If V_n satisfies (4.24) for $n \geq k$, then the V_n defined by (4.17), for $0 \leq n \leq k-1$, also satisfies

$$V_n = c(U_n(\mathcal{F})), \quad 0 \leq n \leq k-1. \tag{4.26}$$

We state as a corollary of Theorem 4.4 the result obtained when the parameter $a = 0$ and the element sets E_j are circular disks given by (4.16).

Theorem 4.5 *Let $\mathcal{F} = \mathcal{F}(\Omega, W) = \mathcal{F}(0, 1, k, \{a_j\}_1^k, \{\rho_j\}_k^\infty)$ be a family of LFASs (4.15), with $a = 0$. Let $F = K(a_j, 1, 0) \in \mathcal{F}$ be given. Let $\{\rho_j\}_{j=k}^\infty$ satisfy*

$$0 < \rho_j \leq \frac{1}{2}, \quad j = k, k+1, k+2, \dots . \tag{4.27}$$

Then: (A) *If there exists an integer $k_0 \geq k$ such that*

$$|h_{k_0}(F)| > \rho_{k_0}, \quad (i.e.,\ h_{k_0}(F) \notin V_{k_0}), \tag{4.28}$$

then F converges to a finite value $v(F) \in \boldsymbol{C}$ and, for all $n \geq k_0 + 1$, the best truncation error bound for $v_n(F)$ with respect to $\mathcal{F}$ is given by

$$\beta_n(F, \mathcal{F}) = \frac{\rho_n \prod_{j=1}^n a_j(F)}{|B_{n-1}(F)|^2 |h_n(F)|(|h_n(F)| - \rho_n)} = \frac{\rho_n}{|h_n(F)| - \rho_n} \cdot |v_n(F) - v_{n-1}(F)|. \tag{4.29}$$

(B) *If $K(a_j/1)$ converges to a finite value $v(K(a_j/1)) = v(F)$ and if $\lim_{j\to\infty} \rho_j = 0$, then there exists an integer $k_0 \geq k$ such that* (4.28) *holds and hence* (4.29) *holds for all $n \geq k_0 + 1$.*

Remark. Corollary 4.1 is an improvement of [3, Theorem 3.2].
Proof. It follows from $a = x_1 = 0$, $x_2 = -1$ and (4.27) that conditions (4.13b,c) and (4.23) hold. The corollary is therefore an immediate consequence of Theorem 4.4. □

4.1.2. *CFs $K(a_j/1)$ and MCFs $K(a_j, 1, w_j)$ with $\lim_{j\to\infty} a_j = \infty$.*
We conclude this section by stating a result (Theorem 4.6) for CFs $K(a_j/1)$ and MCFs $K(a_j, 1, w_j)$ for which

$$\lim_{j\to\infty} a_j = \infty. \tag{4.30}$$

A proof of this result can be found in [24] (see, also [23]). Use is made of the following terminology.

$$P_\alpha := [u \in \boldsymbol{C} : |u| - \operatorname{Re}(ue^{-i2\alpha}) \leq \frac{1}{2}\cos^2\alpha], \text{ for } -\frac{\pi}{2} < \alpha < \frac{\pi}{2}. \tag{4.31}$$

P_α is a region bounded by a parabola ∂P_α with focus at the origin $u = 0$, axis of symmetry along the ray $\arg u = 2\alpha$, and ∂P_α passes through $u = -1/4$. For a sequence $\{c_n\}$ and ρ satisfying

$$c_n \in \boldsymbol{C},\ |c_{n-1}| \leq |1 + c_n|,\ 0 < \rho < |1 + c_n|, \text{ for all } n = 1, 2, 3, \dots, \tag{4.32a}$$

we define $E_n(\{c_j\}, \rho)$, for $n \geq 1$, by

$$E_n(\{c_j\}, \rho) := [u \in \boldsymbol{C} : |u(1 + \bar{c}_n) - c_{n-1}(|1 + c_n|^2 - \rho^2)| + \rho|u| \leq \rho(|1 + c_n|^2 - \rho^2)]. \tag{4.32b}$$

The boundary $\partial E_n(\{c_j\}, \rho)$ is a Cartesian oval (see remark following (4.15)).

Theorem 4.6 *Let α, ρ and R satisfy*

$$-\frac{\pi}{2} < \alpha < \frac{\pi}{2} \quad \text{and} \quad 0 < R < \rho \cos\alpha. \tag{4.33}$$

Let $G = K(a_j/1)$ be a CF whose elements a_j satisfy the conditions (4.30),

$$a_j \in P_\alpha, \qquad j = 1, 2, 3, \ldots, \tag{4.34}$$

and the limit points of $\{a_{j+1} - a_j\}$ all lie in the disk

$$D(\alpha, \rho, R) := [u \in \boldsymbol{C} : |u - 2\rho^2 e^{i2\alpha}| \leq 2R]. \tag{4.35}$$

Then: (A) *$G = K(a_j/1)$ converges to a value*

$$v(G) = \lim_{n\to\infty} v_n(G) \in \widehat{\boldsymbol{C}} = \boldsymbol{C} \cup [\infty]. \tag{4.36}$$

(B) *Let $F = K(a_j, 1, w_j)$ be the MCF whose converging factors w_j are given by*

$$w_j := \sqrt{a_{j+1} + \frac{1}{4}} - \frac{1}{2}, \; j = 1, 2, 3, \ldots, \; (Re\sqrt{\ } > 0). \tag{4.37}$$

If $v(G) \in \boldsymbol{C}$, then

$$\lim_{n\to\infty} \left| \frac{v(G) - v_n(F)}{v(G) - v_n(G)} \right| = 0 \text{ and hence } v(G) = v(F) = \lim_{n\to\infty} v_n(F) \in \boldsymbol{C}. \tag{4.38}$$

(C) *If $v(G) \in \boldsymbol{C}$ and*

$$a_m \in E_m(\{w_j\}, \rho) \text{ and } \rho < |1 + w_m|, \text{ for } m = 1, 2, 3, \ldots,$$

then

$$|v(F) - v_n(F)| \leq 2\rho \prod_{j=1}^{n} \frac{|a_j|}{(|1 + w_j| - \rho)^2}, \; n = 1, 2, 3, \ldots . \tag{4.39}$$

4.2. Limit k-Periodic CFs $K(1/b_j)$ and MCFs $K(1, b_j; w_j)$.

In this section we consider best truncation error bounds for continued fractions

$$\operatorname*{K}_{j=1}^{\infty} \left(\frac{1}{b_j}\right) = \frac{1}{b_1} + \frac{1}{b_2} + \frac{1}{b_3} + \cdots \tag{4.40}$$

and modified continued fractions $K(1, b_j; w_j)$ whose elements b_j satisfy a limit 4-periodic condition $\lim_{n\to\infty} b_n = \infty$ or $\lim_{n\to\infty} b_{4n+i} = \beta_i$ where $1 \leq i \leq 4$. The results in this section are restricted to the case in which $|b_n| \geq 2$ for all sufficiently large n. This condition ensures that $F = K(1/b_j)$ converges to $v(F)$ in the extended complex plane (Theorem 4.35 in [37]).

4.2.1. $K(1/b_j)$, $b_j \to \infty$.

Let $k \geq 0$ be a given non-negative integer; let $\{b_j\}_{j=1}^k$ be a given sequence of complex numbers; and let $\{\rho_n\}_{n=k}^\infty$ be a sequence of positive numbers such that

$$\lim_{n\to\infty} \rho_n = 0 \text{ and } \rho_n + \frac{1}{\rho_{n-1}} \geq 2, \text{ for } n = k+1, k+2, \ldots . \tag{4.41}$$

Let $W := \{0\}$ and let $\Omega = \{\Omega_j\}$ be defined by

$$\Omega_j := \langle 1, E_j, 0, 1\rangle, \; j = 1, 2, 3, \ldots, \tag{4.42}$$

where

$$E_j := \begin{cases} [b_j], \; j = 1, 2, \ldots, k \\ \left[u \in \boldsymbol{C} : |u| \geq \rho_j + \dfrac{1}{\rho_{j-1}}\right], \; j \geq k+1. \end{cases} \tag{4.43}$$

We define a family $\mathcal{F}$ of LFASs by

$$\begin{aligned} \mathcal{F} &:= \mathcal{F}(\Omega, W) := \mathcal{F}(1, \infty, k, \{b_j\}_1^k, \{\rho_j\}_k^\infty) \\ &:= \left[F = \mathop{\mathrm{K}}_{j=1}^{\infty} (1, b_j, 0) : b_j \in E_j \text{ for } j \geq 1\right]. \end{aligned} \tag{4.44}$$

We recall that $\mathrm{K}_{j=1}^{\infty}(1/b_j) = \mathrm{K}_{j=1}^{\infty}(1, b_j, 0)$. A sequence of value regions $\{V_n\}$ with respect to $\mathcal{F}(\Omega, W)$ is given by

$$V_n := \begin{cases} [u : |u| \leq \rho_n], \; n = k, k+1, k+2, \ldots, \\ \frac{1}{b_{n+1} + V_{n+1}} \qquad n = k-1, k-2, \ldots, 1, 0. \end{cases} \tag{4.45}$$

Theorem 4.7 *(Theorem 2.2 in [7]). If the elements b_j of $F = K(1/b_j)$ satisfy $\lim_{n\to\infty} b_j = \infty$, then for the critical tail sequence $\{h_n(F)\}$ we have*

$$\lim_{n\to\infty} h_n(F) = 0 \quad \textit{if} \quad v(F) = \infty \tag{4.46}$$

and

$$\lim_{n\to\infty} h_n(F) = \infty \quad \textit{if} \quad v(F) \neq \infty. \tag{4.47}$$

The following result is subsequently used.

Theorem 4.8 *(Theorem 3.1 in [7]). Let $\mathcal{F} = \mathcal{F}(\Omega, W)$ be a family of LFASs of the form (4.44) and let $F = K(1/b_j) \in \mathcal{F}$ be given. Then:*

(A) *If there exists an integer $k_0 \geq k$ such that $|h_{k_0}(F)| > \rho_{k_0}$, then*

$$|h_n(F)| > \rho_n, \quad n = k_0, k_0+1, \ldots, \tag{4.48}$$

and

$$K(1/b_j) \textit{ converges to a finite value } v(F). \tag{4.49}$$

(B) *If $F = K(1/b_j)$ converges to a finite value $v(F)$, then there exists a $k_0 \geq k$ such that $|h_{k_0}(F)| > \rho_{k_0}$, and hence the assertions of (A) hold and $v(F) \in \boldsymbol{C}$.*

The following truncation error bounds were obtained in Theorems 3.2 and 4.2 in [7].

Theorem 4.9 *Let $\mathcal{F} = (\mathcal{F}(\Omega, W)) = \mathcal{F}(1, \infty, k, \{b_j\}_1^k, \{\rho_j\}_k^\infty)$ be a family of LFASs* (4.44) *and let $F = K(1/b_j) \in \mathcal{F}$ be given. Then*

(A) *If there exists an integer $k_0 \geq k$ such that*

$$|h_{k_0}(F)| > \rho_{k_0}, \tag{4.50}$$

then F converges to a finite value $v(F) \in \boldsymbol{C}$ and, for all $n \geq k_0 + 1$

$$\begin{aligned}|v(F) - v_n(F)| &\leq \frac{\rho_n}{|B_{n-1}(F)|^2 |h_n(F)|(|h_n(F)| - \rho_n)} \\ &= \frac{\rho_n |v_n(F) - v_{n-1}(F)|}{(|h_n(F)| - \rho_n)}.\end{aligned} \tag{4.51}$$

(B) *If $K(1/b_j)$ converges to a finite value $v(F)$, then there exists a $k_0 \geq k$ such that $|h_{k_0}(F)| > \rho_{k_0}$ and hence* (4.51) *holds for $n \geq k_0 + 1$.*

(C) *If (4.50) holds for some integer $k_0 \geq k$, then, for $n \geq k_0 + 1$, the truncation error bound in* (4.51) *is the best bound $\beta_n(F, \mathcal{F})$ for $v_n(F)$ with respect to $\mathcal{F}$.*

4.2.2. $K(1, b_j, w_j)$, *$b_{4j+i} \to \beta_i$ as $j \to \infty$ and $w_{4n+i} = 1/\beta_{i+1}$, $i = 0, 1, 2, 3$, and $m \geq 0$.*

We now consider CFs $K(1/b_j)$ and MCFs $K(1, b_j, w_j)$ for which the elements b_j are complex numbers that satisfy limit 4-periodic properties.

Let $k \geq 0$ be a given non-negative integer; let $\{\beta_j\}_1^4$ satisfy

$$\beta_2 = \beta_4 = \infty, \ \beta_1, \beta_3 \in [u \in \boldsymbol{C} : |u| > 2]; \tag{4.52}$$

and let $\{\rho_j\}_k^\infty$ be a sequence of positive numbers satisfying

$$\lim_{j \to \infty} \rho_j = 0, \tag{4.53}$$

$$\rho_{4j+i} < \left| \frac{1}{\beta_{i+1}} \right|, \text{ for } i = 0, 2 \text{ and } 4j + i \geq k, \tag{4.54}$$

and

$$2 - \frac{|\beta_i|}{|1 + |\beta_i| \rho_{4j+i-1}} \leq \rho_{4j+i} \leq \frac{|\beta_i|^2 \rho_{4j+i-1}}{1 + |\beta_i| \rho_{4j+i-1}}, \tag{4.55}$$

for $i = 1, 3$ and $4j + i \geq k$.

Let $\{E_j\}$ be a sequence of subsets of $\boldsymbol{C}$ defined by

$$E_j := [b_j], \quad j = 1, 2, \ldots, k, \tag{4.56}$$

$$E_{4j+i} := \left[b \in \boldsymbol{C} : \left| b + \frac{1}{\beta_{i+1}} \right| \geq \rho_{4j+i} + \frac{1}{\rho_{4j+i-1}} \right], \ i = 0, 2, \ 4j + i \geq k + 1 \tag{4.57}$$

$$E_{4j+i} := \left[b \in \boldsymbol{C} : \left| \frac{\beta_i}{1 - (|\beta_i| \rho_{4j+i-1})^2} \right| \leq -\rho_{4j+i} + \frac{|\beta_i|^2 \rho_{4j+i-1}}{1 - (|\beta_i| \rho_{4j+i-1})^2} \right], \tag{4.58}$$

$i = 1, 3$, $4j + i \geq k + 1$. Then $\Omega = \{\Omega_j\}$ is defined by

$$\Omega_j := \langle 1, E_j, 0, 1\rangle, \quad j = 1, 2, 3, \ldots . \tag{4.59}$$

It follows from (4.57) and (4.58) that if

$$b_j \in E_j, \quad j = k+1, k+2, k+3, \ldots, \tag{4.60}$$

then

$$\lim_{j\to\infty} b_{4j+1} = \beta_1, \ \lim_{j\to\infty} b_{4j+3} = \beta_3, \ \lim_{j\to\infty} b_{4j+2} = \lim_{j\to\infty} b_{4j+4} = \infty. \tag{4.61}$$

We define a sequence of converging factors $W = \{w_j\}$, for $j = 0, 1, 2, \ldots$, by

$$w_{4m+i} := \lim_{j\to\infty} \frac{1}{b_{4j+i+1}} = \begin{cases} 0, & i = 1, 3, \\ 1/\beta_1, & i = 4, \\ 1/\beta_3, & i = 2. \end{cases} \tag{4.62}$$

A family $\mathcal{F}$ of LFASs is then defined by

$$\begin{aligned} \mathcal{F} &:= \mathcal{F}(\Omega, W) := \mathcal{F}(1, \{\beta_j\}_1^4, k, \{\beta_j\}_1^k, \{\rho_j\}_k^\infty) \\ &:= [F = K(1, b_j, w_j) : b_j \in E_j \text{ for } j \geq 1]. \end{aligned} \tag{4.63}$$

The generating sequence $\{t_j^F(w)\}$ for $F \in \mathcal{F}$ is defined by

$$t_0^F(w) := w, \ t_j^F(w) := \frac{1}{b_j(F) + w}, \ j = 1, 2, 3, \ldots . \tag{4.64}$$

With $\{T_n(F, w)\}$ defined by (1.1e) we have

$$v_n(F) := T_n(F, w_n), \quad n = 1, 2, 3, \ldots \tag{4.65}$$

and

$$v(F) := \lim_{n\to\infty} T_n(F, w_n) = T_n(F, v(F^{(n)})). \tag{4.66}$$

Remarks:

(1) Our choice of $\{w_j\}$ given by (4.62) is motivated by (1.38), (4.61) and

$$\lim_{m\to\infty} v(F^{(4m+i)}) = \lim_{m\to\infty} \frac{1}{b_{4m+i+1}} =: w_{4m+i}, \ m \geq 0, \ i = 0, 1, 2, 3. \tag{4.67}$$

(2) Condition (4.54) ensures $0 \notin V_{4j+i}$ and hence $E_{4j+i+1} \neq \emptyset$, for $i = 0, 2$ and $4j + i \geq k$. Condition (4.55) ensures that $\beta_i \in E_{4j+i}$, for $i = 1, 3$, $4j + i \geq k$, and that $E_j \cap [u \in \mathbf{C} : |u| < 2] = \emptyset$ for $j = k, k+1, \ldots$.

(3) If $K(1/b_j)$ is a continued fraction satisfying (4.61) and $|b_j| \geq 2$ for $j \geq k$ for some positive integer k, then there exists a sequence of positive numbers $\{\rho_j\}_{j=0}^\infty$ satisfying (4.53), (4.54) and (4.55) such that

$$K(1, b_j, w_j) \in \mathcal{F}(1, \{\beta_j\}_1^4, k, \{b_j\}_1^k, \{\rho_j\}_k^\infty)$$

where $\{w_j\}_{j=1}^\infty$ is defined by (4.62).

Theorem 4.10 *(Lemma 2.2 in [8]). If the elements b_j of $F = K(1/b_j)$ satisfy (4.61), $|b_j| \geq 2$ for $j \geq k$ for some positive integer k, and $b_j \neq 0$ for $j = 1, 2, 3, \ldots$, then the critical tail sequence $\{-h_n(F)\}$ satisfies*

$$\lim_{n\to\infty} h_{4n+i}(F) = \begin{cases} 0, & i = 1, 3 \\ -1/\beta_3, & i = 2 \\ -1/\beta_1, & i = 4, \end{cases} \qquad \text{if } v(F) = \infty \tag{4.68}$$

and

$$\lim_{n\to\infty} h_{4n+i}(F) = -\beta_i, \quad \text{if} \quad v(F) \neq \infty. \tag{4.69}$$

We use the following result to obtain truncation error bounds.

Theorem 4.11 *(Theorem 3.2 in [8]). Let $\mathcal{F} = \mathcal{F}(\Omega, W)$ be a family of LFASs of the form (4.63) and let $F = K(1, b_j; w_j) \in \mathcal{F}$ be given. Then:*

(A) *If there exists an integer $k_0 \geq k$ such that*

$$\left| h_{k_0}(F) + \frac{1}{\beta_{(k_0 \bmod 4)+1}} \right| > \rho_{k_0} \tag{4.70}$$

then

$$\left| h_n(F) + \frac{1}{\beta_{(n \bmod 4)+1}} \right| > \rho_n \text{ for } n = k_0, k_0+1, \ldots, \tag{4.71}$$

and $K(1/b_j)$ and $K(1, b_j, w_j)$ converge to the same finite value $v(F) \in \boldsymbol{C}$.

(B) *If $K(1/b_j)$ converges to a finite value f, then there exists a $k_0 \geq k$ such that* (4.70) *holds. Hence $f = v(F) \in \boldsymbol{C}$.*

The following theorem was proven in Theorems 3.3 and 4.1 in [8].

Theorem 4.12 *Let $\mathcal{F} = (\mathcal{F}(\Omega, W)) = \mathcal{F}(1, \{\beta_i\}_1^4, k, \{b_j\}_1^k, \{\rho_j\}_k^\infty)$ be a family of LFASs* (4.63) *and let $F = K(1, b_j, w_j) \in \mathcal{F}$ be given. Then:*

(A) *If there exists an integer $k_0 \geq k$ such that* (4.70) *holds, then $K(1/b_j)$ and F both converge to the same finite value $v(F) \in \boldsymbol{C}$ and for $n \geq k_0$*

$$\begin{aligned} |v(F) - v_n(F)| &\leq \frac{\rho_n}{|B_{n-1}(F)|^2 |w_n + h_n(F)| (|w_n + h_n(F)| - \rho_n)} \\ &= \frac{\rho_n |h_n(F)| |v_n(F) - v_{n-1}(F)|}{|w_n + h_n(F)| (|w_n + h_n(F)| - \rho_n)}. \end{aligned} \tag{4.72}$$

(B) *If $K(1/b_j)$ converges to a finite value $v(K(1/b_j))$, then there exists a $k_0 \geq k$ such that* (4.70) *holds and hence* (4.72) *holds for $n \geq k_0$.*

(C) *If there exists an integer $k_0 \geq k$ such that* (4.70) *holds, then the truncation error bound in* (4.72) *is the best bound $\beta_n(F, \mathcal{F})$ for $v_n(F)$ with respect to $\mathcal{F}$ for $n \geq k_0$.*

5. Asymptotically Best Truncation Error Bounds for LFASs

An approach to truncation error estimates for limit periodic LFT algorithms, differing from the one treated in detail in this article, was explored by one of the authors [59]. Earlier work pointing in the direction can be found in [63], [57] and [60]. An LFT algorithm is the special case of an LFAS (see (1.1a)) where $w_j = 0$, $j \geq 1$.

One starts with formulas based on the invariance of the cross ratio under ℓ.f.t. (see [56]), that is,

$$\left(\frac{S(u)-S(v)}{S(u)-S(w)}\right)\left(\frac{S(w)-S(z)}{S(v)-S(z)}\right) = \left(\frac{u-v}{u-w}\right)\left(\frac{w-z}{v-z}\right). \tag{5.1}$$

In (5.1) we replace S by T_n and set

$$-z = h_n := -T_n^{-1}(\infty). \tag{5.2}$$

This leads to the simplification

$$\frac{T_n(u_n)-T_n(v_n)}{T_n(u_n)-T_n(w_n)} = \left(\frac{u_n-v_n}{u_n-w_n}\right)\left(\frac{w_n+h_n}{v_n+h_n}\right). \tag{5.3}$$

Making suitable substitutions for u_n, v_n, w_n one arrives at

$$\frac{T_n(0)-T_{n+m}(0)}{T_n(0)-T_{n-1}(0)} = \frac{T_m^{(n)}(0)}{-a_n}\left(\frac{-a_n+c_nh_n}{T_m^{(n)}(0)+h_n}\right) \tag{5.4}$$

and

$$\frac{T_{k+1}(0)-T_k(0)}{T_k(0)-T_{k-1}(0)} = \frac{a_{k+1}}{a_k}\left(\frac{a_k-c_kh_k}{a_{k+1}+b_{k+1}h_k}\right). \tag{5.5}$$

Here $\{T_n\}$ is defined in terms of $\{t_n(w)\}$ as in $(1.1b, c, d)$ except that we have dropped the F superscript. $T_m^{(n)}$ is defined as

$$T_m^{(n)}(w) = t_{n+1} \circ \cdots \circ t_{n+m}(w).$$

Combining (5.4) and (5.5) one obtains

$$T_{n+m}(0)-T_n(0) = \frac{T_m^{(n)}(0)(a_{n+1}+b_{n+1}h_n)(a_0b_1-b_0a_1)}{(T_m^{(n)}(0)+h_n)b_0b_1}\prod_{k=1}^{n}\left(\frac{a_k-c_kh_k}{a_{k+1}+b_{k+1}h_k}\right). \tag{5.6}$$

This formula is valid for general $\{T_n\}$ provided the denominator of the right side of (5.6) does not vanish.

The formula (5.6) becomes particularly useful for *limit periodic* LFT algorithms. From now on we shall restrict ourselves to such sequences $\{T_n\}$.

Set

$$\lim_{n\to\infty} t_n(w) =: t(w) =: \frac{a+cw}{b+dw}, \tag{5.7}$$

where we shall assume that

$$a := \lim_{n\to\infty} a_n,\ b := \lim_{n\to\infty} b_n,\ c := \lim_{n\to\infty} c_n,\ d := \lim_{n\to\infty} d_n, \tag{5.8}$$

$a, b, c, d \in \mathbb{C}$. We exclude the cases where $t(w)$ is the identity or parabolic or elliptic. Then, if $t(w)$ is not singular, it has exactly two distinct fixed points x_1 and x_2. If $t(w)$ is singular, we shall denote by x_2 its fixed point and by x_1 the point $-a/b$ for which $t(w)$ is not defined. We also shall assume that both x_1 and x_2 are finite. Next, we introduce

$$r := \frac{dx_1 + b}{dx_2 + b} \tag{5.9}$$

and choose the subscripts so that $|r| < 1$. This can be done since $t(w)$ is assumed not to be elliptic.

It can be shown (the proof is quite intricate) that

$$\lim_{n\to\infty} h_n = -x_1, \tag{5.10}$$

provided $t_n(x_2) \neq x_2$ for all $n > n_0$. It further is true that

$$\lim_{k\to\infty} \frac{a_k - c_k h_k}{a_{k+1} + b_{k+1} h_k} = -r, \tag{5.11}$$

and that there exists a constant M such that

$$\left| \frac{T_m^{(n)}(0)}{T_m^{(n)}(0) + h_n} \right| < M, \tag{5.12}$$

for all $m > 0$ and all $n > n_0$.

In general we only know that such an M exists. However if more information is available about the sequence $\{t_n\}$, then an explicit bound on (5.12) may be obtainable. This is illustrated by our discussion of $K(a_n/1)$ later in this section. The remaining quantities in (5.6) can be easily calculated on the basis of the available information.

For any r' such that

$$|r| < |r'| < 1$$

and $n > n_2 > \max(n_0, n_1)$, the formula (5.6) can be recast into the inequality

$$|T_{n+m}(0) - T_n(0)| < K(r')|r'|^n. \tag{5.13}$$

In general $\lim\limits_{r'\to r} K(r') = \infty$; but if

$$\sum_{n=1}^{\infty} \Delta_n < \infty,$$

where

$$\Delta_n := \max(|a - a_n|, |b - b_n|, |c - c_n|, |d - d_n|), \tag{5.14}$$

then the stronger statement

$$|T_{n+m}(0) - T_n(0)| < K(r)|r|^n \tag{5.15}$$

is valid. Here $K(r) < \infty$.

For pure periodic sequences $\{\overset{p}{T}_n\}$ the truncation error is known to be

$$|\overset{p}{T}_n(0) - w_2| = \left| r^n \frac{w_2}{w_1} (\overset{p}{T}_n(0) - w_1) \right|. \tag{5.16}$$

It follows that the estimates (5.13) and (5.15) can be said to be *asymptotically best.*

The formula

$$\frac{f - T_n(x_2)}{f - T_n(0)} = \frac{f^{(n)} - x_2}{f^{(n)}} \frac{h_n}{h_n + x_2} \tag{5.17}$$

is an easy consequence of (5.3). Here we have set

$$f := \lim_{n \to \infty} T_n(0), \quad f^{(n)} := \lim_{m \to \infty} T_{n+m}^{(n)}(0).$$

(5.17) was initially proved for $K(a_n/1)$ in [61]. Since $f^{(n)} - x_2 \to 0$, it follows from (5.17) that $\{T_n(x_2)\}$ converges to f much faster than $\{T_n(0)\}$ does.

Further analysis shows that $K(r')$ depends on the behavior of $\{\Delta_n\}$, while, clearly, r is completely determined by a, b, c, w in $t(w)$. It can also be shown (the argument is delicate) that $f^{(n)} - x_2$ is roughly proportional to Δ_n.

If $t(w)$ is singular, then $r = 0$ and it follows that the convergence of $\{T_n(0)\}$ is extremely fast. For $K(a_n/1)$, with $a_n \to 0$, this was first observed in [3]. For Schur algorithms with $\gamma_n \to e^{i\theta}$, see [57].

For the special case $K(a_n/1)$, which was analyzed in [63], (5.6) becomes

$$S_{n+m}(0) - S_n(0) = \frac{S_m^{(n)}(0)(a_{n+1} + h_n)(-a_1)}{S_m^{(n)}(0) + h_n} \prod_{k=1}^{n} \left(\frac{a_k}{a_{k+1} + h_k} \right). \tag{5.18}$$

For r we obtain

$$r = \frac{x_1 + 1}{x_2 + 1} = \frac{-x_2}{-x_1} = \frac{x_2}{x_1}. \tag{5.19}$$

Using the fact that

$$\frac{h_k - 1}{h_k} = \frac{a_k}{a_{k-1}} \cdot \frac{a_{k-1}}{a_k + h_{k-1}}$$

we can show that (5.18) is equivalent to (3.3) in [63].

If we assume that for all $n \geq 1$ and some θ, $0 < \theta < 1$

$$a_n \in P(\alpha, \theta) = [w : |w| - \operatorname{Re} we^{-i2\alpha} \leq (\cos\alpha)^2(1 - \theta^2)/2] \tag{5.20}$$

where $2\alpha = \arg(\lim a_n)$, then we can conclude that

$$\operatorname{Re}(e^{-i\alpha} S_m^{(n)}(0)) \geq (\cos\alpha)(1 - \theta)/2. \tag{5.21}$$

Hence an explicit estimate for M in (5.12) can be obtained, since h_n can be computed from the given data.

Using a mixture of methods, which are brought together in [65], one can establish the following explicit results for $K(a_n/1)$:

(A) If $|a_m| \le \min(1/6, \alpha n^{-\rho})$, $\alpha > 0$, $\rho > 0$, $m \ge n \ge n_1$, then there exist constants $K_1 > 0$, $M_1 > 0$ such that for $n > n_1$, $k > 0$

$$|f_{n+k} - f_n| \le K_1 \left(\frac{M_1}{n}\right)^{\rho(n+\frac{3}{2})}.$$

(B) If $\lim a_n = a \in \boldsymbol{C} - (-\infty, -1/4]$, then for every q satisfying

$$\left|\frac{-1+\sqrt{1+4a}}{+1+\sqrt{1+4a}}\right| < q < 1$$

there exists a $K_2 = K_2(q, n_2) > 0$ such that for $n \ge n_2$, $k > 0$

$$|f_{n+k} - f_n| < K_2 q^n.$$

(C) If $a_n \in P(\alpha, \theta)$ (see (5.20)) for some $0 < \theta \le 1$ and $a_n = O(n^\beta)$ for some β, $0 < \beta \le 1$, then there exist $K_3 > 0$, $M_3 > 0$, $E_3 > 0$ and $L_3 > 1$ such that for $n \ge n_3$, $k > 0$

$$|f_{n+k} - f_n| < \begin{cases} a\dfrac{K_3}{n^{E_3}}, & \text{for} \quad \beta = 1, \\[2ex] \dfrac{M_3}{L^{n^{1-\beta}}}, & \text{for} \quad 0 < \beta < 1. \end{cases}$$

(D) For the S-fraction $K(a_n z/1)$ with $a_n > 0$, $|\arg z| < \pi$, let $a_n = O(n^\alpha)$, $0 < \alpha \le 2$. Then there exist constants $K_4 > 0$, $M_4 > 0$, $E_4 > 0$ and $L_4 > 1$ such that for $n \ge n_4$, $k > 0$

$$|f_{n+k}(z) - f_n(z)| < \begin{cases} \dfrac{K_4}{n^{E_4\sqrt{z}}} & \text{for} \quad \alpha = 2 \\[2ex] \dfrac{M_4}{L_4^{n^\delta}}, \ \delta := \dfrac{2-\alpha}{2\sqrt{z}}, & \text{for } 0 < \alpha < 2. \end{cases}$$

References

1. M. Abramowitz and I. A. Stegun, *Handbook of Mathematical Functions with Formulas, Graphs and Mathematical Tables*, National Bureau of Standards, Appl. Math. Ser. 55, U.S. Govt. Printing Office, Washington, D.C. (1964).
2. G. A. Baker, Jr., Best error bounds for Padé approximants to convergent series of Stieltjes, *J. Mathematical Phys.* **10** (1969), 814–820.
3. Christopher Baltus and William B. Jones, Truncation error bounds for limit-periodic continued fractions $K(a_n/1)$ with $\lim a_n = 0$, *Numer. Math.* **46** (1985), 541–569.
4. Christopher Baltus and William B. Jones, A family of best value regions for modified continued fractions, in *Analytic Theory of Continued Fractions* II (ed., W. J. Thron), Lecture Notes in Mathematics **1199**, Springer–Verlag, New York (1986), 1–20.
5. Christopher Baltus and William B. Jones, Truncation error bounds for modified continued fractions with applications to special functions, *Numer. Math.* **55** (1989), 281–307.
6. G. Blanch, Numerical evaluation of continued fractions, *SIAM Rev.* **7** (1964), 383–421.
7. C. M. Craviotto, William B. Jones, W. J. Thron, Best Truncation Error Bounds for Continued Fractions $K(1/b_n)$, $\lim_{n\to\infty} b_n = \infty$, *Continued Fractions and Orthogonal Functions; Theory and Applications* (eds., S. C. Cooper and W. J. Thron), Marcel Dekker, Inc. New York (1994), 115–127.

8. C. M. Craviotto, William B. Jones and W. J. Thron, Truncation Error Bounds for Limit k-Periodic Continued Fractions, (submitted).
9. David A. Field, Estimates of the speed of convergence of continued fraction expansions of functions, *Math. of Comp.* **31** (1977), 495–502.
10. David A. Field, Error bounds for elliptic convergence regions for continued fractions, *SIAM J. Numer. Anal.* **15** (1978), 444–449.
11. David A. Field, Error bounds for continued fractions $K(1/b_n)$, *Numer. Math.* **29** (1978), 261–267.
12. David A. Field and William B. Jones, A priori estimates for truncation error of continued fractions $K(1/b_n)$, *Numer. Math.* **19** (1972), 283–302.
13. John Gill, The use of attractive fixed points in accelerating the convergence of limit-periodic continued fractions, *Proc. Amer. Math. Soc.* **47** (1975), 119–126.
14. John Gill, Enhancing the convergence region of a sequence of bilinear transformation, *Math. Scand.* **43** (1978), 74–80.
15. John Gill, Truncation error analysis for continued fractions $K(a_n/1)$ where $\sqrt{|a_n|} + \sqrt{|a_{n-1}|} < 1$, *Lecture Notes in Math.* **932** (eds., W. B. Jones, W. J. Thron and H. Waadeland), Springer–Verlag, (1982), 71–73.
16. W. B. Gragg, Truncation error bounds for g-fractions, *Numer. Math.* **11** (1968), 370–379.
17. W. B. Gragg, Truncation error bounds for π-fractions, *Bull. Amer. Math. Soc.* **76** (1970), 1091–1094.
18. W. B. Gragg, Truncation error bounds for T-fractions, *Approximation Theory* III (ed., W. Cheney), Academic Press, (1980), 455–460.
19. W. B. Gragg and D. D. Warner, Two Constructive Results in Continued Fractions, *SIAM J. Numer. Anal.* **20** (1983), 1187–1197.
20. T. L. Hayden, Continued fraction approximation to functions, *Numer. Math.* **7** (1965), 292–309.
21. P. Henrici and Pia Pfluger, Truncation error estimates for Stieltjes fractions, *Numer. Math.* **9** (1966), 120–138.
22. K. L. Hillam, Some convergence criteria for continued fractions, Doctoral Thesis, University of Colorado, Boulder (1962).
23. Lisa Jacobsen, William B. Jones and Haakon Waadeland, Further results on the computation of incomplete gamma functions, *Analytic Theory of Continued Fractions* II (ed., W. J. Thron), Lecture Notes in Math. **1199**, Springer–Verlag, New York (1986), 67–89.
24. Lisa Jacobsen, William B. Jones and Haakon Waadeland, Convergence acceleration for continued fractions $K(a_n/1)$ where $a_n \to \infty$, *Rational Approximation and its Applications to Mathematics and Physics* (eds., J. Gilewicz, M. Pindor, W. Siemaszko), Lecture Notes in Mathematics **1237**, Springer–Verlag, New York (1987), 177–187.
25. Lisa Jacobsen and D. R. Masson, On the convergence of limit periodic continued fractions $K(a_n/1)$, where $a_n \to -\frac{1}{4}$, Part II, *Constr. Approx.* **6** (1990), 363–374.
26. Lisa Jacobsen and David R. Masson, A sequence of best parabola theorems for continued fractions, *Rocky Mtn. J. Math.* **21** (1991), 377–385.
27. L. Jacobsen and W. J. Thron, Oval convergence regions and circular limit regions for continued fractions $K(a_n/1)$, *Analytic Theory of Continued Fractions* II (ed., W. J. Thron), Lecture Notes in Mathematics **1199**, Springer–Verlag, New York (1986), 90–126.
28. Lisa Jacobsen, W. J. Thron, Haakon Waadeland, Julius Worpitzky, his contributions to the analytic theory of continued fractions and his times, *Analytic Theory of Continued Fractions* III (ed., Lisa Jacobsen), Lecture Notes in Mathematics **1406**, Springer–Verlag, New York (1989), 25–47.
29. Thomas H. Jefferson, Truncation error estimates for T-fractions, *SIAM J. Numer. Anal.* **6** (1969), 359–364.
30. William B. Jones, Analysis of truncation error of approximations based on the Padé table and continued fractions, *Rocky Mountain J. of Math.* **4** (1974), 241–250.
31. William B. Jones, Schur's algorithm extended and Schur continued fractions, *Nonlinear Numerical Methods and Rational Approximation* (ed., A. Cuyt), D. Reidel Publ. Company, Dordrecht (1988), 281–298.
32. William B. Jones, Olav Njåstad and W. J. Thron, Schur fractions, Perron–Carathéodory fractions and Szegö polynomials, a survey, *Analytic Theory of Continued Fractions* II, (ed., W. J. Thron), Lecture Notes in Math. **1199**, Springer–Verlag, New York (1986), 127–158.

33. William B. Jones, Olav Njåstad and W. J. Thron, Moment theory, orthogonal polynomials, quadrature, and continued fractions associated with the unit circle, *Bull. London Math. Soc.* **21** (1989), 113–152.
34. William B. Jones and R. I. Snell, Truncation error bounds for continued fractions, *SIAM J. Numer. Anal.* **6** (1969), 210–221.
35. William B. Jones and W. J. Thron, A posteriori bounds for the truncation error of continued fractions, *SIAM J. Numer. Anal.* **8** (1971), 693–705.
36. William B. Jones and W. J. Thron, Truncation error analysis by means of approximant systems and inclusion regions, *Numer. Math.* **26** (1976), 117–154.
37. William B. Jones and W. J. Thron, *Continued Fractions: Analytic Theory and Applications*, Encyclopedia of Mathematics and its applications **11**, Addison–Wesley Publ. Company, Reading, Mass. (1980); distributed now by Cambridge University Press, New York.
38. William B. Jones and W. J. Thron, Continued fractions in numerical analysis, *Appl. Numer. Math.* **4** (1988), 143–230.
39. William B. Jones and W. J. Thron, A constructive proof of convergence of the even approximants of positive PC-fractions, *Rocky Mountain J. of Math.* **19** (1989), 199–210.
40. William B. Jones, W. J. Thron and Haakon Waadeland, Truncation Error Bounds for Continued Fractions $K(a_n/1)$ with Parabolic Element Regions, *SIAM J. Numer. Anal.* **20** (1983), 1219–1230.
41. William B. Jones, W. J. Thron and Haakon Waadeland, Value Regions for Continued Fractions $K(a_n/1)$ Whose Elements Lie in Parabolic Regions, *Math. Scand.* **56** (1985), 5–14.
42. R. E. Lane, The value region problem for continued fractions, *Duke Math. J.* **12** (1945), 207–216.
43. L. J. Lange, Divergence, convergence, and speed of convergence of continued fractions $1 + K(a_n/1)$, Doctoral Thesis, University of Colorado, Boulder (1960).
44. W. Leighton and W. J. Thron, Continued fractions with complex elements, *Duke Math. J.* **9** (1942), 763–772.
45. Lisa Lorentzen and Haakon Waadeland, *Continued Fractions with Applications*, Studies in Computational Math., Vol. 3, North–Holland, New York (1992).
46. J. H. McCabe, A continued fraction expansion with a truncation error estimate for Dawson's integral, *Math. Comp.* **28** (1974), 811–816.
47. E. P. Merkes, On truncation errors for continued fraction computations, *SIAM J. Numer. Anal.* **3** (1966), 486–496.
48. Marius Overholt, The values of continued fractions with complex elements, *Padé Approximants and Continued Fractions* (eds., Haakon Waadeland and Hans Wallin), Det. Konkelige Norske Videnskabers Selskab, Skrifter, No. 1 (1983), 109–116.
49. J. F. Paydon and H. S. Wall, The continued fraction as a sequence of linear transformations, *Duke Math. J.* **9** (1942), 360–372.
50. I. Schur, Über Potenzreihen die im Innern des Einheitskreises beschränkt sind, *J. für die reine und angewandte Mathematik* **147** (1917), 205–232 and **148** (1918), 122–145.
51. W. B. Sweezy and W. J. Thron, Estimates of the speed of convergence of certain continued fractions, *SIAM J. Numer. Anal.* **4**, No. 2 (1967), 254–270.
52. W. J. Thron, Twin convergence regions for continued fractions $b_0 + K(1/b_n)$, *Amer. J. Math.* **66** (1944), 428–438.
53. W. J. Thron, On parabolic convergence regions for continued fractions, *Math. Zeitschr.* **69** (1958), 173–182.
54. W. J. Thron, Convergence Regions for Continued Fractions and Other Infinite Processes, *Amer. Math. Monthly* **68** (1961), 734–750.
55. W. J. Thron, A priori truncation error estimates for Stieltjes fractions, in *E. B. Christoffel* (ed., P. L. Butzer and F. Fehér), Birkhäuser Verlag, Basel, (1981), 203–211.
56. W. J. Thron, Continued fraction identities derived from the invariance of the crossratio under linear fractional transformations, *Analytic Theory of Continued Fractions* III (ed., Lisa Jacobsen), Lecture Notes in Mathematics **1406**, Springer–Verlag, New York (1989), 124–134.
57. W. J. Thron, Limit periodic Schur algorithms, the case $|\gamma| = 1$, $\sum d_n < \infty$, *Numer. Algorithms* **3** (1992), 441–450.
58. W. J. Thron, Should the Pringsheim criterion be renamed the Śleszyński criterion?, *Comm. Analytic Theory Cont. Fract.* **1** (1992), 13–18.
59. W. J. Thron, Truncation Error for L.F.T. algorithms $\{T_n(n)\}$, *Continued Fractions and Or-*

thogonal Functions: Theory and Applications, (eds., S. C. Cooper and W. J. Thron), Marcel Dekker, New York (1994), 353–365.

60. W. J. Thron, Truncation Error for limit periodic Schur algorithms, *SIAM J. Math. Analysis*, (to appear).
61. W. J. Thron and Haakon Waadeland, Accelerating convergence of limit periodic continued fractions $K(a_n/1)$, *Numer. Math.* **34** (1980), 155–170.
62. W. J. Thron and Haakon Waadeland, Modifications of continued fractions, a survey, *Analytic Theory of Continued Fractions* (eds., W. B. Jones, W. J. Thron and H. Waadeland), Lecture Notes in Mathematics **932**, Springer–Verlag, New York (1982), 38–66.
63. W. J. Thron and Haakon Waadeland, Truncation Error bounds for limit periodic continued fractions, *Math. of Comp.* **40** (1983), 589–597.
64. Haakon Waadeland, Derivatives of continued fractions with applications to hypergeometric functions, *J. Comp. Appl. Math.* **19** (1987), 161–169.
65. Haakon Waadeland, Computation of Continued Fractions by square root modification: reflection and examples, *Appl. Numer. Math.* **4** (1988), 361–375.
66. J. Worpitzky, Untersuchungen über die Entwickelung der monodromen und monogenen Funktionen durch Kettenbrüche, *Friedrichs-Gymnasium und Realschule*, Jahresbericht, Berlin (1865), 3–39.
67. Peter Wynn, The numerical efficiency of certain continued fraction equations, *Indag. Math.* **24** (1962), 127–148.

TRUNCATION ERROR BOUNDS FOR LIMIT K-PERIODIC CONTINUED FRACTIONS

CATHLEEN CRAVIOTTO, WILLIAM B. JONES and W. J. THRON*
Department of Mathematics, University of Colorado, Boulder, CO 80309-0395, U.S.A.

Abstract. Many special functions of mathematical physics have representations in the form of limit k-periodic continued fractions $f = K(1/b_n)$, where $\lim_{n\to\infty} b_n = b$ or $\lim_{n\to\infty} b_{kn+i} = \beta_i$ where k is a positive integer and $0 \leq i \leq k-1$. For computation of such functions by continued fractions, one needs sharp bounds for the truncation error $|f - f_n|$ resulting when the value f is replaced by the nth approximate f_n. In this paper we present bounds for limit 4-periodic continued fractions $K(1/b_n)$. Some of these bounds are shown to be "best" relative to limited given information.

1. Introduction

Many important special functions of mathematics, statistics and the physical sciences have representations by continued fractions (*CFs*) $\overset{\infty}{\underset{n=1}{K}} (a_n/b_n)$ and modified continued fractions (*MCFs*) $\overset{\infty}{\underset{n=1}{K}} (a_n, b_n; w_n)$ that can be used for computation. When a CF or MCF is used to compute the value of a function $f = f(z)$ at a point $z \in \mathbb{C}$, it is essential to have realistic upper bounds for truncation error $|f - g_n|$ obtained when the value f is replaced by an *nth* approximant g_n. Following the advent of high-speed digital computers, a great deal of research has been devoted to the truncation error analysis of CFs and MCFs (see, e.g., [7] and [2] for recent surveys). Only recently has attention been focussed on the exploitation of limit periodic properties satisfied by many CF expansions and on the determination of "best" error bounds. Results of that type have been obtained for CFs of the form $\overset{\infty}{\underset{n=1}{K}} (a_n/1)$ and MCFs $\overset{\infty}{\underset{n=1}{K}} (a_n, 1; w_n)$ in the papers [1], [2] and [5]. The main purpose of the present paper is to obtain best truncation error bounds for CFs $K(1/b_n)$ and MCFs $K(1, b_n; w_n)$ which have limit periodic behavior. Work in this direction was given in [3] for the case in which $\lim_{n\to\infty} b_n = \infty$.

* Research supported in part by the National Science Foundation under Grant No. DMS-9302584.

A. Cuyt (ed.), Nonlinear Numerical Methods and Rational Approximation II, 375–385.

In the present paper we consider CFs $\overset{\infty}{\underset{n=1}{K}}(1/b_n)$ and MCFs $\overset{\infty}{\underset{n=1}{K}}(1, b_n; w_n)$ for which the elements b_n are complex-valued functions of a complex variable z that satisfy (for fixed z) the limit-4-periodic properties

$$\beta_1 := \lim_{n\to\infty} b_{4n+1}, \quad \beta_3 := \lim_{n\to\infty} b_{4n+3}, \quad \beta_1, \beta_3 \in [u \in \boldsymbol{C} : |u| \geq 2], \tag{1}$$

$$\beta_2 := \lim_{n\to\infty} b_{4n+2} = \infty, \quad \beta_4 := \lim_{n\to\infty} b_{4n+4} = \infty \tag{2}$$

and

$$|b_n| \geq 2 \text{ for } \quad n \geq k, \text{ for some positive integer k.} \tag{3}$$

The complex numbers w_n (which depend upon z) are called the *converging factors* for the MCF $\overset{\infty}{\underset{n=1}{K}}(1, b_n; w_n)$ *Sequences of linear fractional transformations (l.f.t.s)* $\{s_n\}_{n=1}^{\infty}$ *and* $\{S_n^{(m)}\}_{n=0}^{\infty}$ *associated with* $\overset{\infty}{\underset{n=1}{K}}(1/b_n)$ *and* $\overset{\infty}{\underset{n=1}{K}}(1, b_n; w_n)$ are defined by

$$s_n(w) = \frac{1}{b_n + w}, \quad n = 1, 2, 3, \ldots \tag{4}$$

and

$$S_0^{(m)}(w) := w, \quad S_n^{(m)}(w) := S_{n-1}^{(m)}(s_{m+n}(w)), \quad m = 0, 1, 2, \ldots, \ n = 1, 2, 3, \ldots. \tag{5}$$

We adopt the notation $S_n = S_n^{(0)}, n = 1, 2, 3, \ldots$.

The *nth approximants* f_n of $\overset{\infty}{\underset{n=1}{K}}(1/b_n)$ and g_n of $\overset{\infty}{\underset{n=1}{K}}(1, b_n; w_n)$, are defined by

$$f_n := S_n(0) := \frac{1}{b_1} + \frac{1}{b_2} + \frac{1}{b_3} + \cdots + \frac{1}{b_n}, \quad n = 1, 2, 3, \ldots, \tag{6}$$

and

$$g_n := S_n(w_n) := \frac{1}{b_1} + \frac{1}{b_2} + \frac{1}{b_3} + \cdots + \frac{1}{b_n + w_n}, \quad n = 1, 2, 3, \ldots, \tag{7}$$

It can be seen that a MCF $\overset{\infty}{\underset{n=1}{K}}(1, b_n; w_n)$ reduces to its *reference continued fraction*

$$\overset{\infty}{\underset{n=1}{K}}(1/b_n) = \frac{1}{b_1} + \frac{1}{b_2} + \frac{1}{b_3} + \cdots, \tag{8}$$

when $w_n = 0$ for $n \geq 1$. We adopt the notation $K(1/b_n) = \underset{n=1}{\overset{\infty}{K}} (1/b_n)$ and $K(1, b_n; w_n) = \underset{n=1}{\overset{\infty}{K}} (1, b_n; w_n)$. A MCF $\underset{n=1}{\overset{\infty}{K}} (1, b_n; w_n)$ is said to *converge to a value* $g \in \widehat{\mathbf{C}} := \mathbf{C}U[\infty]$ if its sequence of approximants $\{g_n\}_{n=1}^{\infty} = \{S_n(w_n)\}_{n=1}^{\infty}$ converges to g. Condition (3) is sufficient to ensure that the reference continued fraction $\underset{n=1}{\overset{\infty}{K}} (1/b_n)$ converges to a value f in $\widehat{\mathbf{C}}$ (see, e.g., [7]). In the present paper we impose conditions that ensure that the MCF $\underset{n=1}{\overset{\infty}{K}} (1, b_n; w_n)$ and its reference CF $\underset{n=1}{\overset{\infty}{K}} (1/b_n)$ converge to a common finite value

$$f = \lim_{n \to \infty} f_n = \lim_{n \to \infty} S_n(w_n) = \lim_{n \to \infty} g_n \in \mathbf{C}. \tag{9}$$

For a given reference CF $K(1/b_n)$ whose elements b_n satisfy conditions (1), (2), and (3) we define converging factors w_n for the MCF $K(1, b_n; w_n)$ by

$$w_{4n+i} := \lim_{n \to \infty} \frac{1}{b_{4n+i+1}} = \begin{cases} 0 & i = 1, 3 \\ 1/\beta_1 & i = 0 \\ 1/\beta_3 & i = 2. \end{cases} \tag{10}$$

The choice of w_n is motivated by the following argument. For $m = 0, 1, 2, \ldots$, the *mth tail of* $\underset{n=1}{\overset{\infty}{K}} (1/b_n)$ is the CF defined by

$$\underset{n=m+1}{\overset{\infty}{K}} (1/b_n) = \frac{1}{b_{m+1}} + \frac{1}{b_{m+2}} + \cdots. \tag{11}$$

The *nth approximant* $f_n^{(m)}$ *of the mth tail (11)* is defined by

$$f_n^{(m)} := S_n^{(m)}(0), \quad m = 0, 1, 2, \ldots, \quad n = 1, 2, 3, \ldots. \tag{12}$$

We denote the value of the mth tail (11) by

$$f^{(m)} := \lim_{n \to \infty} f_n^{(m)} = \lim_{n \to \infty} S_n^{(m)}(0), \quad m = 0, 1, 2, \ldots, \tag{13}$$

if the limit exists in $\widehat{\mathbf{C}}$. In Section 2 (Lemma 2.1) we show that, if the b_n satisfy (1), (2), and (3) then

$$\lim_{n \to \infty} f^{(4n+i)} = \lim_{n \to \infty} \frac{1}{b_{4n+i+1}}, \quad i = 1, 2, 3, 4. \tag{14}$$

The choice of w_n in (10) is therefore motivated by (14) and the fact that

$$f = S_n(f^{(n)}), \quad n = 1, 2, 3, \ldots. \tag{15}$$

1.1. The Class L4P $[1, \{\beta_j\}_1^4, k, \{b_j\}_1^k, \{\rho_j\}_k^\infty]$

The truncation error bounds obtained in Section 3 apply to all CFs and MCFs contained in the family denoted by L4P $[1, \{\beta_j\}_1^4, k, \{b_j\}_1^k, \{\rho_j\}_k^\infty]$, that consist of all $K(1/b_n)$ and $K(1, b_n; w_n)$ such that the elements b_n satisfy

$$b_n \in E_n, \quad n = 1, 2, 3, \ldots, \tag{16}$$

where

$$E_n := [b_n], \quad n = 1, 2, \ldots, k,$$

and for $4n + i \geq k + 1$

$$E_{4n+i} := \left[b \in \boldsymbol{C} : \left| b + \frac{1}{\beta_{(i \bmod 4)+1}} \right| \geq \rho_{4n+i} + \frac{1}{\rho_{4n+i-1}} \right] \quad \text{for} \quad i = 2, 4, \tag{17}$$

$$E_{4n+i} := \left[b \in \boldsymbol{C} : \left| b - \frac{\beta_i}{1 - (|\beta_i| \rho_{4n+i-1})^2} \right| \leq -\rho_{4n+i} + \frac{|\beta_i|^2 \rho_{4n+i-1}}{1 - (|\beta_i| \rho_{4n+i-1})^2} \right] \tag{18}$$

for $i = 1, 3$, and the converging factors w_n are given by (10). Here k is a given positive integer, $\{\beta_i\}_{i=1}^4$ satisfies

$$\beta_2 = \beta_4 = \infty, \quad \beta_1, \beta_3 \in [u \in \boldsymbol{C} : |u| \geq 2] \tag{19}$$

and $\{\rho_n\}_{n=k}^\infty$ is a given sequence of positive numbers satisfying

$$\lim_{n \to \infty} \rho_n = 0 \tag{20}$$

$$\rho_{4n+i} < \left| \frac{1}{\beta_{i+1}} \right| \quad \text{for} \quad i = 0, 2, \; 4n + i \geq k \tag{21}$$

$$2 - \frac{|\beta_i|}{1 + |\beta_i| \rho_{4n+i-1}} \leq \rho_{4n+i} \leq \frac{|\beta_i|^2 \rho_{4n+i-1}}{1 + |\beta_i| \rho_{4n+i-1}} \tag{22}$$

for $i = 1, 3$, $4n + i \geq k$.

Remarks: 1. Condition (21) ensures $0 \notin V_{4n+i}$ for $i = 2, 4$, $n = 1, 2, 3, \ldots$, and hence $E_{4n+i+1} \neq \emptyset$ for $i = 2, 4$, $n = 1, 2, 3, \ldots$. Condition (22) ensures $\beta_i \in E_{4n+i}$ for $i = 1, 3$ and $n = 0, 1, 2, \ldots$, *and* $E_n \cap [u \in \boldsymbol{C} : |u| < 2] = \emptyset$ for $n = k, k+1, k+2, \ldots$.

2. We write $L4P[1, \{\beta_j\}_{j=1}^4, k, \{b_j\}_1^k, \{\rho_j\}_k^\infty]$ to indicate there exists $\beta_1, \beta_2, \beta_3, \beta_4, k$ satisfying (11) and $\{\rho_j\}_k^\infty$ satisfying (20), (21), and (22).

3. If $\overset{\infty}{\underset{n=1}{K}} (1/b_n)$ is a continued fraction satisfying (1), (2), and (3), then there exists a positive integer k and a sequence of positive numbers $\{\rho_j\}_k^\infty$ satisfying (20), (21), and (22) such that $K(1, b_n; w_n) \in L4P[1, \{\beta_j\}_1^4, k, \{b_j\}_1^k, \{\rho_j\}_k^\infty]$ where $\{w_n\}_{n=1}^\infty$ is defined by (10).

The following formulas and notation are used throughout the paper (see [7] for further details). For $m = 1, 2, 3, \ldots$, the mth tail MCF of $K(1, b_n; w_n)$ is the MCF $\underset{n=m+1}{\overset{\infty}{K}}(1, b_n; w_n)$; $\underset{n=1}{\overset{\infty}{K}}(1, b_n; w_n)$ is its own 0th tail MCF. The nth approximant $g_n^{(m)}$ of the mth tail MCF $\underset{n=m+1}{\overset{\infty}{K}}(1, b_n; w_n)$ is defined for $m = 0, 1, 2, \ldots$, and $n = 1, 2, 3, \ldots$, by

$$g_0^{(m)} := w_m, \quad g_n^{(m)} := S_n^{(m)}(w_{n+n}) =: \frac{1}{b_{m+1}} + \frac{1}{b_{n+2}} + \cdots + \frac{1}{b_{m+n} + w_{n+n}}. \tag{23}$$

2. Preliminaries

It can be seen that for $n = 0, 1, 2, \ldots$, and $n = 1, 2, 3, \ldots$

$$S_{m+n}(w) = S_m(S_n^{(m)}(w)) \tag{24}$$

and thus

$$g_{m+n} := S_{m+n}(w_{m+n}) = S_m(S_n^{(m)}(w_{m+n})) = S_m(g_n^{(m)}). \tag{25}$$

If an mth tail MCF converges to a limit $g^{(m)} := \lim_{n\to\infty} g_n^{(m)}$, it follows that every tail of the MCF converges and

$$g = \lim_{n\to\infty} g_n = S_m(g^{(m)}), \quad n = 1, 2, 3, \ldots. \tag{26}$$

The nth numerator A_n and denominator B_n of $K(1/b_n)$ are defined by the second order linear difference equations

$$A_{-1} := 1, \quad A_0 := 0, \quad A_n := b_n A_{n-1} + A_{n-2} \quad n = 1, 2, 3, \ldots. \tag{27}$$

$$B_{-1} := 0, \quad B_0 := 1, \quad B_n := b_n B_{n-1} + B_{n-2} \quad n = 1, 2, 3, \ldots. \tag{28}$$

These A_n and B_n satisfy the determinant formulas

$$A_n B_{n-1} - A_{n-1} B_n = (-1)^{n-1} \quad n = 1, 2, 3, \ldots, \tag{29}$$

and also

$$S_n(w) = \frac{A_n + wA_{n-1}}{B_n + wB_{n-1}} \quad n = 0, 1, 2, \ldots. \tag{30}$$

It follows from (6), (7) and (30) that

$$f_n = S_n(0) = \frac{A_n}{B_n} \quad n = 0, 1, 2, \ldots, \tag{31}$$

and

$$g_n = S_n(w_n) = \frac{A_n + w_n A_{n-1}}{B_n + w_n B_{n-1}} \quad n = 0, 1, 2, \ldots. \tag{32}$$

The critical tail sequence $\{-h_n\}$ is defined by

$$-h_n := S_n^{-1}(\infty) \quad n = 0, 1, 2, \ldots \tag{33}$$

and satisfies

$$h_0 = \infty, \quad h_n = \frac{B_n}{B_{n-1}} \quad n = 1, 2, 3, \ldots \tag{34}$$

$$h_n = b_n + \frac{1}{h_{n-1}} \quad n = 1, 2, 3, \ldots. \tag{35}$$

The boundary, interior, closure, and diameter of an arbitrary subset A of $\boldsymbol{C}$ are denoted by ∂A, $\mathrm{Int}\,(A)$, $c(A)$, and $\mathrm{diam}\,(A)$ respectively. By $\mathrm{rad}\,(A)$ and $\mathrm{center}\,(A)$ we denote, respectively, the radius and center of a disk A.

We now establish some basic results that are subsequently employed. Lemmas 2.1 and 2.2 describe the convergence of the nth tail sequence $\{f^{(n)}\}_{n=1}^{\infty}$ and the critical tail sequence $\{-h_n\}_{n=0}^{\infty}$. Theorem 2.4 gives sufficient conditions to ensure that $K(1/b_n)$ and $K(1, b_n; w_n)$ converge to the same limit.

Lemma 2.1 *Let $K(1/b_n)$ be a continued fraction satisfying (1), (2), and (3). Then*

$$\lim_{n\to\infty} f^{(4n+i)} = x_1^{(i)} := \lim_{n\to\infty} \frac{1}{b_{4n+i+1}} = \begin{cases} 0 & i = 1, 3 \\ 1/\beta_1 & i = 4 \\ 1/\beta_3 & i = 2. \end{cases} \tag{36}$$

Proof. By (3) and ([7], Theorem 4.35) we have $f^{(n)} \in V := \{z : |z| \leq 1\}$ for all sufficiently large n. Thus

$$\lim_{n\to\infty} |f^{(2n-1)}| \leq \lim_{n\to\infty} \sup \left[\left| \frac{1}{b_{2n} + v} \right| : v \in V \right] = 0,$$

and hence (36) holds for $i = 1, 3$. It follows that for $i = 2, 4$,

$$\begin{aligned} \lim_{n\to\infty} f^{(4n+i)} &= \lim_{n\to\infty} \frac{1}{b_{4n+i+1} + f^{(2(2n+\frac{i+2}{2})-1)}} \\ &= \lim_{n\to\infty} \frac{1}{b_{4n+i+1}} = x_1^{(i)}. \square \end{aligned} \tag{37}$$

By Lemma 2.1 and (10)

$$x_1^{(i)} = w_{4n+i} = \lim_{n\to\infty} \frac{1}{b_{4n+i+1}} = \begin{cases} 0 & i = 1, 3 \\ 1/\beta_1 & i = 4 \\ 1/\beta_3 & i = 2. \end{cases} \tag{38}$$

For $i = 1, 2, 3, 4$ and $n = 0, 1, 2, \ldots$, let $w_n^{(i)}$ be the smaller in modulus of the fixed points of the linear fractional transformation

$$\zeta = s_{4n+(i+1)} \circ s_{4n+(i+2)} \circ s_{4n+(i+3)} \circ s_{4n+i}(\zeta). \tag{39}$$

Then for $i = 1, 2, 3, 4$ and $n = 0, 1, 2, \ldots,$ $w_n^{(i)}$ satisfies the quadratic equation

$$\begin{aligned} 0 =& (\zeta)^2(b_{4n+i+1}b_{4n+i+2}b_{4n+i+3} + b_{4n+i+1} + b_{4n+i+3}) \\ &+\zeta(b_{4n+i+1}b_{4n+i+2}b_{4n+i+3}b_{4n+i} + b_{4n+i+1}b_{4n+i+2} + b_{4n+i+1}b_{4n+i} \\ &+b_{4n+i+3}b_{4n+i} - b_{4n+i+2}b_{4n+i+3}) \\ &+(-b_{4n+i+2}b_{4n+i+3}b_{4n+i} - b_{4n+i+2} - b_{4n+i}) \end{aligned} \tag{40}$$

and we choose $w_n^{(i)}$ to be the root of (39) with the smaller absolute value. Denoting the other root of (39) by $\widetilde{w}_n^{(i)}$, one can show

$$\lim_{n\to\infty} w_n^{(i)} = x_1^{(i)}, \tag{41}$$

and

$$\lim_{n\to\infty} \widetilde{w}_n^{(i)} = x_2^{(i)} := -\lim_{n\to\infty} b_{4n+i} = -\beta_i \quad i = 1, 2, 3, \tag{42}$$

Lemma 2.2 *Let $f \in \widehat{\mathbb{C}}$ denote the value of a CF satisfying (1)-(3) and $b_n \neq 0$ for $n = 1, 2, 3, \ldots$. Let $\{-h_n\}_{n=0}^{\infty}$ be its critical tail sequence.*

(A) *If $f = \infty$, then*

$$\lim_{n\to\infty} h_{4n+i} = -x_1^{(i)} = \begin{cases} 0 & i = 1, 3 \\ -1/\beta_3 & i = 2 \\ -1/\beta_1 & i = 4. \end{cases}$$

(B) *If $f \neq \infty$, then $\lim_{n\to\infty} h_{4n+i} = -x_2^{(i)} = -\beta_i$.*

Proof. Apply [[2],Theorem 2.1] for $K(a_n/1)$ to the equivalent CF $K(1/b_n)$. □

Corollary 2.1 *If $\underset{n=}{\overset{\infty}{K}} (1/b_n)$ is a finite CF satisfying the hypotheses of Lemma 2.2, then*

$$\lim_{n\to\infty} (h_n - b_n) = 0 \tag{43}$$

and

$$\lim_{n\to\infty} \frac{h_n}{b_n} = 1. \tag{44}$$

Proof. We use Lemma 2.2B and (42) to obtain (44) from which (43) follows. □

Our next result follows from ([7], Theorem 3.1).

Theorem 2.1 *Let $\underset{n=1}{\overset{\infty}{K}} (1, b_n; w_n) \in L4P[1, \{\beta_j\}_1^4, k, \{b_j\}_1^k, \{\rho_j\}_k^\infty]$ be given. Then $\underset{n=1}{\overset{\infty}{K}} (1/b_n)$ and $\underset{n=1}{\overset{\infty}{K}} (1, b_n; w_n)$ converge to the same value $f \in \widehat{\mathbb{C}}$.*

3. A Truncation Error Bound

In this section we derive truncation error bounds for the approximants f_n of $K(1/b_n)$ and g_n of the MCF $K(1, b_n; w_n)$. Our method of obtaining bounds in Section 3 and best truncation error bounds in Section 4 makes use of element regions and their corresponding value regions. A sequence $\{E_n\}_{n=1}^{\infty}$ of non-empty subsets of $\boldsymbol{C}$ is called a *sequence of element regions for a MCF* $\underset{n=1}{\overset{\infty}{K}}(1, b_n; w_n)$ if

$$b_n \in E_n \quad \text{for} \quad n = 1, 2, 3, \ldots. \tag{45}$$

A sequence $\{V_n\}_{n=0}^{\infty}$ of subsets of $\widehat{\boldsymbol{C}}$ is called a *sequence of value regions corresponding to a sequence of element regions* $\{E_n\}_{n=1}^{\infty}$ at $\{w_n\}_{n=1}^{\infty}$ if

$$\frac{1}{E_n + w_n} \subseteq V_{n-1} \quad \text{for} \quad n = 1, 2, 3, \ldots, \tag{46}$$

and

$$\frac{1}{E_n + V_n} \subseteq V_{n-1} \quad \text{for} \quad n = 1, 2, 3, \ldots. \tag{47}$$

If $w_n \in V_n$ for $n = 1, 2, 3, \ldots$, then (47) implies (46). In addition if $w_n \in V_n$ for $n = 1, 2, 3, \ldots$, then by (7) and (47)

$$g_n := S_n(w_n) \in S_n(V_n) = S_{n-1}(s_n(V_n)) \subseteq S_{n-1}(V_{n-1}) \text{ for } n = 1, 2, 3, \ldots. \tag{48}$$

It follows that $\{S_n(V_n)\}_{n=1}^{\infty}$ is a nested sequence of non-empty subsets of $\widehat{\boldsymbol{C}}$ satisfying

$$g_{n+m} \in S_n(V_n) \quad \text{for} \quad n = 1, 2, 3, \ldots, \ m = 0, 1, 2, \ldots. \tag{49}$$

Theorem 3.1 *Let $L4P[1, \{\beta_j\}_1^4, k, \{b_j\}_1^k, \{\rho_j\}_k^{\infty}]$ be given, and let $\{E_n\}_{n=1}^{\infty}$ and $\{w_n\}_{n=1}^{\infty}$ be defined by (17), (18), and (10) respectively. Let*

$$V_{4n+i} := \left[v \in \boldsymbol{C} : \left| v - \frac{1}{\beta_{(i \bmod 4)+1}} \right| \leq \rho_{4n+i} \right] \quad i = 1, 2, 3, 4, \ 4n + i \geq k, \tag{50}$$

and

$$V_n := \left[\frac{1}{b_{n+1} + V_{n+1}} \right] \quad n = k-1, k-2, \ldots, 1, 0. \tag{51}$$

Then $\{V_n\}_{n=0}^{\infty}$ is a sequence of value regions corresponding to the sequence of element regions $\{E_n\}_{n=1}^{\infty}$ at $\{w_n\}_{n=1}^{\infty}$.

Proof. For $i = 2, 4$,

$$\frac{1}{V_{4n+i-1}} = \left[v \in \boldsymbol{C} : |v| \geq \frac{1}{\rho_{4n+i-1}} \right],$$

and for $i = 1, 3$,

$$\frac{1}{V_{4n+i-1}} = \left[v \in \boldsymbol{C} : \left| v - \frac{\beta_i}{1 - (|\beta_i| \rho_{4n+i-1})^2} \right| \leq \frac{|\beta_i|^2 \rho_{4n+i-1}}{1 - (|\beta_i| \rho_{4n+i-1})^2} \right].$$

It follows that $E_{4n+i}+V_{4n+i} \subseteq \dfrac{1}{V_{4n+i-1}}$ and hence (47) is satisfied, which completes the proof since $w_n \in V_n$ for $n = 1, 2, 3, \ldots$. □

Theorem 3.2 *Let $K(1, b_n; w_n) \in L4P[1, \{\beta_j\}_1^4, k, \{b_j\}_1^k, \{\rho_j\}_k^\infty]$ be given.*
(A) *If there exists a $k_0 \geq k$ such that*

$$\left| h_{k_0} + \frac{1}{\beta_{(k_0 mod 4)+1}} \right| > \rho_{k_0}, \tag{52}$$

then

$$\left| h_n + \frac{1}{\beta_{(n mod 4)+1}} \right| > \rho_n \quad for \quad n = k_0, k_0+1, k_0+2, \ldots, \tag{53}$$

and $K(1/b_n)$ and $K(1, b_n; w_n)$ converge to the same finite value f.
(B) *If $K(1/b_n)$ converges to a finite value f and $b_n \neq 0$, $n = 1, 2, 3, \ldots$, then there exists a $k_0 \geq k$ such that (52) holds.*

Proof. (A) It follows from (50), (51), and (52) that $-h_{k_0} \notin V_{k_0}$. By (33), $S_{k_0}(-h_{k_0}) = \infty$. Hence $\infty \notin S_{k_0}(V_{k_0})$ and thus $S_{k_0}(V_{k_0})$ is a bounded circular disk. By (5) and Theorem 3.1,

$$S_{k_0+1}(V_{k_0+1}) = S_{k_0}\left(\frac{1}{b_{k_0+1}+V_{k_0+1}}\right) \subseteq S_{k_0}(V_{k_0}).$$

Therefore $S_{k_0+1}(V_{k_0+1})$ is also a bounded circular disk. Since $S_{k_0+1}(-h_{k_0+1}) = \infty$, it follows that $-h_{k_0+1} \notin V_{k_0+1}$. This proves (53) for $n = k_0+1$. By induction one can readily verify (53) and $S_n(V_n) \subseteq S_{n-1}(V_{n-1})$ for $n = k_0+1, k_0+2, \ldots$. Hence $\{S_n(V_n)\}_{n=k_0}^\infty$ is a nested sequence of bounded circular disks and for $n = k_0+1, k_0+2, \ldots$, and $m = 0, 1, 2, \ldots$,

$$g_{n+m} := S_{n+m}(w_{n+m}) \in S_{n+m}(V_{n+m}) \subseteq S_n(V_n) \subseteq S_{k_0}(V_{k_0}). \tag{54}$$

By Theorem 2.1 we know that $\{g_n\}_{n=1}^\infty = \{S_n(w_n)\}_{n=1}^\infty$ and $\{f_n\}_{n=1}^\infty = \{S_n(0)\}_{n=1}^\infty$ both converge to a common limit $f \in \widehat{C}$. It follows from (54) that f is finite.
(B) By (43), (1), (2), and (3) for sufficiently large n,

$$\left| h_n + \frac{1}{\beta_{(n \mathrm{mod} 4)+1}} \right| \geq \left| |h_n - b_n| - |b_n + \frac{1}{\beta_{(n \mathrm{mod} 4)+1}}| \right| \geq 1.$$

Since $\lim_{n\to\infty} \rho_n = 0$, there exists $k_0 \geq k$ such that

$$\left| h_n + \frac{1}{\beta_{(n \mathrm{mod} 4)+1}} \right| > \rho_n \quad \text{for} \quad n = k_0, k_0+1, k_0+2, \ldots. \square$$

Theorem 3.3 *Let a MCF $K(1, b_n; w_n) \in L4P[1, \{\beta_j\}_1^4, k, \{b_j\}_1^k, \{\rho_j\}_k^\infty]$ be given and let f_n and g_n denote the nth approximants, respectively of $K(1/b_n)$ and $K(1, b_n; w_n)$. Then*

(A) *If there exists a* $k_0 \geq k$ *such that*

$$\left| h_{k_0} + \frac{1}{\beta_{(k_0 mod 4)+1}} \right| > \rho_{k_0} \tag{55}$$

then $K(1/b_n)$ *and* $K(1, b_n; w_n)$ *both converge to the same finite value* f *and, for* $n \geq k_0$

$$|f - f_n| \leq \frac{\rho_n}{|B_{n-1}|^2 |h_n| (|h_n| - \rho_n)} \tag{56}$$

and

$$|f - g_n| \leq \frac{\rho_n}{|B_{n-1}|^2 |w_n + h_n| (|w_n + h_n| - \rho_n)} \tag{57}$$

(B) *If* $K(1/b_n)$ *converges to a finite value and if* $b_n \neq 0$ *for* $n = 1, 2, 3, \ldots$, *then there exists a* $k_0 \geq k$ *such that (55) is satisfied and hence the assertions of (A) hold.*

Proof. (A) By Theorem 3.2A, $K(1/b_n)$ and $K(1, b_n; w_n)$ both converge to the same finite value f. Inequalities (56) and (57) follow from ([4], Theorem 2.6A).

(B) Follows directly from Theorem 3.2B. □

4. Best Truncation Error Bounds

In this section we provide best truncation error bounds for $K(1, b_j; w_j) \in L4P[1, \{\beta_j\}_1^4, k, \{b_j\}_1^k, \{\rho_j\}_k^\infty]$. We use the methods and theorems discussed in "A Survey of Truncation Error Analysis for Padé and Continued Fraction Approximants" (in this volume).

We denote by $\mathcal{V}(E)$ the family of all value regions $V = \{V_n\}_{n=0}^\infty$ corresponding to a sequence $E = \{E_j\}_{j=1}^\infty$ of element regions. Given a sequence of element regions $\{E_j\}_{j=1}^\infty$, we define

$$U_j := \left[v : v = \frac{1}{b_{j+1}} + \frac{1}{b_{j+2}} + \cdots + \frac{1}{b_{j+m} + w_{j+m}}, \; b_k \in E_k, \right.$$
$$\left. j+1 \leq k \leq j+m, \quad m = 1, 2, 3, \ldots \right]. \tag{58}$$

It follows from ([4], Theorem 2.1) that

$$\{U_j\}_{j=0}^\infty \in \mathcal{V}(E) \tag{59}$$

and

$$U_j \subseteq V_j \quad \text{whenever} \quad \{V_j\}_{j=0}^\infty \in \mathcal{V}(E) \quad \text{for} \quad j = 0, 1, 2, \ldots. \tag{60}$$

This motivates us to call $\{U_j\}_{j=0}^\infty$ the *best sequence of value regions corresponding to* E.

Let $L4P[1, \{\beta_j\}_1^4, k, \{b_j\}_1^k, \{\rho_j\}_k^\infty]$ be given and here denoted by $\mathcal{F} = L4P[1, \{\beta_j\}_1^4, k, \{b_j\}_1^k, \{\rho_j\}_k^\infty]$. Let a finite $F = K(1, b_j; w_j) \in \mathcal{F}$ be given. For each $n = 1, 2, 3, \ldots$, let

$$\mathcal{F}_n(F) := [G = K(1, c_j; w_j) \in \mathcal{F} : c_j = b_j \text{ for } 1 \leq j \leq n].$$

Let

$$\ell_n(F,\mathcal{F}) := [\lambda \in \boldsymbol{C} : \text{for some } G \in \mathcal{F}_n(F),\ \lambda \text{ is a finite limit point of } \{v_m(G)\}], \tag{61}$$

where $v_m(G)$ denotes the mth approximant of G. Then the set

$$L_n(F,\mathcal{F}) := c(\ell_n(F,\mathcal{F})) \tag{62}$$

is called the nth limit region for $\mathcal{F}_n(F)$. Since the finite value $K(1,b_j;w_j) \in \ell_n \subseteq L_n$ we see $L_n \neq \emptyset$.

The definition of best truncation error bound for g_n with respect to $\{E_j\}_{j=1}^{\infty}$ of element regions is based on the assumption that our usable knowledge about $K(1/b_j)$ consists of i) the values $b_1, b_2, b_3, \ldots, b_n$, ii) $b_j \in E_j$ for $j \geq n+1$. Subject to these restrictions the *best truncation error bound* $\beta_n(F,\mathcal{F})$ for the nth approximant g_n with respect to $\{E_j\}_{j=1}^{\infty}$ is given by

$$\beta_n(F,\mathcal{F}) := [\sup|\lambda - f_n| : \lambda \in L_n(F,\mathcal{F})]. \tag{63}$$

Since $F \in L_n(F,\mathcal{F})$ we have $|F - g_n| \leq \beta_n(F,\mathcal{F})$.

Theorem 4.1 *Let a finite MCF $K(1,b_j;w_j) \in L4P[1,\{\beta_j\}_1^4,k,\{b_j\}_1^k,\{\rho_j\}_k^{\infty}]$ be given and let n be a given positive integer. If there exists a $k_0 \geq k$ such that (55) holds, then the conditions of Theorem 3.3 are satisfied and the truncation error bounds (57) for $n \geq k_0$ are best with respect to the element regions (17) and (18).*

Proof. By Theorem 3.1 and ([7], Theorem 3.1) conditions (2.36) and (2.37) in ([4], Theorem 2.7) are satisfied and hence $c(V_n) = c(U_n(F))$. Let

$$G_j := \frac{1}{b_1} + \frac{1}{b_2} + \cdots + \frac{1}{b_n} + \frac{1}{b_{n+1+4j}} + \frac{1}{b_{n+2+4j}} + \cdots.$$

By Lemma 2.1, $w_{n+k} = lim_{j\to\infty} v(G_j^{(n+k)})$. The result now follows from ([4] , Theorem 2.5).□

References

1. Christopher Baltus and W. B. Jones, Truncation error bounds for limit-periodic continued fractions $K(a_n/1)$ with $\lim a_n = 0$, *Numer. Math.* **46** (1985), 541–569.
2. Christopher Baltus and W. B. Jones, Truncation error bounds for modified continued fractions with applications to special functions, *Numer. Math.* **55** (1989), 281–307.
3. C. M. Craviotto, W. B. Jones and W. J. Thron, Best truncation error bounds for continued fractions $K(1/b_n)$ with $\lim_{n\to\infty} b_n = \infty$, *Continued Fractions and Orthogonal Functions; Theory and Applications* (eds., S. C. Cooper and W. J. Thron).
4. C. Craviotto, W.B. Jones, W.J. Thron, A Survey of Truncation Error Analysis for Padé and Continued Fraction Approximants (in this volume).
5. L. Jacobsen, W. B. Jones, H. Waadeland, Further results on the computation of incomplete gamma functions, Lecture Notes Math **1199** (1986), 90–126.
6. W. B. Jones and W. J. Thron, *Continued Fractions: Analytic Theory and Applications*, in Encyclopedia of Mathematics and its Applications, Addison–Wesley, Reading, MA, 1980.
7. W. B. Jones and W. J. Thron, Continued fractions in numerical analysis, *Applied Numerical Mathematics* **4** (1988), 143–230.

CONTINUED FRACTIONS FOR THE SYMMETRIC STRONG STIELTJES MOMENT PROBLEM

A.K. COMMON*
Institute of Mathematics and Statistics
University of Kent, Canterbury, Kent CT2 7NF, England.

and

J.H. MCCABE
Department of Mathematical and Computational Sciences
University of St. Andrews, St. Andrews
Fife, Scotland KY16 9SS, England.

Abstract.
The properties of the Strong Stieltjes moment problem

$$\mu_n = \int_0^\infty u^n d\psi(u) \; ; \quad n = 0, \pm 1, \pm 2, \ldots$$

with $\psi(u)$ a bounded non-decreasing function of u have been well studied. Here we consider the case when the moments have the SYMMETRY property,

$$\mu_n = \mu_{-n} \; ; \quad n = 1, 2, \ldots$$

It will be demonstrated that continued fractions constructed from series with these moments as coefficients have corresponding symmetry properties. In particular we consider implications for M-fractions, T-fractions and PC-fractions corresponding to the log-normal distribution.

1. Introduction

In previous work by one is us (Common, 1992), it was demonstrated that solutions of the initial value problem for the Discrete Modified KdV equation are related to the subset of solutions of the Strong Stieltjes moment problem

$$\mu_n = \int_0^\infty u^n d\psi(u) \; ; \quad n = 0, \pm 1, \pm 2, \ldots \tag{1}$$

with the SYMMETRY property

$$\mu_n = \mu_{-n} \; ; \quad n = 1, 2, \ldots . \tag{2}$$

When $\psi(u)$ is a bounded non-decreasing function of u with infinitely many points of increase all the corresponding Hankel determinants,

$$H_n^{(m)} = \begin{vmatrix} \mu_m & \mu_{m+1} & \cdots & \mu_{m+n-1} \\ \mu_{m+1} & & & \vdots \\ \vdots & & & \vdots \\ \mu_{m+n-1} & & \cdots & \mu_{m+2n-2} \end{vmatrix} \; ; \; n = 0, 2, 3, \ldots ; \; m = 0, \pm 1, \pm 2, \ldots \tag{3}$$

* UKC/IMS/A93/9b

A. Cuyt (ed.), Nonlinear Numerical Methods and Rational Approximation II, 387–394.

are strictly positive.

Here we consider continued fractions associated with the pair of series

$$\mu_0 z^{-1} + \mu_1 z^{-2} + \cdots + \mu_n z^{-(n+1)} + \cdots \tag{4}$$

and

$$-\mu_{-1} - \mu_{-2} z - \mu_{-3} z^2 - \cdots - \mu_{-n} z^{n-1} - \cdots \tag{5}$$

In §2 we demonstrate that the related $n-d$ table (McCabe, 1983) is symmetric when the coefficients of the series have the SYMMETRY property (2). This result will then be used to obtain a corresponding property of the denominators of related M-fractions. T-fractions corresponding to the Strong Stieltjes moment problem have been investigated by Jones and Thron and Waadeland (1980) and more recently Jones, Njåstad and Thron (1987) have considered Perron-Carathéodory (PC) fractions related to this moment problem. We will investigate in §3 the symmetry of these Strong Stieltjes PC ($SSPC$) fractions when the moments are symmetric. Finally in §4, we consider the example of the log-normal distribution $\psi(u)$ defined by

$$d\psi(u) = \frac{q^{\frac{1}{2}} e^{-\left[\frac{\ln u}{2\mathcal{K}}\right]^2}}{2\mathcal{K}\sqrt{\pi}u}\, du \;\; ; \;\; q = e^{-2\mathcal{K}^2} \tag{6}$$

It will be demonstrated that the $n-d$ table has a particularly simple structure in this case.

2. M-Fractions and the $n-d$ Table

From the two series (4), (5) one may construct the $n-d$ array (McCabe, 1983)

$$\begin{array}{llllllll}
\multicolumn{8}{c}{\cdots\cdots\cdots\cdots\cdots\cdots} \\
d_1^{-3} & n_2^{-3} & \cdots & & & & & \\
d_1^{-2} & n_2^{-2} & d_2^{-2} & n_3^{-2} & \cdots & & & \\
d_1^{-1} & n_2^{-1} & d_2^{-1} & n_3^{-1} & d_3^{-1} & \cdots & & \\
d_1^{0} & n_2^{0} & d_2^{0} & n_3^{0} & d_3^{0} & n_4^{0} & d_4^{0} & \cdots \\
d_1^{1} & n_2^{1} & d_2^{1} & n_3^{1} & d_3^{1} & \cdots & & \\
d_1^{2} & n_2^{2} & d_2^{2} & n_3^{2} & \cdots & & & \\
d_1^{3} & n_2^{3} & \cdots & & & & &
\end{array} \tag{7}$$

from the rhombus rules,

$$n_j^k = d_{j-1}^{k+1} + n_{j-1}^{k+1} - d_{j-1}^k \tag{8}$$

$$d_j^k = \frac{n_j^k * d_{j-1}^{k-1}}{n_j^{k-1}} \tag{9}$$

for $j = 2, 3, \ldots$ and $k = 0, \pm1, \pm2, \ldots$ with starting values,

$$n_1^k = 0 \quad d_1^k = -\mu_{-(k+1)}/\mu_{-k}\ ; \quad k = 0, \pm1, \pm2, \ldots \tag{10}$$

Theorem (2.1)

If the $\{\mu_k; k = 0, \pm1, \pm2, \ldots\}$ satisfy the SYMMETRY condition (2), then

$$d_r^k = \frac{1}{d_r^{-(k+1)}}\ ; \quad r = 1, 2, 3, \ldots \tag{11}$$

$$n_r^k = \frac{n_r^{-(k+1)}}{d_r^{-(k+1)} d_{r-1}^{-(k+1)}}\ ; \quad r = 2, 3, \ldots \tag{12}$$

□

The proof of this theorem is by induction on r and it implies that elements of the kth row of the $n-d$ array (7) are related in a simple way to those in the $-(k+1)^{\text{th}}$ row.

It may be shown that the M-fractions (McCabe, 1983)

$$\frac{-\mu_{-(k+1)}}{1 + d_1^k z} + \frac{n_2^k z}{1 + d_2^k z} + \frac{n_3^k z}{1 + d_3^k z} + \frac{n_4^k z}{1 + d_4^k z} + \cdots; \quad k = 0, \pm1, \pm2, \ldots \tag{13}$$

correspond to the two series

$$\mu_{-k} z^{-1} + \mu_{-k+1} z^{-2} + \mu_{-k+2} z^{-3} + \cdots \tag{14}$$

and

$$-\mu_{-(k+1)} - \mu_{-(k+2)} z - \mu_{-(k+3)} z^2 - \mu_{-(k+4)} z^3 - \cdots \tag{15}$$

Denote the rth denominator of the continued fraction (13) by $B_{r,k}(z)$. Then $B_{r,k}(z)$ is a polynomial of degree r in z and we can form the table

$$\begin{array}{cccc}
\multicolumn{4}{c}{\cdots\cdots\cdots\cdots\cdots\cdots\cdots\cdots\cdots\cdots} \\
B_{1,-2}(z) & B_{2,-2}(z) & B_{3,-2}(z) & B_{4,-2}(z)\ldots \\
B_{1,-1}(z) & B_{2,-1}(z) & B_{3,-1}(z) & B_{4,-1}(z)\ldots \\
\hline
B_{1,0}\ (z) & B_{2,0}\ (z) & B_{3,0}\ (z) & B_{4,0}\ (z)\ldots \\
B_{1,1}\ (z) & B_{2,1}\ (z) & B_{3,1}\ (z) & B_{4,1}\ (z)\ldots \\
\multicolumn{4}{c}{\cdots\cdots\cdots\cdots\cdots\cdots\cdots\cdots\cdots\cdots}
\end{array} \tag{16}$$

Theorem (2.2)

The table of denominator polynomials in (16) is SYMMETRIC about a horizontal line between $B_{1,0}(z)$ and $B_{1,-1}(z)$ in the sense that any polynomial $B_{n,m}(z)$ can be obtained from its image polynomial $B_{n,-(m+1)}(z)$ by reversing the coefficients and *normalising so that* $B_{n,m}(0) = 1$.

Proof. By replaced z by $1/z$ in (13), we obtain after a sequence of similarity transforms,

$$\frac{-\mu_{-(k+1)}z/d_1^k}{1+z/d_1^k} + \frac{n_2^k z/d_1^k d_2^k}{1+z/d_2^k} + \frac{n_3^k z/d_2^k d_3^k}{1+z/d_3^k} + \cdots \tag{17}$$

which on using *Th(2.1)* gives the continued fraction

$$\frac{-\mu_{-(k+1)}d_1^{-(k+1)}z}{1+d_1^{-(k+1)}z} + \frac{n_2^{-(k+1)}z}{1+d_2^{-(k+1)}z} + \frac{n_3^{-(k+1)}z}{1+d_3^{-(k+1)}z} + \cdots \tag{18}$$

This latter continued fraction has denominators in the $-(k+1)^{\text{th}}$ row of table (16) and the result follows. □

In an extended version of this work (Common and McCabe, 1993) we show that when the coefficients $\{\mu_k\}$ in (14), (15) are solutions of the SYMMETRIC Strong Stieltjes moment problem (1), (2) then the denominators $B_{n,m}(z)$ for $m = -1$ and 0 correspond to sets of orthogonal polynomials defined with respect to the symmetric weight function $d\psi(u)$.

3. Perron-Carathéodory Fractions

It is well known that T-fractions have even extensions which are Perron fractions. We now consider this extension in our case of the SYMMETRIC Strong Stieltjes moment problem. In the notation of Jones and Thron (1980), for the series

$$-\mu_0 z - \mu_1 z^2 - \mu_2 z^3 - \cdots - \mu_n z^{n+1} - \cdots \tag{19}$$

and

$$\mu_{-1} + \mu_{-2}z^{-1} + \mu_{-3}z^{-2} + \cdots + \mu_{-n}z^{-n-1} + \cdots \tag{20}$$

there is a corresponding T-fraction,

$$\frac{F_1 z}{1+G_1 z} + \frac{F_2 z}{1+G_2 z} + \frac{F_3 z}{1+G_3 z} + \cdots \tag{21}$$

where

$$F_n = -\frac{H_{n-2}^{(-n+2)}H_n^{(-n+1)}}{H_{n-1}^{(-n+1)}H_{n-1}^{(-n+2)}} \quad ; \quad G_n = -\frac{H_{n-1}^{(-n+1)}H_n^{(-n+1)}}{H_n^{(-n)}H_{n-1}^{(-n+2)}} \quad . \tag{22}$$

An even extension for the above case is given by the following result.

Theorem (3.1)

When the $\{\mu_k; k = 0, \pm1, \pm2, \pm3, \ldots\}$ *are solutions of the SYMMETRIC Strong Stieltjes moment problem, then the T-fraction (21) has an even extension of the form,*

$$\mathop{K}_{n=1}^{\infty}\left[\frac{\alpha_n}{\beta_n}\right] = \frac{\alpha_1}{1} + \frac{\lambda}{\beta_2} + \frac{\alpha_3}{\beta_3} + \frac{\lambda}{\beta_4} + \frac{\alpha_5}{\beta_5} + \cdots \tag{23}$$

where

$$\lambda = -z, \alpha_1 = \mu_o, \alpha_{2n+1} = (1 - \beta_{2n}\beta_{2n+1}) < 0; n = 0, 1, 2, \ldots \tag{24}$$

and

$$\beta_{2n} = \beta_{2n+1} = \frac{H_n^{(-n)}}{H_n^{(-n+1)}} \ ; \ n = 1, 2, \ldots, \ . \tag{25}$$

Proof. The even part of the fraction (23) is the fraction $\underset{n=1}{\overset{\infty}{K}} \left[\frac{a_n}{b_n}\right]$, where

$$a_1 = \mu_0\beta_2, b_1 = \lambda + \beta_2, a_2 = -\lambda(1 - \beta_2\beta_3)\beta_4 \tag{26}$$

$$a_n = -\lambda(1 - \beta_{2n-2}\beta_{2n-1})\beta_{2n-4}\beta_{2n}; n = 2, 3, \ldots \tag{27}$$

$$b_n = \beta_{2n} + \beta_{2n-2}\lambda \ ; \ \ n = 2, 3, 4, \ldots \ . \tag{28}$$

This is equivalent to the fraction $\underset{n=1}{\overset{\infty}{K}} \left[\frac{a_n^*}{b_n^*}\right]$ where

$$a_n^* = \frac{a_n}{\beta_{2n}\beta_{2n-2}} \ , \quad b_n^* = \frac{b_n}{\beta_{2n}} \ ; \quad n = 1, 2, 3, \ldots; \ \ \beta_0 = 1 \ . \tag{29}$$

Using expressions (25) for the β_n's in (26) to (29)

$$a_1^* = \mu_o \ , \quad b_1^* = 1 + \lambda\frac{H_1^{(0)}}{H_1^{(-1)}} = 1 + \lambda\frac{\mu_0}{\mu_1} \tag{30}$$

$$a_n^* = -\lambda\left[\frac{H_{n-2}^{(-n+2)}H_{n-1}^{(-n+2)}}{H_{n-2}^{(-n+3)}H_{n-1}^{(-n+1)}} - \frac{H_{n-1}^{(-n+1)}H_{n-2}^{(-n+2)}}{H_{n-1}^{(-n+2)}H_{n-2}^{(-n+3)}}\right] \tag{31}$$

$$b_n^* = 1 + \lambda\frac{H_{n-1}^{(-n+1)}H_n^{(-n+1)}}{H_{n-1}^{(-n+2)}H_n^{(-n)}} \ . \tag{32}$$

In the SYMMETRIC case

$$H_n^{(-m)} = H_n^{(m-2n+2)} \tag{33}$$

and using this in Jacobi's identity,

$$\left[H_{n-1}^{(-n+2)}\right]^2 - \left[H_{n-1}^{(-n+1)}\right]^2 = -H_{n-2}^{(-n+3)}H_n^{(-n+1)} \ . \tag{34}$$

Substituting in (31),

$$a_n^* = \lambda\frac{H_{n-2}^{(-n+2)}H_n^{(-n+1)}}{H_{n-1}^{(-n+1)}H_{n-1}^{(-n+2)}} \ . \tag{35}$$

Comparing (35), (32) with (22), we see that when $\lambda = -z$, the T-fraction (21) is equivalent to the EVEN part of the PC fraction (23). Finally we can use (34) with $n \to n+1$ to write

$$\alpha_n = \frac{(H_n^{(-n+1)})^2 - (H_n^{(-n)})^2}{(H_n^{(-n)})^2} = -\frac{H_{n-1}^{(-n+2)}H_{n+1}^{(-n)}}{(H_n^{(-n)})^2} \tag{36}$$

and this proves (24). □

Continued fractions of the form (23), but with $\beta_{2n} \neq \beta_{2n+1}$, have been constructed from the general Strong Stieltjes moment problem by Jones, Njåstad and Thron (1987) and they called them *SSPC* fractions. We see that when the SYMMETRY property (2) of the coefficients of the corresponding series holds, then $\beta_{2n} = \beta_{2n+1}$ for $n = 1, 2, \ldots$. We call this sub-class of *SSPC* fractions *SSSPC* fractions.

PC fraction of the general form (23) with $\alpha_{2n+1} = 1 - \beta_{2n}\beta_{2n+1}$ were introduced by Jones, Njåstad and Thron (1986). They defined in particular positive PC fractions corresponding to the case where the general complex β_n satisfied the relations $\beta_{2n} = \overline{\beta}_{2n+1}$ and $|\beta_{2n}| < 1$; $n = 1, 2, \ldots$. We see these fractions, when the β_n are real, complement our SSSPC fractions since the latter have $\beta_{2n} = \beta_{2n+1}$ and $|\beta_{2n}| > 1$; $n = 1, 2, \ldots$.

4. The Log-Normal Distribution

The Log-Normal distribution is that where

$$d\psi(u) = \frac{q^{1/2} e^{-\left[\frac{\ln u}{2\mathcal{K}}\right]^2}}{2\mathcal{K}\sqrt{\pi} u}\, du \; ; \quad q = e^{-2\mathcal{K}^2} \tag{37}$$

and then

$$\mu_k = \mu_{-k} = q^{(1-k^2)/2} \quad ; \quad k = 0, 1, 2, \ldots . \tag{38}$$

Therefore this distribution corresponds to a SYMMETRIC Strong Stieltjes moment problem. The corresponding Hankel determinants are then given by the following algebraic functions of q:

$$H_k^{(m)} = \left[q^{-(m+k-2)^2/2-(m+k-2)}\right]^k q^{-k(k^2-1)/6} \prod_{j=1}^{k-1} (1 - q^j)^{k-j} . \tag{39}$$

Substituting in (24), (25)

$$\alpha_{2n+1} = (1 - q^{-n}) < 0 \; ; \; \beta_{2n} = \beta_{2mn+1} = \frac{H_n^{(-n)}}{H_n^{(-n+1)}} = q^{-n/2} \; ; \; n = 1, 2, \ldots \tag{40}$$

The corresponding *SSSPC* fraction is

$$\mathop{K}_{n=1}^{\infty} \left(\frac{\alpha_n}{\beta_n}\right) = \frac{q^{\frac{1}{2}}}{1} + \frac{\lambda}{q^{-\frac{1}{2}}} + \frac{(1-q^{-1})}{q^{-\frac{1}{2}}} + \frac{\lambda}{q^{-1}} + \frac{(1-q^{-2})}{q^{-1}} + \cdots \tag{41}$$

and has even contraction which is the T-fraction,

$$\mathop{K}_{n=1}^{\infty} \left(\frac{a_n}{b_n}\right) = \frac{q^{1/2}\lambda}{1 - \lambda q^{1/2}} - \frac{q^{-1/2}(1-q)\lambda}{1 - \lambda q^{1/2}} - \frac{q^{-3/2}(1-q^2)\lambda}{1 - \lambda q^{1/2}} - \cdots \tag{42}$$

which has been given previously by Cooper, Jones and Thron (1990).

The $n-d$ array (7) in this case has the form,

$$\begin{array}{llllll}
\multicolumn{6}{c}{\cdots\cdots\cdots\cdots\cdots\cdots\cdots\cdots\cdots\cdots\cdots\cdots\cdots\cdots\cdots} \\
(k=-3) & -q^{5/2} & -q^{3/2}\,(1-q) & -q^{5/2} & -q^{1/2}(1-q^2) & \ldots \\
(k=-2) & -q^{3/2} & -q^{1/2}\,(1-q) & -q^{3/2} & -q^{-1/2}(1-q^2) & \ldots \\
(k=-1) & -q^{1/2} & -q^{-1/2}(1-q) & -q^{1/2} & -q^{-3/2}(1-q^2) & \ldots \\
(k=0\ ) & -q^{-1/2} & -q^{-3/2}(1-q) & -q^{-1/2} & -q^{-5/2}(1-q^2) & \ldots \\
(k=1\ ) & -q^{-3/2} & -q^{-5/2}(1-q) & -q^{-3/2} & -q^{-7/2}(1-q^2) & \ldots \\
(k=2\ ) & -q^{-5/2} & -q^{-7/2}(1-q) & -q^{-5/2} & -q^{-9/2}(1-q^2) & \ldots \\
(k=3\ ) & -q^{-7/2} & -q^{-9/2}(1-q) & -q^{-7/2} & -q^{-11/2}(1-q^2) & \ldots \\
\multicolumn{6}{c}{\cdots\cdots\cdots\cdots\cdots\cdots\cdots\cdots\cdots\cdots\cdots\cdots\cdots\cdots\cdots}
\end{array} \tag{43}$$

It is seen that this table has a particularly simple structure since the entries in the $(k+1)^{\text{th}}$ row are just q^{-1} times the corresponding entries in the kth row.

The T-fraction (42) corresponds to the $(k=-1)$ row of this table while

$$\underset{n=1}{\overset{\infty}{K}}\left(\frac{a'_n}{b'_n}\right)=\frac{q^{-1/2}\lambda}{1-\lambda q^{-1/2}}-\frac{q^{-3/2}(1-q)\lambda}{1-\lambda q^{-1/2}}-\frac{q^{-5/2}(1-q^2)\lambda}{1-\lambda q^{-1/2}}-\cdots \tag{44}$$

corresponds to the $(k=0)$ row. Now $\underset{n=1}{\overset{\infty}{K}}\left(\frac{a_n}{b_n}\right)$ can be obtained by substituting $\lambda\to\lambda q$ in $\underset{n=1}{\overset{\infty}{K}}\left(\frac{a'_n}{b'_n}\right)$. Therefore the denominator polynomials $B_{m,k}(\lambda)$ defined in §2 satisfy

$$B_{n,-1}(\lambda)=B_{n,0}(q\lambda)\ ;\quad n=0,1,2,\ldots \tag{45}$$

and more generally

$$B_{n,k}(\lambda)=B_{n,0}(q^{-k}\lambda)\ ;\quad n=0,1,2,\ldots,\ k=0,\pm1,\pm2,\ldots \tag{46}$$

Therefore in the Log-normal case, the table (16) of denominators for the family of M-fractions (13) may be constructed in a simple way from those in the $(k=0)$ row. It of course also has the previously determined reflection symmetry about a horizontal line between the $(k=0$ and $(k=-1)$ rows.

References

1. Common, A.K.: 1992, 'A Solution of the Initial Value Problem for Half-Infinite Integrable Lattice Systems', *Inverse Problems* Vol. no. **8**, pp. 393-408.
2. Common, A.K. and McCabe, J.H.: 1993, 'The Symmetric Strong Moment Problem', University of Kent preprint IMS/A93/3. Submitted to *Journ.Comput.Appl.Maths.* for publication.
3. Cooper, S.C., Jones, W.B. and Thron, W.J.: 1990, 'Orthogonal Laurent-Polynomials and Continued Fractions Associated with Log-Normal Distributions', *Journ.Comput.Appl.Maths* Vol. no. **32**, pp. 39-46.
4. Jones, W.B. and Thron, W.J.: 1980, 'Continued Fractions: Analytic Theory and Applications', Additon-Wesley, Reading MA.
5. Jones, W.B., Thron, W.J. and Waadeland, H.: 1980, 'A Strong Stieltjes Moment Problem', *Trans.Amer.Math.Soc.* **261**, pp. 503-528.
6. Jones, W.B., Njåstad, O. and Thron, W.J.: 1986, 'Continued Fractions Associated with Trigonometric and Other Strong Moment Problems', *Constr.Approx.* **2**, pp. 197-211.
7. Jones, W.B., Njåstad, O. and Thron, W.J.: 1987, 'Perron-Carathéodory Continued Fractions', in *Rational Approximation and its Applications to Mathematics and Physics*, Lecture Notes in Maths. Vol. no. **1237** (J. Gilewicz et.al. eds.).
8. McCabe, J.H.: 1983, 'The Quotient-Difference Algorithm and the Padé Table: An Alternative Form and a General Continued Fraction', *Maths. of Computation* Vol. no. **41** pp. 183-197.

OBSERVATIONS ON INDETERMINATE STIELTJES MOMENT PROBLEMS

CATHERINE M. BONAN-HAMADA, WILLIAM B. JONES and W. J. THRON*
Department of Mathematics, University of Colorado, Boulder, CO 80309-0395, U.S.A.

Abstract. This paper is concerned with indeterminate classical Stieltjes moment problems (CSMPs) and indeterminate strong Stieltjes moment problems (SSMPs). We are particularly interested in what we call the natural solutions to each type of moment problem. Known results are summarized and we compare Stieltjes' results on indeterminate CSMPs to analogous results for a special class of indeterminate SSMPs. For the indeterminate CSMP each natural solution is a step function whose points of increase have a limit point at ∞. Each natural solution to an indeterminate SSMP is a step function whose points of increase have limit points only at 0 and/or ∞. Every convex linear combination of the natural solutions to an indeterminate Stieltjes moment problem (classical and strong) is also a solution to that problem. It is shown that there exist indeterminate Stieltjes moment problems (both classical and strong) which have step function solutions which are not convex linear combinations of the natural solutions. Finally, it is shown how to construct new moment problems and their solutions from known moment problems and their solutions so that the moments of both problems are related by "shifts" and "flips."

Key words: Classical Stieltjes moment problems, strong Stieltjes moment problems, continued fractions, indeterminate moment problems, natural solutions

1. Introduction

For each interval (a, b) such that $-\infty \le a < b \le \infty$ let $\Psi(a,b)$ denote the family of distribution functions $\psi(t)$ (i.e., real-valued, bounded, non-decreasing functions with infinitely many points of increase) on $a < t < b$. The *classical Stieltjes moment problem (CSMP) for a sequence* $\{\mu_n\}_{n=0}^{\infty}$ is to find necessary and sufficient conditions for there to exist a $\varphi \in \Psi(0,\infty)$ such that

$$\mu_n = \int_0^\infty (-t)^n d\varphi(t), \quad n = 0, 1, 2, \dots. \tag{1}$$

A distribution function $\varphi(t) \in \Psi(0,\infty)$ satisfying (1) is called *a solution to the CSMP for* $\{\mu_n\}_{n=0}^{\infty}$. The CSMP was posed and solved by T. J. Stieltjes in 1894 [7].

The *strong Stieltjes moment problem (SSMP) for a bisequence* $\{c_n\}_{n=-\infty}^{\infty}$ consists of finding necessary and sufficient conditions for there to exist a $\psi \in \Psi(0,\infty)$ such that

$$c_n = \int_0^\infty (-t)^n d\psi(t), \quad n = 0, \pm 1, \pm 2, \dots. \tag{2}$$

* Research supported in part by the National Science Foundation under Grant No. DMS-9302584.

A. Cuyt (ed.), *Nonlinear Numerical Methods and Rational Approximation II*, 395–407.

A distribution function $\psi(t)$ satisfying (2) is called a *solution to the SSMP for* $\{c_n\}_{n=-\infty}^{\infty}$. It is also said that $\psi(t)$ has $\{c_n\}_{n=-\infty}^{\infty}$ as its moment sequence. The SSMP was posed and solved in 1980 [5].

A solvable moment problem is said to be *determinate* if there exists only one solution. Otherwise it is called *indeterminate*. This paper will deal primarily with indeterminate CSMPs and SSMPs.

2. Natural Solutions to Indeterminate CSMPs and SSMPs

In his famous paper [7] published in 1894, Stieltjes proved a great many results about moment theory. Here we summarize those that pertain to the present paper.

Theorem 2.1 (Stieltjes) *The CSMP for a sequence $\{\mu_k\}_{k=0}^{\infty}$ has a solution iff there exists a continued fraction of the form*

$$\frac{1}{b_1} + \frac{1}{b_2 z} + \frac{1}{b_3} + \frac{1}{b_4 z} + \cdots, \quad b_n > 0, \ n = 1, 2, 3, \ldots, \tag{3}$$

corresponding to the formal Laurent series (fLs)

$$L_\infty(z) := \sum_{k=0}^{\infty} \mu_k z^{-k} \tag{4}$$

in the sense that

$$\Lambda_\infty\left(\frac{A_n(z)}{B_n(z)}\right) - L_\infty(z) = O\left(\left(\frac{1}{z}\right)^n\right), \quad n = 1, 2, 3, \ldots, \tag{5}$$

where $A_n(z)$ and $B_n(z)$ denote the (polynomial) nth numerator and denominator, respectively, of (3).

Here $\Lambda_\infty(R(z))$ denotes the Laurent series expansion of $R(z)$ about $z = \infty$ and $O\left(\left(\frac{1}{z}\right)^n\right)$ denotes a fLs of increasing powers of $\frac{1}{z}$ starting with a power $\geq n$.
Remark: Stieltjes actually worked with continued fractions of the form

$$\frac{1}{b_1 z} + \frac{1}{b_2} + \frac{1}{b_3 z} + \frac{1}{b_4} + \cdots.$$

We chose to present a completely analogous statement of his results using the continued fraction (3) in order to exhibit the similarities between the CSMP and the SSMP.

Theorem 2.2 (Stieltjes) *The following three statements are equivalent:*

(i) *The CSMP for a given sequence $\{\mu_k\}_{k=0}^{\infty}$ is indeterminate.*

(ii) *The corresponding continued fraction (3) is divergent for all $z \in S_\pi := \{z \in \mathbb{C} : |\arg z| < \pi\}$.*

(iii)

$$\sum_{n=0}^{\infty} b_n < \infty. \tag{6}$$

Theorem 2.3 (Stieltjes) *If the CSMP for $\{\mu_k\}_{k=0}^{\infty}$ is indeterminate, then:*

(A) *For $\sigma = 0, 1$, the sequences $\{A_{2n+\sigma}(z)\}_{n=1}^{\infty}$ and $\{B_{2n+\sigma}(z)\}_{n=1}^{\infty}$ converge uniformly on compact subsets of $\mathbf{C}$ to entire functions $A^{(\sigma)}(z)$ and $B^{(\sigma)}(z)$, respectively. Moreover, for $z \in \mathbf{C}$*

$$B^{(1)}(z)A^{(0)}(z) - B^{(0)}(z)A^{(1)}(z) = 1.$$

(B) *For $\sigma = 0, 1$, the zeros $t_{2n+\sigma,k}$, $k = 1, 2, \ldots, 2n+\sigma$, of $B_{2n+\sigma}(z)$ are negative and simple and there exists a $\beta > 0$, independent of n and σ, such that*

$$t_{2n+\sigma,2n+\sigma} < t_{2n+\sigma,2n+\sigma-1} < \cdots < t_{2n+\sigma,1} < -\beta < 0. \tag{7}$$

(C) *For each $\sigma = 0, 1$ and each $k = 1, 2, 3, \ldots$, the sequence $\{t_{2n+\sigma,k}\}_{n=k}^{\infty}$ is convergent and the limits*

$$t_k^{(\sigma)} := \lim_{n\to\infty} t_{2n+\sigma,k} \tag{8}$$

satisfy

$$\cdots < t_3^{(\sigma)} < t_2^{(\sigma)} < t_1^{(\sigma)} \leq -\beta < 0 \tag{9}$$

and

$$\lim_{k\to\infty} t_k^{(\sigma)} = -\infty. \tag{10}$$

(D) *For $\sigma = 0, 1$ the set of zeros of $B^{(\sigma)}(z)$ is given by $\{t_1^{(\sigma)}, t_2^{(\sigma)}, t_3^{(\sigma)}, \ldots\}$ and each zero is simple.*

(E) *For $\sigma = 0, 1$ and $n = 1, 2, 3, \ldots$*

$$\frac{A_{2n+\sigma}(z)}{B_{2n+\sigma}(z)} = \sum_{k=1}^{2n+\sigma} \frac{z\lambda_{2n+\sigma,k}}{z - t_{2n+\sigma,k}}, \quad \lambda_{2n+\sigma,k} := \frac{A_{2n+\sigma}(t_{2n+\sigma,k})}{t_{2n+\sigma,k}B'_{2n+\sigma}(t_{2n+\sigma,k})} > 0. \tag{11}$$

(F) *For $\sigma = 0, 1$ and $k = 1, 2, 3, \ldots$, $\{\lambda_{2n+\sigma,k}\}_{n=1}^{\infty}$ converges to*

$$\lambda_k^{(\sigma)} := \lim_{n\to\infty} \lambda_{2n+\sigma,k} \quad \textit{where} \quad \sum_{k=1}^{\infty} \lambda_k^{(\sigma)} = \frac{1}{b_1} = \mu_0 > 0. \tag{12}$$

Each step function $\varphi^{(\sigma)}(t)$, $\sigma = 0, 1$, with a positive jump $\lambda_k^{(\sigma)}$ at $t = -t_k^{(\sigma)} > 0$, $k = 1, 2, 3, \ldots$, is a solution to the CSMP for $\{\mu_k\}_{k=0}^{\infty}$.

(G) *For $\sigma = 0, 1$,*

$$F^{(\sigma)}(z) := \frac{A^{(\sigma)}(z)}{B^{(\sigma)}(z)} = z\int_0^{\infty} \frac{d\varphi^{(\sigma)}(t)}{z+t} = \sum_{k=1}^{\infty} \frac{z\lambda_k^{(\sigma)}}{z - t_k^{(\sigma)}}. \tag{13}$$

is holomorphic in $S := \mathbf{C} - [t_1^{(\sigma)}, t_2^{(\sigma)}, t_3^{(\sigma)}, \ldots]$, and the series is uniformly convergent on every compact subset of S.

We note that throughout this paper the superscript (σ) is an index and is not to be interpreted as a derivative.

The distribution functions $\varphi^{(0)}(t)$ and $\varphi^{(1)}(t)$ are called the *natural solutions to the indeterminate CSMP for* $\{\mu_k\}_{k=0}^{\infty}$. Every convex linear combination $\varphi(\alpha,t) := (1-\alpha)\varphi^{(0)}(t)+\alpha\varphi^{(1)}(t)$, $0 \le \alpha \le 1$, of $\varphi^{(0)}(t)$ and $\varphi^{(1)}(t)$ is a solution to the CSMP for $\{\mu_k\}_{k=1}^{\infty}$. An indeterminate CSMP may also have other solutions not of the form $\varphi(\alpha,t)$ (see, e.g., Section 3).

The situation for the SSMP is more complicated than that for the CSMP but there are a number of similarities. In the next three theorems we summarize those results obtained in [5] which we use in this paper.

Theorem 2.4 *The SSMP for a sequence* $\{c_k\}_{k=-\infty}^{\infty}$ *has a solution iff there exists a positive T-fraction*

$$\frac{F_1 z}{1+G_1 z} + \frac{F_2 z}{1+G_2 z} + \frac{F_3 z}{1+G_3 z} + \cdots, \; F_n > 0, \; G_n > 0, \; n = 1,2,\ldots, \tag{14}$$

corresponding to the pair $(L_0(z), L_\infty(z))$ *of fLs*

$$L_0(z) := \sum_{k=1}^{\infty} -c_{-k} z^k \quad \text{and} \quad L_\infty(z) := \sum_{k=0}^{\infty} c_k z^{-k} \tag{15}$$

in the sense that for $n = 1,2,3,\ldots$,

$$\Lambda_0\left(\frac{P_n(z)}{Q_n(z)}\right) - L_0(z) = O(z^{n+1}) \text{ and } \Lambda_\infty\left(\frac{P_n(z)}{Q_n(z)}\right) - L_\infty(z) = O\left(\left(\frac{1}{z}\right)^n\right) \tag{16}$$

where $P_n(z)$ *and* $Q_n(z)$ *denote the (polynomial) nth numerator and denominator, respectively, of (14).*

Theorem 2.5 *The following statements are equivalent:*

(i) *The SSMP for a given bisequence* $\{c_k\}_{k=-\infty}^{\infty}$ *is indeterminate.*

(ii) *The corresponding positive T-fraction (14) is divergent for all* $z \in S_\pi := \{z \in \mathbf{C} : |\arg z| < \pi\}$.

(iii)

$$\sum_{n=1}^{\infty} e_n < \infty \quad \text{and} \quad \sum_{n=1}^{\infty} d_n < \infty \tag{17}$$

where for $n = 1,2,3,\ldots$,

$$e_1 := \frac{1}{F_1}, \; e_{2n-1} := \frac{\prod_{k=1}^{n-1} F_{2k}}{\prod_{k=1}^{n} F_{2k-1}}, \; e_{2n} := \frac{\prod_{k=1}^{n} F_{2k-1}}{\prod_{k=1}^{n} F_{2k}}, \; d_n := G_n e_n. \tag{18}$$

Theorem 2.6 *If the SSMP for* $\{c_k\}_{k=-\infty}^{\infty}$ *is indeterminate, then:*

(A) *The subsequences* $\{P_{2n}(z)/Q_{2n}(z)\}_{n=1}^{\infty}$ *and* $\{P_{2n+1}(z)/Q_{2n+1}(z)\}_{n=0}^{\infty}$ *converge uniformly on compact subsets of the cut plane* $S_\pi := \{z \in \mathbb{C} : |\arg z| < \pi\}$ *to different holomorphic functions* $G^{(0)}(z)$ *and* $G^{(1)}(z)$, *respectively.*

(B) *There exist distinct distribution functions* $\psi^{(0)}(t)$ *and* $\psi^{(1)}(t)$ *that are solutions of the SSMP for* $\{c_k\}_{k=-\infty}^{\infty}$ *and satisfy*

$$G^{(\sigma)}(z) := \lim_{n\to\infty} \frac{P_{2n+\sigma}(z)}{Q_{2n+\sigma}(z)} = z\int_0^\infty \frac{d\psi^{(\sigma)}(t)}{z+t}, \quad z \in S_\pi, \ \sigma = 0, 1. \tag{19}$$

We will call $\psi^{(0)}(t)$ and $\psi^{(1)}(t)$ the *natural solutions to the SSMP for* $\{c_k\}_{k=-\infty}^{\infty}$. As in the CSMP, every convex linear combination $\psi(\alpha, t) := (1-\alpha)\psi^{(0)}(t) + \alpha\psi^{(1)}(t)$, $0 \le \alpha \le 1$, of $\psi^{(0)}(t)$ and $\psi^{(1)}(t)$ is a solution to the SSMP for $\{c_k\}_{k=-\infty}^{\infty}$. Although the existence of natural solutions $\psi^{(0)}(t)$ and $\psi^{(1)}(t)$ to each indeterminate SSMP is established in [5], it was not determined whether the $\psi^{(\sigma)}(t)$ are step functions or not.

In [1] a constructive approach is used to show that the natural solutions to a special class ($G_n = G > 0$, $n = 1, 2, 3, \dots$) of indeterminate SSMPs are step functions whose points of increase have limit points at both 0 and ∞. In order to establish the similarities between this class of indeterminate SSMPs and the indeterminate CSMPs, we use an equivalence transformation ([4], Theorem 2.6), with $r_0 = 1$, $r_{2n-1} = e_{2n-1}/z$ and $r_{2n} = e_{2n}$, $n = 1, 2, \dots$, to see that the positive T-fraction (14) is equivalent to the continued fraction

$$\frac{1}{\frac{e_1}{z} + d_1} + \frac{1}{e_2 + d_2 z} + \frac{1}{\frac{e_3}{z} + d_3} + \frac{1}{e_4 + d_4} + \cdots, \tag{20}$$

where the e_n and d_n are given in (18). The nth numerator $U_n(z)$ and denominator $V_n(z)$ of (20) are related to $P_n(z)$ and $Q_n(z)$ by the equations

$$U_{2n+\sigma}(z) = \left(\prod_{k=1}^{2n+\sigma} e_k\right)\frac{P_{2n+\sigma}(z)}{z^{n+\sigma}}, \quad V_{2n+\sigma}(z) = \left(\prod_{k=1}^{2n+\sigma} e_k\right)\frac{Q_{2n+\sigma}(z)}{z^{n+\sigma}} \tag{21}$$

where $\sigma = 0, 1$ and $n = 1-\sigma, 2-\sigma, 3-\sigma, \dots$. We know that a SSMP is indeterminate if and only if the corresponding positive T-fraction (14) and hence the equivalent fraction (20) diverge. In the following theorem we summarize some of the results contained in [1] in such a manner as to emphasize the similarities and differences with Theorem 2.3.

Theorem 2.7 *If the SSMP for* $\{c_k\}_{k=-\infty}^{\infty}$ *is indeterminate and if* $G_n = G > 0$ *for* $n = 1, 2, 3, \dots$, *then:*

(A) *For* $\sigma = 0, 1$, *the sequences* $\{U_{2n+\sigma}(z)\}_{n=1}^{\infty}$ *and* $\{V_{2n+\sigma}(z)\}_{n=1}^{\infty}$ *of Laurent polynomials converge uniformly on compact subsets of the punctured plane* $\mathbb{C} - \{0\}$ *to holomorphic functions* $U^{(\sigma)}(z)$ *and* $V^{(\sigma)}(z)$, *respectively. Moreover, for* $z \in \mathbb{C} - \{0\}$,

$$U^{(1)}(z)V^{(0)}(z) - U^{(0)}(z)V^{(1)}(z) = 1.$$

(B) *For* $\sigma = 0, 1$, *the zeros* $x_{2n+\sigma,p}$, $-n+1-\sigma \le p \le n$, *of* $V_{2n+\sigma}(z)$ *are negative, simple and we have*

$$x_{2n,-n+1} < x_{2n,-n+2} < \cdots < x_{2n,0} < -\frac{1}{G} < x_{2n,1} < \cdots < x_{2n,n} < 0, \tag{22}$$

$$x_{2n+1,-n} < x_{2n+1,-n+1} < \cdots < x_{2n+1,0} = -\frac{1}{G} < x_{2n+1,1} < \cdots < x_{2n+1,n} < 0. \tag{23}$$

(C) *For each* $\sigma = 0, 1$ *and each* $p = 0, \pm 1, \pm 2, \ldots$, *the sequence* $\{x_{2n+\sigma,p}\}_{n=|p|+1}^{\infty}$ *is convergent and the limits*

$$x_p^{(\sigma)} := \lim_{n\to\infty} x_{2n+\sigma,p} \tag{24}$$

satisfy

$$\cdots < x_{-3}^{(\sigma)} < x_{-2}^{(\sigma)} < x_{-1}^{(\sigma)} < x_0^{(\sigma)} < x_1^{(\sigma)} < x_2^{(\sigma)} < x_3^{(\sigma)} < \cdots < 0 \tag{25}$$

and

$$\lim_{p\to\infty} x_p^{(\sigma)} = 0 \quad \text{and} \quad \lim_{p\to-\infty} x_p^{(\sigma)} = -\infty. \tag{26}$$

(D) *For* $\sigma = 0, 1$, *the set of zeros of* $V^{(\sigma)}(z)$ *is given by* $\{\ldots, x_{-2}^{(\sigma)}, x_{-1}^{(\sigma)}, x_0^{(\sigma)}, x_1^{(\sigma)}, x_2^{(\sigma)}, \ldots\}$ *and each zero is simple.*

(E) *For* $\sigma = 0, 1$, *and* $n = 1, 2, 3, \ldots$

$$\frac{U_{2n+\sigma}(z)}{V_{2n+\sigma}(z)} = \sum_{p=-n+1-\sigma}^{n} \frac{z a_{2n+\sigma,p}}{z - x_{2n+\sigma,p}}, \quad a_{2n+\sigma,p} := \frac{U_{2n+\sigma}(x_{2n+\sigma,p})}{x_{2n+\sigma,p} V'_{2n+\sigma}(x_{2n+\sigma,p})} > 0. \tag{27}$$

(F) *For* $\sigma = 0, 1$ *and* $p = 0, \pm 1, \pm 2, \ldots$, $\{a_{2n+\sigma,p}\}_{n=|p|+1}^{\infty}$ *converges to*

$$a_p^{(\sigma)} := \lim_{n\to\infty} a_{2n+\sigma,p} \quad \text{where} \quad \sum_{p=-\infty}^{\infty} a_p^{(\sigma)} = \frac{F_1}{G} = c_0 > 0. \tag{28}$$

Each step function $\psi^{(\sigma)}(t)$, $\sigma = 0, 1$, *with a positive jump* $a_p^{(\sigma)}$ *at* $t = -x_p^{(\sigma)} > 0$, $p = 0, \pm 1, \pm 2, \ldots$, *is a solution to the SSMP for* $\{c_k\}_{k=-\infty}^{\infty}$.

(G) *For* $\sigma = 0, 1$, *and* $z \in S := \mathbb{C} - [\ldots, x_{-1}^{(\sigma)}, x_0^{(\sigma)}, x_1^{(\sigma)}, \ldots]$,

$$G^{(\sigma)}(z) := \frac{U^{(\sigma)}(z)}{V^{(\sigma)}(z)} = z \int_0^{\infty} \frac{d\psi^{(\sigma)}(t)}{z+t} = \sum_{p=-\infty}^{\infty} \frac{z a_p^{(\sigma)}}{z - x_p^{(\sigma)}}. \tag{29}$$

the convergence of the series being uniform on compact subsets of S.

Remark: The authors of [1] proved part (A) of Theorem 2.7 with a slightly weaker hypothesis than that given above. In ([1], Theorem 3.3), it was shown that part (A) of Theorem 2.7 is true of the conditions (17) and $|G_n| \leq M$, $n = 1, 2, \ldots$, are satisfied. It is shown in Theorem 2.8 that part (A) of Theorem 2.7 is in fact true with the hypothesis that condition (17) holds.

Theorem 2.8 *For a given sequence* $\{c_k\}_{k=-\infty}^{\infty}$, *if (17) holds, then for* $\sigma = 0, 1$, *the sequences* $\{U_{2n+\sigma}(z)\}_{n=1}^{\infty}$ *and* $\{V_{2n+\sigma}(z)\}_{n=1}^{\infty}$ *of Laurent polynomials converge uniformly on compact subsets of the punctured plane* $\mathbb{C} - \{0\}$ *to holomorphic functions* $U^{(\sigma)}(z)$ *and* $V^{(\sigma)}(z)$, *respectively. Moreover for* $z \in \mathbb{C}$,

$$U^{(1)}(z)V^{(0)}(z) - U^{(0)}(z)V^{(1)}(z) = 1. \tag{30}$$

Proof. Define $b_{2n+1}(z) := \frac{e_{2n+1}}{z} + d_{2n+1}$, $n = 0, 1, 2, \ldots$, $b_{2n}(z) := e_{2n} + d_{2n}z$, $n = 1, 2, \ldots$, $0 < E := \sum_{n=1}^{\infty} e_n < \infty$ and $0 < D := \sum_{n=1}^{\infty} d_n < \infty$. Then from the difference equations (see, e.g., [4], p. 20) for the continued fraction (20) we have for $n = 1, 2, 3, \ldots$,

$$U_{-1}(z) := 1, \quad U_0(z) := 0, \quad V_{-1}(z) := 0, \quad V_0(z) := 1. \tag{31}$$

$$U_n(z) = b_n(z)U_{n-1}(z) + U_{n-2}(z) \tag{32}$$

$$V_n(z) = b_n(z)V_{n-1}(z) + V_{n-2}(z). \tag{33}$$

Let K be a compact subset of $\mathbb{C} - \{0\}$. Then there exists a constant $N(K)$ so that $\frac{1}{N(K)} \leq |z| \leq N(K)$ for all $z \in K$. Thus

$$\sum_{n=1}^{\infty} |b_{2n}(z)| = \sum_{n=1}^{\infty} |e_{2n} + d_{2n}z| \leq \sum_{n=1}^{\infty} (e_{2n} + d_{2n}|z|)$$

$$\leq \sum_{n=1}^{\infty} e_n + |z| \sum_{n=1}^{\infty} d_n \leq E + N(K)D < \infty$$

for all $z \in K$. Here we use the fact that for $n = 1, 2, \ldots$, $e_n > 0$ and $d_n > 0$. Similarly, $\sum_{n=0}^{\infty} |b_{2n+1}(z)| \leq N(K)E + D < \infty$ for all $z \in K$. Consequently for $z \in K$,

$$\sum_{n=1}^{\infty} |b_n(z)| = \sum_{n=1}^{\infty} |b_{2n}(z)| + \sum_{n=0}^{\infty} |b_{2n+1}(z)| \leq (E + D)(1 + N(K)) =: S(K) < \infty.$$

Using the difference equations (31), (32), (33) one obtains by induction

$$|U_n(z)| \leq \prod_{k=1}^{n} (1 + |b_k(z)|) \text{ and } |V_n(z)| \leq \prod_{k=1}^{n} (1 + |b_k(z)|), \; n = 1, 2, \ldots. \tag{34}$$

Since $\sum_{n=1}^{\infty} |b_n(z)| \leq S(K)$ for each $z \in K$, it follows from (34) that $|U_n(z)|$ and $|V_n(z)|$ are both bounded above for all $z \in K$ and $n = 1, 2, \ldots$, by $e^{S(K)}$. Thus we have $\sum_{n=1}^{\infty} |b_n(z)U_{n-1}(z)| \leq S(K)e^{S(K)} < \infty$ and $\sum_{n=1}^{\infty} |b_n(z)V_{n-1}(z)| \leq S(K)e^{S(K)} < \infty$. Consequently the series $\sum_{n=1}^{\infty} b_n(z)U_{n-1}(z)$ and $\sum_{n=1}^{\infty} b_n(z)V_{n-1}(z)$ converge uniformly and absolutely on K. Using induction on the difference equations (31), (32), (33) one sees that for $\sigma = 0, 1$ and $n = 1, 2, 3, \ldots$, $U_{2n+\sigma}(z) = \sigma + \sum_{k=1}^{n} b_{2k+\sigma}(z)U_{2k+\sigma-1}(z)$

and $V_{2n+\sigma}(z) = 1 - \sigma + \sum_{k=0}^{n} b_{2k+\sigma}(z)V_{2k+\sigma-1}(z)$. Hence for $\sigma = 0,1$, $\{U_{2n+\sigma}(z)\}_{n=1}^{\infty}$ and $\{V_{2n+\sigma}(z)\}_{n=1}^{\infty}$ converge uniformly on K and $U^{(\sigma)}(z) := \lim_{n\to\infty} U_{2n+\sigma}(z)$ and $V^{(\sigma)}(z) := \lim_{n\to\infty} V_{2n+\sigma}(z)$. Equation (30) follows from the determinant formula for continued fractions (see, e.g., [4], p. 20). □

In an unpublished personal communication [6] Olav Njåstad has shown that each natural solution $\psi^{(\sigma)}(t)$, $\sigma = 0,1$, to an indeterminate SSMP is a step function having points of increase at the negatives of the poles of $G^{(\sigma)}(z)$. We give a small extension of this result (Theorem 2.9) by investigating the possible limit points of the poles of $G^{(\sigma)}(z)$. In ([1], Theorem 3.1) it is shown that for $n = 1,2,\ldots$, the n zeros of $V_n(z)$ are simple and negative. Thus by Hurwitz' theorem [8], the zeros of $V^{(\sigma)}(z)$ are negative. From (19), (21) and Theorem 2.8 we see that for $|\arg z| < \pi$

$$G^{(\sigma)}(z) := \lim_{n\to\infty} \frac{P_{2n+\sigma}(z)}{Q_{2n+\sigma}(z)} = \lim_{n\to\infty} \frac{U_{2n+\sigma}(z)}{V_{2n+\sigma}(z)} = \frac{U^{(\sigma)}(z)}{V^{(\sigma)}(z)}.$$

It follows from (30) that $U^{(\sigma)}(z)$ and $V^{(\sigma)}(z)$ have no common zeros and so each natural solution $\psi^{(\sigma)}(t)$, $\sigma = 0,1$, to an indeterminate SSMP is a step function with points of increase at the negatives of the zeros of $V^{(\sigma)}(z)$. Since the set of points of increase of each natural solution $\psi^{(\sigma)}(t) \in \Psi(0,\infty)$, $\sigma = 0,1$, is denumerable (countably infinite), the set of zeros of each $V^{(\sigma)}(z)$, $\sigma = 0,1$, is also denumerable. Using Hurwitz' theorem, [8] Rouché's theorem [2] and the fact that the zeros of $V_n(z)$ are simple, one can show that the zeros of $V^{(\sigma)}(z)$ are simple. We know that for $\sigma = 0,1$, $V^{(\sigma)}(z)$ is continuous on $\mathbb{C} - \{0\}$ and $V^{(\sigma)}(z) \not\equiv 0$. Hence one can see by the identity theorem [3] that 0 and $-\infty$ are the only possible limit points for the zeros of $V^{(\sigma)}(z)$. We summarize these results in the following theorem.

Theorem 2.9 *Each natural solution $\psi^{(\sigma)}(t)$, $\sigma = 0,1$, to an indeterminate SSMP is a step function whose points of increase have limit points only at 0 and/or ∞. The points of increase occur at the negatives of the zeros (all are negative) of $V^{(\sigma)}(z)$.*

In Theorem 2.7 we saw an example of an indeterminate SSMP such that the points of increase of each natural solution have limit points at both 0 and ∞. There exist SSMPs which have step function solutions where the points of increase have a limit point at ∞ but not at 0 (see, e.g., Theorem 3.2). If we apply Theorem 4.1 to one of these SSMPs we arrive at a SSMP which has a step function solution where the points of increase have a limit point at 0 but not at ∞.

Let the n zeros of $V_n(z)$ be denoted by $\zeta_{n,\nu}$, $\nu = 1,2,\ldots,n$ where $\zeta_{n,n} < \zeta_{n,n-1} < \cdots < \zeta_{n,1} < 0$. By the Hurwitz theorem [8] we know that each zero of $V^{(\sigma)}(z)$ is the limit of a sequence of zeros $\{\zeta_{2n+\sigma,\nu_{2n+\sigma}}\}_{n=1}^{\infty}$ of the $V_{2n+\sigma}(z)$. We hope to one day be able identify such sequences. Along these lines we have been able to prove the following:

Theorem 2.10 *Let $\{c_k\}_{k=-\infty}^{\infty}$ be a sequence whose SSMP is indeterminate. Let $\zeta_{n,\nu}$, $\nu = 1,2,\ldots,n$ denote the n zeros of $V_n(z)$ arranged so that $\zeta_{n,n} < \zeta_{n,n-1} < \cdots < \zeta_{n,1} < 0$. Then $\lim_{n\to\infty} \zeta_{n,1} = 0$ and $\lim_{n\to\infty} \zeta_{n,n} = -\infty$.*

Proof. In ([1], Theorem 3.1) it is shown that for $n = 1, 2, \ldots$, the n zeros, $\zeta_{n,\nu}$, $\nu = 1, 2, \ldots, n$, of $V_n(z)$ are the same as the n zeros of $Q_n(z)$ and are simple, negative and separate the zeros of $V_{n+1}(z)$. For $\nu = 2, 3, \ldots$, set $\gamma_\nu := e_{\nu-1}e_\nu$ and for $n = 2, 3, \ldots$, set $\rho_n := \min\limits_{2\le\nu\le n} \gamma_\nu$. Then $\rho_n = \gamma_{\mu_n}$ for some $\mu_n \in \{2, 3, \ldots, n\}$ and $\sum\limits_{n=2}^{\infty} \rho_n < \infty$ since $\sum\limits_{n=1}^{\infty} e_n < \infty$ and $0 < \rho_n \le \gamma_n < e_n$ for n sufficiently large. Using induction on the three-term recursion relations (see, e.g., [4], p. 20) for the positive T-fraction (14) we see that

$$Q_n(z) = \prod_{\nu=1}^{n}\left(1 - \frac{z}{\zeta_{n,\nu}}\right) = 1 + \left(G_1 + \sum_{\nu=2}^{n}(G_\nu + F_\nu)\right) z + \cdots + \left(\prod_{\nu=1}^{n} G_\nu\right) z^n.$$

Comparing the coefficients of z we arrive at the formula

$$\sum_{\nu=1}^{n} \frac{-1}{\zeta_{n,\nu}} = G_1 + \sum_{\nu=2}^{n}(G_\nu + F_\nu). \tag{35}$$

One easily verifies using (18) that

$$F_1 = \frac{1}{e_1},\ G_1 = \frac{d_1}{e_1},\quad F_n = \frac{1}{e_{n-1}e_n} \text{ and } G_n = \frac{d_n}{e_n},\quad n = 2, 3, \ldots. \tag{36}$$

Suppose that there exists a constant M independent of n so that

$$0 < \frac{-1}{\zeta_{n,n}} < \cdots < \frac{-1}{\zeta_{n,2}} < \frac{-1}{\zeta_{n,1}} < M$$

for all n. From (35) and (36) we have

$$M \cdot n > \sum_{\nu=1}^{n} \frac{-1}{\zeta_{n,\nu}} = G_1 + \sum_{\nu=2}^{n}(G_\nu + F_\nu) = \frac{d_1}{e_1} + \sum_{\nu=2}^{n}\left(\frac{d_\nu}{e_\nu} + \frac{1}{e_{\nu-1}e_\nu}\right)$$

$$> \sum_{\nu=2}^{n} \frac{1}{e_{\nu-1}e_\nu} > \frac{1}{e_{\mu_n-1}e_{\mu_n}} = \frac{1}{\rho_n}.$$

Thus $\rho_n > \dfrac{1}{M \cdot n}$ from which it follows that $\sum\limits_{n=2}^{\infty} \rho_n$ diverges giving us a contradiction. Hence the set $\left\{\dfrac{-1}{\zeta_{n,\nu}} : n, \nu = 1, 2, \ldots\right\}$ is unbounded and in particular $\lim\limits_{n\to\infty} \zeta_{n,1} = 0$. If we replace z by $\dfrac{1}{z}$ in the positive T-fraction (14), use an equivalence transformation (see, e.g., [4], p. 33) with $r_0 = 1$, $r_n = \dfrac{z}{G_n}$, $n = 1, 2, \ldots$, and then multiply the resulting continued fraction by z we obtain a positive T-fraction

$$\frac{\widehat{F}_1 z}{1 + \widehat{G}_1 z} + \frac{\widehat{F}_2 z}{1 + \widehat{G}_2 z} + \frac{\widehat{F}_3 z}{1 + \widehat{G}_3 z} + \cdots \tag{37}$$

where for $n = 2, 3, \ldots$

$$\widehat{F}_1 := \frac{F_1}{G_1}, \quad \widehat{G}_1 := \frac{1}{G_1}, \quad \widehat{F}_n := \frac{F_n}{G_{n-1}G_n}, \quad \widehat{G}_n := \frac{1}{G_n}. \tag{38}$$

The nth numerator $\widehat{P}_n(z)$ and denominator $\widehat{Q}_n(z)$ of (37) are given by

$$\widehat{P}_n(z) = \frac{z^{n+1}}{\prod_{k=1}^n G_k} P_n\left(\frac{1}{z}\right), \quad \widehat{Q}_n(z) = \frac{z^n}{\prod_{k=1}^n G_k} Q_n\left(\frac{1}{z}\right). \tag{39}$$

Thus for $\nu = 1, 2, \ldots, n$, the zeros $\hat{\zeta}_{n,\nu}$ of $\widehat{Q}_n(z)$ are related to the zeros $\zeta_{n,\nu}$ of $Q_n(z)$ by

$$\hat{\zeta}_{n,\nu} = \frac{1}{\zeta_{n,n-\nu+1}}. \tag{40}$$

Using equations (38) and (18) one shows that

$$\hat{e}_1 := \frac{1}{\widehat{F}_1} = \frac{G_1}{F_1} = G_1 e_1, \quad \hat{e}_{2n} := \prod_{k=1}^n \widehat{F}_{2k-1} \Big/ \prod_{k=1}^n \widehat{F}_{2k} = e_{2n} G_{2n} = d_{2n},$$

$$\hat{e}_{2n+1} := \prod_{k=1}^{n-1} \widehat{F}_{2k} \Big/ \prod_{k=1}^n \widehat{F}_{2k-1} = e_{2n+1} d_{2n+1} = d_{2n+1}, \hat{d}_n := \widehat{G}_n \hat{e}_n = \frac{d_n}{G_n} = e_n.$$

That is, for $n = 1, 2, \ldots$, $\hat{e}_n = d_n$ and $\hat{d}_n = e_n$. Thus $\sum_{n=1}^{\infty} \hat{e}_n < \infty$ and $\sum_{n=1}^{\infty} \hat{d}_n < \infty$ and so by the first part of this proof, $\lim_{n\to\infty} \hat{\zeta}_{n,1} = 0$ from which it follows that $\lim_{n\to\infty} \zeta_{n,n} = \lim_{n\to\infty} \frac{1}{\hat{\zeta}_{n,1}} = -\infty$. □

3. Related Indeterminate CSMPs and SSMPs

We call the CSMP for $\{\mu_n\}_{n=0}^{\infty}$ and the SSMP for $\{c_n\}_{n=-\infty}^{\infty}$ *related* if $\mu_n = c_n$ for $n = 0, 1, 2, \ldots$. Clearly any solution to the SSMP for $\{c_n\}_{n=-\infty}^{\infty}$ is also a solution to the related CSMP with moments $\{c_n\}_{n=0}^{\infty}$.

As stated in Section 2, if $\varphi^{(0)}(t)$ and $\varphi^{(1)}(t)$ are the natural solutions to the indeterminate CSMP for $\{\mu_n\}_{n=0}^{\infty}$, then every convex linear combination $\varphi(\alpha, t) := (1-\alpha)\varphi^{(0)}(t) + \alpha\varphi^{(1)}(t)$, $0 \le \alpha \le 1$, is also a solution to the CSMP for $\{\mu_n\}_{n=0}^{\infty}$. In the following theorem it is shown that there exist indeterminate CSMPs which have step function solutions that are not of the form $\varphi(\alpha, t)$.

Theorem 3.1 *Suppose that $\{c_n\}_{n=-\infty}^{\infty}$ is a bisequence that leads to an indeterminate SSMP. Let $\psi^{(0)}(t)$ and $\psi^{(1)}(t)$ denote the natural solutions to the SSMP for $\{c_n\}_{n=-\infty}^{\infty}$ and suppose that for $\sigma = 0, 1$, the points of increase of $\psi^{(\sigma)}(t)$ have a limit point at 0. Then $\psi^{(0)}(t)$ and $\psi^{(1)}(t)$ are distinct solutions to the related CSMP for $\{c_n\}_{n=0}^{\infty}$ and they are not convex linear combinations of the natural solutions to the CSMP for $\{c_n\}_{n=0}^{\infty}$*

Proof. Clearly $\psi^{(0)}(t)$ and $\psi^{(1)}(t)$ are solutions to the CSMP for $\{c_n\}_{n=0}^{\infty}$. Let $\varphi^{(0)}(t)$ and $\varphi^{(1)}(t)$ be the natural solutions to the CSMP for $\{c_n\}_{n=0}^{\infty}$. We recall from Theorem 2.3(F) that for each $\sigma = 0, 1$, $\varphi^{(\sigma)}(t)$ is a step function on $(0, \infty)$ with jump $\lambda_k^{(\sigma)} > 0$ at $t = -t_k^{(\sigma)} > 0$ where $0 < \beta \leq -t_k^{(\sigma)} < -t_{k+1}^{(\sigma)}$, $k = 1, 2, \ldots$. Thus for each α, $0 \leq \alpha \leq 1$, $\varphi(\alpha, t) := (1-\alpha)\varphi^{(0)}(t) + \alpha\varphi^{(1)}(t) = 0$ for $0 \leq t < \beta$. But $\psi^{(\sigma)}(t) > 0$ for $0 < t < \beta$ since for $\sigma = 0, 1$, the points of increase of $\psi^{(\sigma)}(t)$ have a limit point at 0. Hence $\psi^{(\sigma)}(t)$ is not of the form $\varphi(\alpha, t)$ for $0 \leq \alpha \leq 1$. □

The next theorem is a connection between step function solutions of CSMPs and SSMPs.

Theorem 3.2 *If a CSMP has a step function solution $\varphi(t)$ with jump $\lambda_k > 0$ at $t = t_k \geq \beta > 0$, $k = 1, 2, \ldots$, where $\lim_{k\to\infty} t_k = \infty$ and $\sum_{k=1}^{\infty} \lambda_k < \infty$ and such that for $n = 0, 1, 2, \ldots$, $c_n = \int_0^{\infty} (-t)^n d\varphi(t) = \sum_{k=1}^{\infty} (-t_k)^n \lambda_k$, then:*

(A) *For $n = -1, -2, -3, \ldots$, the integrals*

$$\int_0^{\infty} (-t)^n d\varphi(t) = \sum_{k=1}^{\infty} (-t_k)^n \lambda_k$$

exist.

(B) *$\varphi(t)$ is a solution of the SSMP for $\{c_n\}_{n=-\infty}^{\infty}$, where*

$$c_n := \int_0^{\infty} (-t)^n d\varphi(t), \quad n = -1, -2, \ldots .$$

Proof. (A) Since $0 < \beta \leq t_k$ and $\lim_{k\to\infty} t_k = \infty$, there exists a number B such that $0 < (t_k)^n \leq B$ for all $k \geq 1$ and $n \leq -1$. Therefore since $\sum_{k=1}^{\infty} \lambda_k < \infty$, we see that for each fixed $n \in \{-1, -2, -3, \ldots\}$,

$$\left| \int_0^{\infty} (-t)^n d\varphi(t) \right| = \left| \sum_{k=1}^{\infty} (-t_k)^n \lambda_k \right| \leq B \sum_{k=1}^{\infty} \lambda_k < \infty.$$

(B) That $\varphi(t)$ is a solution of the SSMP for $\{c_n\}_{n=-\infty}^{\infty}$ follows from the definition of the c_n. □

4. Shifts and Flips in Moment Sequences

In Theorem 4.1 we assume that $\psi(t) \in \Psi(0, \infty)$ is a step function such that the jumps occur at a bisequence of points $\{t_j\}_{j=-\infty}^{\infty}$ with $0 < \cdots < t_{-1} < t_0 < t_1 < \cdots$. Completely analogous theorems for which $\psi(t)$ is a step function such that the jumps occur at a sequence of points $\{t_j\}_{j=1}^{\infty}$ with $0 < t_1 < t_2 < t_3 \cdots$ or $0 < \cdots < t_3 < t_2 < t_1$ are omitted.

Theorem 4.1 *Suppose that $\psi(t) \in \Psi(0,\infty)$ is a step function solution to the SSMP for a bisequence $\{c_n\}_{n=-\infty}^{\infty}$ and that $\psi(t)$ has jump $a_j > 0$ at $t = t_j$, $j = 0, \pm 1, \pm 2, \ldots$, where $\sum_{j=-\infty}^{\infty} a_j < \infty$ and $0 < \cdots < t_{-1} < t_0 < t_1 < \cdots$, so that for $k = 0, \pm 1, \pm 2, \ldots$,*

$$c_k = \int_0^\infty (-t)^k d\psi(t) = \sum_{j=-\infty}^{\infty} (-t_j)^k a_j.$$

Then:

(A) *For $\alpha \in \mathbb{R}$ and $n = 0, \pm 1, \pm 2, \ldots$, the series $\gamma_n^{(\alpha)} := (-1)^n \sum_{j=-\infty}^{\infty} \frac{a_j}{(t_j)^{n+\alpha}}$ converge. If $\theta_\alpha(\tau)$ is the step function with jump $\lambda_j := \frac{a_j}{(t_j)^\alpha} > 0$ at $\tau_j := \frac{1}{t_j} > 0$, $j = 0, \pm 1, \pm 2, \ldots$, then $\theta_\alpha(\tau) \in \Psi(0,\infty)$ and $\theta_\alpha(\tau)$ is a solution to the SSMP for $\{\gamma_n^{(\alpha)}\}_{n=-\infty}^{\infty}$.*

(B) *For $\alpha \in \mathbb{R}$ and $n = 0, \pm 1, \pm 2, \ldots$, the series $\eta_n^{(\alpha)} := (-1)^n \sum_{j=-\infty}^{\infty} (t_j)^{n+\alpha} a_j$ converge. If $\psi_\alpha(t)$ is the step function with jump $\lambda_j := (t_j)^\alpha a_j > 0$ at $t = t_j$, $j = 0, \pm 1, \pm 2, \ldots$, then $\psi_\alpha(t) \in \Psi(0,\infty)$ and $\psi_\alpha(t)$ is a solution to the SSMP for $\{\eta_n^{(\alpha)}\}_{n=-\infty}^{\infty}$.*

Remark. In Theorem 4.1(A) if $\alpha = m \in \mathbb{Z}$, then

$$\gamma_n^{(m)} = (-1)^m \sum_{j=-\infty}^{\infty} (-t_j)^{-n-m} a_j = (-1)^m c_{-n-m}.$$

In Theorem 4.1(B) if $\alpha = m \in \mathbb{Z}$, then

$$\eta_n^{(m)} = (-1)^m \sum_{j=-\infty}^{\infty} (-t_j)^{n+m} a_j = (-1)^m c_{n+m}.$$

Proof. Fix $\alpha \in \mathbb{R}$ and pick $k \in \mathbb{Z}$ so that $k - 1 < \alpha \le k$. Then

$$\begin{aligned} 0 < |\gamma_n^{(\alpha)}| &= \left| \sum_{j=-\infty}^{\infty} (-1)^n (t_j)^{-n-\alpha} a_j \right| \le \sum_{j=-\infty}^{\infty} (t_j)^{-n-\alpha} a_j \\ &< \sum_{j=-\infty}^{\infty} ((t_j)^{-n-k} + (t_j)^{-n-k+1}) a_j = \int_0^\infty (t^{-n-k} + t^{-n-k+1}) d\psi(t) \\ &= (-1)^{-n-k} c_{-n-k} + (-1)^{-n-k+1} c_{-n-k+1} < \infty. \end{aligned}$$

Thus $\gamma_n^{(\alpha)}$ exists. Clearly $\theta_\alpha(\tau)$ is a real-valued, non-decreasing function with infinitely many points of increase. Since

$$0 < \sum_{j=-\infty}^{\infty} \lambda_j = \sum_{j=-\infty}^{\infty} (t_j)^{-\alpha} a_j \leq \sum_{j=-\infty}^{\infty} ((t_j)^{-k} + (t_j)^{-k+1}) a_j = \int_0^{\infty} (t^{-k} + t^{-k+1}) d\psi(t)$$

$$= (-1)^{-k} c_{-k} + (-1)^{-k+1} c_{-k+1} < \infty,$$

we see that $\theta_\alpha(\tau)$ is bounded and hence $\theta_\alpha(\tau) \in \Psi(0,\infty)$. Moreover,

$$\gamma_n^{(\alpha)} := \sum_{j=-\infty}^{\infty} \frac{(-1)^n a_j}{(t_j)^{n+\alpha}} = \sum_{j=-\infty}^{\infty} \left(-\frac{1}{t_j}\right)^n \frac{a_j}{(t_j)^\alpha}$$

$$= \sum_{j=-\infty}^{\infty} (-\tau_j)^n \lambda_j = \int_0^{\infty} (-\tau)^n d\theta_\alpha(\tau).$$

Hence $\theta_\alpha(\tau)$ is a solution to the SSMP for $\{\gamma_n^{(\alpha)}\}_{n=-\infty}^{\infty}$.

(B) A proof is similar to that given for (A) and hence is omitted. □

Results analogous to Theorem 4.1 can be obtained using step function solutions to a CSMP. In this case, however, we must restrict $\alpha > 0$.

References

1. Catherine Bonan-Hamada, William B. Jones and W. J. Thron, A class of indeterminate strong Stieltjes moment problems with discrete distributions, submitted.
2. John B. Conway, *Functions of One Complex Variable*, Springer-Verlag, New York, 1973.
3. Einar Hille, *Analytic Function Theory*, Vol. I, Blaisdell, New York, 1959.
4. William B. Jones and W. J. Thron, *Continued Fractions: Analytic Theory and Applications*, in Encyclopedia of Mathematics and Its Applications, Vol. 11, Addison–Wesley, Reading, Mass., 1980. Distributed now by the Cambridge University Press.
5. William B. Jones, W. J. Thron and Haakon Waadeland, A strong Stieltjes moment problem, *Trans. Amer. Math Soc.* **261** (1980), 503–528.
6. Olav Njåstad, personal communication, 1993.
7. T. J. Stieltjes, Recherches sur les fractions continues, *Ann. Fac. Sci. Toulouse* (1894), J. 1–122; **9** (1894), A, 1–47; *Oeuvres*, Vol. 2, pp. 402–566. Also published in *Mémoires présentés par divers savants a l'Académie de sciences de l'Institut National de France*, Vol. **33**, pp. 1–196. This paper and an English translation are now also available in *Oeuvres Complétes - Collected Papers* (G. van Dijk, ed.), Springer–Verlag, 1993, vol.2, pp. 406–570; pp. 609–745 (English).
8. W. J. Thron, *Introduction to the Theory of a Complex Variable*, Wiley, New York, 1953.

A FAMILY OF CLASSICAL DETERMINATE STIELTJES MOMENT PROBLEMS WITH DISCRETE SOLUTIONS

WILLIAM B. JONES and W. J. THRON
Department of Mathematics, University of Colorado, Boulder, Colorado 80309–0395, U.S.A.

and

NANCY J. WYSHINSKI*
Department of Mathematics, Trinity College, Hartford, Connecticut 06106, U.S.A.

Abstract. It follows from a theorem attributed to Favard [1] that the sequence of n-th denominators $Q_n(a,z)$ of a continued fraction of the form

$$\frac{zb_1}{z+b_2} - \frac{b_2b_3}{z+b_3+b_4} - \frac{b_4b_5}{z+b_5+b_6} - \cdots - \frac{b_{2n-2}b_{2n-1}}{z+b_{2n-1}+b_{2n}} - \cdots$$

where

$$b_n := \frac{1}{4[4(n+a)^2-1]}, \quad n \geq 1, \quad a > -\frac{1}{2},$$

forms an orthogonal polynomial sequence (OPS) with respect to a distribution function $\psi(a,t)$ on $\mathbf{R}$. It was recently shown [4] that the $Q_n(a,z)$ satisfy the following asymptotic formulas

$$\lim_{n\to\infty} \frac{Q_n(a,z)}{z^n} = {}_0F_1\left(a+\frac{3}{2}; \frac{1}{16z}\right) = \Gamma(a+3/2)(4\sqrt{-z})^{a+\frac{1}{2}} J_{a+\frac{1}{2}}\left(\frac{1}{2\sqrt{-z}}\right),$$

where the convergence is uniform on compact subsets of $0 < |z| < \infty$. $J_\nu(w)$ denotes the Bessel function of the first kind of order ν. In this paper we prove, using continued fraction methods, that the distribution function $\psi(a,t)$ is a step function with jumps $\lambda_k(a) = -2q_k(a)/(2a+1)$ at the zeros $q_k(a)$, $k = 1,2,3,\cdots$, of $J_{a+\frac{1}{2}}\left(\frac{1}{2\sqrt{-z}}\right)$. The connection with Lommel polynomials is discussed.

1. Introduction

In this paper we investigate the n-th numerators $P_n(a,z)$ and n-th denominators $Q_n(a,z)$ of continued fractions of the form

$$\frac{zb_1}{z+b_2} - \frac{b_2b_3}{z+b_3+b_4} - \frac{b_4b_5}{z+b_5+b_6} - \cdots - \frac{b_{2n-2}b_{2n-1}}{z+b_{2n-1}+b_{2n}} - \cdots \tag{1.1a}$$

where

$$b_n := \frac{1}{4[4(n+a)^2-1]}, \quad n \geq 1, \quad a > -\frac{1}{2}. \tag{1.1b}$$

The polynomials $P_n(a,z)$ and $Q_n(a,z)$ are defined by the difference equations

$$P_0(a,z) := 0, \quad P_1(a,z) := b_1 z, \quad Q_0(a,z) := 1, \quad Q_1(a,z) := z + b_2, \tag{1.2a}$$

* Research supported in part by the National Science Foundation under Grant Nos. DMS-9103141 and INT-9113400.

A. Cuyt (ed.), Nonlinear Numerical Methods and Rational Approximation II, 409–422.

$$P_n(a,z) = (z + b_{2n-1} + b_{2n})P_{n-1}(a,z) - b_{2n-2}b_{2n-1}P_{n-2}(a,z), \quad n \geq 2, \quad (1.2b)$$

$$Q_n(a,z) = (z + b_{2n-1} + b_{2n})Q_{n-1}(a,z) - b_{2n-2}b_{2n-1}Q_{n-2}(a,z), \quad n \geq 2. \quad (1.2c)$$

The continued fraction (1.1) corresponds to a unique formal Laurent series

$$L = \sum_{k=0}^{\infty} c_k(a) z^{-k}$$

in the sense that for each $n = 1, 2, 3, \cdots$,

$$\frac{P_n(a,z)}{Q_n(a,z)} = \sum_{k=0}^{2n-1} c_k(a) z^{-k} + O\left(\left(\frac{1}{z}\right)^{2n}\right)$$

(see, e.g., Section 7.2.2 in [5]). The *classical Stieltjes moment problem (CSMP)* for a sequence $\{c_k(a)\}_{k=0}^{\infty}$ is said to be *determinate* if there exists a unique distribution function $\psi(t)$ (i.e., bounded, non-decreasing function with infinitely many points of increase on $(0, \infty)$) such that

$$c_k(a) = \int_0^{\infty} (-t)^k d\psi(t), \quad k = 0, 1, 2, \cdots.$$

Since the real J-fraction (1.1) is obtained as the even part of a convergent S-fraction, Section 2.4.2 in [5], it is known that the CSMP for $\{c_k(a)\}_{k=0}^{\infty}$ is determinate. We show in the main result of this paper (Theorem 1.1) that the unique solution to the CSMP for $\{c_k(a)\}_{k=0}^{\infty}$ is a step function $\psi(a,t)$ and we express its jumps $\lambda_k(a)$ and points of jump explicitly in terms of the zeros $q_k(a)$ of the Bessel function $J_{a+\frac{1}{2}}\left(\frac{1}{2\sqrt{-z}}\right)$.

From [4] we obtain the following asymptotic formulas

$$\begin{aligned} \lim_{n\to\infty} \frac{P_n(a,z)}{z^n} &= z\left[{}_0F_1\left(a+\tfrac{1}{2};\tfrac{1}{16z}\right) - {}_0F_1\left(a+\tfrac{3}{2};\tfrac{1}{16z}\right)\right] \\ &= z\left[\Gamma(a+1/2)(4\sqrt{-z})^{a-\frac{1}{2}} J_{a-\frac{1}{2}}\left(\tfrac{1}{2\sqrt{-z}}\right)\right. \qquad (1.3a) \\ &\quad \left. -\Gamma(a+3/2)(4\sqrt{-z})^{a+\frac{1}{2}} J_{a+\frac{1}{2}}\left(\tfrac{1}{2\sqrt{-z}}\right)\right], \end{aligned}$$

$$\lim_{n\to\infty} \frac{Q_n(a,z)}{z^n} = {}_0F_1\left(a+\frac{3}{2};\frac{1}{16z}\right) = \Gamma(a+3/2)(4\sqrt{-z})^{a+\frac{1}{2}} J_{a+\frac{1}{2}}\left(\frac{1}{2\sqrt{-z}}\right), \quad (1.3b)$$

where the convergence is uniform on every compact subset of $0 < |z| < \infty$. These results were derived using separate convergence theorems ([8], [11]) for continued fractions; explicit formulas were found for the coefficients of the polynomials $Q_n(a,z)$ using formulas of Euler-Minding ([6]). The symbol ${}_0F_1$ denotes the confluent hypergeometric function

$${}_0F_1(c;z) := \sum_{n=0}^{\infty} \frac{1}{(c)_n} \frac{z^n}{n!}, \quad c \in \mathbf{C} \setminus [0, -1, -2, \cdots],$$

where $(c)_0 := 1$ and $(c)_n := c(c+1)(c+2)\cdots(c+n-1)$, $n \geq 1$. These functions are related to Bessel functions $J_\nu(z)$ of the first kind of order ν by the relations

$$J_\nu(z) = \frac{(\frac{1}{2}z)^\nu}{\Gamma(\nu+1)} {}_0F_1\left(\nu+1; -\frac{1}{4}z^2\right), \quad z \in \mathbf{C}.$$

After simplifying we can write

$$\lim_{n\to\infty} \frac{P_n(a,z)}{Q_n(a,z)} = z \ll \left[\frac{1}{\Gamma(a+\frac{1}{2})(4\sqrt{-z})} \frac{J_{a-\frac{1}{2}}(\frac{1}{2\sqrt{-z}})}{J_{a+\frac{1}{2}}(\frac{1}{2\sqrt{-z}})} - 1\right]. \tag{1.4}$$

The main result of this paper is given by the following theorem.

Before stating this result we let $q_k(a), k = 1,2,3,\cdots$, denote the zeros of the Bessel function $J_{a+\frac{1}{2}}\left(\frac{1}{2\sqrt{-z}}\right)$, all negative, simple and satisfying, for $k \geq 1$,

$$q_1(a) < q_2(a) < q_3(a) < \cdots < q_k(a) < 0. \tag{1.5}$$

Let

$$\lambda_k(a) := \frac{-2q_k(a)}{2a+1}, \; k = 1,2,3,\cdots, \tag{1.6}$$

and let $\psi(a,t)$ denote the step function with jump $\lambda_k(a)$ at $-q_k(a)$, $k = 1,2,3,\cdots$; hence

$$\psi(a,t) = \begin{cases} \sum\limits_{k=n+1}^{\infty} \lambda_k(a), & -q_{n+1}(a) < t \leq -q_n(a), \\ \sum\limits_{k=1}^{\infty} \lambda_k(a), & -q_1(a) < t. \end{cases} \tag{1.7a}$$

The distribution function $\psi(a,t)$ has support on $(0,\infty)$. The closely related distribution function $\varphi(a,t)$ defined by

$$\varphi(a,t) := c - \psi(a,-t), \quad \left(c := \sum_{k=1}^{\infty} \lambda_k(a)\right), \tag{1.7b}$$

has its support on the interval $(-\infty, 0)$. The convergence of the series $\sum\limits_{k=1}^{\infty} \lambda_k(a)$ is established in the following:

Theorem 1.1 *Let $a > -\frac{1}{2}$ be given and let $\{P_n(a,z)\}$ and $\{Q_n(a,z)\}$ be the polynomial sequences defined by (1.2). Let $\{q_k(a)\}, \{\lambda_k(a)\}, \psi(a,t)$ and $\varphi(a,t)$ be defined above in (1.6), (1.7) and (1.8). Let $L := \sum\limits_{k=0}^{\infty} c_k(a) z^{-k}$ denote the unique formal Laurent series corresponding to the continued fraction (1.1) at $z = \infty$. Then:*

(A) The positive numbers $\lambda_k(a)$ satisfy

$$\sum_{k=1}^{\infty} \lambda_k(a) = b_1 \quad (b_1 \text{ defined by } (1.1b)). \tag{1.8}$$

(B)

$$\lim_{n\to\infty} \frac{P_n(a,z)}{Q_n(a,z)} = z\int_0^\infty \frac{d\psi(a,t)}{z+t} = z\int_{-\infty}^0 \frac{d\varphi(a,t)}{z-t}$$

$$= z\sum_{k=1}^{\infty} \frac{\lambda_k(a)}{z-q_k(a)} = z \ll \left[\frac{1}{\Gamma(a+\frac{1}{2})(4\sqrt{-z})} \frac{J_{a-\frac{1}{2}}(\frac{1}{2\sqrt{-z}})}{J_{a+\frac{1}{2}}(\frac{1}{2\sqrt{-z}})} - 1\right],$$

the convergence being uniform on all compact subsets of $\mathbf{C} \setminus [q_1(a), q_2(a), q_3(a), \cdots]$.

(C) The step function $\psi(a,t)$ *is the unique solution to the CSMP for* $\{c_k(a)\}_{k=0}^{\infty}$.

(D) $\{Q_n(a,z)\}$ *is an orthogonal polynomial sequence with respect to the distribution function* $\varphi(a,t)$.

If $a = 0$, Theorem 1.1 reduces to the following example (from [4]) of a particular orthogonal polynomial sequence. In that case the continued fraction (1.1) converges to the meromorphic function

$$\begin{aligned}\lim_{n\to\infty} \frac{P_n(0,z)}{Q_n(0,z)} &= z\frac{{}_0F_1(\frac{1}{2};\frac{1}{16z})}{{}_0F_1(\frac{3}{2};\frac{1}{16z})} - z = \frac{\sqrt{z}}{2}\coth\left(\frac{1}{2\sqrt{z}}\right) - z \\ &= z\sum_{k=1}^{\infty} \frac{\lambda_k}{z+t_k} = z\int_0^\infty \frac{d\psi(0,t)}{z+t}, \quad |\arg z| < \pi,\end{aligned} \tag{1.9}$$

where

$$\lambda_k := \frac{1}{2\pi^2 k^2},\ t_k := \frac{1}{4\pi^2 k^2} \text{ and } \psi(0,t) := \begin{cases} \sum_{k=n+1}^{\infty} \lambda_k, & t_{n+1} < t \le t_n, \\ \sum_{k=1}^{\infty} \lambda_k, & t_1 < t. \end{cases} \tag{1.10}$$

The distribution function $\psi(0,t)$ is therefore a step function given explicitly by (1.10b). The moments with respect to this distribution function can be expressed in terms of the Bernoulli numbers B_m by

$$\int_0^\infty t^n \, d\psi(0,t) = \frac{|B_{2n+2}|}{(2n+2)!}, \qquad n = 0,1,2,\ldots, \tag{1.11}$$

and

$$\lim_{n\to\infty} \frac{Q_n(0,z)}{z^n} = {}_0F_1\left(\frac{3}{2};\frac{1}{16z}\right) = 2\sqrt{z}\,\sinh\left(\frac{1}{2\sqrt{z}}\right), \quad 0 < |z| < \infty. \tag{1.12}$$

We conclude this introduction by stating further notation and known results that are subsequently used. First we introduce

$$U_n(a,z) := z^{-n}P_n(a,1/z) \text{ and} V_n(a,z) := z^{-n}Q_n(a,1/z),\ n = 0,1,2,\cdots. \tag{1.13}$$

If $A_n(a,\zeta)$ and $B_n(a,\zeta)$ denote the n-th numerator and denominator of the S-fraction

$$1+\mathop{\rm K}_{n=1}^{\infty}\left(\frac{b_n\zeta}{1}\right)=1+\frac{b_1\zeta}{1}\,+\,\frac{b_2\zeta}{1}\,+\,\frac{b_3\zeta}{1}\,+\,\cdots, \tag{1.14}$$

where the b_n are defined by (1.1b), then by (1.3) in [4] we obtain

$$A(a,\zeta):=\lim_{n\to\infty}A_n(a,\zeta)={}_0F_1\left(a+\frac{1}{2};\frac{\zeta}{16}\right),\ \zeta\in\mathbf{C}, \tag{1.15a}$$

$$B(a,\zeta):=\lim_{n\to\infty}B_n(a,\zeta)={}_0F_1\left(a+\frac{3}{2};\frac{\zeta}{16}\right),\ \zeta\in\mathbf{C}, \tag{1.15b}$$

the convergence being uniform on all compact subsets of $\mathbf{C}$. Therefore $A(a,\zeta)$ and $B(a,\zeta)$ are entire functions of ζ. By a theorem due to Maillet (see, e.g., the remark on page 149 and footnote on page 150 in [6]), *if $A_n(\zeta)$ and $B_n(\zeta)$ denote the n-th numerator and denominator of an S-fraction*

$$\mathop{\rm K}_{n=1}^{\infty}\left(\frac{a_n\zeta}{1}\right),\ a_n>0,\ \textit{satisfying}\ \sum_{n=1}^{\infty}a_n<\infty, \tag{1.16}$$

then

$$A(\zeta):=\lim_{n\to\infty}A_n(\zeta)\ \textit{and}\ B(\zeta):=\lim_{n\to\infty}B_n(\zeta), \tag{1.17}$$

are entire functions of ζ with no common zeros; the convergence in (1.17) is uniform on all compact subsets of $\mathbf{C}$. It follows from this and (1.1b) that the entire functions (1.15) satisfy the condition

$$A(a,\zeta)\ \text{and}\ B(a,\zeta)\ \text{have no common zero.} \tag{1.18}$$

It is readily shown that, for $n=0,1,2,\cdots$,

$$U_n(a,z):=z^{-n}P_n(a,1/z)=z\Big(A_{2n}(a,z)-B_{2n}(a,z)\Big) \tag{1.19a}$$

and

$$V_n(a,z):=z^{-n}Q_n(a,1/z)=B_{2n}(a,z), \tag{1.19b}$$

and hence

$$U(a,z):=\lim_{n\to\infty}U_n(a,z)=z\Big(A(a,z)-B(a,z)\Big) \tag{1.20a}$$

and

$$V(a,z):=\lim_{n\to\infty}V_n(a,z)=B(a,z). \tag{1.20b}$$

It follows immediately from (1.15), (1.18) and (1.20) that the entire functions $U(a,z)$ and $V(a,z)$ satisfy

$$U(a,z)\ \text{and}\ V(a,z)\ \text{have no common zero.} \tag{1.21}$$

The following remarks about notation are in order: If we denote by $P_n^*(a,z)$ and $Q_n^*(a,z)$ the polynomials $P_n(a,z)$ and $Q_n(a,z)$, respectively, defined in [4], then P_n^*

and Q_n^* are related to the polynomials P_n and Q_n defined by (1.2) in the following way:

$$P_n^*(a,z) = z^{-1}P_n(a,z) + Q_n(a,z) \text{ and } Q_n^*(a,z) = Q_n(a,z), \; n \geq 0. \tag{1.22}$$

Finally we state, for later reference, a theorem of Hurwitz found in [10]:

Theorem 1.2 (Hurwitz) *Let $f_n(z)$ be holomorphic and single-valued for each integer $n \geq 0$ in an open region R, and let $\{f_n(z)\}$ converge uniformly to $f(z) \not\equiv c$ (c a constant) in every compact subset of R. Let $a \in R$. Then $f(a) = 0$ if and only if there exists a sequence $\{z_n\}$ with $\lim_{n\to\infty} z_n = a$ and there exists n_0 such that $f_n(z_n) = 0$ for each $n > n_0$.*

2. Zeros

From the difference equations (1.2) we see that, for $n \geq 0$, $Q_n(a,z)$ is a monic polynomial in z of degree n; for $n \geq 1$, $P_n(a,z)$ is a polynomial in z of degree n with leading coefficient b_1. In the following lemmas we consider properties of the zeros of $Q_n(a,z)$ and $P_n(a,z)$.

Lemma 2.1 *(A) For $n \geq 1$, $Q_n(a,z)$ is a polynomial of degree n of the form*

$$Q_n(a,z) = z^n + \cdots + b_2 b_4 \cdots b_{2n}.$$

(B) The n zeros $q_k^{(n)}(a), k = 1,2,\cdots,n$, of $Q_n(a,z)$ are all negative and simple and we write

$$q_1^{(n)}(a) < q_2^{(n)}(a) < \cdots < q_n^{(n)}(a) < 0. \tag{2.1}$$

(C) The $n-1$ zeros $q_k^{(n-1)}$ of $Q_{n-1}(a,z)$ separate the zeros of $Q_n(a,z)$; i.e,

$$q_1^{(n)}(a) < q_1^{(n-1)}(a) < q_2^{(n)}(a) < q_2^{(n-1)}(a) < \cdots < q_{n-1}^{(n-1)}(a) < q_n^{(n)}(a) < 0, \tag{2.2}$$

Proof. Adhering to the notation introduced in Lemma 2.2 in [4], we may write

$$Q_n(a,z) = \sum_{j=0}^{n} q_{n,j} z^{n-j},$$

where

$$q_{n,0} := 1,$$

$$q_{n,j} := \sum_{i_j=2}^{2n-2j+2} b_{i_j} \sum_{i_{j-1}=i_j+2}^{2n-2j+4} b_{i_{j-1}} \cdots \sum_{i_2=i_3+2}^{2n-2} b_{i_2} \sum_{i_1=i_2+2}^{2n} b_{i_1}, \; j = 1,2,\cdots,n.$$

Thus the coefficients $q_{n,j}$ are positive for all $0 \leq j \leq n$. Using induction and the difference equations (1.2) one can verify (A), (B) and (C). □

Remark. The zeros $v_k^{(n)}(a)$ of $V_n(a,z)$ are given by

$$v_k^{(n)}(a) = \frac{1}{q_k^{(n)}(a)}, \; k = 1,2,\cdots,n.$$

Lemma 2.2 *(A) For each $k = 1, 2, 3, \cdots$, the sequence $\{v_k^{(n)}(a)\}_{n=k}^{\infty} = \left\{\frac{1}{q_k^{(n)}(a)}\right\}_{n=k}^{\infty}$ is monotonically increasing and bounded above and hence converges to a limit*

$$v_k(a) := \lim_{n\to\infty} v_k^{(n)}(a) = \lim_{n\to\infty} \frac{1}{q_k^{(n)}(a)} =: \frac{1}{q_k(a)}. \tag{2.3}$$

(B) Moreover,

$$\cdots \le v_4(a) \le v_3(a) \le v_2(a) \le v_1(a) \le 0. \tag{2.4}$$

Proof. By Lemma 2.1, we have

$$v_1^{(1)}(a) < v_1^{(2)}(a) < v_1^{(3)}(a) < 0 \cdots < 0 \quad \Rightarrow \quad v_1(a) \le 0,$$

and

$$v_2^{(2)}(a) < v_2^{(3)}(a) < v_2^{(4)}(a) < 0 \cdots < 0 \quad \Rightarrow \quad v_2(a) \le 0.$$

But for $n \ge 2$,

$$v_2^{(n)}(a) < v_1^{(n-1)}(a) \quad \Rightarrow \quad v_2(a) \le v_1(a) \le 0.$$

An induction argument completes the assertion of the lemma. □

Lemma 2.3 *For $n \ge 1$, $P_n(a, z)$ is a polynomial of degree n such that $P_n(a, 0) = 0$. The n zeros $p_k^{(n)}(a)$, $k = 1, 2, \cdots, n$, of $P_n(a, z)$ are simple and non-positive and we write*

$$p_1^{(n)}(a) < p_2^{(n)}(a) < \cdots < p_n^{(n)}(a) = 0, \tag{2.5}$$

Proof. Again, adhering to the notation introduced in Lemma 2.2 in [4], we may write

$$P_n(a, z) = \sum_{j=1}^{n} p_{n,j} z^{n-j+1},$$

where

$$p_{n,1} := b_1,$$

$$p_{n,j} := b_1 \sum_{i_{j-1}=3}^{2n-2j+4} b_{i_{j-1}} \sum_{i_{j-2}=i_{j-1}+2}^{2n-2j+6} b_{i_{j-2}} \cdots \sum_{i_2=i_3+2}^{2n-2} b_{i_2} \sum_{i_1=i_2+2}^{2n} b_{i_1}, \; j = 2, 3, \cdots, n.$$

Thus the coefficients $p_{n,j}$ are positive for all $1 \le j \le n$ (the constant term is zero). The proof of this lemma is similar to that of Lemma 2.1. □

Remark. The zeros $u_k^{(n)}(a)$ of $U_n(a, z)$ are given by

$$u_k^{(n)}(a) = \frac{1}{p_k^{(n)}(a)}, \; k = 1, 2, \cdots, n-1.$$

Lemma 2.4 *(A) For each $k = 1, 2, 3, \cdots$, the sequence $\{u_k^{(n)}(a)\}_{n=k}^{\infty} = \left\{\frac{1}{p_k^{(n)}(a)}\right\}_{n=k}^{\infty}$ is monotonically increasing and bounded above and hence converges to a limit.*

$$u_k(a) := \lim_{n\to\infty} u_k^{(n)}(a) = \lim_{n\to\infty} \frac{1}{p_k^{(n)}(a)} =: \frac{1}{p_k(a)}. \tag{2.6}$$

(B) Moreover,

$$\cdots \le u_4(a) \le u_3(a) \le u_2(a) \le u_1(a) \le 0. \tag{2.7}$$

Proof. A proof is similar to that of Lemma 2.2 and hence omitted. □

Lemma 2.5 *For $n \ge 2$, the $n-1$ negative zeros $p_k^{(n)}(a), k = 1, 2, \cdots, n-1$ of $P_n(a, z)$ separate the n zeros $q_k^{(n)}(a), k = 1, 2, \cdots, n$ of $Q_n(a, z)$ and hence*

$$q_1^{(n)}(a) < p_1^{(n)}(a) < q_2^{(n)}(a) < p_2^{(n)}(a) < \cdots < p_{n-1}^{(n)}(a) < q_n^{(n)}(a) < p_n^{(n)}(a) = 0, \tag{2.8}$$

Proof. Using the determinant formula

$$P_n(a, z)Q_{n-1}(a, z) - P_{n-1}(a, z)Q_n(a, z) = z \prod_{i=1}^{n-1} b_i \tag{2.9}$$

and induction, we see that $P_n(a, z)$ and $Q_n(a, z)$ have no common zero. ($Q_n(a, 0) > 0$.)

Next, consider equation (2.9) and let $n = 2$ and $z = q_i^{(2)}(a)$, $i = 1, 2$. By Lemma 2.1, $Q_1(a, q_1^{(2)}(a)) > 0$ and thus $P_2(a, q_1^{(2)}(a)) < 0$; similarly, $Q_1(a, q_2^{(2)}(a)) < 0$ and it follows that $P_2(a, q_2^{(2)}(a)) > 0$. Together these inequalities imply that $q_1^{(2)}(a) < p_1^{(2)}(a) < q_2^{(2)}(a)$. Continuing in this manner, one can show by induction that the lemma holds. □

Lemma 2.6 *For $k = 1, 2, 3, \cdots$,*

$$v_k(a) \le u_{k-1}(a) \le v_{k-1}(a) \le \cdots. \tag{2.10}$$

Proof. From Lemmas 2.2, 2.4 and 2.5, it follows that

$$v_n^{(n)}(a) < u_{n-1}^{(n)}(a) < v_{n-1}^{(n)}(a) < \cdots v_2^{(n)}(a) < u_1^{(n)}(a) < v_1^{(n)}(a) < 0.$$

Thus

$$v_{k+1}^{(n)}(a) < u_k^{(n)}(a) < v_k^{(n)}(a)$$

for $n = k+1, k+2, \cdots$. Taking the limit as $n \to \infty$ yields the result (2.10). □

Theorem 2.1 *(A) For each $k = 1, 2, 3, \cdots$,*

$$U(a, u_k(a)) = 0 \text{ and } V(a, v_k(a)) = 0, \text{ anda } U(a, u_0(a)) = 0, \text{ where } u_0(a) := 0. \tag{2.11a}$$

(B) For each $k = 1, 2, 3, \cdots$,

$$v_k(a) < u_{k-1}(a) < v_{k-1}(a) < \cdots < u_1(a) < v_1(a) < u_0(a) = 0. \tag{2.11b}$$

(C) All zeros of $U(a,z)$ are contained in $\{u_k(a)\}_{k=0}^{\infty}$, and all zeros of $V(a,z)$ are contained in $\{v_k(a)\}_{k=1}^{\infty}$.

Proof. (A) Since $U_n(a, u_k^{(n)}(a)) = 0$ for $1 \le k \le n$, and $u_k(a) = \lim_{n\to\infty} u_k^{(n)}(a)$, it follows from (1.20a) and Hurwitz' theorem that $U(a, u_k(a)) = 0, k = 1, 2, 3, \cdots$. A similar argument shows that $V(a, v_k(a)) = 0, k = 1, 2, 3, \cdots$.

(B) All inequalites in (2.11b) except $v_1(a) < 0$ follow from Lemma 2.6, (2.11a) and the fact stated in (1.21) that $U(a,z)$ and $V(a,z)$ have no common zero. The inequality $v_1(a) < 0$ is a consequence of the fact that $V(a,0) = B(a,0) = 1$; here we have used (1.20b) and (1.15b).

(C) We recall from Lemmas 2.2 and 2.4 that, for each $k = 1, 2, 3, \cdots$, the sequences $\{u_k^{(n)}\}_{n=k}^{\infty}$ and $\{v_k^{(n)}\}_{n=k}^{\infty}$ are strictly monotonically increasing, bounded above and converge to finite (negative) limits $u_k(a)$ and $v_k(a)$, respectively. Suppose that $V(a,z)$ has a zero v_0 in the interval $(v_1(a), 0)$. Then by Hurwitz' theorem there exists a sequence $\{z_n\}$ with $\lim_{n\to\infty} z_n = v_0$ and there exists an n_0 such that $V_n(a, z_n) = 0$ for $n > n_0$. This is impossible since for all $n \ge 1$, all zeros of $V_n(a,z)$ are less than $v_1(a)$. Hence $V(a,z)$ has no zero in the interval $(v_1(a), 0)$. Similarly $U(a,z)$ has no zero in the interval $(u_1(a), v_1(a))$, since for all $n \ge 1$, all zeros of $U_n(a,z)$ are less than $u_1(a)$. By similar reasoning one can show that $V(a,z)$ has no zero in the interval $(v_2(a), v_1(a))$ and that $U(a,z)$ has no zero in the interval $(u_2(a), u_1(a))$. Continuing in this manner, one can show that for every $k = 1, 2, 3, \cdots$, $U(a,z)$ has no zero in the interval $(u_{k+1}(a), u_k(a))$ and $V(a,z)$ has no zero in the interval $(v_{k+1}(a), v_k(a))$. This completes the proof. □

Lemma 2.7 *$P_n(a,z)/Q_n(a,z)$ has a partial fraction representation of the form*

$$\frac{P_n(a,z)}{Q_n(a,z)} = z \sum_{k=1}^{n} \frac{\lambda_k^{(n)}(a)}{z - q_k^{(n)}(a)}, \tag{2.12}$$

where

$$\lambda_k^{(n)}(a) := \frac{P_n(a, q_k^{(n)}(a))}{q_k^{(n)}(a) Q_n'(a, q_k^{(n)}(a))} > 0, \ \textit{and} \ \sum_{k=1}^{n} \lambda_k^{(n)}(a) = b_1, \quad n \ge 1. \tag{2.13}$$

Proof. Since the zeros of $Q_n(a,z)$ are all simple, we know that $P_n(a,z)/Q_n(a,z)$ can be expressed in the form (2.12). Moreover,

$$\begin{aligned}\lambda_k^{(n)}(a) &:= \frac{P_n(a, q_k^{(n)}(a))}{q_k^{(n)}(a) Q_n'(a, q_k^{(n)}(a))} = \frac{b_1 q_k^{(n)}(a) \prod_{i=1}^{n-1} (q_k^{(n)}(a) - p_i^{(n)}(a))}{q_k^{(n)}(a) \prod_{i=1, i\ne k}^{n} (q_k^{(n)}(a) - q_i^{(n)}(a))} \\ &= \frac{b_1 \prod_{i=1}^{k-1} (q_k^{(n)}(a) - p_i^{(n)}(a)) \prod_{i=k}^{n-1} (p_i^{(n)}(a) - q_k^{(n)}(a))}{\prod_{i=1}^{k-1} (q_k^{(n)}(a) - q_i^{(n)}(a)) \prod_{i=k+1}^{n} (q_i^{(n)}(a) - q_k^{(n)}(a))}.\end{aligned}$$

Positivity of $\lambda_k^{(n)}(a)$ now follows from Lemmas 2.1, 2.3 and 2.5.

Finally,

$$b_1 = \lim_{z\to\infty} \frac{P_n(a,z)}{Q_n(a,z)} = \lim_{z\to\infty} z\sum_{k=1}^{n} \frac{\lambda_k^{(n)}(a)}{z - q_k^{(n)}(a)}$$

$$= \lim_{z\to\infty} \sum_{k=1}^{n} \frac{\lambda_k^{(n)}(a)}{1 - (q_k^{(n)}(a)/z)} = \sum_{k=1}^{n} \lambda_k^{(n)}(a). \quad \square$$

3. Step Function.

For each $n = 1, 2, 3, \cdots$, we define $\psi_n(a,t)$ to be the step function with jump $\lambda_k^{(n)}$ at $t = -q_k^{(n)}(a)$, $k = 1, 2, \ldots, n$ and $\psi_n(a,t) = 0$ for $t < q_1^{(n)}(a)$; that is,

$$\psi_n(a,t) := \begin{cases} 0, & t \le -q_n^{(n)}(a), \\ \sum_{j=1}^{n-k} \lambda_j^{(n)}(a), & -q_{k+1}^{(n)}(a) < t \le -q_k^{(n)}(a), \ 1 \le k \le n-1, \\ b_1, & t > -q_1^{(n)}(a). \end{cases} \tag{3.1}$$

For $k = 1, 2, 3, \cdots$,

$$\lambda_k(a) := \frac{-2q_k(a)}{2a+1} = \frac{\lim_{n\to\infty} \frac{P_n(a,q_k^{(n)}(a))}{(q_k^{(n)}(a))^n}}{\lim_{n\to\infty} q_k^{(n)}(a)[\frac{d}{dz}\,(\frac{Q_n(a,z)}{z^n})]|_{z=q_k^{(n)}(a)}}$$

$$= \frac{\lim_{n\to\infty} \frac{P_n(a,q_k^{(n)}(a))}{(q_k^{(n)}(a))^n}}{\lim_{n\to\infty} \frac{Q_n'(a,q_k^{(n)}(a))}{(q_k^{(n)}(a))^{n-1}}} = \lim_{n\to\infty} \frac{\frac{P_n(a,q_k^{(n)}(a))}{(q_k^{(n)}(a))^n}}{\frac{Q_n'(a,q_k^{(n)}(a))}{(q_k^{(n)}(a))^{n-1}}} \tag{3.2}$$

$$= \lim_{n\to\infty} \frac{P_n(a, q_k^{(n)}(a))}{q_k^{(n)}(a) Q_n'(a, q_k^{(n)}(a))} = \lim_{n\to\infty} \lambda_k^{(n)}(a).$$

By Helly's Selection Principle ([1], p. 53) there exists a subsequence $\{\psi_{n_j}(a,t)\}$ converging to a non-decreasing function $\widehat{\psi}(a,t)$, i.e.,

$$\lim_{j\to\infty} \psi_{n_j}(a,t) =: \widehat{\psi}(a,t), \text{ for all } t \in \mathbf{R}.$$

Lemma 3.1 *Let $\widehat{\psi}(a,t)$ be obtained from Helly's Selection Principle as described above. Then $\widehat{\psi}(a,t)$ is the step function with jump $\lambda_k(a)$ at $-q_k(a)$, $k = 1, 2, 3, \ldots$, and without loss of generality $\widehat{\psi}(a,t) = 0$ for $t < q_1(a)$. Moreover, $\widehat{\psi}(a,t) = \psi(a,t)$ (see (1.7a)).a*

Proof. Consider the interval $-q_2(a) < t < -q_1(a)$. Let $\epsilon > 0$ and $-q_2(a) < t_2 < t_1 < -q_1(a)$ be given. Since $-q_1^{(n)}(a) \uparrow -q_1(a)$ as $n \to \infty$, there exists j_0 such that

$$t_1 < -q_1^{(n_j)}(a) < -q_1(a)$$

for $j > j_0$. Thus $\psi_{n_j}(a,t)$ is constant on $-q_2(a) < t < -q_1^{n_j}(a)$ for $j \geq j_0$. Choose $j_0(\epsilon, t_1, t_2)$ such that

$$|\psi_{n_j}(a,t_1) - \widehat{\psi}(a,t_1)| < \epsilon, \text{ for } j \geq j_0(\epsilon, t_1, t_2)$$

and

$$|\psi_{n_j}(a,t_2) - \widehat{\psi}(a,t_2)| < \epsilon, \text{ for } j \geq j_0(\epsilon, t_1, t_2).$$

It follows that

$$\begin{aligned} |\widehat{\psi}(a,t_1) - \widehat{\psi}(a,t_2)| &\leq |\widehat{\psi}(a,t_1) - \psi_{n_j}(a,t_1)| + |\psi_{n_j}(a,t_2) - \widehat{\psi}(a,t_2)| \\ &< 2\epsilon \end{aligned}$$

for $j \geq j_0(\epsilon, t_1, t_2)$. Thus

$$\widehat{\psi}(a,t_1) = \widehat{\psi}(a,t_2)$$

and $\widehat{\psi}(a,t)$ is constant in this subinterval. The proof for the interval $(-q_3(a), -q_2(a))$ is similiar and therefore this process may be continued. Since by (3.2)

$$\lambda_k(a) = \lim_{n\to\infty} \lambda_k^{(n)}(a),$$

$\lambda_k(a)$ is the jump of $\widehat{\psi}(a,t)$ at $q_k(a)$. Therefore $\widehat{\psi}(a,t) = \psi(a,t)$ by (1.7a). □

4. Proof of Theorem 1.1

(A): By (1.6) and (3.2)

$$\lambda_k(a) := \frac{-2q_k(a)}{2a+1} = \lim_{n\to\infty} \lambda_k^{(n)}(a) > 0, \; k = 1,2,3,\ldots. \tag{4.1}$$

By (2.13)

$$\sum_{k=1}^{m} \lambda_k^{(n)}(a) \leq b_1, \text{ for } 1 \leq m \leq n, \; n = 1,2,3,\cdots.$$

Therefore,

$$\lim_{n\to\infty} \sum_{k=1}^{m} q_k^{(n)}(a) = \sum_{k=1}^{m} \lim_{n\to\infty} q_k^{(n)}(a) = \sum_{k=1}^{m} \lambda_k(a) \leq b_1, \; m = 1,2,3,\ldots.$$

It follows that

$$\sum_{k=1}^{\infty} \lambda_k(a) =: b_1^* \leq b_1. \tag{4.2}$$

In a later step of the proof we show that $b_1^* = b_1$.

(B): By Lemma 3.1, (3.1), (2.12), (4.2) and an extension of Helly's 2nd Theorem ([1], p. 54)

$$\begin{aligned}\lim_{j\to\infty}\frac{P_{n_j}(a,z)}{Q_{n_j}(a,z)} &= \lim_{j\to\infty} z\sum_{k=1}^{n_j}\frac{\lambda_k^{(n_j)}(a)}{z-q_k^{(n_j)}(a)} = \lim_{j\to\infty} z\int_0^\infty \frac{d\psi_{n_j}(a,t)}{z+t} \\ &= z\int_0^\infty \frac{d\psi(a,t)}{z+t} := z\sum_{k=1}^{\infty}\frac{\lambda_k}{z-q_k(a)},\end{aligned} \tag{4.3}$$

the convergence being uniform on every compact subset of $\mathbf{C}\backslash[q_1(a), q_2(a), q_3(a), \cdots]$. (B) follows from (1.4) and (4.3).

(C) follows from the convergence of the S-fraction (1.16) and Theorems 9.4A, 9.5A and 9.6A in [5].

(D) is a consequence of Theorem 7.16 and (7.2.51) in [5].

Finally we show that in (4.2) $b_1^* = b_1$. By (4.3)

$$\begin{aligned}\sum_{k=1}^{\infty}\frac{\lambda_k(a)}{1-\frac{q_k(a)}{z}} &= z\int_0^\infty \frac{d\psi(a,t)}{z+t} = \int_0^\infty \left(1-\frac{t}{z+t}\right)d\psi(a,t) \\ &= c_0(a) - \frac{1}{z}\int_0^\infty \frac{t\,d\psi(a,t)}{1+\frac{t}{z}},\end{aligned} \tag{4.4}$$

where the series converges uniformly on compact subsets of $\mathbf{C}\backslash[q_1(a), q_2(a), q_3(a), \cdots]$. It is readily seen that

$$\lim_{\substack{x\to\infty\\x>0}}\sum_{k=1}^{\infty}\frac{\lambda_k(a)}{1-\frac{q_k(a)}{x}} = \sum_{k=1}^{\infty}\lambda_k(a) =: b_1^*, \tag{4.5a}$$

and

$$\lim_{\substack{x\to\infty\\x>0}}\frac{1}{x}\int_0^\infty \frac{t\,d\psi(a,t)}{1+\frac{t}{x}} = 0. \tag{4.5b}$$

Since the continued fraction (1.1) corresponds to the series $\sum_{k=0}^{\infty} c_k(a)z^{-k}$ at $z=\infty$, it follows that $c_0(a) = b_1$. Combining this with (4.4) and (4.5) yields

$$\sum_{k=1}^{\infty}\lambda_k(a) = c_0(a) = b_1. \Box$$

5. Lommel Polynomials

The Bessel functions of the first kind of order ν, defined by

$$J_\nu(z) := \left(\frac{z}{2}\right)^\nu \sum_{k=0}^{\infty}\frac{(-1)^k\left(\frac{z}{2}\right)^{2k}}{k!\,\Gamma(\nu+k+1)}, \quad \nu\in C, \quad z\in C, \tag{5.1}$$

satisfy the second order linear difference equations ([12], p. 294)

$$J_{\nu+1}(z) = \frac{2\nu}{z}J_\nu(z) - J_{\nu-1}(z). \tag{5.2}$$

These equations may be used to express $J_{\nu+m}(z)$ in terms of $J_\nu(z)$ and $J_{\nu-1}(z)$ in the form ([12], p. 295)

$$J_{\nu+m}(z) = J_\nu(z)R_{m,\nu}(z) - J_{\nu-1}(z)R_{m-1,\nu+1}(z), \tag{5.3}$$

where the coefficients of $J_\nu(z)$ and $J_{\nu-1}(z)$ (denoted above by $R_{m,\nu}(z)$ and $R_{m-1,\nu+1}(z)$, respectively) are polynomials in $\frac{1}{z}$ called *Lommel polynomials.* Lommel polynomials are known to satisfy the recurrence relation ([12], p. 299)

$$R_{0,\nu}(z) := 1, \quad R_{1,\nu}(z) := \frac{2\nu}{z}, \tag{5.4a}$$

$$R_{m+1,\nu}(z) = \frac{2(\nu+m)}{z} R_{m,\nu}(z) - R_{m-1,\nu}(z), \quad m \geq 1, \ z \in \mathbf{C} \setminus \{0\}. \tag{5.4b}$$

Schwartz ([7], p. 63) proved that Lommel polynomials are orthogonal with respect to a distribution function $\psi(t)$ and explicitly computed this distribution function. Denoting the positive zeros of $J_\nu(\frac{1}{z})$ by j_k, $j_{k+1} < j_k$, and using "Stieltjes inversion formula," he showed that $\psi(t)$ is a step function with jumps j_k^2 at $\pm j_k$, $k \geq 1$ and that the interval of orthogonality is $(-j_1, j_1)$. Defining modified Lommel polynomials by

$$h_{m,\nu}(z) := R_{m,\nu}\left(\frac{1}{z}\right), \quad m \geq 0, \quad z \in \mathbf{C}, \tag{5.5}$$

and using a known asymptotic formula for $J_\nu(1/z)$ [12], Dickenson [2] rediscovered Schwartz's results. Ismail [3] gave the q-extension of modified Lommel polynomials and established their orthogonality.

The $Q_n(a,z)$ are related to Lommel polynomials by

$$R_{2n,\nu+1}\left(\frac{2}{\sqrt{z}}\right) = (-16)^n(\nu+1)_{2n}\, Q_n\left(\nu - \frac{1}{2}, -\frac{z}{16}\right), \quad n \geq 1, \quad \nu > 0.$$

Acknowledgements

The authors wish to thank Mourad Ismail for pointing out a connection with Lommel polynomials and the referee for the helpful comments.

References

1. Chihara, T. S., *Introduction to Orthogonal Polynomials*, (Mathematics and Its Applications Ser.), Gordon (1978).
2. Dickenson, David, On Lommel and Bessel Polynomials, *Proc. Amer. Math. Soc.* 5 (1954), 946-956.
3. Ismail, Mourad E.H., The Zeros of Basic Bessel Functions, the Functions $J_{\nu+ax}(x)$, and Associated Orthogonal Polynomials, *J. Math. Analy.* 86 (1982), 1-19.
4. Jones, William B., Arne Magnus, John H. McCabe, and Nancy J. Wyshinski, Asymptotic Properties of a Family of Orthogonal Polynomial Sequences, *J. Comp. and Applied Math.* 32 (1990), 143-151.
5. Jones, William B. and W. J. Thron, *Continued Fractions: Analytic Theory and Applications*, Encyclopedia of Mathematics and its Applications, 11, Addison–Wesley Publishing Company, Reading, MA (1980), distributed now by Cambridge University Press, New York.
6. Perron, O, *Die Lehre Von Den Ketterbrüchen*, (1929).

7. Schwartz, H.M., A Class of Continued Fractions, *Duke Math. J.* 6 (1940), 48-65.
8. Ślezyński, Convergence of continued fractions, *Zap. Mat. Otodiel. Novoross. Obshch. estest. (Odessa)* **8** (1888), 97-127 (in Russian).
9. Szegö, Gabor, *Orthogonal Polynomials*, Colloquium Publications, Vol. 23, Amer. Math. Soc., New York (1939).
10. Thron, W.J., *The Theory of Functions of A Complex Variable*, Wiley, New York (1953).
11. Thron, W.J., Some Results on Separate Convergence of Continued Fractions, *Comput. Methods and Function Theory*, Proceedings Valparaiso 1989, Lecture Notes in Mathematics **1435**, Springer-Verlag, New York, (1990), 191-200.
12. Watson, G.N., *A Treatise on the Theory of Bessel Functions*, Cambridge, University Press (1948).

CONVERGENCE CRITERIA OF TWO-DIMENSIONAL CONTINUED FRACTIONS

KRYSTYNA KUCHMINS'KA *
Applied Problems of Mechanics and Mathematics Institute
Ukrainian Academy of Sciences
3-b Naukova St., 290601 LVIV, MSP, UKRAINE

Abstract. Some well known convergence criteria for continued fractions are generalized on the two-dimensional case. Using the difference formula for approximants of the two-dimensional continued fraction and the majorant method some estimates of the convergence speed are obtained.

Key words: Two-dimensional continued fraction, convergence, majorant fraction

1. Introduction

We define a two-dimensional continued fraction (shortly, TDCF) by using compositions of non-singular linear fractional transformations. Let a_i, b_i, a_{ij}, b_{ij} $(i, j \in N)$ be complex numbers, $a_i \neq 0$, $a_{ij} \neq 0$, and we consider the following transformations

$$\begin{aligned} t_i(x,y,z) &= \frac{a_{i-1}}{b_{i-1} + a_{i,i-1}x + a_{i-1,i}y + z}, \quad i = 1, 2, \ldots \\ t_{ij}(u) &= \frac{1}{b_{ij} + a_{i+1,j}u}(i > j), \quad t_{ij}(u) = \frac{1}{b_{ij} + a_{i,j+1}u}(i < j), i, j = 0, 1, \ldots \end{aligned} \tag{1}$$

and compositions of such transformation

$$\begin{aligned} T_{k-i,i} &= t_{i+1,i}t_{i+2,i}\cdots t_{k-i,i}(0), \\ T_{i,k-i} &= t_{i,i+1}t_{i,i+2}\cdots t_{i,k-i}(0), k > i, k \geq 1, i = 0, ..., \left[\frac{k}{2}\right] - 1, \end{aligned}$$

[k] means the entire part of the number k.

Then we can define the TDCF as following sequence $\{f_k\}$, f_k is the element of the extended complex plane

$$f_k = t_1(T_{k-1,0}; T_{0,k-1}; t_2(T_{k-2,1}; T_{1,k-2}; \cdots$$
$$t_{[k/2]}(T_{[(k+1)/2],[(k-2)/2]}; T_{[(k-2)/2],[(k+1)/2]}; t_{[(k+1)/2]}(0,0,0)\ldots), k = 1, 2, \ldots, \tag{2}$$

$t_{[(k+1)/2]}(0,0,0) = 0$, if $[(k+1)/2] = [k/2]$; $T_{0,0} = 0$, $f_1 = \frac{a_0}{b_0}$.

* Research supported in part by the Fundamental Investigation Foundation of the Science and Technology State Committee of Ukraine and the International Science Foundation (USA).

A. Cuyt (ed.), Nonlinear Numerical Methods and Rational Approximation II, 423–431.

From this definition we find the following representation of f_k

$$\cfrac{a_0}{b_0+\cfrac{a_{10}}{b_{10}+\genfrac{}{}{0pt}{}{\vdots}{+\cfrac{a_{k-1,0}}{b_{k-1,0}}}}+\cfrac{a_{01}}{b_{01}+\genfrac{}{}{0pt}{}{\vdots}{+\cfrac{a_{0,k-1}}{b_{0,k-1}}}}+\cfrac{a_1}{b_1+\cfrac{a_{21}}{b_{21}+\genfrac{}{}{0pt}{}{\vdots}{+\cfrac{a_{k-3,1}}{b_{k-3,1}}}}+\cfrac{a_{12}}{b_{12}+\genfrac{}{}{0pt}{}{\vdots}{+\cfrac{a_{1,k-3}}{b_{1,k-3}}}}+\begin{matrix}\vdots\end{matrix}}}, \tag{3}$$

$$\vdots\; + \cfrac{a_{[(k-1)/2]}}{b_{[(k-1)/2]}+\dfrac{a_{k-1-[(k-1)/2],[(k-1)/2]}}{b_{k-1-[(k-1)/2],[(k-1)/2]}}+\dfrac{a_{[(k-1)/2],k-1-[(k-1)/2]}}{b_{[(k-1)/2],k-1-[(k-1)/2]}}}$$

where $a_{k-1-[(k-1)/2],[(k-1)/2]} = a_{[(k-1)/2],k-1-[(k-1)/2]} = 0,$ if k is odd, or shortly for an infinite TDCF (f):

$$f = \mathop{D}_{k=0}^{\infty} \frac{a_k}{\Phi_k}, \quad \Phi_k = b_k + \mathop{D}_{j=1}^{\infty} \frac{a_{k+j,k}}{b_{k+j,k}} + \mathop{D}_{j=1}^{\infty} \frac{a_{k,k+j}}{b_{k,k+j}}. \tag{4}$$

Finite TDCFs of the form (4)

$$f_1 = \frac{A_1}{B_1} = \frac{a_0}{b_0}, \quad f_m = \frac{A_m}{B_m} = \mathop{D}_{k=0}^{[(m-1)/2]} \frac{a_k}{\Phi_k^{(m-1-2k)}}, \tag{5}$$
$$\Phi_k^{(m)} = b_k + \mathop{D}_{j=1}^{m} \frac{a_{k+j,k}}{b_{k+j,k}} + \mathop{D}_{j=1}^{m} \frac{a_{k,k+j}}{b_{k,k+j}}$$

are called m-th approximants, A_m and B_m are called the numerator and the denominator of f_m, respectively [3,4].

Introducing recursiveness by the notations

$$Q_i^{(s-1-2i)} = \Phi_i^{(s-1-2i)} + \frac{a_{i+1}}{Q_{i+1}^{(s-3-2i)}}, \quad i = 0, 1, \ldots, [(s-1)/2]$$
$$Q_j^{(1)} = \Phi_j^{(1)}, \quad Q_j^{(0)} = b_j, \quad j = [(s-1)/2], s = 1, 2, \ldots \tag{6}$$

for the TDCF (4) we determine equations

$$f_n - f_{2m+1} = \sum_{i=0}^{m} \frac{(-1)^i(\Phi_i^{(2m-2i)} - \Phi_i^{(n-1-2i)}) \prod_{j=0}^{i} a_i}{\prod_{j=0}^{i}(Q_j^{(2m-2j)} Q_j^{(n-1-2j)})} + $$
$$+ \frac{(-1)^{m+1} \prod_{j=0}^{m+1} a_i}{\prod_{j=0}^{m} Q_j^{(2m-2j)} \prod_{j=0}^{m+1} Q_j^{(n-1-2j}} r_{m,n}, \tag{7}$$

where

$$r_{m,n} = \begin{cases} 0 \text{ if } n = 2m+2 \\ 1 \text{ if } n > 2m+2. \end{cases}$$

Equations (7) be called the difference formulas for approximants f_k of the TDCF [3,4]. These formulas play the main role in the investigation of the TDCF convergence.

2. Śleszyński-Pringsheim-like Criteria

A two-dimensional continued fraction is said to converge if its sequence of approximants (5) converges to a point in the extended complex plane.
A two-dimensional continued fraction is said to converge absolutely if the series $\sum_{n=0}^{\infty} |f_{n+1} - f_n|$ composed of its approximants (5) converges.

Theorem 1 *Let the elements of the TDCF*

$$\underset{i=0}{\overset{\infty}{D}} \frac{c_i}{\Phi_i}, \quad \Phi_i = 1 + \underset{p=1}{\overset{\infty}{D}} \frac{c_{i+p,i}}{1} + \underset{p=1}{\overset{\infty}{D}} \frac{c_{i,i+p}}{1} \tag{8}$$

be complex numbers and satisfy the inequalities

$$|c_i| \leq \frac{1}{3} g_i(1 - g_{i-1}), \quad |c_0| \leq g_0,$$
$$|c_{i+1,i}| \leq \frac{1}{3} g_{i+1,i}(1 - g_i), \quad |c_{i,i+1}| \leq \frac{1}{3} g_{i,i+1}(1 - g_i), \tag{9}$$
$$|c_{ij}| \leq g_{ij}(1 - g_{i-1,j})(i > j), \quad |c_{ij}| \leq g_{ij}(1 - g_{i,j-1})(i < j), i, j = 0, 1, \ldots,$$

where

$$g_i, g_{ij} \in R, \quad 0 \leq g_i < 1, \quad 0 \leq g_{ij} < 1, \quad g_{-1} = 0, \tag{10}$$

or

$$g_i, g_{ij} \in R, \quad 0 < g_i \leq 1, \quad 0 < g_{ij} \leq 1 \tag{11}$$

Then the TDCF (8) converges absolutely and its value region is the disk $|z| \leq 1$.

Proof. We show that the TDCF

$$\frac{-g_0}{\hat{\Phi}_0 + \underset{i=1}{\overset{\infty}{D}} \frac{-\frac{1}{3} g_i(1 - g_{i-1})}{\hat{\Phi}_i}}, \tag{12}$$

$$\hat{\Phi}_i = 1 - \frac{\frac{1}{3} g_{i+1,i}(1 - g_i)}{1 + \underset{p=i+2}{\overset{\infty}{D}} \frac{-g_{pi}(1 - g_{p-1,i})}{1}} - \frac{\frac{1}{3} g_{i,i+1}(1 - g_i)}{1 + \underset{p=i+2}{\overset{\infty}{D}} \frac{-g_{ip}(1 - g_{i,p-1})}{1}}$$

is majorant one for (8).

Let f_n and $\hat{f}_n$ be n-th approximants of TDCFs (8) and (12), respectively:

$$f_n = \underset{i=0}{\overset{[(n-1)/2]}{D}} \frac{c_i}{\Phi_i^{(n-1-2i)}}, \quad \hat{f}_n = \frac{-g_0}{\hat{\Phi}_0^{(n-1)} + \underset{i=1}{\overset{[(n-1)/2]}{D}} \frac{-\frac{1}{3} g_i(1 - g_{i-1})}{\hat{\Phi}_i^{(n-1-2i)}}}.$$

Introducing recurrent relations similarly as (6):

$$\hat{Q}_i^{(s-1-2i)} = \hat{\Phi}_i^{(s-1-2i)} - \frac{1}{3} g_{i+1}(1 - g_i)/\hat{Q}_{i+1}^{(s-3-2i)},$$

$$Q_{1j}^{(s-1-2i)} = 1 + c_{i+j+1,i}/Q_{1,j+1}^{(s-1-2i)}, \quad Q_{2j}^{(s-1-2i)} = 1 + c_{i,i+j+1}/Q_{2,j+1}^{(s-1-2i)},$$
$$\hat{Q}_{1j}^{(s-1-2i)} = 1 - g_{i+j+1,i}(1 - g_{i+j,i})/\hat{Q}_{1,j+1}^{(s-1-2i)},$$
$$\hat{Q}_{2j}^{(s-1-2i)} = 1 - g_{i,i+j+1}(1 - g_{i,i+j})/\hat{Q}_{2,1+j}^{(s-1-2i)},$$
$$i = 0, \ldots, [(s-1)/2]; \quad j = 0, \ldots, s-1-2i,$$

using the method of estimates of [5] we can prove inequalities

$$\begin{aligned} |Q_i^{(s-1-2i)}| &\geq \hat{Q}_i^{(s-1-2i)} \geq g_i, \\ |Q_{1j}^{(s-1-2i)}| &\geq \hat{Q}_{1j}^{(s-1-2i)} \geq g_{i+j,i}, \\ |Q_{2j}^{(s-1-2i)}| &\geq \hat{Q}_{2j}^{(s-1-2i)} \geq g_{i,i+j} \end{aligned} \tag{13}$$

under the restriction (10) and

$$\begin{aligned} |Q_i^{(s-1-2i)}| &\geq \hat{Q}_i^{(s-1-2i)} > g_i, \\ |Q_{1j}^{(s-1-2i)}| &\geq \hat{Q}_{1j}^{(s-1-2i)} > g_{i+j,i}, \\ |Q_{2j}^{(s-1-2i)}| &\geq \hat{Q}_{2j}^{(s-1-2i)} > g_{i,i+j} \end{aligned} \tag{14}$$

under (11).

In addition the fulfilment of two last inequalities from (13),(14) and conditions (9) allows to estimate the following difference

$$|\Phi_i^{(s-1-2i)} - \Phi_i^{(s-2i)}| \leq \hat{\Phi}_i^{(s-1-2i)} - \hat{\Phi}_i^{(s-2i)}.$$

Using the difference equations (7), we arrive at

$$|f_{r+1} - f_r| \leq \hat{f}_r - \hat{f}_{r+1}. \tag{15}$$

The sequence $\{\hat{f}_r\}$ is monotonic decreasing and by virtue of inequalities (13),(14), bounded below

$$\hat{f}_n = \frac{-g_0}{\hat{Q}_0^{(n-1)}} \geq \frac{-g_0}{g_0} = -1,$$

hence there exists the limit of $\hat{f}_n$, so $\{\hat{f}_n\}$ converges to a finite value.

The absolute convergence of the TDCF (8) follows from the inequality (15). Since

$$|f_n| = \frac{|c_0|}{|Q_0^{(n-1)}|} \leq \frac{g_0}{\hat{Q}_0^{(n-1)}} \leq 1,$$

then the value region of the TDCF (8),i.e. the closure of the set of all of its approximants, belongs to the disk $|z| \leq 1$.

We show that the value region of the TDCF (8) coincides with that disk. If c is an arbitrary complex number such that $|c| < 1$, then the first approximant of the TDCF (8) with $c_0 = c$ and arbitrary complex numbers c_i,c_{ij} $(i, j = 0, 1, \ldots, i \neq j)$ satisfying (9) takes the value c. If $c = 1$, then the TDCF (8) with $c_0 = c/4$, $c_{ij} = -1/4$ $(i \neq j)$, $c_{i,i+1} = c_{i+1,i} = -1/8$, $c_i = 1/16$ $(i, j = 0, 1, \ldots)$ satisfies the restrictions of (9) under $g_{ij} = 1/2$, $g_i = 1/4$, $g_0 = 1/4$ and takes the value c.

Theorem 2 *The TDCF (4) with complex elements satisfying*

$$|b_i| \geq |a_i| + 3, \quad |b_{ij}| \geq |a_{ij}| + 1, \quad i \neq j; \quad i, j = 0, 1, \ldots \tag{16}$$

converges absolutely and its value region is the disk $|z| \leq 1$.

Proof. By means of equivalence transformations we can lead the TDCF (4) to (8), which has such elements

$$c_0 = a_0/b_0, \quad c_{i+1,i} = a_{i+1,i}/(b_i b_{i+1,i}), \quad c_{i,i+1} = a_{i,i+1}/(b_i b_{i,i+1}),$$
$$c_{ij} = a_{ij}/(b_{ij} b_{i-1,j}), (i > j+1); \quad c_{ij} = a_{ij}/(b_{ij} b_{i,j-1})(j > i+1);$$
$$c_i = a_i/(b_i b_{i-1}), \quad i \neq j; \quad i, j = 0, 1, \ldots.$$

Let $g_i = |a_i|/(|a_i| + 3)$, $g_{ij} = |a_{ij}|/(|a_{ij}| + 1)$, $i \neq j$, then we can use the theorem 1.

REMARK. We can show that the value region of the TDCF from the theorem 2 coincides with the disk $|z| \leq 1$. If $|c| < 3$, then the first approximant with elements $a_0 = 3c/(3 - |c|)$, $b_0 = 3|c|/(3 - |c|) + 3$ takes the value $c/3$. Let $c = 3\exp(i\phi)$ $(-\pi < \phi \leq \pi)$, then we consider the TDCF satisfying (16)

$$\frac{c_0 \exp(i\phi)}{\hat{\Phi}_0 + \overset{\infty}{\underset{i=1}{D}} \frac{-c_i}{\hat{\Phi}_i}}, \hat{\Phi}_i = c_i + 3 + \overset{\infty}{\underset{p=1}{D}} \frac{-c_{i+p,i}}{c_{i+p,i} + 1} + \overset{\infty}{\underset{p=1}{D}} \frac{-c_{i,i+p}}{c_{i,i+p} + 1}.$$

Putting all $c_i > 0$, $c_{ij} = c_{ji} = 1$ $(i \neq j; i, j = 0, 1, \ldots)$, we obtain the fraction

$$\cfrac{c_0 \exp(i\phi)/(c_0 + 1)}{1 - \cfrac{c_1/(c_0 + 1)(c_1 + 1)}{1 - \cfrac{c_2/(c_1 + 1)(c_2 + 1)}{1 - \ddots}}}, \tag{17}$$

or

$$\frac{g_0 c/3}{1 - \overset{\infty}{\underset{i=1}{D}} \frac{g_i(1 - g_{i-1})}{1}}, \; g_i = \frac{c_i}{1 + c_i}, \; i = 0, 1, \ldots$$

the value of which according to [8] is equal to

$$\frac{c}{3}(1 - (1 + \sum_{k=0}^{\infty} \prod_{i=0}^{k} \frac{g_i}{1 - g_i})^{-1}).$$

Since $\sum_{k=o}^{\infty} \prod_{i=0}^{k} \frac{g_i}{1-g_i} = \sum_{k=0}^{\infty} \prod_{i=0}^{k} c_i$, then (17) takes the value $c/3$, if the series $\sum_{k=0}^{\infty} \prod_{i=0}^{k} c_i$ is convergent.

Theorem 3 *The TDCF (4) with complex elements converges absolutely, if*

$$|b_i| \geq 3|a_i| + 1, \quad |b_0| \geq |a_0| + 1/3,$$
$$|b_{i+1,i}| \geq 3|a_{i+1,i}| + 1, \quad |b_{i,i+1}| \geq 3|a_{i,i+1}| + 1, \tag{18}$$
$$|b_{ij}| \geq |a_{ij}| + 1, |i - j| > 1.$$

Its value region is the disk $|z| \leq 1$.

Proof. We can reduce the TDCF (4) to the TDCF (12); if we choose

$$g_i = \frac{3|a_i|}{3|a_i|+1}, \quad g_{i+1,i} = \frac{3|a_{i+1,i}|}{3|a_{i+1,i}|+1}, \quad g_{i,i+1} = \frac{3|a_{i,i+1}|}{3|a_{i,i+1}|+1},$$
$$g_{ij} = \frac{|a_{ij}|}{|a_{ij}|+1}, \quad |i-j|>1, \quad i,j=0,1,\ldots,$$

then we obtain $0 \le g_i < 1$, $0 \le g_{ij} < 1$ $(i \ne j; i,j = 1,2,\ldots)$ and conditions of the theorem 1 are satisfying.

Theorems 2 and 3 generalize Śleszyński-Pringsheim's [7] criterion.

3. Worpitzky-like Criterion

Worpitzky's Criterion is also well known one for the classical continued fractions [2,6]. This criterion was generalized for branched continued fractions [1] and we propose our result for the TDCF (8).

Theorem 4 *Let the complex elements of the TDCF (8) satisfy conditions*

$$|c_i| \le \alpha = \frac{1}{2}t(1-t), \quad |c_{ij}| \le \alpha = \frac{1}{2}t(1-t),$$
$$c_0 = 1, \quad 0 \le t \le \frac{1}{2}, \quad i,j = 1,2,\ldots. \tag{19}$$

Then:

(i) the TDCF (8) converges absolutely;

(ii) there exist positive constants A and B such that

$$|f_n - f_{2m+1}| \le$$
$$A\frac{(1-2t)(t(1-t))^{m+1}((1-t)^{[(n-1)/2]-m} - t^{[(n-1)/2]-m})}{2^{m+1}(1-2t(1-t))^{m+3/2}((1-t)^{[(n-1)/2]+2} - t^{[n-1)/2]+2})((1-t)^{m+2} - t^{m+2})},$$

if $0 \le t < 1/2$, *or*

$$B\frac{2\sqrt{2}([(n-1)/2]-m)}{([(n-1)/2]+2)(m+2)}, \quad t = 1/2 \tag{20}$$

($n > 2m+1$, f_m is the m-th approximant of the TDCF (8));

(iii) the value of the TDCF (8) and all of its approximants belong to the disk

$$|z-(1-g^2)^{-1}| \le g(1-g^2)^{-1}, \quad g = t - \frac{1}{2}(\sqrt{1-2t(1-t)}-1); \tag{21}$$

(iv) the limit constant $\alpha = \frac{1}{8}$ is the best one,i.e. it can not be increased without losing convergence of the TDCF (8).

Proof. First we show that a periodic fraction

$$(\hat{\Phi}_0 + \mathop{D}_{i=1}^{\infty} \frac{\frac{1}{2}t(1-t)}{\hat{\Phi}_i})^{-1}, \hat{\Phi}_i = 1 - \cfrac{t(1-t)}{1 - \cfrac{\frac{1}{2}t(1-t)}{1 - \cfrac{\frac{1}{2}t(1-t)}{1- \ddots}}} \tag{22}$$

is the majorant one for the TDCF (8). Similarly as inequalities (13),(14) were obtained, we can prove the following

$$|Q_{1j}^{(s-1-2i)}| \geq \hat{Q}_{1,2,j}^{(s-1-2i)} \geq \frac{1}{2}(1+\sqrt{1-2t(1-t)}),$$
$$|Q_{2j}^{(s-1-2i)}| \geq \hat{Q}_{1,2,j}^{(s-1-2i)} \geq \frac{1}{2}(1+\sqrt{1-2t(1-t)}), \tag{23}$$
$$|Q_{i}^{(s-1-2i)}| \geq \hat{Q}_{i}^{(s-1-2i)} \geq \frac{1}{2}(\sqrt{1-2t(1-t)}+\sqrt{1-4t(1-t)}).$$

where

$$\hat{Q}_m^{(s-1-2m)} = \hat{\Phi}_m^{(s-1-2m)} - \frac{\frac{1}{2}t(1-t)}{\hat{Q}_{m+1}^{(s-3-2m)}}$$

and $\hat{Q}_{1,2,j}^{(s-1-2m)}$ means the fraction $Q_{1j}^{(s-1-2m)}$ or $Q_{2j}^{(s-1-2m)}$ with the value $-\frac{1}{2}t(1-t)$ instead of $c_{i+p,i}$ or $c_{i,i+p}$, $(p = j+1, \ldots, s-1-2m)$. Note that all $Q_i^{(s-1-2i)} \neq 0$ and $\hat{Q}_i^{(s-1-2i)} > 0$.

Let $\frac{C_n}{D_n}$ be the n-th approximant of the TDCF (22), then using (7),(23) we obtain

$$|\frac{A_n}{B_n} - \frac{A_{2m+1}}{B_{2m+1}}| \leq \frac{C_n}{D_n} - \frac{C_{2m+1}}{D_{2m+1}} \quad (n > 2m+1), \tag{24}$$

$\frac{A_n}{B_n}$ means the n-th approximant of the TDCF (8). The sequence $\left\{\frac{C_n}{D_n}\right\}$ is monotonic increasing and $\frac{C_n}{D_n}$ satisfies the inequality

$$\frac{C_n}{D_n} \leq \frac{2}{\sqrt{1-2t(1-t)}+\sqrt{1-4t(1-t)}}.$$

Furthermore from

$$\sum_{r=0}^{m} |\frac{A_{r+1}}{B_{r+1}} - \frac{A_r}{B_r}| \leq \sum_{r=0}^{m} (\frac{C_{r+1}}{D_{r+1}} - \frac{C_r}{D_r}) = \frac{C_{m+1}}{D_{m+1}}, \quad C_0 = 0, \quad D_0 = 1$$

we obtain

$$\sum_{r=0}^{\infty} |\frac{A_{r+1}}{B_{r+1}} - \frac{A_r}{B_r}| \leq \lim_{m\to\infty} \frac{C_{m+1}}{D_{m+1}} \leq \frac{2}{\sqrt{1-2t(1-t)}+\sqrt{1-4t(1-t)}}.$$

The first part of the statement is valid.

To prove the next parts of the theorem we find the majorant continued fraction for (22).

Let us consider a continued fraction

$$\frac{a}{1-} \; \frac{k(1-k)}{1-} \; \frac{k(1-k)}{1-\cdots}, \quad 0 \leq k \leq 1/2 \tag{25}$$

where a is a constant, and study its properties; we denote the m-th numerator and denominator of the m-th approximant g_m of (25) as P_m , Q_m, respectively, then

$$Q_m = (1-k)^m + k(1-k)^{m-1} + \cdots + k^m, \quad P_m = aQ_{m-1}, \quad m = 1, 2, \ldots \tag{26}$$

and all $Q_m > 0$ for any $0 \le k \le 1/2$.

Using the mathematical induction method and conditions of the theorem we can prove, if $k = \frac{1}{2}(1 - \frac{1-2t}{\sqrt{1-2t(1-t)}})$, $0 \le t \le 1/2$ and $a = (1-2t(1-t))^{-\frac{1}{2}}$ that

$$\hat{Q}_m^{(s-1-2m)} \ge \sqrt{1-2t(1-t)}d_{[(s-1)/2]-m}, \quad m \le [(s-1)/2], \tag{27}$$

where d_m is the m-th approximant of the continued fraction

$$1 - \frac{k(1-k)}{1-} \quad \frac{k(1-k)}{1-\cdots}. \tag{28}$$

It's well known that

$$g_n - g_m = \frac{k^m(1-k)^m}{\sqrt{1-2t(1-t)}\prod_{k=0}^{m} d_{n-k-1}\prod_{k=0}^{m-1} d_{m-k-1}} \quad (n > m)$$

By (27) for any $n > 2m+1$ we have after some evaluations:

$$\frac{C_n}{D_n} - \frac{C_{2m+1}}{D_{2m+1}} \le \hat{A}(g_{[(n-1)/2]+2} - g_{m+1})$$

Applying (26) and this fact that $d_k = (g_{k+1}\sqrt{1-2t(1-t)})^{-1} = \frac{Q_{k+1}}{Q_k}$ we find

$$\frac{C_n}{D_n} - \frac{C_{2m+1}}{D_{2m+1}} \le \hat{A}\frac{k^{m+1}(1-k)^{m+1}Q_{[(n-1)/2]-m-1}}{\sqrt{1-2t(1-t)}Q_{[(n-1)/2]+1}Q_{m+1}}. \tag{29}$$

For $t = 1/2$ we have $k = 1/2$, $Q_m = 2^{-m}(m+1)$. For $0 \le t < 1/2$ we set $t = 1/x$ in (26) and returning to t we obtain

$$Q_m = ((1-t)^{m+1} - t^{m+1})(1-2t)^{-1}.$$

Put these values for Q_m in (29) we obtain (20). A constant $\hat{A}$ takes two different values, depending on t ($\hat{A} = A$, if $0 \le t < 1/2$; $\hat{A} = B$, if $t = 1/2$).

Representing the TDCF (8) in the form $z = \frac{1}{1+w}$, where

$$w = \Phi_0 - 1 + \underset{i=1}{\overset{\infty}{D}} \frac{c_i}{\Phi_i}$$

and $|w| \le t - \frac{1}{2}(\sqrt{1-2t(1-t)} - 1)$ in the same way as in classical one [8] we find (21).

The constant $\alpha = \frac{1}{8}$ is the best one because the TDCF (8) with $c_i = -c$, $c_{ij} = -c$, $c > 1/8$ is divergent.

The value region $|z - \frac{1}{1-g^2}| \le \frac{g}{1-g^2}$ is the best one, because the value of the TDCF (8) with $c_i = c_{ij} = -\frac{1}{8}$ is equal to $2\sqrt{2}$.

References

1. Bodnar D.I.: 1986, 'Branched continued fractions,' Naukova Dumka, Kiev (in Russian).

2. Jones W.B., Thron W.J.: 1980, 'Continued fractions: analytic theory and applications', *Encyclopedia of Mathematics and its Applications*, Addison-Wesley Publishing Company, Massachusets.
3. Kuchminskaya Kh.(Kuchmins'ka): 1984, 'On sufficient conditions of convergence for two-dimensional continued fractions', *Mat. Metody i Fiz.-Meh. Polya*, **20**, pp. 19–23 (in Russian).
4. Kuchminskaya Kh.: 1987, 'On approximation of functions by two-dimensional continued fractions', *Rational Approximation and its Applications in Mathematics and Physics (Eds. J.Gilewicz, M.Pindor, W.Siemaszko), Lecture Notes in mathematics*, Springer-Verlag, **1237**, pp. 207–216.
5. Kuchminskaya Kh.: 1989, 'Convergence criteria for two-dimensional continued fractions with complex elements', *Metody issledovaniya differencialnyh i integralnyh operatorov*, Naukova Dumka, Kiev, pp.122–127 (in Russian).
6. Lorentzen L., Waadeland H.: 1992, 'Continued fractions with applications', North-Holland, Elsevier Science Publishers B.V., Amsterdam.
7. Thron W.: 1992, 'Should the Pringsheim criterion be renamed the Śleszyński criterion ?', *Communucations in the analytic theory of continued fractions*, **1**, pp. 13–20.
8. Wall H.S.: 1948, 'Analytic theory of continued fractions', Van Nostrand.

FIRST-ORDER LINEAR RECURRENCE SYSTEMS AND GENERAL N-FRACTIONS

PAUL LEVRIE
Departement Computerwetenschappen, K.U.Leuven, Celestijnenlaan 200A, B-3001 Heverlee, and Katholieke Industriële Hogeschool Antwerpen, Salesianenlaan 30, B-2660 Hoboken, Belgium

and

MARC VAN BAREL and ADHEMAR BULTHEEL
Departement Computerwetenschappen, K.U.Leuven, Celestijnenlaan 200A, B-3001 Heverlee, Belgium

Abstract. In this paper we define a natural generalization of ordinary continued fractions and n-fractions (de Bruin [2]). This so-called general n-fraction is associated with a first-order recurrence system, and we look at some of its applications: computation of eigenvectors, stable computation of non-dominant solutions of recurrence systems, and calculation of vector rational interpolants.

Key words: continued fraction, n-fraction, recurrence system

1. Introduction

It is well known ([5], [8]) that there is a close relationship between ordinary continued fractions and second-order linear recurrence relations: the sequence of numerators (denominators) of a continued fraction is a solution of such a recurrence relation, and if the continued fraction converges, its value is related to the minimal solution of this recurrence relation. A similar relationship exists between some generalizations of continued fractions and higher-order linear recurrence relations, for instance n-fractions, introduced by de Bruin [2], are natural generalizations of ordinary continued fractions, and they are related to $(n+1)$-st-order linear recurrence relations. Both ordinary continued fractions and n-fractions have many interesting applications in the field of applied mathematics: calculation of special functions [15], (vector) Padé approximation [2], (vector) rational interpolation [7],... .
In this paper we look at a continued fraction-like construction which has continued fractions and n-fractions as special cases. The idea is based on the following observation: an ordinary continued fraction $K(a_k/b_k)$ is of the form

$$K(a_k/b_k) = \frac{a_0|}{|b_0} + \frac{a_1|}{|b_1} + \dots + \frac{a_k|}{|b_k} + \dots .$$

Its k-th approximant is given by A_k/B_k where both numerator and denominator are solutions of the following second-order recurrence relation:

$$y_k = b_k y_{k-1} + a_k y_{k-2}, \quad k = 0, 1, 2, \dots .$$

A. Cuyt (ed.), Nonlinear Numerical Methods and Rational Approximation II, 433–446.

Another way to calculate A_k and B_k is by using a product of 2×2-matrices:

$$\begin{pmatrix} A_{k-1} & B_{k-1} \\ A_k & B_k \end{pmatrix} = \begin{pmatrix} 0 & 1 \\ a_k & b_k \end{pmatrix} \begin{pmatrix} 0 & 1 \\ a_{k-1} & b_{k-1} \end{pmatrix} \cdot \ldots \cdot \begin{pmatrix} 0 & 1 \\ a_0 & b_0 \end{pmatrix}$$

(This can already be found in Milne-Thomson [11]). This matrix-product is closely related to the first-order recurrence system:

$$\begin{pmatrix} y_k^{(1)} \\ y_k^{(2)} \end{pmatrix} = \begin{pmatrix} 0 & 1 \\ a_k & b_k \end{pmatrix} \begin{pmatrix} y_{k-1}^{(1)} \\ y_{k-1}^{(2)} \end{pmatrix}, \quad k = 0, 1, \ldots .$$

Note that the matrix of this recurrence system has a special form. The generalization of a continued fraction we will discuss in this paper is in the same way associated with a first-order recurrence system, but the matrix of this system will be a full $p \times p$-matrix.

In the second section we define this *general n-fraction* by its sequence of approximants and we show that its approximants may be calculated using the composition of linear fractional transformations with some of the coefficients being matrices. Next we define convergence and we prove a Pincherle [4]-Van der Cruyssen [14]-like theorem that relates the convergence of the general n-fractions to the existence of (in some way) non-dominant solutions of the associated recurrence system.

In the third section we look at some of the applications. These include: the calculation of eigenvectors of a matrix (a scaled version of the power method), the (stable) calculation of certain non-dominant solutions of first-order recurrence systems, and the solution of a simultaneous rational interpolation problem.

2. General n-Fractions: Definitions and Notations

The general n-fraction associated with the first-order recurrence system

$$Y_k = \mathcal{A}_k Y_{k-1} \quad , \quad k = 0, 1, \ldots,$$

or, written componentwise,

$$\begin{pmatrix} y_k^{(1)} \\ y_k^{(2)} \\ \vdots \\ y_k^{(n+1)} \end{pmatrix} = \begin{pmatrix} c_k^{(1,1)} & c_k^{(1,2)} & \cdots & c_k^{(1,n)} & d_k^{(1)} \\ \vdots & \vdots & & \vdots & \vdots \\ c_k^{(n,1)} & c_k^{(n,2)} & \cdots & c_k^{(n,n)} & d_k^{(n)} \\ a_k^{(1)} & a_k^{(2)} & \cdots & a_k^{(n)} & b_k \end{pmatrix} \begin{pmatrix} y_{k-1}^{(1)} \\ y_{k-1}^{(2)} \\ \vdots \\ y_{k-1}^{(n+1)} \end{pmatrix} \tag{1}$$

is defined by its sequence of n-tuples of approximants

$$\{\frac{\mathcal{G}_k^{(n+1,j)}}{\mathcal{G}_k^{(n+1,n+1)}}, \; j = 1, \ldots, n\}$$

where $(\mathcal{G}_k^{(1,j)} \ldots \mathcal{G}_k^{(n+1,j)})^\tau$, $(j = 1, \ldots, n+1)$ (τ = transpose) is the solution of (1) with initial value

$$(\mathcal{G}_{-1}^{(1,j)} \ldots \mathcal{G}_{-1}^{(n+1,j)})^\tau = (0 \ldots 0 \; 1 \; 0 \ldots 0)$$

with the 1 in the j-th position.

Example. Let

$$\begin{pmatrix} c_k^{(1,1)} & \cdots & c_k^{(1,n)} \\ \vdots & & \vdots \\ c_k^{(n,1)} & \cdots & c_k^{(n,n)} \end{pmatrix} = \begin{pmatrix} 0 & 1 & 0 & \cdots & 0 \\ 0 & 0 & 1 & \cdots & 0 \\ \vdots & \vdots & \vdots & \ddots & \vdots \\ 0 & 0 & 0 & \cdots & 1 \\ 0 & 0 & 0 & \cdots & 0 \end{pmatrix} \quad \text{and} \quad \begin{pmatrix} d_k^{(1)} \\ \vdots \\ d_k^{(n-1)} \\ d_k^{(n)} \end{pmatrix} = \begin{pmatrix} 0 \\ \vdots \\ 0 \\ 1 \end{pmatrix}. \tag{2}$$

If we put $z_{k+j} = y_k^{(j)}$ in (1), then the first-order system (1) is equivalent with the $(n+1)$-th-order linear recurrence relation

$$z_{k+n+1} = b_k z_{k+n} + a_k^{(n)} z_{k+n-1} + \cdots + a_k^{(1)} z_k.$$

In this case the general n-fraction reduces to the n-fraction (see de Bruin [2]) associated with this recurrence relation.

Using the notation

$$\begin{pmatrix} C_k & D_k \\ A_k & B_k \end{pmatrix}$$

for the matrix of the system (1), we shall from now on assume that it is **non-singular** for all k.

The approximants of the general n-fraction associated with the system (1) may be calculated from the composition of linear fractional transformations :

$$\begin{aligned} s_k(W) &= \frac{A_k + WC_k}{B_k + WD_k} && (k = 0, 1, ...) \\ S_0(W) &= s_0(W) \qquad \text{and} \quad S_k(W) = S_{k-1}(s_k(W)) && (k = 1, 2, ...) \end{aligned} \tag{3}$$

where $W = (W^{(1)} \ldots W^{(n)})$ is a $1 \times n$–matrix (note: the denominator in the definition of s_k is a scalar).

We have the following theorem :

Theorem 1

$$S_k(W) = \frac{\tilde{A}_k + W\tilde{C}_k}{\tilde{B}_k + W\tilde{D}_k}$$

with

$$\begin{pmatrix} \tilde{C}_k & \tilde{D}_k \\ \tilde{A}_k & \tilde{B}_k \end{pmatrix} = \begin{pmatrix} C_k & D_k \\ A_k & B_k \end{pmatrix} \begin{pmatrix} C_{k-1} & D_{k-1} \\ A_{k-1} & B_{k-1} \end{pmatrix} \cdot \ldots \cdot \begin{pmatrix} C_0 & D_0 \\ A_0 & B_0 \end{pmatrix}$$

Proof : By induction on k, using simple algebra.

If

$$\begin{pmatrix} \tilde{C}_k & \tilde{D}_k \\ \tilde{A}_k & \tilde{B}_k \end{pmatrix} = \begin{pmatrix} \tilde{c}_k^{(1,1)} & \tilde{c}_k^{(1,2)} & \cdots & \tilde{c}_k^{(1,n)} & \tilde{d}_k^{(1)} \\ \vdots & \vdots & & \vdots & \vdots \\ \tilde{c}_k^{(n,1)} & \tilde{c}_k^{(n,2)} & \cdots & \tilde{c}_k^{(n,n)} & \tilde{d}_k^{(n)} \\ \tilde{a}_k^{(1)} & \tilde{a}_k^{(2)} & \cdots & \tilde{a}_k^{(n)} & \tilde{b}_k \end{pmatrix}$$

it is easy to see from (1) and (3) that $\mathcal{G}_k^{(n+1,j)} = \tilde{a}_k^{(j)}$ and $\mathcal{G}_k^{(n+1,n+1)} = \tilde{b}_k$. Hence

$$S_k(0) = \frac{\tilde{A}_k}{\tilde{B}_k} = \left(\frac{\tilde{a}_k^{(1)}}{\tilde{b}_k} \ \frac{\tilde{a}_k^{(2)}}{\tilde{b}_k} \ \ldots \ \frac{\tilde{a}_k^{(n)}}{\tilde{b}_k} \right)$$

is an n-tuple of approximants of the general n-fraction associated with (1). The general n-fraction associated with (1) is said to converge if

$$\lim_{k\to\infty} \frac{\tilde{A}_k}{\tilde{B}_k} \in \mathbb{C}^n.$$

We have the following generalization of a result by Van der Cruyssen [14]:

Theorem 2 *The recurrence system (1) has a fundamental system of solutions*

$$\{(X_k^{(1,j)} \ \ldots \ X_k^{(n+1,j)})^T, \ j = 1, \ldots, n+1\}$$

satisfying

$$(a) \quad \det \begin{pmatrix} X_{-1}^{(1,1)} & \cdots & X_{-1}^{(1,n)} \\ \vdots & & \vdots \\ X_{-1}^{(n,1)} & \cdots & X_{-1}^{(n,n)} \end{pmatrix} \neq 0$$

$$(b) \quad \lim_{k\to\infty} \frac{X_k^{(n+1,j)}}{X_k^{(n+1,n+1)}} = 0 \quad \textit{for} \quad j = 1, \ldots, n$$

iff the general n-fraction associated with (1) converges, i.e.

$$\lim_{k\to\infty} \frac{\tilde{A}_k}{\tilde{B}_k} \in \mathbb{C}^n.$$

Proof : Let us first assume that (a) and (b) are satisfied. Since

$$\{(\mathcal{G}_k^{(1,j)} \ \ldots \ \mathcal{G}_k^{(n+1,j)})^T, \ j = 1, \ldots, n+1\} \quad \text{and} \quad \{(X_k^{(1,j)} \ \ldots \ X_k^{(n+1,j)})^T, \ j = 1, \ldots, n+1\}$$

are two fundamental systems of solutions of (1) there exists a constant non-singular $(n+1) \times (n+1)$–matrix F for which

$$\begin{pmatrix} \mathcal{G}_k^{(1,1)} & \cdots & \mathcal{G}_k^{(1,n)} & \mathcal{G}_k^{(1,n+1)} \\ \vdots & & \vdots & \vdots \\ \mathcal{G}_k^{(n,1)} & \cdots & \mathcal{G}_k^{(n,n)} & \mathcal{G}_k^{(n,n+1)} \\ \mathcal{G}_k^{(n+1,1)} & \cdots & \mathcal{G}_k^{(n+1,n)} & \mathcal{G}_k^{(n+1,n+1)} \end{pmatrix} = \begin{pmatrix} X_k^{(1,1)} & \cdots & X_k^{(1,n)} & X_k^{(1,n+1)} \\ \vdots & & \vdots & \vdots \\ X_k^{(n,1)} & \cdots & X_k^{(n,n)} & X_k^{(n,n+1)} \\ X_k^{(n+1,1)} & \cdots & X_k^{(n+1,n)} & X_k^{(n+1,n+1)} \end{pmatrix} \cdot F \ .$$

If we write F in block form :

$$F = \begin{pmatrix} F^{(1,1)} & F^{(1,2)} \\ F^{(2,1)} & F^{(2,2)} \end{pmatrix}$$

with $F^{(1,1)}$ an $n \times n$–matrix, we have

$$(\mathcal{G}_k^{(n+1,1)} \; ... \; \mathcal{G}_k^{(n+1,n)}) = (X_k^{(n+1,1)} \; ... \; X_k^{(n+1,n)}) \cdot F^{(1,1)} + X_k^{(n+1,n+1)} \cdot F^{(2,1)}$$

and

$$\mathcal{G}_k^{(n+1,n+1)} = (X_k^{(n+1,1)} \; ... \; X_k^{(n+1,n)}) \cdot F^{(1,2)} + X_k^{(n+1,n+1)} \cdot F^{(2,2)} .$$

Hence

$$\left(\frac{\mathcal{G}_k^{(n+1,1)}}{\mathcal{G}_k^{(n+1,n+1)}} \; \cdots \; \frac{\mathcal{G}_k^{(n+1,n)}}{\mathcal{G}_k^{(n+1,n+1)}} \right) = \frac{\left(\frac{X_k^{(n+1,1)}}{X_k^{(n+1,n+1)}} \; \cdots \; \frac{X_k^{(n+1,n)}}{X_k^{(n+1,n+1)}} \right) \cdot F^{(1,1)} + F^{(2,1)}}{\left(\frac{X_k^{(n+1,1)}}{X_k^{(n+1,n+1)}} \; \cdots \; \frac{X_k^{(n+1,n)}}{X_k^{(n+1,n+1)}} \right) \cdot F^{(1,2)} + F^{(2,2)}}$$

and we have immediately from (b) that

$$\lim_{k\to\infty} \left(\frac{\mathcal{G}_k^{(n+1,1)}}{\mathcal{G}_k^{(n+1,n+1)}} \; \cdots \; \frac{\mathcal{G}_k^{(n+1,n)}}{\mathcal{G}_k^{(n+1,n+1)}} \right) = \frac{F^{(2,1)}}{F^{(2,2)}}$$

if $F^{(2,2)} \neq 0$. To prove this observe that

$$\begin{pmatrix} \mathcal{G}_k^{(1,n+1)} \\ \vdots \\ \mathcal{G}_k^{(n,n+1)} \end{pmatrix} = \begin{pmatrix} X_k^{(1,1)} & \cdots & X_k^{(1,n)} \\ \vdots & & \vdots \\ X_k^{(n,1)} & \cdots & X_k^{(n,n)} \end{pmatrix} \cdot F^{(1,2)} + \begin{pmatrix} X_k^{(1,n+1)} \\ \vdots \\ X_k^{(n,n+1)} \end{pmatrix} \cdot F^{(2,2)}$$

and for $k = -1$ this reduces to

$$\begin{pmatrix} 0 \\ \vdots \\ 0 \end{pmatrix} = \begin{pmatrix} X_{-1}^{(1,1)} & \cdots & X_{-1}^{(1,n)} \\ \vdots & & \vdots \\ X_{-1}^{(n,1)} & \cdots & X_{-1}^{(n,n)} \end{pmatrix} \cdot F^{(1,2)} + \begin{pmatrix} X_{-1}^{(1,n+1)} \\ \vdots \\ X_{-1}^{(n,n+1)} \end{pmatrix} \cdot F^{(2,2)} .$$

$F^{(2,2)} = 0$ together with (a) would imply that $F^{(1,2)} = (0 \; ... \; 0)^\tau$ which is impossible since F is non-singular.

Let us now assume that the general n-fraction associated with (1) converges and that

$$\lim_{k\to\infty} \left(\frac{\mathcal{G}_k^{(n+1,1)}}{\mathcal{G}_k^{(n+1,n+1)}} \; \cdots \; \frac{\mathcal{G}_k^{(n+1,n)}}{\mathcal{G}_k^{(n+1,n+1)}} \right) = (t^{(1)} \; ... \; t^{(n)}) = T .$$

The column-vectors of the matrix

$$\begin{pmatrix} \tilde{C}_k - \tilde{D}_k \cdot T & \tilde{D}_k \\ \tilde{A}_k - \tilde{B}_k \cdot T & \tilde{B}_k \end{pmatrix} = \begin{pmatrix} \tilde{C}_k & \tilde{D}_k \\ \tilde{A}_k & \tilde{B}_k \end{pmatrix} \begin{pmatrix} I_n & 0 \\ -T & 1 \end{pmatrix}$$

(I_n is the identity matrix) form a fundamental system of solutions of (1) satisfying (b) since

$$\lim_{k\to\infty} \frac{\tilde{A}_k - \tilde{B}_k \cdot T}{\tilde{B}_k} = \lim_{k\to\infty} \frac{\tilde{A}_k}{\tilde{B}_k} - T = 0$$

and

$$\det \begin{pmatrix} \tilde{C}_{-1} - \tilde{D}_{-1} \cdot T & \tilde{D}_{-1} \\ \tilde{A}_{-1} - \tilde{B}_{-1} \cdot T & \tilde{B}_{-1} \end{pmatrix} = 1 \neq 0 \, .$$

□

Note that as a consequence a solution of (1) from the subspace spanned by the columns of the matrix

$$\begin{pmatrix} \tilde{C}_k - \tilde{D}_k \cdot T \\ \tilde{A}_k - \tilde{B}_k \cdot T \end{pmatrix}$$

has initial values $(Z_{-1}^{(1)} \;\ldots\; Z_{-1}^{(n)} \; Z_{-1}^{(n+1)})^\tau$ which satisfy

$$Z_{-1}^{(n+1)} = -t^{(n)} Z_{-1}^{(n)} - \ldots - t^{(1)} Z_{-1}^{(1)} \, . \tag{4}$$

In 1917 O. Perron [12] proved the following generalization to first-order recurrence systems of the well-known Poincaré-Perron theorem for linear recurrence relations of Poincaré-type. We shall use this result later on.

Theorem 3 *If the recurrence system (1) has the property that :* $\lim_{k\to\infty} \mathcal{A}_k = \mathcal{A}$ *, with for all k* $\det \mathcal{A}_k \neq 0$*, and if the eigenvalues of* $\mathcal{A}$ *are all different in modulus :*

$$|\alpha_1| < |\alpha_2| < \cdots < |\alpha_{n+1}|, \tag{5}$$

then for each $j \in \{1, 2, ..., n+1\}$ *the recurrence system (1) has a solution*

$$(X_k^{(1,j)} \;\ldots\; X_k^{(n+1,j)})^\tau$$

for which

$$\lim_{k\to\infty} \frac{X_{k+1}^{(i,j)}}{X_k^{(i,j)}} = \alpha_j$$

for all $i \in \{1, 2, ..., n+1\}$ *for which the eigenvector* $(P^{(1,j)}, ..., P^{(n+1,j)})^\tau$ *corresponding to the eigenvalue* α_j *has i-th component different from zero, i.e.* $P^{(i,j)} \neq 0$.

3. Applications

3.1. Numerical Calculation of Non-Dominant Solutions of a Recurrence System

Theorem 2 provides us with a method to calculate solutions of the recurrence system (1) which in a certain sense are non-dominant, and cannot be calculated numerically from (1) using forward recurrence:
if the condition (b) of Theorem 2 holds, then a solution of (1) from the subspace

spanned by $\{(X_k^{(1,j)} \ldots X_k^{(n+1,j)})^\tau, j = 1, ..., n\}$ has $(n+1)$–st component dominated by the $(n+1)$–st component of $(X_k^{(1,n+1)} \ldots X_k^{(n+1,n+1)})^\tau$. To see this we look at the following example. If we take $n = 1$ and

$$\begin{array}{ll} C_k = (2 \cdot 8^{k+1} - 1)/(8^{k+1} - 1)\ , & D_k = -2^{k+1}/(8^{k+1} - 1) \\ A_k = -3 \cdot 4^{k+1}/(8^{k+1} - 1)\ , & B_k = (4 \cdot 8^{k+1} - 1)/(8^{k+1} - 1) \end{array}$$

in (1), then this recurrence system has the solutions $(2^{k+2}\ 1)^\tau$, and $(1\ 4^{k+2})^\tau$. Hence every solution of the system can be written as :

$$r\begin{pmatrix} 2^{k+2} \\ 1 \end{pmatrix} + s\begin{pmatrix} 1 \\ 4^{k+2} \end{pmatrix} .$$

If we want to calculate the solution $(2^{k+2}\ 1)^\tau$ using forward recurrence, a rounding error in the initial value will result in an $s \neq 0$, and so the second component of $(1\ 4^{k+2})^\tau$ will eventually become dominant in the calculated solution.

Let us now assume that the conditions (a) and (b) of Theorem 2 hold, but with (a) replaced by

$$\det \begin{pmatrix} X_k^{(1,1)} & \cdots & X_k^{(1,n)} \\ \vdots & & \vdots \\ X_k^{(n,1)} & \cdots & X_k^{(n,n)} \end{pmatrix} \neq 0$$

for all k. Then the general n-fraction associated with the recurrence system

$$Y_k = \mathcal{A}_{k+p} Y_{k-1}$$

converges for each $p = 0, 1, \ldots$. Let its value be $T_p = (t_p^{(1)}\ \ldots\ t_p^{(n)})$.
A solution of the subspace spanned by $\{(X_k^{(1,j)} \ldots X_k^{(n+1,j)})^\tau, j = 1, ..., n\}$ satisfies

$$\begin{aligned} Z_{k-1}^{(n+1)} &= -T_k \cdot \begin{pmatrix} Z_{k-1}^{(1)} \\ \vdots \\ Z_{k-1}^{(n)} \end{pmatrix} \\ \begin{pmatrix} Z_k^{(1)} \\ \vdots \\ Z_k^{(n)} \end{pmatrix} &= (C_k - D_k \cdot T_k) \begin{pmatrix} Z_{k-1}^{(1)} \\ \vdots \\ Z_{k-1}^{(n)} \end{pmatrix} \end{aligned} \qquad (6)$$

for $k = 0, 1, \ldots$. (This follows immediately from (4)). We can use this to calculate solutions of this subspace which are dominant in this subspace : for some index N we calculate $S_N(0)$ using (3). This gives us an approximation $T_0(N) = S_N(0)$ for the value T_0 of the general n-fraction associated with (1), and also approximations $T_k(N)$ for T_k $(k = 1, 2, ..., N)$. If the solution we want to calculate is $(Z_k^{(1)}\ \ldots\ Z_k^{(n)}\ Z_k^{(n+1)})^\tau$ where $Z_{-1}^{(i)}$, $i = 1, ..., n$, are given initial values, we get an approximation for this solution from :

$$Z_{k-1}^{(n+1)} = -T_k(N) \cdot \begin{pmatrix} Z_{k-1}^{(1)} \\ \vdots \\ Z_{k-1}^{(n)} \end{pmatrix}$$
$$\begin{pmatrix} Z_{k}^{(1)} \\ \vdots \\ Z_{k}^{(n)} \end{pmatrix} = (C_k - D_k \cdot T_k(N)) \begin{pmatrix} Z_{k-1}^{(1)} \\ \vdots \\ Z_{k-1}^{(n)} \end{pmatrix} \tag{7}$$

for $k = 0, 1, ..., N$. For the example above we get with $N = 80$ the results given in the following table (for the solution $(2^{k+2}\ \ 1)^\tau$). In the table the results (in double precision) using forward recursion and using (7) are compared. We note that other

TABLE I

k	relative error in $Z_k^{(1)}$ using (7)	relative error in $Z_k^{(1)}$ forward	relative error in $Z_k^{(2)}$ using (7)	relative error in $Z_k^{(2)}$ forward
-1	0.0E+00	0.0E+00	0.111022E-15	0.111022E-15
10	0.0E+00	0.0E+00	0.000000E+00	0.000000E+00
20	0.0E+00	0.0E+00	0.000000E+00	0.113687E-12
30	0.0E+00	0.0E+00	0.000000E+00	0.119209E-06
40	0.0E+00	0.0E+00	0.000000E+00	0.125000E+00
50	0.0E+00	0.0E+00	0.000000E+00	0.131072E+06
60	0.0E+00	0.0E+00	0.909495E-12	0.137439E+12
70	0.0E+00	0.0E+00	0.953674E-06	0.144115E+18
80	0.0E+00	0.0E+00	0.100000E+01	0.151116E+24

authors have already given algorithms to compute non-dominant solutions of (1): Mattheij in the papers [9], [10], and Zahar [16]. Our method is equivalent with Zahar's method, the differences being that Zahar's algorithm is more general and it is a forward-backward algorithm, while ours is a backward-forward algorithm. However, it is possible to adapt our algorithm to the more general setting of Zahar's work. We also note that the theoretical method behind these two algorithms is known in the literature as *method of embedding* (see [1]).

3.2. The 1-Periodic Case

Let us assume that for all k $\mathcal{A}_k = \mathcal{A}$ with $\det\mathcal{A} \neq 0$, that the eigenvalues α_j of $\mathcal{A}$ satisfy (5) and that the corresponding eigenvectors $(P^{(1,j)}, ..., P^{(n+1,j)})^\tau$ satisfy

$$\det \begin{pmatrix} P^{(1,1)} & \dots & P^{(1,n)} \\ \vdots & & \vdots \\ P^{(n,1)} & \dots & P^{(n,n)} \end{pmatrix} \neq 0 \,,$$

and that $P^{(n+1,n+1)} \neq 0$. The conditions of Theorem 3 are satisfied with

$$X_k^{(i,j)} = \alpha_j^{k+1} P^{(i,j)}$$

for $i, j = 1, ..., n+1$ and the general n-fraction associated with (1) converges to $T = (t^{(1)} \; ... \; t^{(n)})$ where $(t^{(1)} \; ... \; t^{(n)} \; 1)^\tau$ is an eigenvector of $\mathcal{A}^\tau$ corresponding to the eigenvalue α_{n+1} : if the general n-fraction converges to T, then it is easy to see from (3) that

$$T = \frac{A+TC}{B+TD} = \frac{(T \;\; 1)\begin{pmatrix} C \\ A \end{pmatrix}}{(T \;\; 1)\begin{pmatrix} D \\ B \end{pmatrix}}$$

and hence with $\lambda = (T \;\; 1)\begin{pmatrix} D \\ B \end{pmatrix}$

$$(\lambda T \;\; \lambda) = (T \;\; 1)\begin{pmatrix} C & D \\ A & B \end{pmatrix}$$

or

$$\begin{pmatrix} C & D \\ A & B \end{pmatrix}^\tau \begin{pmatrix} T^\tau \\ 1 \end{pmatrix} = \lambda \begin{pmatrix} T^\tau \\ 1 \end{pmatrix} .$$

Furthermore $(T \;\; 1)$ satisfies (4) :

$$(T \;\; 1)\begin{pmatrix} P^{(1,j)} \\ \vdots \\ P^{(n+1,j)} \end{pmatrix} = 0 \tag{8}$$

for $j = 1, ..., n$. Now the only eigenvector of $\mathcal{A}^\tau$ satisfying (8) is the one corresponding to the eigenvalue α_{n+1}.

It is easy to see that for the periodic case the algorithm presented in the previous section is just a scaled version of the power method for the computation of the dominant eigenvalue (and the corresponding eigenvector) of a matrix.

3.3. The Limit-Periodic Case

Let us assume that $\lim_{k\to\infty} \mathcal{A}_k = \mathcal{A}$ with $\det \mathcal{A}_k \neq 0$ for all k, that the eigenvalues α_j of $\mathcal{A}$ satisfy (5) and that the corresponding eigenvectors $(P^{(1,j)}, ..., P^{(n+1,j)})^\tau$ satisfy

$$\det \begin{pmatrix} P^{(1,1)} & \dots & P^{(1,n)} \\ \vdots & & \vdots \\ P^{(n,1)} & \dots & P^{(n,n)} \end{pmatrix} \neq 0 , \tag{9}$$

and that $P^{(n+1,j)} \neq 0$ for $j = 1, ..., n+1$. Then the conditions of Theorem 3 are satisfied and the recurrence system (1) has a fundamental system of solutions $\{(X_k^{(1,j)} \; ... \; X_k^{(n+1,j)})^\tau$ for which

$$\lim_{k\to\infty} \frac{X_{k+1}^{(n+1,j)}}{X_k^{(n+1,j)}} = \alpha_j$$

and hence

$$\lim_{k\to\infty} \frac{X_k^{(n+1,j)}}{X_k^{(n+1,n+1)}} = 0$$

as a consequence of (5). Furthermore from the proof of Theorem 3 (see [12]) it follows that

$$\lim_{k\to\infty} \frac{X_k^{(i,j)}}{X_k^{(n+1,j)}} = \frac{P^{(i,j)}}{P^{(n+1,j)}} . \tag{10}$$

From the determinantal condition (9) together with (10) we get that

$$\det \begin{pmatrix} X_k^{(1,1)} & \cdots & X_k^{(1,n)} \\ \vdots & & \vdots \\ X_k^{(n,1)} & \cdots & X_k^{(n,n)} \end{pmatrix} \neq 0$$

from some $k = k_0$ on, hence the general n-fraction associated with $Y_k = \mathcal{A}_{k+p} Y_{k-1}$ will converge for $k = k_0, k_0 + 1, \ldots$ to the value T_p. Since

$$T_p = \frac{A_p + T_{p+1} C_p}{B_p + T_{p+1} D_p}$$

(as is easy to prove using Theorem 1), and (from (4))

$$t_p^{(j)} = - \frac{\det \begin{pmatrix} X_{p-1}^{(1,1)} & \cdots & X_{p-1}^{(j-1,1)} & X_{p-1}^{(n+1,1)} & X_{p-1}^{(j+1,1)} & \cdots & X_{p-1}^{(n,1)} \\ \vdots & & \vdots & \vdots & \vdots & & \vdots \\ X_{p-1}^{(1,n)} & \cdots & X_{p-1}^{(j-1,n)} & X_{p-1}^{(n+1,n)} & X_{p-1}^{(j+1,n)} & \cdots & X_{p-1}^{(n,n)} \end{pmatrix}}{\det \begin{pmatrix} X_{p-1}^{(1,1)} & \cdots & X_{p-1}^{(n,1)} \\ \vdots & & \vdots \\ X_{p-1}^{(1,n)} & \cdots & X_{p-1}^{(n,n)} \end{pmatrix}} ,$$

we get using (10) that $\lim_{p\to\infty} t_p^{(j)} = \tilde{t}^{(j)}$ with $\tilde{T} = (\tilde{t}^{(1)} \ \ldots \ \tilde{t}^{(n)})$ satisfying

$$\tilde{T} = \frac{A + \tilde{T}C}{B + \tilde{T}D} \quad \text{and} \quad P^{(n+1,j)} = -\tilde{t}^{(n)} P^{(n,j)} - \ldots - \tilde{t}^{(1)} P^{(1,j)} .$$

Now it is easy to see that in this case too $\lim_{p\to\infty} T_p$ is the eigenvector of $\mathcal{A}^T$ corresponding to the eigenvalue α_{n+1}.

We note that this result can be used to accelerate the convergence of the algorithm discussed in section 3.1 in the limit-periodic case. For similar results on ordinary continued fractions and n-fractions we refer to [8], [3], [6].

3.4. A New Approach to the Simultaneous Rational Interpolation Problem

In their paper [13] Van Barel and Bultheel describe an algorithm to solve the vector case of the rational interpolation problem. We shall reformulate this algorithm and show that the solution of the problem may be given in the form of a general n-fraction (for more details we refer to [13]).

Let us first state the problem:

Definition 1 (proper simultaneous rational interpolation problem)
Find all polynomial $(n+1)$-tuples $(p^{(1)}, p^{(2)}, \ldots, p^{(n+1)})$ such that
– *the following rational interpolation conditions are satisfied:*

$$(p^{(1)}(z) \ \cdots \ p^{(n)}(z))/p^{(n+1)}(z) = F(z) + (O(\pi_\sigma(z)) \ \cdots \ O(\pi_\sigma(z)))$$

with $F(z)$ a vector Newton series using the sequence of points $z_1, z_2, \ldots$, and $\pi_\sigma(z) = \prod_{j=1}^{\sigma}(z - z_j)$,
– *the following degree conditions are satisfied:*

$$\deg(p^{(1)} \ \cdots \ p^{(n+1-r)} \ p^{(n+2-r)} \ \cdots \ p^{(n+1)}) \leq (\mu \ \cdots \ \mu \ \overbrace{\mu+1 \ \cdots \ \mu+1}^{r}) \tag{11}$$

with $n\sigma + 1 = (\mu+1)(n+1) + r$ and $0 \leq r < n+1$.

Note that the number of parameters, i.e. the total number of coefficients $(\mu+1)(n+1) + r$, is one more than the number of conditions $n\sigma$. Increasing the order σ by one, i.e. $\sigma' = \sigma + 1$, transforms μ and r as follows: if $r = 0$ then $\mu' = \mu$ and $r' = n$; if $0 < r < n+1$ then $\mu' = \mu + 1$ and $r' = (r+n)\mathbf{mod}(n+1)$.

Definition 2 (linearized simultaneous rational interpolation problem)
Find all polynomial $(n+1)$-tuples $(p^{(1)}, p^{(2)}, \ldots, p^{(n+1)})$ such that:
– *the following linearized rational interpolation conditions are satisfied:*

$$(p^{(1)}(z) \ \cdots \ p^{(n)}(z) \ p^{(n+1)}(z)) \begin{pmatrix} -I_n \\ F(z) \end{pmatrix} = (O(\pi_\sigma(z)) \ \cdots \ O(\pi_\sigma(z))) \ , \tag{12}$$

with the same notation as before,
– *the same degree conditions as for the proper case are satisfied.*

Note that this linearized problem has at least one solution with one degree of freedom (which disappears when the polynomial tuple is considered as a vector rational function). When $p^{(n+1)}(z_j) \neq 0$, for $j = 1, 2, \ldots, \sigma$, the solution of the linearized problem is also a solution of the proper problem. It is easy to see that all solutions of the proper problem form a subset of all solutions of the linearized problem. Hence, we first compute all solutions of the linearized problem.

Theorem 4 *All polynomial $(n+1)$-tuples satisfying the linearized rational interpolation conditions (12) form a submodule S_σ of the* $\mathbf{F}[z]$*-module* $\mathbf{F}[z]^{n+1}$ *with a set of $(n+1)$ basis vectors.*

Proof : see [13].
Let us give an efficient algorithm to compute these basis vectors in a recursive way, i.e. for increasing σ.

– **Initialization:** The rows of the $(n+1) \times (n+1)$ identity matrix form a set of basis vectors for S_0.
– **Recursive step:** Suppose we have a basis for S_σ. We can compute a basis for $S_{\sigma+1}$ as follows:

- Using the basis vectors $g_\sigma^{(1)}, \ldots, g_\sigma^{(n+1)}$ for S_σ, we can determine the corresponding residual vectors $r_\sigma^{(1)}, \ldots, r_\sigma^{(n+1)}$:

$$\begin{pmatrix} g_\sigma^{(1)}(z) \\ \vdots \\ g_\sigma^{(n+1)}(z) \end{pmatrix} \begin{pmatrix} -I_n \\ F(z) \end{pmatrix} = \begin{pmatrix} r_\sigma^{(1)} \\ \vdots \\ r_\sigma^{(n+1)} \end{pmatrix} \pi_\sigma(z) + \begin{pmatrix} O(\pi_{\sigma+1}(z)) \cdots O(\pi_{\sigma+1}(z)) \\ \vdots \qquad\qquad \vdots \\ O(\pi_{\sigma+1}(z)) \cdots O(\pi_{\sigma+1}(z)) \end{pmatrix}$$

  where we have written the polynomial $(n+1)$-tuples rowwise.
- The set of homogeneous linear equations

$$\left(v_{\sigma+1}^{(1)}\ v_{\sigma+1}^{(2)} \cdots v_{\sigma+1}^{(n+1)}\right) \begin{pmatrix} r_\sigma^{(1)} \\ r_\sigma^{(2)} \\ \vdots \\ r_\sigma^{(n+1)} \end{pmatrix} = 0 \tag{13}$$

  always has at least one non-zero solution (which can always be multiplied by a scalar). Let us assume that there always exists a solution $v_{\sigma+1}$ of (12) with $v_{\sigma+1}^{(1)} \neq 0$, $\sigma = 0, 1, \ldots$ (for the case $v_{\sigma+1}^{(1)} = 0$ for some σ we refer to [13]).
- Once we have determined the non-zero solution $v_{\sigma+1}$ of (12) it is easy to show that the following set of polynomial $(n+1)$-tuples forms a basis for $S_{\sigma+1}$:

$$\begin{pmatrix} g_{\sigma+1}^{(1)}(z) \\ \vdots \\ g_{\sigma+1}^{(n+1)}(z) \end{pmatrix} = \mathcal{A}_\sigma \begin{pmatrix} g_\sigma^{(1)}(z) \\ \vdots \\ g_\sigma^{(n+1)}(z) \end{pmatrix}$$

  or

$$G_{\sigma+1} = \mathcal{A}_\sigma G_\sigma \tag{14}$$

  with

$$\mathcal{A}_\sigma = \begin{pmatrix} 0 & z - z_{\sigma+1} & \cdots & 0 & 0 \\ 0 & 0 & \cdots & 0 & 0 \\ \vdots & \vdots & & \vdots & \vdots \\ 0 & 0 & \cdots & 0 & z - z_{\sigma+1} \\ v_{\sigma+1}^{(1)} & v_{\sigma+1}^{(2)} & \cdots & v_{\sigma+1}^{(n)} & v_{\sigma+1}^{(n+1)} \end{pmatrix}.$$

  From the fact that $(r_0^{(1)}\ r_0^{(2)} \cdots r_0^{(n)})^\tau = -I_n$, it can easily be shown that $r_\sigma^{(n+1)} \neq 0$, $v_{\sigma+1}^{(n+1)} \neq 0$ and $r_\sigma^{(1)}, \ldots, r_\sigma^{(n)}$ are linearly independent, i.e. the solution of (13) is unique except for a scalar multiple.

The set S_σ of all solutions of (12) is generated by the $n+1$ polynomial $(n+1)$-tuples $g_\sigma^{(1)}, g_\sigma^{(2)}, \ldots, g_\sigma^{(n+1)}$. We want to extract only those solutions satisfying the additional degree conditions (11). It is easy to prove that the last basis vector $g_\sigma^{(n+1)}$ satisfies the degree conditions with equality for at least one component. On the other hand, the other basis vectors do not satisfy the degree conditions. It is also true that the last component of $g_\sigma^{(n+1)}$ is different from zero in the points z_j

$(j = 1, 2, \ldots, \sigma)$. Hence, the linearized as well as the proper simultaneous rational interpolation problem has one and just one solution, the last basis vector $g_\sigma^{(n+1)}$. Now it is easy to see from (14) that, in the notation of the second paragraph,

$$\tilde{A}_k = (g_{k+1}^{(1)}, g_{k+1}^{(2)}, \ldots, g_{k+1}^{(n)}) \ , \quad \tilde{B}_k = g_{k+1}^{(n+1)}$$

and hence that the solution of our first problem is given by the approximants of the general n-fraction associated with this recurrence system. Because the matrix of this system has a special form, it is easy to transform it in the form of an n-fraction (see (2)). This is done by transforming the general n-fraction represented by $\cdots \mathcal{A}_\sigma \mathcal{A}_{\sigma-1} \cdots \mathcal{A}_1 \mathcal{A}_0$ into the following form $\cdots W_\sigma W_{\sigma-1} \cdots W_1 W_0$ where all W_σ, $\sigma > 0$ have the form

$$\begin{pmatrix} 0 & 1 & \cdots & 0 & 0 \\ 0 & 0 & \cdots & 0 & 0 \\ \vdots & \vdots & & \vdots & \vdots \\ 0 & 0 & \cdots & 0 & 1 \\ w_{\sigma+1}^{(1)} & w_{\sigma+1}^{(2)} & \cdots & w_{\sigma+1}^{(n)} & w_{\sigma+1}^{(n+1)} \end{pmatrix}.$$

Suppose we have already transformed the first σ blocks, i.e.

$$\mathcal{A}_{\sigma-1} \cdots \mathcal{A}_0 = E_{\sigma-1} W_{\sigma-2} \cdots W_0 \tag{15}$$

with $E_{\sigma-1}$ having the following form

$$E_{\sigma-1} = \begin{pmatrix} 0 & u_\sigma^{(1)} & \cdots & 0 & 0 \\ 0 & 0 & \cdots & 0 & 0 \\ \vdots & \vdots & & \vdots & \vdots \\ 0 & 0 & \cdots & 0 & u_\sigma^{(n)} \\ s_\sigma^{(1)} & s_\sigma^{(2)} & \cdots & s_\sigma^{(n)} & s_\sigma^{(n+1)} \end{pmatrix}$$

with $s_\sigma^{(j)}$ and $u_\sigma^{(j)}$ polynomials. These polynomials together with the polynomials $w_\sigma^{(j)}$ can be determined by using the equality $\mathcal{A}_\sigma E_{\sigma-1} = E_\sigma W_{\sigma-1}$ giving the following equations

$$\begin{aligned} w_\sigma^{(i)} &= s_\sigma^{(i)}, \quad i = 1, 2, \ldots n+1 \\ s_{\sigma+1}^{(i)} &= u_\sigma^{(i)} v_{\sigma+1}^{(i)}, \quad i = 1, 2, \ldots, n+1 \\ u_{\sigma+1}^{(i)} &= u_\sigma^{(i+1)} \cdot (z - z_{\sigma+1}), \quad i = 1, 2, \ldots, n \end{aligned}$$

with $u_\sigma^{(n+1)} = 1$ and initial values $u_1^{(i)} = z - z_1$, $i = 1, 2, \ldots, n$ and $s_1^{(i)} = v_1^{(i)}$, $i = 1, 2, \ldots, n+1$. Using these recurrence relations, it is easy to prove that

$$\begin{aligned} u_\sigma^{(i)}(z) &= \textstyle\prod_{j=0}^{n-i} (z - z_{\sigma-j}), \quad i = 1, 2, \ldots, n \\ w_\sigma^{(i)}(z) &= u_{\sigma-1}^{(i)}(z) v_\sigma^{(i)}, \quad i = 1, 2, \ldots, n+1 \\ s_\sigma^{(i)}(z) &= w_\sigma^{(i)}(z), \quad i = 1, 2, \ldots, n+1 \end{aligned}$$

where the factors $z - z_k = 1$ for $k \leq 0$. Because we can always scale the initial solution v_1 such that $v_1^{(n+1)} = 1$, the solutions of the proper (and linearized) simultaneous rational interpolation problem can also be written as the convergents of the n-fraction associated with (1).
A similar n-fraction was found in [7] using another method.

References

1. Agarwal R.P.: 1992, *Difference equations and inequalities: theory, methods and applications*, Marcel Dekker, Inc., New York.
2. de Bruin M.G.: 1974, *Generalized continued fractions and a multidimensional Padé Table*, Doctoral Thesis, Amsterdam.
3. de Bruin M.G., Jacobsen L.: 1987, 'Modification of Generalised Continued Fractions I', Lecture Notes in Math. (J. Gilewicz et al. Eds), Springer-Verlag, Berlin, pp. 161–176.
4. Gautschi W.: 1967, 'Computational Aspects of Three-Term Recurrence Relations', *SIAM Review* Vol. **9**, pp. 24–82.
5. Jones W.B., Thron W.J.: 1980, *Continued Fractions: Analytic Theory and Applications*, Encyclopedia of Mathematics **11**, Addison-Wesley Publishing Company, Reading, Mass..
6. Levrie P., Jacobsen L.: 1988, 'Convergence Acceleration for Generalized Continued Fractions', *Trans. Amer. Math. Soc.* Vol. **305** no. **1**, pp. 263–275.
7. Levrie P., Bultheel A.: 1993, 'A note on Thiele n-fractions', *Numer. Algorithms* Vol. **4**, pp. 225–239.
8. Lorentzen L., Waadeland H.: 1992, *Continued Fractions with Applications*, North-Holland, Amsterdam.
9. Mattheij R.M.M.: 1980, 'Characterizations of Dominant and Dominated Solutions of Linear Recursions', *Numer. Math.* Vol. **35**, pp. 421–442.
10. Mattheij R.M.M.: 1982, 'Stable Computation of Solutions of Unstable Linear Initial Value Recursions', *BIT* Vol. **22**, pp. 79–93.
11. Milne-Thomson L.M.: 1933, *The calculus of finite differences*, MacMillan.
12. Perron O.: 1917, 'Über Systeme von Linearen Differenzengleichungen erster Ordnung', *J. Reine Angew. Math.* Vol. **147**, pp. 36–53.
13. Van Barel M., Bultheel A.: 1990, 'A New Approach to th Rational Interpolation Problem: the Vector Case', *J. Comp. Appl. Math.* Vol. **33**, pp. 331–346.
14. Van der Cruyssen P.: 1979, 'Linear Difference Equations and Generalized Continued Fractions', *Computing* Vol. **22**, pp. 269–278.
15. Wimp J.: 1984, *Computation with Recurrence Relations*, Pitman Advanced Publishing Program, Pitman Publishing Inc., Boston.
16. Zahar R.V.M.: 1990, 'A Generalization of Olver's Algorithm for Linear Difference Systems', in *Asymptotic and Computational Analysis* (R. Wong ed.), Marcel Dekker, Inc., New York, pp. 535–551.